ELSEVIER'S

DICTIONARY OF HORTICULTURE

EDITORIAL STAFF

J. NIJDAM

*Former Deputy Head of the Horticultural College
at Utrecht (final wording)*

A. DE JONG

*Ministry of Agriculture and Fisheries at The Hague,
Secretary of the Working Group Horticultural Dictionary*

ELSEVIER'S
DICTIONARY OF
HORTICULTURE

in nine languages
English, French, Dutch, German,
Danish, Swedish, Spanish, Italian, Latin

COMPILED UNDER THE AUSPICES OF
THE MINISTRY OF AGRICULTURE AND FISHERIES
AT THE HAGUE, THE NETHERLANDS

ELSEVIER PUBLISHING COMPANY
AMSTERDAM – LONDON – NEW YORK

1970

ELSEVIER PUBLISHING COMPANY
335 Jan van Galenstraat
P.O. Box 211, Amsterdam, The Netherlands

ELSEVIER PUBLISHING CO. LTD.
Barking, Essex, England

AMERICAN ELSEVIER PUBLISHING COMPANY, INC.
52 Vanderbilt Avenue
New York, New York 10017

Library of Congress Card Number: 72-103349
Standard Book Number: 444-40812-6

Printed in The Netherlands

PREFACE

This 9-language dictionary is the successor to the 8-language Horticultural Dictionary published in 1961 by the Dutch State Publishing Company. The steadily increasing international contact, stimulated by the need to keep up with horticultural developments in other countries, has underlined the need for a new edition of the original dictionary, which has been out of print for quite some time.

As regards format and arrangement, this edition differs considerably from the previous one, mainly as a result of expansion of the material. To begin with, the number of terms has been increased by about 40%. Further, in response to demand, Italian has been added as ninth language. Another difference is that English has now been chosen as the basic language; this was decided in view of the world-wide interest shown in the previous edition.

Despite this enlargement of the scope, the dictionary has of course its restrictions. The technical terms chosen are derived from general horticulture; terms relating to viniculture, for example, will therefore not be found here. Highly specialized terms, which one would expect to find only in a specialized dictionary relating to one of the subsections of horticulture, have not been included, nor have those terms that one can find in any good general dictionary.

Many people have collaborated enthusiastically in the preparation of this dictionary. In particular, special mention should be made of Mr. J. Nijdam, who again accepted the role of coordinating editor. The members of the working group, consisting of representatives of all sectors of horticulture, had the task of chosing the terms to be included; they fulfilled this task with devotion and skill. Finally, the invaluable help of a great many colleagues throughout the world must be acknowledged. The reward of all concerned will be the conviction that this dictionary will serve a useful purpose and be widely used in horticulture circles everywhere.

CONTENTS

INSTRUCTIONS – MODE D'EMPLOI – GEBRUIKSAANWIJZING
HINWEISE FÜR DIE BENUTZUNG
INSTRUKTIONER – INSTRUKTIONER
INSTRUCCIONES – ISTRUZIONI

The words have been arranged alphabetically on the basis of their English spelling. Alphabetical indexes are provided for each language, each word being given the reference number under which the equivalents in the other languages will be found in the main list.

Les mots sont classés par ordre alphabétique d'après leur orthographe anglais. Des index alphabétiques pour chaque langue permettent, grâce à des numéros de référence pour chaque mot, de trouver dans la liste principale les termes équivalents dans les autres langues.

De woorden zijn alfabetisch gerangschikt volgens hun schrijfwijze in het Engels. In de alfabetische registers van de verschillende talen zijn de woorden voorzien van een nummer dat correspondeert met de nummering in de basistabel.

Die Wörter sind alphabetisch nach ihrer englischen Schreibweise angeordnet. Für jede Sprache wird ein alphabetisch geordnetes Wörterregister gegeben, und jedes Wort ist mit einer Ordnungsnummer gekennzeichnet, unter der die entsprechenden Wörter in den anderen Sprachen im Hauptwörterregister zu finden sind.

Ordene er ordnet alfabetisk efter deres engelske stavemåde. Der findes alfabetiske indholdsfortegnelser for hvert sprog, og hvert ord er forsynet med et nummer, under hvilket man i hovedlisten kan finde det tilsvarende ord på de andre sprog.

Orden äro arrangerade i alfabetisk ordning på grundval av deras engelska stavning. Alfabetiska förteckningar finnes för varje språk, och för varje ord gives det referensnummer under vilket motsvarigheterna på de andra språken förekommer på huvudförteckningen.

La palabras se colocan en orden alfabético en base a su ortografía inglés. Se proveen indices alfabéticos para cada lenguaje y en los mismos se ha asignado a cada palabra el número de referencia bajo el que se encuentra en la lista principal su equivalente en las otras lenguas.

Le parole sono disposte in ordine alfabetico sulla base della loro ortografia inglese. Si provvedono indici alfabetici per ciascuna lingua, nei quali ogni parola ha il numero di riferimento sotto il quale può trovarsi il suo equivalente nell'elenco principale.

IX

ABBREVIATIONS – ABRÉVIATIONS – AFKORTINGEN
ABKÜRZUNGEN – FORKORTELSER – FÖRKORTNINGAR
ABREVIACIONES – ABBREVIAZIONI

	ENGLISH	FRANÇAIS	NEDERLANDS	DEUTSCH
adj.	adjective	adjectif	bijvoeglijk naamwoord	Adjektiv
subst.	substantive	substantif	zelfstandig naamwoord	Substantiv
m	masculine	masculin	mannelijk	männlich
f	feminine	feminin	vrouwelijk	weiblich
n	neuter	neutre	onzijdig	sächlich
pl	plural	pluriel	meervoud	Plural
bot.	botanical	botanique	botanisch	botanisch
(US)	term mainly used in North America	terme utilisé principalement en Amérique du Nord	term gebruikelijk in Noord-Amerika	in den USA üblicher Ausdruck
(UK)	term mainly used in Great Britain	terme utilisé principalement en Grande-Bretagne	term gebruikelijk in Groot-Brittannië	in England üblicher Ausdruck
fr	French	français	Frans	französisch
ne	Dutch	néerlandais	Nederlands	holländisch
de	German	allemand	Duits	deutsch
da	Danish	danois	Deens	dänisch
sv	Swedish	suédois	Zweeds	schwedisch
es	Spanish	espagnol	Spaans	spanish
it	Italian	italien	Italiaans	italienisch

	DANSK	SVENSKA	ESPAÑOL	ITALIANO
adj.	tillægsord	adjektiv	adjetivo	aggettivo
subst.	navneord	substantiv	substantivo	sostantivo
m	hankøn	maskulinum	masculino	maschile
f	hunkøn	femininum	femenino	femminile
n	intetkøn	neutrum	neutro	neutro
pl	flertal	pluralis	plural	plurale
bot.	botanisk	botanisk	botánico	botanico
(US)	udtryk der hovedsagelig anvendes i Nordamerika	term använd huvudsakligen i Nordamerika	término empleado principalmente en los Estados Unidos	termine impiegato principalmente nell'America del Nord
(UK)	udtryk der hovedsagelig anvendes i Storbritannien	term använd huvudsakligen i Storbritannien	término empleado principalmente en el Reino Unido	termine impiegato principalmente nel Regno Unito
fr	fransk	franska	francés	francese
ne	hollandsk	holländska	holandés	olandese
de	tysk	tyska	alemán	tedesco
da	dansk	danska	danés	danese
sv	svensk	svenska	sueco	svedese
es	spansk	spanska	español	spagnuolo
it	italiensk	italienska	italiano	italiano

COLLABORATORS

Of the many persons who gave their help in the translation or compiling of the terms in the various fields of horticulture, the following ones deserve special mention:

GENERAL TERMS AND CO-ORDINATION FOR THE RELATIVE LANGUAGES
D. Akenhead, Commonwealth Bureau of horticulture and plantation crops, East Malling Research Station, near Maidstone, Kent, England (English); E. Blankholm, Aagaard, Denmark (Danish); J. Douglas, Instituut voor de Veredeling van Tuinbouwgewassen, Wageningen, The Netherlands (Italian); J. C. Garnaud, Comité national interprofessionnel de l'horticulture et des pépinières, 42 Ave. de Normandie, 94 – Halles de Rungis, France (French); Dr. K. W. Müller, Staatliche Lehr- und Forschungsanstalt für Gartenbau, Weihenstephan, W. Germany (German); M. del Pozo Ibanez, Ing. Agr., Avenido de América 50, Madrid, Spain (Spanish); R. Sabarte Belacortu, Dijkgraafseweg 1, Wageningen, The Netherlands (Spanish); S. Willhammar, Civilhortonom, Statens Plantskolenämnd, Åkarp, Sweden (Swedish).

ARBORICULTURE
Prof. Asger Klougart, København, Denmark; Dr. H. Rademacher, Direktor der Lehr- und Versuchsanstalt für Gartenbau, Aurich (Ostfr.), W. Germany; J. Segers, Pépinières Minier, 74 Rue Volnay, 49 – Angers, France; S. Willhammar, Statens Plantskolenämnd, Åkarp, Sweden.

CULTIVATION UNDER GLASS
Dieter Jansen, Taunusstrasse 20, Wiesbaden, W. Germany; R. de Keyzer, Stookte 1, Wetteren, Belgium; Prof. Asger Klougart, The Royal Veterinary and Agricultural College, København, Denmark; S. Willhammar, Statens Plantskolenämnd, Åkarp, Sweden.

DISEASES AND PESTS
General terms on diseases E. Blankholm, Aagaard, Denmark; Lic. Agro. Harald Eriksson, Alnarpsinstitutets Trädgårdsavdelning, Alnarp, Sweden; A. Faivre-Amiot, Station centrale de pathologie végétale, 78 – Versailles, France; C. Schoen, Plantenziektenkundige Dienst, Wageningen, The Netherlands; S. Willhammar, Statens Plantskolenämnd, Åkarp, Sweden.

Bacteria, Fungus diseases Dr. Fr. Hejndorf, Statens Plantepathologiske For-søg, Lyngby, Denmark; Dr. F.J. Moore, Ministry of Agriculture, Fisheries and Food, Plant Pathology Laboratory, Harpenden, Herts., England; Dr. Chr. Stark, Pflanzenschutzamt Hamburg, 2 Hamburg 36, W. Germany; R. Tramier, Station de Botanique et de Pathologie Végétale, 06 – Antibes, France; S. Will-hammar, Statens Plantskolenämnd, Åkarp, Sweden.

Eelworms Dr. E. Gadea, Instituto de Biologia Aplicada, Universidad de Barcelona, Barcelona, Spain; E. Johansson, Statens Växtskyddsanstalt, Solna 7, Sweden; Dr. A. Scognamiglio, Osservatorio per le Malattie delle Piante, Pes-cara, Italy; Dr. B. Weischer, Institut für Hackfruchtkrankheiten und Nemato-denforschung, Münster, Westf., W. Germany.

Insects and mites Dr. F.H. Jacob, Plant Pathology Laboratory, Harpenden, Herts., England; Dr. K. Lindhardt, Statens Plantepathologiske Forsøg, Lyngby, Denmark; Prof. Dr. Minos Martelli, Direttore dell' Istituto di Entomologia Agraria, Osservatorio per le Malattie delle Piante, Milano, Italy; Dr. Olof Ry-berg, Sergelsväg 14a, Malmö, Sweden; Dr. G. Schmidt, Biologische Bundes-anstalt für Land- und Forstwirtschaft, Berlin-Dahlem, W. Germany.

Viruses Dr. L. Bos, Instituut voor Plantenziektenkundig Onderzoek, Wage-ningen, The Netherlands; F.C. Brawden, Director Rothamsted Exp. Station, Harpenden, Herts., England; Dr. P. Cornuet, Institut National de la Recherche Agronomique, Station Centrale de Pathologie Végétale, 78 – Versailles, France; Dr. P. Grandcini, Stazione Sperimentale di Maiscultura, Bergamo, Italy; Prof. Dr. Max Klinkovski, Institut für Phytopathologie, Ascherleben, E. Germany.

Weeds Lic. Agro. Harald Eriksson, Alnarpsinstitutets Trädgårdsavdelning, Alnarp, Sweden; P.J. Quillon, Directeur agronomique de la Société Borax Français, 78 – Saint-Germain-en Laye, France; H.A. Robers, National Vege-table Research Station, Wellesbourne, Warwick, England; Dr. A. Scognamiglio, Osservatorio per le Malattie delle Piante, Pescara, Italy.

FLORICULTURE, BOTANY, GARDEN DESIGN

Prof. R. Bossard, Ecole Nationale Supérieure d'Horticulture, 78 – Versailles, France; W.C. Krause, Editor of "The Grower", 49 Doughty Street, London W.C. 1, England; Prof. Asger Klougart, The Royal Veterinary and Agricultural College, København, Denmark; Prof. G. Puccini, Stazione Sperimentale, Sanremo, Italy; S. Willhammar, Statens Plantskolenämnd, Åkarp, Sweden; Karin Bjursell-Willhammar, Falkvägen 27, Åkarp, Sweden.

FLOWER BULB CULTIVATION

H. Groet, Coöp. Agr. de la Prod. des Bulbes Les Noëls, 41 – Vineuil, France; Dr. D.E. Horton, County Horticultural Adviser N.A.A.S., Wellington Rd.,

Kirton, Boston Lincs., England; Dr. H. Rademacher, Lehr- und Versuchsanstalt für Gartenbau, 296 Aurich (Ostfr.), W. Germany; E. Rasmussen, Lumbyvejen, Aarslev, Denmark.

FRUIT GROWING

Sv. Joh. Jørgensen, Danmarks Erhvervs Frugtavleren Forening, Vendegarde 74, Odense, Denmark; A. Felderer, Landwirtschaftsinspectorat, Siegeplatz 48/1, Bolzano, Italy.

HORTICULTURAL ENGINEERING

Dr. R. Bohn, Zentralverband des Deutschen Gemüse-, Obst- und Gartenbaues, Koblenzerstrasse 33, Bonn a/Rh., W. Germany; Dr. J. Cardúz, Instituto de Edafologia y Biologia Vegetal, Universidad de Barcelona, Barcelona 14, Spain; Svend Aa. Christensen, Produktivitetsudvalget for Gartneri og Frugtavl, Anker Heegaardsgade 2, København, Denmark; Prof. Dr. Ghisleni, Istituto di Orticoltura e Floricoltura, Torino, Italy; Lennart Gröné, Lantbrukshögskolan, Alnarp, Sweden; Mr. Harris, National Institute of Agricultural Engineering, Wrest Park, Silsoe (Bedford), England; J. Lemoyne de Forges, Ecole Nationale d'Horticulture, 78 – Versailles, France.

MUSHROOM CULTIVATION

G. Derks, c/o F.lli Sartor, Funghi del Montello, Treviso, Italy; J. Roca Dumora, Plaza de Catlanuna 6, Barcelona 2, Spain; S. Willhammar, Statens Plantskolenämnd, Åkarp, Sweden.

PROCESSING, PRESERVATION, TRANSPORT

Prof. Dr. G. Bünemann, Techn. Universität, Berlin, W. Germany; Pedro Caldentey, Ministerio de Agricultura, Secretaria General Tecnica Información de Precios y Mercados, Madrid 7, Spain; Dr. J.C. Fidler, Ditton Laboratory, Maidstone, Kent, England; T. Johnsson, Institutionen för Frükt- och bärodling, Alnarp, Sweden; P. Moll's Rasmussen, Statens Forsøgstation Blangstedgaard, Odense, Denmark; Dr. K. Stoll, Eidg. Versuchsanstalt für Obst-, Wein- und Gartenbau, 8820 Wädenswil, Switzerland.

SOIL SCIENCE AND FERTILIZING

Prof. M. de Boodt, Rijkslandbouwhogeschool, afd. Bodemkunde en Geologie, Gent, Belgium; Dr. J. Cardúz, Instituto de Edafologia y Biologia Vegetal, Universidad de Barcelona, Barcelona 14, Spain; H. Fabian, An der Dekanei 12, 309 Verden-Aller, W. Germany; Fr. Heick, Det Danske Hedeselskab, Viborg, Denmark; A. Massacesi, Commissario Straordinario dell' Istituto Sperimentale per lo Studio e la Difesa del Suelo, Borgo Pinti 80, Firenze, Italy.

TRADE AND ECONOMICS

Dr. Massimo Bartolelli, Centro de Specializzazione e Ricerche Economico-Agrario per il Mezzogiorno, Università di Napoli, Portici (Napoli), Italy; Mårten Carlsson, Lantbrukshögskolan, University lecturer, Alnarp, Sweden; Dr. R.R.W. Folley, Department of Agricultural Economics, Wye College, University of London, near Ashford, Kent, England; Javier Gros, Ing. Agr., Centro Hortifruticola del Ebro, Zaragoza, Spain; M. Petit, Institut National de la Recherche Agronomique, Paris – 16, France; Dr. A. Staunig, Erhvervsraadet for Gartneri og Frugtavl, Anker Heegaardsgade 2, København, Denmark; Prof. Dr. H. Storck, Institut für Wirtschaftslehre des Gartenbaues, 805 Freising-Weihenstephan, W. Germany.

VEGETABLE GROWING IN THE OPEN

E. Blankholm, Aagaard, Denmark; H. Duggen, Kronshagenerweg 130, Kiel, W. Germany; G. Flament, Société Vilmorin-Andrieux, Paris, France; Prof. Asger Klougart, The Royal Veterinary and Agricultural College, København, Denmark; W. Mollemans, Tuinbouwschool Anderlecht, Brussel, Belgium.

MEASURES, WEIGHTS AND OTHER UNITS

LENGTH MEASURES
1 inch (in) = 2.54 cm
1 foot (ft) = 12 in = 30.48 cm
1 yard (yd) = 3 ft = 36 in = 91.44 cm
1 mile = 1760 yds = 1609.32 m

SURFACE MEASURES
1 acre = 4 roods = 160 sq. perches = 4840 sq. yards = 40.46 a
1 ha = 100 a = 10.000 m² = 2.47 acres = 4 Morgen (Germany)
1 a = 1/100 ha = 0.0247 acres = 119.6 sq. yds = 1076 sq. ft

MEASURES OF CAPACITY
1 (UK) gal. (gallon) = 4 liq. quarts = 8 liq. pints = 4.54 liter
1 (US) gal. = 4 liq. quarts = 8 liq. pints = 3.78 liter
1 register ton (UK) = 100 cu. ft = 2.8317 m³
1 barrel (dry) (US) = 3.281 bushels = 105 dry quarts = 0.115 m³
1 bushel (UK) = 0.125 quarter = 36.367 dm³ (liter)
1 bushel (US) = 4 pecks = 32 dry quarts = 35.238 dm³
1 cu. ft = 28.316 dm³
1 cu. in = 16 cm³
1 peck (UK) = 2 gallons = 9.09 dm³
1 hl = 100 l (dm³) = 2.75 (UK) bushels = 2.837 (US) bushels
1 l = 1 dm³ = 1.75 pints = 2.2 (UK) gallons
1 m³ (cbm) = 10 hl = 1000 l = 1.3 cu. yd
1 dm³ (liter) = 1000 cm³ = 61 cu. in = 0.035 cu. ft
1 cm³ (ccm) = 0.061 cu. in

WEIGHTS
(t = ton; lgt. = longton; sht. = shortton; cwt. = hundredweight; lb. = pound; oz. = ounce; kg = kilogram; gr = gram; dz. = Doppelzentner)

1 t = 1000 kg = 0.984 (UK) ton = 1.10 (US) ton
1 lgt. = 2204 lbs. = 1016.06 kg
1 sht. = 2000 lbs. = 907.185 kg
1 cwt. long = 112 lbs. = 50.80 kg

1 cwt. short (US) = 1 central = 100 lbs.   = 45.36 kg
1 lb.   = 16 oz.   = 0.45 kg
1 oz.   = 28.34 gr
1 kg   = 1000 gr   = 2.20 lbs   = 35.27 oz.
1 dz.   = 100 kg   = 2 Zentner   = 1.96(UK)cwt. = 2.2(US)cwt.

TEMPERATURE

Ratio F : C : R = 9 (+ 32) : 5 : 4
212 °F   = 100 °C   = 80 °R
0 °F   = –17.8 °C = –14.2 °R
0 °C   = 32 °F

PRESSURE

1 oz./sq. in.   = 44 mm WK
1 in. of water   = 25.4 mm WK
1 lb./sq. in.   = 0.0703 kg/cm^2
1 lb./sq. ft   = 4.88 kg/cm^2
1 in. mercury   = 345 mm WK
1 mm WK   = 0.0227 oz./sq.in. = 0.0394 in. of water = 0.0029 in. mercury
1 kg/cm^2 = 14.2 lb./sq. in.   = 0.205 lb./sq. ft

WORK AND POWER

1 horsepower (HP) = 1.0138 pk
1 pk   = 0.986 HP
1 ft lb.   = 0.1388 kgm
1 kgm   = 7.23 ft lb.
1 HP   = 746 W   = 76 kg/sec. = 33000 ft lb./min. = 550 ft lb/ sec.
1 kW   = 1.359 pk   = 1.34 HP
1 kWh   = 3411 BTU   = 860 kcal.

HEAT UNITS

1 BTU (British thermal unit)   = 1 °F per lb.   = 0.252 kcal.
1 BTU per cu. ft   = 0.555 kcal./kg
1 BTU per lb.   = 8.9 kcal./m^3
1 BTU per sq. ft   = 2.71 kcal./m^2
1 BTU per sq. ft per hour per °F = 4.87 kcal./m^2 h °C
1 BTU per ft per hour per °F   = 1.49 kcal./m h °C
1 BTU per 1 in. per hour per °F = 17.85 kcal./m h °C

1 kcal.   = 3.97 BTU
1 kcal./kg   = 1.80 BTU/lb.

1 kcal./m^3	= 0.1121 BTU/cu. ft
1 kcal./m^2	= 0.369 BTU/sq. ft
1 kcal./m^2 h °C	= 0.206 BTU/sq. ft h °F
1 kcal./m h °C	= 0.671 BTU/ft h °F
1 kcal./m h °C	= 0.056 BTU/in.h °F

RADIATION QUANTITIES

Quantity	Units (ISO-symbol)	Other units		
radiant flux	W	1.163	W	= 1 kcal./h
irradiance	W/m^2	4.186.10^4	W/m^2	= 1 cal./cm^2 s
radiant intensity	W/sr	4.186	W/sr	= 1 cal./sr.s
radiance	W/sr. m^2	4.186.10^4	W/sr. m^2	= 1 cal./sr.cm^2 s

PHOTOMETRIC QUANTITIES

N.B.   This system is derived from the physiology of the human eye and is not suited to the plant physiology.

luminous flux	lm = cd/sr.		
illumination	lm/m^2 = lux	10^4 lux	= 1 phot.
luminous intensity	cd (candela)		
luminance	cd/m^2	0.3183 cd/m^2	= 1 asb (apostilb)
		10^4 cd/m^2	= 1 sb (stilb)
		3.183.10^3 cd/m^2	= 1 la (lambert)

Sr (steradian) is the solid angle in the centre of a sphere, which encloses a surface on this sphere equivalent to the square of the radius.

BASIC TABLE

A

1
ABDOMEN
fr abdomen *m*
ne achterlijf *n*
de Hinterleib *m*
da bagkrop
sv bakkropp
es abdomen *m*
it addome *m*

2
A-BLADE
fr sarcleuse *f* à patte de
    canard
ne ganzevoet (werktuig)
de Gänsefuss *m*
da gåsefod
sv gåsfot
es escardadera *f*
it lama *f*

3
ABSOLUTE HUMIDITY
fr humidité *f* absolue
ne absolute vochtigheid
de absolute Feuchtigkeit *f*;
    absolute Feuchte *f*
da absolut fuktighed
sv absolut fuktighet
es humedad *f* absoluta
it umiditá *f* assoluta

4
ABSORB, TO
fr absorber
ne absorberen
de absorbieren
da absorbere
sv absorbera
es absorber
it assorbire

5
ABSORPTION
fr absorption *f*
ne absorptie

de Absorption *f*
da absorbtion
sv absorption
es absorción *f*
it assorbimento *m*

6
ABSORPTION CAPACITY
fr pouvoir *m* absorbant
ne absorptievermogen *n*
de Absorptionsvermögen *m*
da absorbtionsevne
sv absorptionsförmåga
es poder *m* de absorción
it capacitá *f* d' assorbimento

7
ABSORPTION OF FOOD
fr absorption *f* de substances
    nutritives
ne voedselopneming
de Nahrungsaufnahme *f*
da næringsoptagelse
sv näringsupptagning
es absorción *f* de substancias
    nutritivas
it assorbimento *m* nutritivo

8
ABUNDANT
fr dru; bien venant;
    luxuriant
ne welig
de geil
da frodig
sv frodig
es exuberante
it abbondante

ACACIA, false,   1284

9
ACALYPHA
fr Acalypha *m*
ne kattestaart
de Acalyphe *f*
da kattesvans
sv kattsvans

es acalifa *f*
it Acalypha *f*
    Acalypha L.

10
ACAULESCENT;
STEMLESS
fr acaule
ne stengelloos
de stengellos
da stængelløs
sv stjälklos
es acaule
it acaule

11
ACCLIMATIZE TO;
TO HARDEN OFF
fr acclimater
ne acclimatiseren
de akklimatisieren;
    eingewöhnen
da acklimatisere
sv acklimatisera
es aclimatizar
it acclimare

12
ACCOUNTING (FARM)
fr comptabilité *f*
ne boekhouding
de Buchführung *f*
da regnskabsføring
sv bokföring; redovisning
es contabilidad *f*
it contabilità *f* (aziendale)

13
ACCOUNTING BUREAU
fr organisme *m* comptable
ne boekhoudbureau *n*
de Buchstelle *f*
da regnskabsbureau *n*
sv bokföringskontor *n*;
    bokföringsbyrå *n*
es central *f* contable
it ufficio *m* di contabilità

14
ACCOUNTING YEAR
fr exercice *m*
ne boekjaar *n*
de Wirtschaftsjahr *n*
da regnskabsår *n*
sv bokföringsår *n*
es ejercicio *m*
it annata *f* contabile

15
ACETIC ACID
fr acide *m* acétique
ne azijnzuur
de Essigsäure *f*
da eddikesyre
sv ättiksyra
es ácido *m* acético
it acido *m* acetico

16
ACHENE
fr akène *m*; achaine *m*
ne dopvrucht
de Schliessfrucht *f*
da nødfrugt
sv nötfrukt
es aquenio *m*
it achenio *m*

17
ACID
fr acide
ne zuur
de sauer
da sur
sv sur
es ácido
it acido

ACID, gibberellic, 1571
—, humic, 1846
—, hydrochloric, 1858

18
ACID HUMUS
fr humus *m* acide
ne zure humus
de saurer Humus *m*
da sur humus
sv sur humus
es humus *m* ácido
it umus *m* acido

19
ACIDIFY, TO
fr acidifier
ne verzuren
de versauern
da gøre sur
sv försura
es acidificar
it acidificare

20
ACIDITY
fr degré *m* d'acidité
ne zuurgraad
de Säuregrad *m*
da surhedsgrad
sv surhetsgrad
es grado *m* de acidez *m*
it grado *m* d'acidità

21
ACONITE; MONKSHOOD
fr aconit *m*
ne monnikskap
de Eisenhut *m*
da stormhat
sv stormhatt
es acónito *m*
it aconito *m*
   Aconitum L.

22
ACTIVE ELEMENT
fr matière *f* active
ne actieve stof
de Wirkstoff *m*; activer
   Stoff *m*
da aktive bestanddele
sv verksam substans
es material *m* activo
it sostanza *f* attiva

23
**ACTUAL PRODUCTION
UNIT (FRUIT GROWING)**
fr peuplement *m*
ne plantopstand
de Pflanzenbestand *m*
da plantebestand
sv trädbestånd *n*
es plantación farborea
it inventario *m* di piante

24
**ACYTHOPEUS
ATERRIMUS**
fr Acythopeus *m*
ne orchideeënsnuitkever
de mattschwarzer
   Orchideenrüssler *m*
da orkidéersnudebille
sv orchidévivel
es curculióndio *m* de las
   orquídeas
it Acythopeus *m*
   Acythopeus aterrimus
   Waterh.

25
ADDED VALUE
fr valeur *f* ajoutée
ne toegevoegde waarde
de Wertschöpfung *f*
da tillæ gsbeløb *n*
sv nettovärdeökning
   (mervärde *n*)
es valor *m* añadido
it valore *m* aggiunto

26
**ADDITIONAL
LIGHTING;
SUPPLEMENTARY
LIGHTING**
fr exposition *f* supplémen-
   taire; éclairage *m* supplé-
   mentaire; éclairage *m*
   additionnel
ne bijbelichting
de Beibeleuchtung *f*;
   Nebenbeleuchtung *f*
da tilskudsbelysning
sv tillskottsbelysning
es iluminación *f* accesoria
it illuminazione faggiuntiva

ADELGES, 3587
—, douglas fir, 2733
—, larch, 4203
—, pine bark, 2733
—, silver fir, 3411
—, spruce pineapple, 3587

27
ADHESIVE
fr adhésif *m*; agglutinant *m*
ne hechtmiddel *m*

de Klebemittel *n*
da klæbemiddel *n*
sv vätningsmedel *n*
es adhesivo *m*
it mezzo adesivo *m*

28
ADSORB, TO
fr adsorber
ne adsorberen
de adsorbieren
da adsorbere
sv adsorbera
es adsorber
it adsorbire

29
ADSORPTION
fr adsorption *f*
ne adsorptie
de Adsorption *f*
da adsorbtion
sv adsorption
es adsorción *f*
it adsorbimento *m*

30
ADSORPTION CAPACITY
fr pouvoir *m* adsorbant
ne adsorptievermogen *n*
de Adsorptionsvermögen *n*
da adsorbtionsevne
sv adsorptionsförmåga
es poder *m* de adsorción
it capacità *f* d'adsorbimento

ADULT INSECT, 1872

ADVANCE, TO, 1425

31
ADVANCES ON
CURRENT ACCOUNT
fr avances *f pl* en compte
courant
ne rekening-courant krediet
de Kontokorrentkredit *m*
da mellemregning
sv löpande kredit
es créditos *m pl* en cuenta
corriente
it attivo *m* del conto corrente

32
ADVENTITIOUS BUD
fr bourgeon *m* adventif
ne adventiefknop
de Adventivknospe *f*
da adventivknop
sv adventivknopp
es yema *f* adventicia
it gemma *f* avventizia

33
ADVICE
fr instruction *f*
ne voorlichting
de Beratung *f*
da vejledning
sv rådgivning
es asesoramiento *m*
it informazione *f*;
istruzione *f*

34
ADVISER
fr conseiller *m*
ne consulent
de Berater *m*
da konsulent
sv konsulent
es asesor *m*
it consultore *m*

35
ADVISORY SERVICE;
EXTENSION SERVICE
(US)
fr service *m* d'informations
ne voorlichtingsdienst
de Beratungsdienst *m*
da konsulenttjeneste
sv upplysningsverksamhet;
informationstjänst
es servicio *m* de extensión
it servizio *m* d'informazione

AEOLIAN SOIL, 2206

36
AERIAL
fr aérien
ne bovengronds
de oberirdisch
da overjordisk
sv ovanjordisk
es aéreo,
it aereo

37
AERIAL ROOT
fr racine *f* aérienne
ne luchtwortel
de Luftwurzel *f*
da luftrod
sv luftrot
es raíz *f* aérea
it radice *f* aerea

38
AERIAL ROOTLET
fr racine *f* aérienne
ne hechtwortel
de Haftwurzel *f*
da hefterod
sv häftrot
es fibrilla *f* radiciforme;
garfio *m* radiciforme
it radice *f* abbarbicante

39
AERIAL SURVEY
fr cartographie *f* aérienne
ne luchtkartering
de Luftkartierung *f*
da luftkortlægning
sv luftkartering;
flymarkkartering
es cartografía *f* aérea
it cartografia *f* aerea

40
AEROBIC
fr aérobie
ne aërobe
de aerobisch
da ærob
sv aerob
es aerobio
it aerobico

41
AEROBIC BACTERIA
fr bactéries *f pl* aérobies
ne aërobe bacteriën
de Aeroben *f pl*;
aerobe Bakterien *f pl*
da ærobe bakterier
sv aeroba bakterier
es bacterias *f pl* aerobias
it batteri *m pl* aerobici

42
AEROSOL
fr aérosol *m*
ne aerosol
de Aerosol *n*
da ærosol
sv aerosol
es aerosol *m*
it aerosol *m*

43
AFFOREST, TO
fr boiser
ne bebossen
de aufforsten
da plante skov; skovdække
sv plantera skog på
es repoblación *f* forestal
it rimboscare

AFRICAN HEMP, 3535

44
AFRICAN MARIGOLD;
FRENCH MARIGOLD
fr tagète *f*; oeillet *m* d'Inde;
rose *f* d'Inde
ne afrikaan
de Samtblume *f*
da fløjelsblomst; Tagetes
sv sammetsblomma
es damasquina *f*
it rosa *f* indica
Tagetes L.

45
AFTER-RIPENING
fr maturation *f*
complémentaire
ne narijping
de Nachreifung *f*
da eftermodning
sv eftermognad
es maduración *f* artificial
it maturazione *f*
complementare

46
AGEING
fr vieillissement *m*
ne veroudering
de Alterung *f*

da aldring
sv åldring
es envejecimiento *m*
it invecchiamento *m*

47
AGGREGATE;
GENERATOR
fr groupe *m* (électrogène)
ne aggregaat *n*
de Aggregat *n*
da aggregat *n*
sv aggregat
es generador *m* eléctrico,
grupo *m* electrógeno
it aggregato *m*

48
AGITATOR
fr agitateur *m*
ne roerder
de Rührwelle *f*
da omrører
sv omrörare
es agitador *m*
it agitatore *m*

49
AGRICULTURAL
fr agricole; agraire
ne agrarisch
de agrarisch; landwirt-
schaftlich
da agrar; landbrugs-
sv lantbruks-
es agrario
it agricolo

AGRICULTURAL
CHALK, 1671

50
AIR, TO;
TO VENTILATE
fr ventiler; aérer
ne luchten; ventileren
de lüften
da lufte; løfte
sv lufta; ventilera
es airear; ventilar
it ventilare; arieggiare

51
AIR BLAST ATOMIZING
BURNER
fr brûleur *m* à pulvérisation
pneumatique
ne luchtverstuivingsbrander
de Druckluftzerstäuber-
brenner *m*
da luftforstøvningsfyr *n*
sv pressluftbrännare
es quemador *m* de pulverización neumático
it bruciatore *m* ad atomizzatore pneumatico

52
AIR-CONDITIONING
fr climatisation *f*;
conditionnement *m* d'air
ne klimatisering
de Klimatisierung *f*
da klimaregulering
sv luftkonditionering
es acondicionamiento *m* del
aire; climatización *f*
it condizionamento *m* dell'
aria

53
AIR COOLER
fr refroidisseur *m* à air
ne luchtkoeler
de Luftkühler *m*
da luftkøler
sv luftkylare
es refrigerador *m* de aire
it refrigerante *m* d'aria

54
AIR DISCHARGE
fr sortie *f* de l'air
ne luchtafvoer
de Luftabzug *f*
da luftaftræk *n*
sv luftavförsel
es salida *f* del aire
it condotta *f* aerea

55
AIR DISCHARGE
CHANNEL; AIRDUCT
fr canal *m* d'évacuation de
l'air
ne luchtafvoerkanaal *n*

de Luftabzugkanal *m*
da aftrækskanal
sv luftväxlingskanal
es canal *m* de aireación
it canale *m* di scolo d'aria

56
AIR DRIED
fr séché à l'air
ne luchtdroog
de lufttrocken
da lufttørret
sv lufttorr
es secado al aire
it seccato all'aria

AIRDUCT, 55

57
AIR LAYERING
fr marcottage *m* aérien
ne marcotteren
de abmoosen; markottieren
da aflægning
sv avläggning
es acodo *m* al aire;
acodo *m* alto
it margottare

58
AIRMIXING FAN
fr ventilateur *m* de brassage
ne mengventilator
de Umwälzventilator *m*
da ventilator "vifte"
sv blandningsventilator
es ventilador *m*
it ventilatore *m* miscelatore

59
AIR POLLUTION
fr pollution *f* atmosphérique
ne luchtverontreiniging
de Luftverschmutzung *f*
da luftforurening
sv luftförorening
es impurificación *f* del aire
it insudiciamento *m* d'aria

60
AIR SUPPLY; FLOW
fr entrée *f* de l'air
ne luchttoevoer
de Luftzufuhr *f*

da lufttilførsel
sv lufttillförsel
es entrada *f* del aire
it approvvigionamento *m*
aereo

61
AIRTIGHT
fr hermétique
ne luchtdicht
de luftdicht
da lufttæt
sv lufttät; hermetisk
es hermético
it ermetico

62
AIR WASHER
fr laveur *m* d'eau
ne luchtwasser
de Luftwäscher *m*
da luftvasker
sv lufttvättare
es lavador *m* de aire
it lavatore *m* d'aria

63
ALBUMIN
fr albumine *f*
ne eiwit
de Eiweiss *n*
da æggehvidestof *n*
sv äggviteämne
es albúmina *f*
it albume *m*

64
ALDER
fr aune *m*; aulne *m*
ne els
de Erle *f*
da el
sv al
es aliso *m*
it alno *m*; ontano *m*
Alnus Mill.

65
ALDER BUCKTHORN
fr bourdaine *f*
ne vuilboom
de Faulbaum *m*
da tørstetræ *n*
sv brakved

es frangula *f*; arraclán *m*
it alno *n* nero
Rhamnus frangula L.

66
ALDER LEAF BEETLE
fr galéruque *f* de l'aulne
ne elzehaan
de blauer Erlenblattkäfer *m*
da ellebladbille
sv blå allövsbaggen
es crisomela *f* del aliso;
galerucela *f* del aliso
it crisomela *f* dell'ontano
Agelastica alni L.

67
ALDER SAWFLY
CATERPILLAR;
HAZEL SAWFLY
fr *Croesus m* septentrionalis
ne elzebastaardrups
de breitfüssige Birkenblattwespe *f*
da bladhvepselarve
sv bredfotad bladstekel
es falsa oruga *f* del aliso
it tentredine *f* dell'ontano
Croesus septentrionalis L.

68
ALKALINE
fr alcalin
ne basisch; alkalisch
de alkalisch; basisch
da basisk, alkalisk
sv basisk, alkalisk
es básico; alcalino
it basico

69
ALLOTMENT
fr jardin *m* familial
ne volkstuin
de Kleingarten *m*;
Schrebergarten *m*
da kolonihave
sv koloniträdgård
es jardín *m* popular
it giardino *m* popolare

70
ALMOND
fr amande *f*
ne amandel

de Mandel *f*
da mandel
sv mandel
es almendra *f*
it mandorla *f*

ALMOND, dwarf flower-
ing, 1181
—, flowering, 1398

71
ALMOND TREE
fr amandier *m*
ne amandelboom
de Mandelbaum *m*
da mandeltræ *n*
sv mandelträd *n*
es almendro *m*
it mandorlo *m*
Prunus amygdalus
Batsch.

72
ALOE
fr aloès *m*
ne Aloë
de Aloe *f*
da aloe
sv aloe
es áloe *m*; acíbar *m*
it aloe *m*
Aloë L.

73
ALPINE PLANT;
ROCKPLANT
fr plante *f* alpine;
plante *f* de rocaille
ne rotsplant
de Alpenpflanze *f*
da stenplante *mf*;
alpeplante *mf*
sv stenpartiväxt; alpväxt
es planta *f* alpina
it pianta *f* alpina

74
ALTERNARIA BLIGHT
(CARROT)
fr alternariose *f* des feuilles
ne loofverbruining
de Laubverfärbung *f*
da farveskifte af gulerodsblad
sv skiftning av lövfärg

es oscurecimiento *m* de las
hojas
it alternariosi *f* delle foglie
Alternaria dauci (Kühn)
Graves et Skolko

75
ALTERNARIA BLIGHT
(POTATO)
fr alternariose *f*
ne alternariaziekte
de Dürrfleckenkrankheit *f*;
Kohlschwärze *f*
da kartoffelskimmel
sv torrfläcksjuka
es alternariosis *f*
it peronospora *f* del
pomodoro
Alternaria solani
(E. et M.) J. et Gr.

76
ALTERNATE (bot.)
fr alterne
ne verspreid
de wechselständig
da spredt
sv strödda blad
es alternado
it alternato

77
ALUM
fr alun *m*
ne aluin
de Alaun *m*
da alun
sv alun
es alumbre *m*
it allume *m*

78
AMARYLLIS
fr amaryllis *f* de serre
ne amaryllis
de Amaryllis *f*
da ridderstjerne
sv amaryllis
es amarilis *f*
it amarilli *f*
Hippeastrum Herb.

79
AMBROSIA BEETLE;
(SMALL) SHOT-HOLE
BORER
fr xylébore *m* (disparate)
ne houtkever
de kleiner Holzbohrer *m*;
ungleicher Holzbohrer *m*
da barkbille
sv brun vedborre
es barrenillo *m* (dispar)
it sileboro *m*; bostrico *m*
Xyleborinus saxeseni
Ratz.;
Xyleborus xylographus
(UK)

80
AMELIORATE, TO;
TO IMPROVE;
TO BREED
fr améliorer; sélectionner
ne veredelen
de veredeln
da forædle
sv förädla
es mejorar
it migliorare

81
AMENTACEOUS PLANT;
AMENTIFEROUS
PLANT;
CATKIN BEARER
fr amentacée *f*;
plante *f* portant des
chatons
ne katjesdrager
de Kätzchenträger *m*
da rakletræ *n*
sv hängeväxt
es planta *f* amentifera
it pianta *f* amentacea

82
AMERICAN BIRD
CHERRY
fr capulin *m*
ne Amerikaanse vogelkers
de amerikanische Trauben-
kirsche *f*
da glansbladet hæk
sv glanshägg
es estaquilla *f*
it ciliegio *m* selvatico
Prunus serotina Ehrh.

83
**AMERICAN
CRANBERRY**
fr canneberge *f*
ne Amerikaanse veenbes
de amerikanische
    Moosbeere *f*
da Amerikansk tranebær *n*
sv amerikanskt tranbär *n*
es arándano *m* americano
it ossicocco *m* palustre
    americano
    Vaccinium macrocarpon
    Ait.

84
**AMERICAN GOOSE-
BERRY MILDEW**
fr Oïdium *m* du groseiller
ne Amerikaanse kruisbesse-
    meeldauw
de amerikanischer Stachel-
    beermehltau *m*
da stikkelsbærdræber
sv amerikansk krusbär-
    smjöldagg
es mildio *m* del grosellero
    espinoso
it mal *m* bianco dell'uva
    spina
    Sphaerotheca morsuvae
    (Schw.) Berk. et Curt.

85
AMMONIA (NH₄OH)
fr ammoniaque *f*
ne ammonia
de Salmiakgeist *m*
da ammoniak
sv ammoniumhydroxid
es hidróxido *m* amónico
it ammoniaca *f* liquida

86
AMMONIUM (NH₄)
fr ammonium *m*
ne ammonium *n*
de Ammonium *n*
da ammonium
sv ammonium
es amonio *m*
it ammonium *m*

87
AMMONIUM ALUM
fr alun *m* ammoniacal
ne ammoniakaluin
de Ammoniakalaun *m*
da ammoniakalun *n*
sv ammoniakalun
es alumbre *m* amoniacal
it allume *m* ammoniacato

88
AMMONIUM NITRATE
fr nitrate *m* d'ammonium;
    salpêtre *m* d'ammonium;
    salpêtre *m* ammoniacal
ne ammonsalpeter
de Ammonsalpeter *m*
da ammonsalpeter *n*
sv ammonsalpeter
es nitrato *m* amónico
it salnitro *m* ammoniacato

89
AMORTIZATION
fr amortissement *m*
ne aflossing
de Amortisation *f*;
    Tilgung *f*
da amortisation; afvikling
sv amortering; avbetalning
es amortización *f*
it ammortamento *m*

90
AMPHIBIOUS BISTORT
fr renouée *f* amphibie
ne veenwortel
de Wasserknöterich *m*
da vand-pileurt
sv vattenpilört
es corregüela *f*
it poligono *m*
    Polygonum amphibium L.

91
ANAEROBIC
fr anaérobie
ne anaërobe
de anaerob
da anærob
sv anaerob
es anaerobio
it anaerobe

92
ANALYSIS
fr analyse *f*
ne analyse
de Analyse *f*
da analyse
sv analys
es análisis *m*
it analisi *f*

ANALYSIS, leaf,   2069

93
ANALYSIS REPORT
fr résultats *m pl* d'analyse
ne analyserapport *n*
de Analysenergebnis *n*
da analyseresultat *n*
sv analysrapport
es boletín *m* de análisis
it rapporto *m* analisi

94
ANCHOR, TO
fr ancrer
ne verankeren
de verankern
da forankre
sv förankra
es anclar
it ancorare

95
ANCHUSA
fr buglosse *f*
ne ossetong
de Ochsenzunge *f*
da oksetunge
sv oxtunga
es buglosa *f*; lengua *f* de buey
it ancusa *f*
    Anchusa L.

96
ANDROMEDA
fr andromède *f*
ne rotsheide
de Lavendelheide *f*
da Rosmarinlyng
sv rosling
es Andromeda *f*
it Andromeda *f*
    Andromeda L.

97
ANEMONE
fr anémone *f*
ne anemoon
de Anemone *f*
da Anemone
sv anemon; sippa
es anemona *f*
it Anemone *m*
Anemone L.

98
ANGELICA
fr angélique *f*
ne Angelica
de Angelika *f*; Engelwurz *f*
da kvan
sv kvanne
es angélica *f*
it Angelica *f*
Angelica archangelica L.

99
ANGIOSPERMOUS
fr angiosperme
ne bedektzadig
de bedecktsamig
da dækfret
sv angiospermae; gömfröig
es angiosperma
it angiosperme

100
ANGLE BLADE
fr rasette *f* latérale
ne hoekschoffel
de Winkelschar *f*
da winkelhakke
sv vinkelskär *n*
es escardadera *f*
(para escardar esquinas)
it lama *f* ad angolo

101
ANGLE OF
DIVERGENCE
fr divergence *f* du faisceau
lumineux
ne spreidingshoek
de Streuwinkel *m*
da spredningsvinkel
sv spridningsvinkel
es divergencia *f* del haz
luminoso
it angolo *m* di divergenza

102
ANGLE SHADES MOTH
fr noctuelle *f*
ne agaatvlinder
de braune Achateule *f*
da brun agatugle
sv tandfly
es mariposa *f* de ágata
it Trigonophora *f*
Trigonophora meticulosa
L.; Plogophora m. (UK)

103
ANGULAR
fr anguleux; angulaire
ne kantig
de eckig; kantig
da kantet
sv kantig
es anguloso; angular
it angoloso; angolato

104
ANGULAR LEAF SPOT
(CUCUMBER)
fr taches *f pl* angulaires
ne bacterievlekkenziekte
de eckige Blattflecken-
krankheit *f*
da pletbakteriose
sv bladbakterios
es punteado *m* de las hojas
it batteriosi *f*
Pseudomonas lachrymans

105
ANHYDROUS AMMONIA
(NH₃)
fr ammoniac *m*
ne ammoniak
de Ammoniak *m*
da ammoniak-base
sv ammoniak
es amoníaco *m*
it ammoniaca *f*

106
ANNUAL
fr annuel
ne eenjarig
de einjährig
da enårig
sv ettårig
es anual
it annuo

107
ANNUAL COSTS
fr coûts *m pl* annuels
ne jaarkosten
de Jahreskosten *f pl*;
Jahresaufwand *m*
da årsomkostninger
sv årskostnad
es costes *m pl* anuales
it costi *m pl* annuali

108
ANNUAL KNAWEL
fr gnavelle *f* annuelle
ne eenjarige hardbloem
de einjähriger Knäuel *m*
da enårig knavel
sv grönknavel
es escleranto *m* anual
it centigrani *f*
Scleranthus annuus L.

109
ANNUAL MEADOW
GRASS
fr paturin *m* annuel
ne straatgras *n*
de einjähriges Rispengras *n*
da enårig rapgræs *n*
sv gröe
es espiguilla *f*, Poa *f*
it Poa *f* annua
Poa annua L.

110
ANNUAL RING
fr couche *f* annuelle
ne jaarring
de Jahresring *m*
da årring
sv årsring
es anillo *m* anual
it circhio *m* annuale

111
ANNUAL SOWTHISTLE
fr laiteron *m* potager
ne melkdistel
de Kohl-Gänsedistel *f*
da almindelig svinemællo
sv mjölktistel
es cerraja *f*
it cicerbita *f*
Sonchus oleraceus L.

112
ANNUAL WEED
fr mauvaise herbe *f* à graines
   (annuellebisannuelle)
ne zaadonkruid *n*
de Samenunkraut *n*
da frøukrudt
sv fröogräs *n*
es mala hierba *f*
it malerba *f* perenne

113
ANNUITY
fr annuité *f*
ne annuïteit
de Annuität *f*
da annuitet
sv annuitet
es anualidad *f*
it annualità *f*

114
ANT
fr fourmi *f*
ne mier
de Ameise *f*
da myre
sv myra
es hormiga *f*
it formica *f*
   Formicidae

115
ANTAGONISM
fr antagonisme *m*
ne antagonisme
de Antagonismus *m*
da antagonisme
sv antagonism
es antagonismo *m*
it antagonismo *m*

116
ANTHER
fr anthère *f*
ne helmknop
de Staubbeutel *m*
da støvknap
sv ståndarknapp
es antera *f*
it antera *f*

117
ANTHER SMUT
fr charbon *m*
ne brand; brandziekte

de Antherenbrand *m*
da brand
sv sot, ståndarsot
es carbón *m*
it carbone *m*
   Ustilago violacea (Pers)
   Roussel

118
ANTHRACNOSE
fr anthracnose *f*
ne brandvlekkenziekte
de Brandfleckenkrankheit *f*
da skivesvamp *mf*
sv gurkröta
es antracnosis *f*
it antracnosi *f*
   Glomerella lagenarium
   (Pass.)

119
ANTHRACNOSE (BEAN);
BEAN SPOT DISEASE
fr anthracnose *f*
ne vlekkenziekte
de Brennfleckenkrankheit *f*
da bønnesyge
sv bönfläcksjuka
es antracnosis *f*
it antracnosi *f*
   Colletotrichum linde-
   muthianum (Sacc. et
   Magn.) Bri. et Cav.

120
ANTHRACNOSE
(DIGITALIS)
fr taches *f pl* foliaires
ne bladvlekkenziekte
de Blattfleckenkrankheit *f*
da bladpletsyge
sv bladfläcksjuka
es punteado *m* de las hojas
it antracnosi *f*
   Colletotrichum fuscum
   Laub.

ANTHRACNOSE,
rose, 2095

121
ANTHURIUM
fr Anthurium *m*
   scherzerianum
ne flamingoplant;
   Anthurium

de Flamingoblume *f*
da flamingoblomst
sv flamingoblomma
es anturio *m*
it anturia *f*
   Anthurium scherzerianum
   Schott.

122
ANTI-GAME PRODUCTS
fr produits *m pl* de protection contre le gibier
ne wildafweermiddelen
de Wildabwehrmittel *n pl*
da vildtafværgemidler *n pl*
sv avskräckingsmedel *n pl*
es medios *m pl* protectores
   contra los daños de la
   caza
it mezzi *m pl* protettivo
   contro i selvatici

123
ANTIRRHINUM;
SNAPDRAGON
fr muflier *m*
ne leeuwebek
de Löwenmaul *n*
da løvemund
sv lejongap
es antirrino *m*;
   boca *f* de dragón
it antirrino *m*
   Antirrhinum L.

124
ANTISEPTIC
fr antiseptique
ne bederfwerend;
   antiseptisch
de fäulnisverhindernd
da antiseptisk
sv antiseptisk
es antiséptico *m*
it antisettico

APEX OF A LEAF, 2111

125
APHID; PLANT LOUSE
fr pou *m*; puceron *m*
ne luis; bladluis
de Laus *f*; Blattlaus *f*
da lus; bladlus

sv lus; bladlus
es piojo *m*; áfido *m*;
   pulgón *m*
it afide *m*; pidocchio *m*
   Aphididae

APHID, artichoke
tuber, 2142
—, beech, 285
—, cabbage, 510
—, Cooley spruce gall, 2733
—, currant, 3060
—, currant root, 946
—, currant sowthistle, 948
—, dock, 1095
—, Eastern spruce gall, 3587
—, European walnut, 1245
—, green peach, 2625
—, large willow, 2037
—, lettuce root, 2142
—, pea, 2622
—, peach-potato, 2625
—, pear-bedstraw, 2639
—, pear-grass, 2640
—, raspberry, 3034
—, red currant blister, 3060
—, rubus, 3215
—, shallot, 3362
—, spruce, 3586
—, strawberry, 3680
—, woolly, 4202
—, woolly larch, 4203
see also: apple aphid;
         bean aphid;
         potato aphid

126
APPLE
fr pomme *f*
ne appel
de Apfel *m*
da æble *n*
sv äpple *n*
es manzana *f*
it mela *f*

127
APPLE AND PEAR
BRYOBIA
fr bryobe *m* précieux
ne harlekijnmijt
de braune Spinnmilbe *f*
da stikkelsbærmide
sv kvalster *n*

es ácaro *m* arlequinado del
   manzano
it briobia *f* delle piante da
   frutto
   Bryobia rubrioculus
   Scheuten; Bryobia
   praetiosa

APPLE-AND-THORN
SKELETONIZER, 132

APPLE APHID, 1650
—, green, 1650
—, rosy, 3193
—, woolly, 4202

128
APPLE BLOSSOM
WEEVIL
fr anthonome *m* du
   pommier
ne appelbloesemkever
de Apfelblütenstecher *m*
da æblesnudebille
sv äppelblomvivel
es gorgojo *m* del manzano
it antonomo *m* del melo
   Anthonomus pomorum L.

129
APPLE BUD WEEVIL
fr anthonome *m* du
   poirier
ne pereknopkever
de Birnenknospenstecher *m*
da pæresnudebille
sv päronblomvivel
es antónomo *m* del peral;
   gorgojo *m* del peral,
   picudo *m* del peral
it antonomo *m* del pero
   Anthonomus cinctus Koll.;
   Anthonomus pyri (UK)

130
APPLE FRUIT MOTH
(LARVA = APPLE FRUIT
MINER)
fr teigne *f* du pommier
ne lijsterbesmot
de Ebereschenmotte *f*
da rønnebærmøl *n*
sv rönnbärsmal *n*

es minadora *f* del manzano;
   polilla *f* de las hojas del
   manzano
it tignola *f* del sorbo
   Argyresthia conjugella
   Zell.

131
APPLE JUICE
fr jus de pommes;
   cidre *m* doux
ne appelsap *n*; appelmost
de Apfelsaft *m*; Süssmost *m*
da æblemos
sv äppelmust
es jugo *m* de manzana
it succo *m* di mela

132
APPLE LEAF
SKELETONIZER;
APPLE-AND-THORN
SKELLETONIZER (US)
fr teigne *f* des feuilles du
   pommier
ne skeletteermot
de Apfelblattmotte *f*
da æblebladmøl *n*
sv bredvingad äppelmal
es polilla *f* de las hojas del
   manzano
it tignola *f* arrotolatrice
   delle foglie dei peri
   Simaethis pariana Clerck;
   Antophila p. (US)

133
APPLE SAUCE
fr purée *f* de pommes
ne appelmoes
de Apfelmus *n*
da æblemos *f*
sv äppelmos *n*
es pasta *f* de manzanas
it passata *f* di mela

134
APPLE SAWFLY;
EUROPEAN APPLE
SAWFLY (US)
fr hoplocampe *m* des
   pommes
ne appelzaagwesp
de Apfelsägewespe *f*

da æblehveps
sv äppelstekel
es hoplocampa *f* del
     manzano;
     avispa *f* del manzano
it tentredine *f* delle mele
     Hoplocampa testudinea
     Kl.

135
APPLE SUCKER
fr psylle *m* du pommier
ne appelbladvlo
de gemeiner Apfelblatt-
     sauger *m*
da æblebladloppe
sv äppelbladloppa
es psila *f* del manzano
it psilla *f* del melo
     Psylla mali Schmidb.

136
APPLE SYRUP
fr sirop *m* de pommes
ne appelstroop
de Apfelsirup *m*;
     Apfelwein *m*
da æblesirup
sv äppelmarmelad
es jarabe *m* de manzanas
it sciroppo *m* di mela;
     vino *m* di mela

137
APPLE TREE
fr pommier *m*
ne appelboom
de Apfelbaum *m*
da æbletræ *n*
sv äppelträd *n*
es manzano *m*
it melo *m*
     Malus L.

138
APPLE WEEVIL
fr Phyllobius *m*
ne entkever
de Pfropfreiskäfer *m*
da øresnudebille
sv fårade öronvivel
es gorgojo *m*;
     curculionido *m*
it fillobio *m*
     Phyllobius oblongus L.

139
APRICOT
fr abricot *m*
ne abrikoos
de Aprikose *f*
da abrikos
sv apriks
es albaricoque *m*
it albicocca *f*

140
APRICOT TREE
fr abricotier *m*
ne abrikozeboom
de Aprikosenbaum *m*
da abrikostræ *n*
sv aprikosträd *n*
es albaricoquero *m*
it albicocco *m*
     Prunus armeniaca L.

141
AQUATIC PLANT
fr plante *f* aquatique
ne waterplant
de Wasserpflanze *f*
da vandplante
sv vattenväxt
es planta *f* acuática
it pianta *f* aquatica

142
ARABLE FARMING
fr grande culture *f*
     agriculture *f*
ne akkerbouw
de Ackerbau *m*;
     Landwirtschaft *f*
da agerbrug *n*
     landbrug *n*
sv jordbruk; lantbruk
es labranza *f*
it agricoltura *f*

143
ARABLE LAND
fr terre *f* arable
ne bouwland *n*
de Ackerland *n*
da agerjord
sv åkerjord
es tierra *f* labrantía,
     tierra *f* de labor
it terra *f* arabile

144
ARBORICULTURE
fr arboriculture *f*
ne boomteelt
de Gehölzzucht *f*;
     Baumzucht *f*
da trædyrkning
sv odling av träd och
     buskar
es arboricultura *f*
it arboricoltura *f*

145
ARBOR VITAE
fr thuja *m*
ne levensboom
de Lebensbaum *m*
da livstræ *n*; Thuja, tuja
sv tuja
es tuya *f*
it albero *m* della vita
     Thuya L.

146
ADRIS BRUNNIVENTRIS
fr tenthrède *f* de la tige du
     rosier
ne dalende rozescheutboor-
     der
de abwärtssteigender Rosen-
     triebbohrer *m*
da rosenborer
sv rosenskottstekel
es barreno *m* de los retoños
     del rosal
it tentredine *f* minatrice dei
     getti delle rose
     Ardis brunniventris Htg.

147
AREA
fr superficie *f*
ne areaal *n*
de Areal *n*; Anbaufläche *f*
da areal *n*
sv areal; yta
es área *f*
it area *f*

ARGE ROSAE,   2036

148
ARID
fr sec; aride
ne dor

de dürr
da tør
sv torr
es árido; seco
it arido

149
ARMILLARIA ROOT
ROT
fr armillaire f de miel;
pourridié m des racines
ne honingzwam
de Hallimasch m
da honningsvamp
sv honungsskivling
es agárico m;
moho m de las raíces
it chiodini f; famigliola f;
marciume m radicale
Armillaria mellea
(Vahl) Quél.

150
ARNICA
fr Arnica m;
tabac-des-Vosges m
ne wolverlei
de Bergwohlverleih m
da guldblomme
sv slåttergubbe
es Arnica m;
tabaco m de montaña
it Arnica f
Arnica montana L.

151
AROLLA PINE
fr pin m cembro
ne alpenden
de Zirbelkiefer f; Arve f
da cembrafyr
sv cembratall
es pino m de los Alpes
it Pinus f cembra
Pinus cembra L.

152
AROMATIC
fr aromatique
ne aromatisch
de aromatisch
da aromatisk
sv aromatisk
es aromático
it aromatico

153
AROMATIC HERB
fr herbe f aromatique
ne specerijkruid n
de Gewürzkraut n
da krydderurt
sv kryddväxt
es hierba f aromática
it erba f aromatica

154
ARRANGEMENT OF
LEAVES;
PHYLLOTAXIS
fr disposition f des feuilles
ne bladstand
de Blattstellung f
da bladstilling
sv bladställning
es disposición f de las hojas
it posizione f della foglia

155
ARSENICAL COMPOUND
fr composé m arsénié
ne arseenverbinging
de Arsenverbindung f
da arsénforbindelse mf
sv arsenikförening
es compuesto m arsenical
it composto m arsenicale

156
ARTHROPODS
fr arthropodes m pl
ne geleedpotigen
de Gliedertiere n pl;
Gliederfüssler m pl
da leddyr n pl
sv leddjur n pl
es artrópodos m pl
it artropodi m pl
Arthropoda

ARTICHOKE, globe, 1597
—, Jerusalem, 1977
—, tuber aphid, 2142

157
ARTICULATE
fr articulé
ne geleed
de gegliedert
da leddelt

sv ledad
es articulado
it articolato

158
ARTIFICIAL BEE
fr abeille f mécanique
ne kunstbij
de Kunstbiene f
da kunstbi; vibrator
sv konstbi n; vibrator
es abeja f mecánica
it ape f artificiale

159
ARTIFICIAL LIGHT
fr lumière f artificielle
ne kunstlicht n
de künstliches Licht n
da kunstlys n
sv artificiellt ljus n
es luz f artificial
it luce f artificiale

160
ARTIFICIAL LIGHTING;
ILLUMINATION
fr éclairage m
ne belichting
de Belichtung f;
Beleuchtung f
da belysning
sv belysning
es iluminación f
it illuminazione f

161
ARTIFICIAL MANURE
fr engrais m artificiel
ne kunstmeststof
de Handelsdünger m
da kunstgødning
sv handelsgödselmedel
es estiércol m artificial
it concime m chimico

162
ARUM LILY
fr arum m
ne aäronskelk
de Zimmercalla f
da hvid kalla
sv kalla

es cala f; aro m
it aro m
Zantedeschia aethiopica
Spreng.

163
ASCENDING (bot.)
fr ascendant
ne opstijgend
de aufsteigend
da opstigende
sv uppstigande
es ascendente
it ascensionale

164
ASEXUAL
fr asexué parthénocarpe
ne ongeslachtelijk
de ungeschlechtlich
da ukønnet; vegetativ
sv vegetativ
es asexual
it agamica

165
ASEXUAL
MULTIPLICATION
fr multiplication f asexuée
ne vegetatieve vermeerdering
de vegetative Vermehrung f
da vegetativ formering
sv vegetativ förökning
es multiplicación f asexual
it propagazione f agamica

166
ASH
fr frêne m
ne es
de Esche f
da ask
sv ask
es fresno m
it frassino m
Fraxinus L.

ASH, mountain, 2404

167
ASP
fr tremble m
ne ratelpopulier
de Zitterpappel f

da bævreasp
sv asp
es álamo m temblón
it tremula f; alberella f
Populus tremula L.

168
ASPARAGUS
fr asperge f
ne asperge
de Spargel m
da asparges
sv sparris
es espárrago m
it sparago m
Asparagus officinalis L.

169
ASPARAGUS BEETLE
fr criocère f de l'asperge
ne aspergekever
de Spargelhähnchen n
da aspargesbille
sv sparrisbagg
es criócero m del espárrago
it criocera f dell'asparago
Crioceris spp.;
Crioceris asparagi (UK)

170
ASPARAGUS FLY
fr mouche f de l'asperge
ne aspergevlieg
de Spargelfliege f
da aspargesflue
sv sparrisfluga
es mosca f del espárrago
it mosca f dell'asparago
Platyparea poeciloptera
Schrank

171
ASPARAGUS KNIFE
fr gouge f à asperge
ne aspergemes n
de Spargelmesser n
da aspargeskniv
sv sparriskniv; sparrisjärn n
es cuchillo m para
cortar espárragos
it coltello m a sparagio

172
ASPARAGUS RIDGING
PLOUGH
fr butteuse f à asperge
ne aspergeploeg
de Spargeldammpflug m
da aspargesplov
sv sparrissängplov
es aporcadora f de espárragos
it aratro m rincalzatora per
asperagi

173
ASPERATE
fr à feuilles rugueuses
ne ruwbladig
de rauhblättrig
da rubladet
sv strävbladig
es asperifolia
it a foglia rugosa

174
ASPERGILLUS
fr aspergillus m
ne aspergillus
de Aspergillus m
da vandkandeskimmel;
aspergillus
sv aspergillus;
borstmögel
es aspergilo m
it aspergillo m

175
ASSETS
fr actif m
ne bezittingen (activa)
de Aktiva m pl
da aktiver
sv tillgångar
es activo m
it attività f

176
ASSIMILATE, TO
fr assimiler
ne assimileren
de assimilieren
da assimilere
sv assimilera
es asimilar
it assimilare

ASSIMILATION, 3786

177
ASSIMILATION
PRODUCTS
fr produits *m pl* de
l'assimilation
ne assimilaten *npl*
de Assimilate *m pl*
da assimilater
sv assimilat
es productos *m pl* de
asimilación
it prodotti *m pl*
dell'assimilazione

178
ASSORTMENT
fr assortiment *m*
ne sortiment *n*
de Sortiment *n*
da sortiment *n*
sv sortiment *n*
es surtido *m*
it assortimento *m*

179
ASTER;
MICHAELMAS DAISY
fr Aster *m*
ne Aster
de Staudenaster *f*
da høstaster; Aster
sv Aster; höstaster
es estraña *f*;
reina margarita *f*
it astero *m*
Aster L.

ASTER, China, 662
—, perennial, 2671

180
ATMOSPHERE
fr atmosphère *f*
ne atmosfeer
de Atmosphäre *f*
da atmosfære
sv atmosfär
es atmósfera *f*
it atmosfera *f*

181
ATMOSPHERIC
HUMIDITY
fr humidité *f*
atmosphérique
ne luchtvochtigheid

de Luftfeuchtigkeit *f*
da luftfugtighed
sv luftfuktighet
es humedad *f* del aire
it umidità dell'aria

182
ATMOSPHERIC
PRESSURE
fr pression *f* atmosphérique;
pression *f* de l'air
ne luchtdruk
de Luftdruck *m*
da lufttryk *n*
sv lufttryck *n*
es presión *f* atmosférica
it pressione *f* atmosferica

183
ATOMIC WEIGHT
fr poids *m* atomique
ne atoomgewicht *n*
de Atomgewicht *n*
da atomvægt
sv atomvikt
es peso *m* atómico
it peso *m* atomico

184
ATOMIZE, TO
fr atomiser
ne vernevelen
de vernebeln
da forstøve; tågesprøjte
sv dimspruta; ,,dimma"
es atomizar
it nebulizzare

185
ATOMIZING CUP
fr coupelle *f*
ne verstuiverbeker
de Drehbecher *m*
da forstøverbæger *n*
sv förångningsgryta
es vaso *m* de vaporización
it capsula *f* di polverizzazione

186
ATTACHMENT (bot.)
fr attache *f*
ne aanhechting
de Anheftung *f*

da bladfæste *n*
sv fäste *n*
es fijación *f*
it attaccare *m*

187
ATTACK
fr attaque *f*; atteinte *f*
ne aantasting
de Befall *m*
da angreb *n*
sv angrepp
es ataque *m*
it attacco *m*

188
ATTACK, TO
fr atteindre; attaquer;
infester
ne aantasten
de berühren; antasten
da angribe
sv angripa
es infestar; atacar
it attaccare

189
AUCTION CLOCK
fr cadran *m*
ne veilingklok
de Auktionsuhr *f*
da auktionsur *n*;
hollandsk ur *n*
sv auktionsur *n*
es reloj *m* de subasta
it indicatore *m* all'asta;
mostra *f* all'asta

190
AUCTION COSTS
fr coût *m* du veiling
ne veilingkosten
de Versteigerungskosten *f pl*
da auktionsomkostninger
sv auktionskostnader;
försäljningskostnader
es costes *m pl* de subasta
it costi *m pl* della vendita
all'asta

191
AUCTION CREDIT
fr crédit *m* du veiling
ne veilingkrediet

de Versteigerungskredit *m*
da auktionskredit
sv auktionskredit
es crédito *m* de mercado-
   subasta
it credito *m* dell'asta

192
AUCTION MARKET;
CLOCK AUCTION
fr vente *f* aux enchères
   décroissantes; veiling
ne veiling
de Versteigerung *f*;
   Auktion *f*
da auktion
sv auktion
es subasta *f*; mercado *m*
it asta *f* pubblica

193
AUCTION VALUE
fr valeur *m* au veiling
ne veilingwaarde
de Versteigerungswert *m*
da salgsværdi
sv försäljningsvärde *n*
es valor *m* de subasta
it valore *m* all'asta

194
AUCUBA MOSAIC
fr mosaïque *f* aucuba;
   taches *f pl*
ne aucubabont
de Aukubamosaik *n*
da aucubamosaik
sv aucubamosaik
es mosaico *m* aucuba
it mosaico *m* aucuba
   Solanum-virus 8

AUCUBA MOSAIC,
potato, 2865

195
AUSTRIAN PINE
fr pin *m* noir
ne Oostenrijkse den
de österreichische
   Schwarzkiefer *f*
da østrigsk fyr
sv svarttall; österrikisk tall

es pino *m* negro de Austria
it pino *m* nigro
   Pinus nigra cv. austriaca
   (Hoess) A. et G.

196
AUTOMATIC SPRINKLE-
INSTALLATION
fr arrosoir *m* automatique
ne regenautomaat
de Beregnungsautomat *m*
da vandingsautomat
sv dyanläggning;
   droppbevattning
es pluviómata *f*
it automa *m* di pioggia

197
AUTOMATION
fr automatisation *f*
ne automatisering
de Automatisierung *f*
da automatisering
sv automatisering
es automatización *f*
it automazione *f*

AUTUMN CROCUS, 2312

198
AVAILABLE
fr assimilable
ne opneembaar
de aufnehmbar
da optagelig
sv upptagbar
es asimilable
it assimilabile

199
AVENS; GEUM
fr benoîte *f*
ne nagelkruid *n*
de Nelkenwurz *f*
da nellikerod
sv nejlikrot
es cariofilada *f*;
   hierba *f* de San Benito
it erba *f* benedetta
   Geum L.

200
AVENUE TREE
fr arbre *m* d'alignement
ne laanboom

de Alleebaum *m*
da allétræ *n*
sv alléträd *n*
es árbol *m* de paseo,
   árbol *m* de avenida
it albero *m* da viale

201
AVERAGE PRICE
fr prix *m* moyen
ne middenprijs
de Mittelpreis *m*
da gennemsnitspris
sv medelpris *n*
es precio *m* medio
it prezzo medio *m*

202
AVERAGE YIELD
fr rendement *m* moyen
ne gemiddelde opbrengst
de Durchschnittsertrag *m*
da gennemsnitsudbytte
sv medelintäkt;
   medelavkastning
es rendimiento *m* medio
it produzione *f* media

AWL-SHAPED, 3710

203
AXIL
fr aisselle *f*
ne oksel
de Achsel *f*
da aksel
sv axel; bladveck
es axila *f*
it ascella *f*

AXIL, leaf, 2071

204
AXILLARY
fr axillaire
ne okselstandig
de achselständig
da akselstillet
sv utgående från bladvecken
es axilar
it ascellare

205
AXILLARY BUD
fr bourgeon *m* axillaire
ne okselknop

de Achselknospe *f*
da akselknop
sv sidoknopp
es botón *m* axilar
it gemma *f* ascellare

**206
AXIS (bot.)**
fr axe *m*
ne as; spil
de Ackse *m*
da akse
sv axel
es eje *m*; árbol *m*
it perno *m*

AXIS, main, 2241

**207
AZALEA**
fr azalée *f*
ne azalea
de Azalee *f*
da azalea
sv azalea

es azalea *f*
it azalea *f*
Rhododendron spp.

AZALEA, Japanese, 1971

**208
AZALEA GALLMITE**
fr phytopte *f* de l'azalée
ne azaleagalmijt
de Azaleagallmilbe
da azalea galmide
sv azaleagallkvalster
es ácoro *m* de la azalea
it eriofide *m* dell'azalea
Phyllocoptes azaleae Nal.

**209
AZALEA LEAF GALL**
fr cloque *f* de l'azalée;
galles *f pl* foliaires de
l'azalée
ne oortjeziekte
de Ohrläppchenkrankheit *f*;
Klumpenblätterkrank-
heit *f*; Löffelkrankheit *f*

da (azalea-) klumpblad
sv azaleasvulst
es agallas *f pl* foliares del
arándano
it malattia *f* orecchietta;
galla *f* fogliale dell'azalea
Exobasidium japonicum
Shir.

**210
AZALEA LEAF MINER**
fr teigne *f* des azalées
ne azaleamot
de Azaleenmotte *f*
da azaleamøl *n*
sv azaleamal *n*
es polilla *f* minadora de la
azalea
it minatrice *f* delle foglie
dell'azalea;
falena *f* d'azalea
Gracillaria azaleella
Brants;
Caloptilia azaleella (UK)

B

BACCA, 306

211
BACCIFEROUS;
BERRYING
fr baccifère; à baies
ne besdragend
de beerentragend
da bærbærende
sv bärgivande
es bacifero
it baccellino

212
BACCIFORM;
BERRY-SHAPED
fr bacciforme
ne besvormig
de beerenförmig
da bærformet
sv bärformig
es baciforme
it bacciforme

213
BACTERIAL BLIGHT
fr maladie f bactérienne des
    taches
ne bacterievlekkenziekte
de Bakterienfleckenkrank-
    heit f
da brunbakteriose
sv blad-och stjälkbakterios
es manchas f pl bacterianas
it malattia f petecchiale di
    batterio
    Xanthomonas spp.

214
BACTERIAL BLIGHT
(BEGONIA)
fr maladie f bactérienne des
    taches
ne bacterieziekte
de „Ölfleckenkrankheit" f
da brunbakteriose
sv blad-och stjälkbakterios
es punteado m
it batteriosi f
    Xanthomonas begoniae
    (Takim.) Dows.

215
BACTERIAL BLIGHT
(FORSYTHIA; SYRINGA);
BACTERIAL ROT;
LILAC BLIGHT
fr maladie f bactérienne;
    bacteriose f;
    taches noires
ne zwart n
de Bakterien-Triebfäule f;
    Fliederseuche
da Syrénbakteriose
sv Syrenbakterios
es bacteriosis f de las lilas;
    negrón m
it tumori f dei lilla;
    batteriosi f
    Pseudomonas syringae
    v. Hall

216
BACTERIAL BLIGHT
(PELARGONIUM)
fr maladie f bactérienne
ne bacterievlekkenziekte
de bakterielle Blattflecken-
    und Stengelfäule f
da pletbakteriose
sv pelargonbakterios
es punteado m
it batteriosi f
    Xanthomonas pelargonie
    (Brown.) St. et Burkh.

217
BACTERIAL BLIGHT
(WALNUT)
fr maladie f bactérienne;
    bactériose f
ne bacterieziekte; vruchtrot
de Walnussbrand m
da valnødbakteriose
sv valnötsbakterios;
    fruktröta
es enfermedad f bacteriana;
    podredumbre f
it mal m secco
    Pseudomonas juglandis
    Pierce

218
BACTERIAL BLOSSOM
BLIGHT (PIRUS)
fr bactériode f des fleurs
ne bloesemsterfte
de Bakterienbrand m
da syrénbakteriose
sv syrenbakterios
es moho m
it batteriosi
    Pseudomonas syringae
    v. Hall

219
BACTERIAL BLOSSOM
BLIGHT (PEAR);
LILAC BLIGHT (PEAR);
BACTERIAL ROT OF
AVOCADO PEAR
fr bactériose f des bour-
    geons;
    blast m des citrus;
    taches f pl noires de l'avo-
    catier
ne knopsterfte
de Bakterienbrand m;
    „Knospensterben" n
da syrénbakteriose
es marchitez f bacteriana de
    los agrios; podredumbre
    f bacteriana de los agrios;
    muerte f del capullo
it tumori f dei lilla;
    tumori f dei frutti del
    limone;
    batteriosi f dei rami e
    frutti degli agrumi
    Pseudomonas syringae
    v. Hall

BACTERIAL BLOTCH,
224

220
BACTERIAL CANKER
(CHERRY)
fr chancre m bactérien
ne bacteriekanker
de Rindenbrand m;
    Bakterienbrand m

da bakteriekræft
sv stam-och bladbakterios
es cáncer *m* bacteriano
it cancro *m* batterico
   Pseudomonas mors
   prunorum Worm.

221
BACTERIAL CANKER
(PEACH)
fr chancre *m* bactérien
ne bacteriekanker
de Bakterienbrand *m*
da bakteriekræft
sv stam-och bladbakterios
es cáncer *m* bacteriano
it cancro *m* batterico
   Pseudomonas syringae
   v. Hall

222
BACTERIAL LEAF SPOT
(AFRICAN MARIGOLD)
fr maladie *f* bactérienne
ne bacterievlekkenziekte
de bakterielle Blattfleckenfäule *f*
da pletbakteriose
sv bladbakterios
es punteado *m* de las hojas
it batteriosi *f*
   Pseudomonas tagetes

223
BACTERIAL LEAF SPOT
(CABBAGE;
CAULIFLOWER)
fr maladie *f* bactérienne
ne bacterievlekkenziekte
de bakterielle Blattfleckenkrankheit *f*
da pletbakteriose
sv fläckbakterios
es punteado *m* de las hojas
it batteriosi *f*
   Pseudomonas maculicola
   Mc Cullock

BACTERIAL ROT, 215

BACTERIAL ROT OF
AVOCADO PEAR, 219

224
BACTERIAL SPOT;
BACTERIAL BLOTCH
fr taches *f pl* bactériennes;
   la goutte *f*
ne bacterievlekken
de Bakterienflecken *m pl*
da bakterieplet
sv bakteriefläck
es manchas *f pl* bacterianas
it maculatura *f* batterica
   Pseudomonas tolaesi
   Paine

225
BACTERICIDAL
fr bactéricide
ne bacteriëndodend
de bakterientötend
da bakteriedræbende
sv bakteriedödande
es bactericido
it battericido

226
BACTERIOLOGY
fr bactériologie *f*
ne bacteriologie
de Bakteriologie *f*
da bakteriologi
sv bakteriologi
es bacteriología *f*
it batteriologia *f*

227
BACTERIOPHAGE
fr bactériophage
ne bacteriofaag
de bakteriophag
da bakteriofag
sv bakteriofag
es bacteriófago
it batteriofago

228
BACTERIUM
fr bactérie *f*
ne bacterie
de Bakterium *n*
da bakterie
sv bakterie
es bacteria *f*
it batterio *m*

229
BADLY COLOURED;
DISCOLOURED
fr de couleur *f* fausse
ne miskleurig
de fehlfarbig
da misfarvet; fejlfarvet
sv missfärgad
es descolorido
it scolorato

230
BAG
fr sac *m*
ne zak
de Sack *m*
da sæk
sv säck
es saco *m*
it sacco *m*

231
BAIT
fr appât *m*
ne lokmiddel *n*
de Lockmittel *n*
da lokkemad
sv lockbete
es cebo *m*
it esca *f*

232
BALANCE
fr solde *m*
ne saldo *n*
de Saldo *m*
da saldo
sv saldo *n*
es saldo *m*
it saldo *m*

233
BALANCE PLOUGH
fr charrue *f* balance
ne kipploeg
de Kippflug *m*
da svingplov
sv balansplog
es arado *m* volquete
it aratro *m* a bilanciere

234
BALCONY PLANT
fr plante *f* à balcon
ne balkonplant

de Balkonpflanze *f*
da altanplante
sv balkongväxt
es planta *f* de balcón
it pianta *f* a balcone

235
BALE OF STRAW
fr ballot *m* de paille
ne strobaal
de Strohballen *m*
da halmballe
sv halmbal; ströbal
es fardo *m* de paja
it balla *f* di paglia

BALLAST MATERIAL,
1897

236
BALL VALVE
fr soupape *f* à boulet
ne kogelafsluiter
de Kugelventil *n*
da kugleventil
sv kulventil
es válvula *f* de bola
it valvola *f* a sfera

237
BALM
fr mélisse *f* officinale;
    mélisse *f* citronnelle
ne citroenkruid *n*
de Zitronenmelisse *f*
da citronmelisse
sv citronmelisse
es melisa *f*; toronjil *m*
it cedrina *f*
    Melissa officinalis L.

238
BALSAM; IMPATIENS;
SWEET SULTAN
fr balsamine *f*; impatiente *f*
ne vlijtig liesje *n*; balsamien
de fleissiges Lieschen *n*;
    Balsamine *f*
da flittig lise; balsamin
sv flitiga Lisa; balsamin
es balsamina *f* nicaragua;
    miramelindo
it balsamina *f*
    Impatiens sultani Hook.

239
BAMBOO
fr bambou *m*
ne bamboe
de Bambus *m*
da bambus
sv bambu
es bambú *m*
it bambu *m*
    Bambusa Schreb.

240
BAMBOO CANE
fr canne *f* de bambou;
    bambou *m*
ne tonkinstok
de Bambusrohr *n*;
    Bambusstock *m*
da tonkinstok
sv tonkinkäpp; bambukäpp
es caña *f* de bambú
it mazza *f* di bambu;
    bambu *m*

241
BANANA
fr banane *f*
ne banaan
de Banane *f*
da banan
sv banan
es plátano *m*
it banana *f*

242
BANANA TREE
fr bananier *m*
ne banaan
de Bananenbaum *n*
da banan
sv banan
es plátano *m*
it banano *m*
    Musa L.

243
BANDED ROSE SAWFLY;
CURLED ROSE SAWFLY
(US)
fr mouche *f* à scie;
    tenthrède *f*
ne aardbeibladwesp;
    klikkeboorder
de weissgegürtelte Rosen-
    blattwespe *f*

da bladhveps;
    Emphytus cinctus
sv vitgördlad rosenblad-
    stekel
es avispa *f* de las hojas del
    fresal
    Allantus cinctus L.;
    Emphytus c.

244
BANK-BED
(CUCUMBER GROWING)
fr ados *m pl*;
    levée *f* de terre
ne wal
de Wall *m*; Hügel *m*
da vold; forhøjning; rabat
sv (gurk-) bädd
es caballón *m*
it alzamento *m* di terra

245
BANK LOAN
fr crédit *m* bancaire
ne bankkrediet *n*
de Bankkredit *m*
da bankkredit
sv bankkredit; banklån *n*
es crédito *m* bancario
it credito *m* bancario

246
BANK MOWER
fr faucheuse *f* de talus
ne taludmaaier
de Böschungsmäher *m*
da skrånings-slåtmaskine
sv slåttermaskin för slänt
es segadora *f* en pendiente
it falciatrice *f* per lavori in
    pendio

247
BAR; ROD (WINDOWS)
fr croisillon *m*
ne roede
de Sprosse *f*
da sprosse
sv spröjs
es barra *f*; listón *m*;
it steccone *m*

248
BARBED WIRE
fr fil *m* de fer barbelé
ne prikkeldraad *n*

de Stacheldraht *m*
da pigtråd
sv taggträd
es alambre *m* de púas
it filo *m* di ferro spinato

249
BARBERRY
fr épine-vinette *f*
ne Berberis (zuurbes)
de Berberitze *f*
da Berberis
sv Berberis
es agracejo *m*
it berbero *m*; crespino *m*
Berberis L.

250
BARK
fr écorce *f*
ne schors
de Borke *f*; Rinde *f*
da bark
sv bark
es corteza *f*
it corteccia *f*

BARK, inner, 1915

251
BARK BEETLE
fr bostryche *f*; scolyte *m*
ne bastkever; schorskever
de Borkenkäfer *m*;
Splintkäfer *m*
da barkbille
sv splintborre; barkborre
es barrenillo *m*;
escolítido *m*
it scolito *m*
Scolytus spp.

BARK BEETLE, elm, 1214
—, fruit, 2032
—, large fruit, 2032
—, small elm, 3441

BARK NECROSIS, 551

252
BARN; SHED
fr grange *f*
ne schuur
de Scheune *f*
da skur *n*; lade

sv lada
es cobertizo *m*
it tettoia *f*

253
BARNYARD GRASS
fr pied-de-coq *m*
ne hanepoot
de Hühnerhirse *f*
da hanespore
sv hönshirs
es pata *f* de gallo
it giavone *m*
Echinochloa crus-galli
(L. P.B.)

254
BARROW, TO;
TO WHEEL
fr brouetter
ne kruien
de schieben
da køre; trille
sv skjuta
es acarrear en carretilla
it trasportare colla carriuola

255
BASAL PLATE
fr plateau *m* du bulbe
ne bolbodem
de Zwiebelboden *m*
da rodkage
sv lökbotten
es base *f* del bulbo
it base *f* del bulbo

256
BASAL ROT
fr fusariose *f* du narcisse
ne bolrot *n*
de Zwiebelfäule *f*
da løgråd
sv lökrota
es fusariosis *f* de los bulbos
it putrefazione *f* basale
Fusarium oxysporum
Schl.

257
BASIC DRESSING;
BASIC MANURING
fr fumure *f* d'investissement
ne voorraadbemesting

de Vorratsdüngung *f*
da grundgødning
sv grundgödsling
es abonado *m* de almacenamiento
it concimazione *f* di
provvisione

258
BASIC MANURE
fr fumure *f* de fond
ne basisbemesting
de Grunddüngung *f*
da grundgødning
sv grundgödsling
es abonado *m* básico;
abonado *m* de fondo
it concimazione *f* fondamentale

259
BASIC SLAG
fr scories *f pl* Thomas;
phospate *m* Thomas
ne Thomasslakkenmeel *n*
de Thomasschlackenmehl *n*;
Thomasphosphat *n*;
Thomasmehl *n*
da Thomasslagge
sv Thomasfosfat *n*
es escorias *f pl* Thomas
it scorie Thoma *f pl*

260
BASIL
fr basilic *m*
ne basilicum
de Basilikum *n*;
Basilienkraut *n*
da basilikum
sv basilika
es albahaca *f*
it basilico *m*
Ocimum basilicum L.

261
BASIN SOIL
fr sol *m* de cuvette
ne komgrond
de Bassinboden *m*
da bassinjord
sv svämlera; bassänglera
es suelo *m* limoso

262
BASKET; HAMPER
fr panier *m*; corbeille *f*
ne mand
de Korb *m*
da kurv
sv korg
es cesto *m*
it cesta *f*

263
BASKET WILLOW;
OSIER
fr saule *m*;
  osier *m*
ne teenwilg
de Korbweide *f*
da kurvpil
sv korgvide
es mimbre *m*
it salcio *m* da vimine;
  vinco *m*
  Salix spp.

264
BAST
fr liber *m*
ne bast
de Bast *m*; Borke *f*
da bast
sv bast *n*
es corteza *f*
it libro *m*; corteccia *f*

265
BASTARD
fr bâtard *m*
ne bastaard
de Bastard *m*
da bastard
sv bastard
es bastardo *m*
it bastardo *m*

266
BASTARDINDIGO
fr faux-indigo *m*
ne bastaardindigo
de Bastardindigo *m*
da særkrone
sv amorpha; segelbuske
es índigo *m* bastardo
it indaco *m* bastardo
  Amorpha L.

267
BAST MAT
fr natte *f*; rabane *f*
ne moscovische mat
de Bastmatte *f*
da bastmåtte
sv bastmatta
es estera *f* rusa
it stoia *f* moscovia

BATCH GROUND, 790

268
BAY; LAUREL
fr laurier *m*
ne laurier
de Lorbeer *m*
da laurbær *n*
sv lager
es laurel *m*
it alloro *m*, lauro *m*
  Laurus L.

269
BAY (GLASSHOUSE)
fr chapelle *f* de serre
ne kap
de Dach *n*
da tag *n*
sv tak *n*
es cubierta *f* de invernadero
it tetto *m*

BAY, rose, 2519

270
BEAK SHAPED
fr en forme de rostre
ne snavelvormig
de schnabelförmig
da næbformet
sv näbblik
es rostrado
it rostriforme

271
BEAN
fr haricot *m*
ne boon
de Bohne *f*; Speckbohne;
  Fleischbohne
da bønne; brydbønne
sv böna; vaxböna
es judía *f* verde
it fagiolino *m*; fava *f*
  Phaseolus spp.

BEAN, broad, 434
—, kidney, 1996
—, runner, 3218
see also: French bean;
  slicing bean;
  snap bean

BEAN APHID, 325
—, black bean, 325

272
BEAN BEETLE;
BROAD BEAN WEEVIL
(US)
fr bruche *f* des fèves
ne tuinboonkever
de Pferdebohnenkäfer *m*
da bønnefrøbille
sv bönsmyg; bönbagge
es gorgojo *m* de las habas
it tonchio *m* delle fave;
  bruco *m* delle fave
  Bruchus rufimanus Boh.

273
BEAN BLACK ROOT
fr jambe *f* noire
ne zwarte-vaatziekte
de Schwarzbeinigkeit *f*
da 'black root'
sv 'black root'
es pie *m* negro
it black root;
  annerimento *m* basale

274
BEAN HARVESTER
fr récolteuse *f* de haricots
  verts
ne bonenplukmachine
de Bohnenpflückmaschine *f*
da bønneplukkemaskine
sv bönskördemaskin
es cosechadora *f* de judías
  verdes
it raccoglitrice *f* per fagiolini

275
BEAN SEED FLY;
SEED-CORN MAGGOT
(US)
fr mouche *f* grise des semis
ne bonevlieg
de Gemüsewurzelfliege *f*
da bønneflue

sv bönstjälkfluga
es mosca *f* de las judías
it mosca *f* grigia dei cereali
Chortophila cilicrura
Rond;
Delia platura (UK);
Hylemya platura (US)

BEAN SPOT DISEASE,
119

276
BEANSTICK
fr rame *f*; perche *f*
ne bonestaak
de Bohnenstange *f*
da bønnestage
sv bönstör
es rodrigón *m*;
tutor *m* para judías
it pertica *f*

277
BEAN STIPPLE STREAK
fr moucheture *f* nécrotique
du haricot;
bigarrure *f* du haricot
ne stippelstreep (boon)
de 'Stippelstreep'
(Gartenbohne)
da bønne-nekrose
sv stippelstreep (böna)
es Nicotiana-virus *m*
it striatura *f* necrotica
(fagiolo)
Nicotiana-virus

BEAN WEEVIL, 1134
—, broad, 272

278
BEAN YELLOW MOSAIC
fr mosaïque *f* jaune du haricot
ne scherpmozaïek *n* (boon)
de Bohnengelbmosaik *n*
da bønne-gulmosaik
sv gul bönmosaik
es mosaico *m* amarillo de las
judías
it mosaico *m* giallo del
fagiolo

279
BED
fr planche *f*; corbeille *f*;
parterre *f*; plate-bande *f*
ne bed *n*
de Beet *n*
da bed
sv säng; bädd
es platabanda *f*; cama *f*;
almácigo *m*
it aiuola *f*

280
BED CULTURE
fr culture *f* en ados
ne beddenteelt
de Beetkultur *f*;
Beetanbau *m*
da bedkultur
sv bäddodling
es cultivo *m* en camas
it coltura *f* su quadri

281
BEDDING SYSTEM
fr culture *f* sur rangs
ne rijenteelt
de Reihenanbau *m*
da rækkerne på langs
sv odling i rader
es cultivo *m* en hileras
it coltivazione *f* in file

282
BEDEGUAR GALL
WASP;
MOSSY-ROSEGALL
WASP (US)
fr bedeguar *m*
ne bedeguar-gal
de gemeine Rosengallwespe *f*
da bedeguar; rosengalhveps
sv sömntornstekel
es rhodites *m* del rosal
it rodite *m* della rosa
Rhoditus rosae L.;
Diplolepus r. (UK)

283
BEE
fr abeille *f*
ne bij
de Biene *f*
da bi

sv bi
es abeja *f*
it ape *f*
Apis mellifera L.;
Apis mellifica L.

BEE, artificial, 158
—, bumble, 492
—, humble, 492

284
BEECH
fr hêtre *m*
ne beuk
de Buche *f*
da bøg
sv bok
es haya *m*
it faggio *m*
Fagus silvatica L.

285
BEECH APHID
fr puceron *m* du hêtre
ne beukebladluis
de Buchenblattlaus *f*
da bøgebladlus
sv bokbladlus
es pulgón *m* de las hojas
it afide *m* colonoso dei
faggio
Phyllaphis fagi L.

BEECH COCCUS, 287

286
BEECH LEAF MINER
fr orcheste *m* du hêtre
ne beukespringkever
de Buchenspringrüssler *m*
da bøgeloppe
sv bokbladminerare
es escarabajo *m* saltarín
it orcheste *m* del faggio
Rhynchaenus fagi L.

287
BEECH SCALE;
BEECH COCCUS
fr cochenille *f* du hêtre
ne beukewolluis
de Buchenwollaus *f*
da bøgeuldlus
sv bokullus

es pulgón *m* lanoso del haya
it cocciniglia *f* del faggio
  Cryptococcus fagi Brsp.

288
BEE COLONY
fr colonie *f* d'abeilles
ne bijenvolk *n*
de Bienenvolk *m*
da bifamilie
sv bisamhälle *n*
es colonia *f* de abejas
it famiglia *f* d'api;
  alveare *m*

289
BEE-HIVE
fr ruche *f*
ne bijenkorf
de Bienenkorb *m*
da bistade *n*, bikube
sv bikupa
es colmena *f*
it arnia *f*

290
BEEKEEPER
fr apiculteur *m*
ne bijenhouder
de Bienenzüchter *m*
da biavler
sv biodlare
es apicultor *m*
it apicoltore *m*

291
BEEKEEPING
fr apiculture *f*
ne bijenteelt
de Bienenzucht *f*
da biavl
sv biodling
es apicultura *f*
it apicoltura *f*

292
BEET; RED BEET
fr betterave *f* rouge;
  betterave *f* potagère
ne kroot
de rote Rübe *f*; rote Beete *f*;
  Salatrübe *f*
da rødbede
sv rödbeta
es remolacha *f* comestible

it barbabietola *f* da orta
  Beta vulgaris L.;
  cv. Rubra L.

BEET, spinach,   3552

BEET BEETLE,   2726

293
BEET CYST NEMATODE
fr anguillule *f* de la betterave
ne bietecysteaaltje *n*
de Rübennematode *m*
da roecystennematod
sv betnematod
es nematodo *m* de la remolacha
it nematode *m* della
  barbabietola da zucchero
  Heterodera schachtii

294
BEETLE
fr coléoptère *m*,
  escarbot *m*
ne tor; kever
de Käfer *m*
da bille
sv skalbagge
es coleóptero *m*;
  escarabajo   *m*
it scarabeo *m*;
  scarafàggio *m*,
  coleottero *m*

BEETLE, ambrosia,   79
—, asparagus,   169
—, beet,   2726
—, blossom,   373
—, carrion,   591
—, click,   704
—, Colorado,   753
—, Colorado potato, 753
—, flea, 1375
—, lily, 2167
—, longhorn, 2209
—, pea,   2623
—, pigmy mangold,   2726
—, pulse,   2975
—, raspberry, 3035
—, seed,   2975
—, strawberry seed,   3685
see also: bark beetle;
        bean beetle;
        leaf beetle;
        willow beetle

295
BEET LEAF MINER
fr mouche *f* de la betterave
ne bietevlieg
de Rübenfliege *f*
da bedeflue
sv betfluga
es mosca *f* de la remolacha
it mosca *f* della barbabietola
  Pegomya hyoscyami
  Panz.;
  Pegomya betae (UK)

296
BEET MOSAIC
fr mosaïque *f* de la betterave
  (rouge)
ne bietemozaiek *n*
de Rübenmosaik *n* (rote
  Rübe)
da bede-mosaik
sv betmosaik
es mosaico *m* de la remolacha
it mosaico *m* della barbabietola

297
BEET YELLOWS
fr jaunisse *f* de la betterave
ne vergelingsziekte (biet)
de Vergilbungskrankheit *f*
  der Rübe
da virusgulsot
sv virusgulsot (betor)
es amarilleamiento *m* de la
  remolacha
it giallume *m* della barbabietola
  Beta-virus 4

298
BEGONIA
fr bégonia *m*
ne Begonia
de Begonie *f*
da begonie; skævblad *n*
sv Begonia
es Begonia *f*
it Begonia *f*
  Begonia L.

BEGONIA, tuberous,   3946

299
BEGONIA REX
fr bégonia *m* rex
ne bladbegonia
de Blattbegonie *f*
da bladbegonie
sv bladbegonia
es Begonia *f* rex;
Begonia *f* de hojas ornamentales
it Begonia *f* rex
Begonia rex hybriden

300
BELL (HYACINTH)
fr fleuron *m*
ne nagel
de einzelne Hyazinthen-
blüte *f*
da klokke
sv hyacintblomma
es uña
it campanula *f*

BELL, TO, 4156

BELLBINE, 1768

301
BELLFLOWER
fr campanule *f*
ne klokje *n*
de Glockenblume *f*
da klokkeblomst
sv klocka
es campanula *f*
it Campanula *f*
Campanula L.

302
BELLOWS; PUFFER
fr soufreuse *f* à soufflet
ne blaasbalg
de Blasebalg *m*
da blæsebalg
sv blåsbälg
es soplete *m* de azufre;
espolvoreador *m* de
azufre
it impolveratrice *f* a
soffietto

303
BELL SHAPED
fr campanulé
ne klokvormig
de glockig; glockenförmig
da klokkeformet
sv klockformig
es campaniforme
it campanaceo

BENCH, 3599

304
BENCH GRAFTING
fr roncer
ne zetten
de veredeln mit Reis
da pode
sv ympa
es injertar
it innestare

305
BERRIES
fr baies *f pl*
ne besvruchten
de Beerenfrüchte
da bærfrugt
sv bärfrukt
es bayas *f pl*
it bacce *f pl*

306
BERRY; BACCA
fr baie *f*
ne bes
de Beere *f*
da bær *n*
sv bär *n*
es baya *f*
it bacca *f*

BERRY, June, 3456

BERRYING, 211

BERRY-SHAPED, 212

BESOM, 2052

307
BHUTAN PINE
fr pin *m* de l'Himalaja
ne tranenden
de Tränenkiefer *f*
da tårefyr

sv himalajatall
es pino *m* del Himalaya
it Pinus *f* griffithii
Pinus griffithii
M. Clelland

308
BIBIONID;
MARCH FLY (US)
fr bibion *m*
ne rouwvlieg
de Haarmücke *f*
da hårmyg
sv hårmygga
es bibio *m* de las huertas
it bibio *m*
Bibionidae

309
BIENNIAL
fr bisannuel
ne tweejarig
de zweijährig
da toårig
sv tvåårig
es bienal
it biennale

310
BIENNIAL BEARING
fr alternance *f*
ne beurtjaren
de Alternanz *f*; abwechselnde
Erträge *m pl*
da hvert andet års bæring
sv vart annat års bärning;
oregelbunden fruktbarhet
es producción *f* alterna
it alternanza *f* di produzione

311
BILBERRY;
WHORTLEBERRY
fr myrtille *f*
ne blauwe bosbes;
gewone bosbes
de Heidelbeere *f*;
Bickbeere *f*;
Blaubeere *f*
da blåbær
sv blåbär *n*
es arándano *m*
it mirtillo *m*;
mirtillo *m* nero
Vaccinium myrtillus L.

BILLHOOK,   676

312
BIND IN, TO
fr entailler
ne insnoeren
de einschnüren
da indsnøre
sv insnöra
es cinchar,
    estrangular
it allacciare

BINDWEED, black,   328
—, field,   1329
—, hedge, 1768

313
BIOLOGICAL
COMMUNITY
fr biocénose f
ne levensgemeenschap
de Lebensgemeinschaft f
da livsfællesskab n
sv livsgemenskap n
es comunidad f biológica
it compagnia f di vita

314
BIOLOGICAL CONTROL
fr prophylaxie f biologique
ne biologische bestrijding
de biologische Bekämpfung f
da biologisk bekæmpelse
sv biologisk bekämpning
es defensa f biológica;
    control m biológico
it lotta f biologica

315
BIOLOGY
fr biologie f
ne biologie
de Biologie f
da biologi
sv biologi
es biología f
it biologia f

316
BIPINNATE
fr bipenné
ne dubbel gevind
de doppelt gefiedert

da dobbeltfinnet
sv dubbelt parbladig
es bipinado
it bipennato

317
BIRCH
fr bouleau m
ne berk
de Birke f
da birk
sv björk
es abedul m
it betulla f
    Betula L.

BIRD-CHERRY, American,
82
—, European,   1243

318
BIRD SCARER
DETONATOR
fr détonateur m épouvan-
    tail
ne knalapparaat n
de Knallapparat m
da fuglekanon
sv knallapparat
es detonador m
it sparentapasseri m,
    detonatore m

319
BIRTHWORT
fr aristoloche f
ne pijpbloem
de Osterluzei f;
    Pfeifenwinde f
da stangerod;
    pibetobaksplante
sv pipranka
es artistoloquia m
it Aristolochia f
    Aristolochia L.

320
BISEXUAL
fr bissexué
ne tweeslachtig
de zwittrig
da tvekønnet
sv tvåkönad
es bisexual
it androgino; bisessuale

BISHOP'S WEED,   1668

321
BITTER
fr amer
ne bitter
de bitter
da bitter
sv bitter; besk
es amargo
it amaro

322
BITTERFREE
fr non amer
ne bittervrij
de bitterfrei
da bitterfri
sv bitterfri
es exento de amargor
it esente dall'amarezza

323
BITTER PIT
fr bitter pit m;
    maladie f des taches
    amères
ne stip n
de Stippigkeit f
da priksyge
sv pricksjuka
es corazón m amargo;
    bitter pit m;
    nódulos m pl amargos
it maculatura f amara;
    macchie f d'amaro

324
BITTER ROT (APPLE)
fr gloéosporiose f
ne gloeosporiumrot n
de Gloeosporium-Fäule f
da gloeosporium
sv gloeosporiumröta
es podredumbre f gelatinosa
it marciume m amaro
    Pezicula malicorticis
    (J.) Nannf.

325
BLACK BEAN APHID;
BEAN APHID (US)
fr puceron m noir des fèves;
    puceron m du haricot
ne zwarte boneluis

de schwarze Blattlaus *f*;
  schwarze Rübenlaus
da bedebladlus
sv bönbladlus;
  betbladlus
es pulgón *m* negro;
  pulgón *m* de las habas
it pidocchio delle fave;
  afide *m* nero della fava
  Aphis fabae Scop.;
  Doralis fabae

326
BLACKBERRY
fr mûre *f* sauvage
ne braam
de Brombeere *f*
da brombær *n*
sv björnbär *n*
es zarzamora *f*
it mora *f*; prugnola *f*

327
BLACKBERRY BUSH
fr mûrier *m* sauvage
ne braamstruik
de Brombeerstrauch *m*
da brombærbusk
sv björnbärsbuske
es zarza *f*
it moro *m* prugnole
  Rubus spp.

328
BLACK BINDWEED
fr vrillée *f*;
  faux-liseron *m*
ne zwaluwtong
de Winden-Knöterich *m*
da snerle-pileurt
sv åkerbinda
es pimentilla *f* trepadora
it convolvulo *m* nero
  Polygonum convolvulus L.

329
BLACKBIRD
fr merle *m*
ne merel
de Amsel *f*
da solsort
sv koltrast
es mirlo *m*
it merlo *m*
  Turdus merula L.

330
BLACK BODY
RADIATION
fr rayonnement *m* d'origine
  thermique
ne temperatuurstraling
de Wärmestrahlung *f*;
  Temperaturstrahlung *f*
da varmestråling
sv värmestrålning
es radiación *f* de origen
  térmico
it irradiazione *f* del corpo
  nero

331
BLACK CANKER
(SALIX)
fr chancre *m*
ne zwarte kanker
de 'schwarzer Krebs' *m*
da 'sort kræft'
sv pilkräfta
es cáncer *m* negro (Salix)
it cancro *m* del salice
  Physalospora myabeana
  Fuk.

332
BLACK CURRANT
fr cassis *m*
ne zwarte bes
de schwarze Johannisbeere *f*
da solbær *n*
sv svarta vinbär *n*
es grosella *f* negra
it ribes nero *m*

333
BLACK CURRANT BUSH
fr cassisier *m*
ne zwarte-bessestruik
de schwarzer Johannisbeerstrauch *m*
da solbærbusk *n*
sv svarta vinbär *n*
es grosellero *m* negro
it arbusto *m* di ribes nero
  Ribes nigrum L.

334
BLACK CURRANT LEAF
MIDGE
fr cécidomyie *f*
ne bessebladgalmug

de Gallmücke
da galmyg
sv gallmygga
es mosquito *f* del grosellero
it cecidomia *f* del ribis
  Dasyneura tetensi Ruebs.

335
BLACK CURRANT
REVERSION
fr feuilles *f pl* étroites
  (cassis)
ne brandnetelbladziekte
  (zwarte bes)
de viröser Atavimus *m*
  (schwarze Johannisbeere)
da ribbesvind *n*
sv reversion
es Ribes-virus I (grosellero)
it reversione *f* da virus
  (ribes nero)

336
BLACK FEN;
BLACK PEAT
fr tourbe *f* noire
ne zwartveen *n*
de Schwarztorf *f*;
  Schwarzmoor *n*
da tørvemose
sv svarttorv
es turba *f* negra
it torba *f* nera

337
BLACK LEG (BETA)
fr pied *m* noir
ne wortelbrand
de Wurzelbrand *m*
da rodbrand
sv rotbrand
es chamuscodo *m* de las
  raíces
it mal *m* del cuore;
  mal *m* del piede
  Pleospora betae Björling

338
BLACK LEG (POTATO)
fr jambe *f* noire;
  pied *m* noir
ne zwartbenigheid
de Schwarzbeinigkeit *f*

da sortbensyge
sv stjälkbakterios
es pie *m* negro
it marciume *m* pedale
Erwinia atroseptica
(van Hall) Jenn.

339
BLACK LEG (TULIPA)
fr Sclerotium *m*
ne zwartbenigheid
de Sclerotium *n*;
Schwarzbeinigkeit *f*
da sortbensyge
sv stjälkbakterios
es pie *m* negro
it sclerozio *m*
Sclerotium wakkerii
Boer. & Posth.

340
BLACK MEDICK
fr luzerne *f* lupaline;
minette *f*
ne hopperupsklaver
de Hopfenklee *m*;
Hopfenluzerne *f*
da humle sneglebælg
sv humleluzern
es lupulina *f*
it lupinello *m*
Medicago lupulina L.

341
BLACK MOULD;
SOOTY MOULD
fr fumagine *f*
ne roetdauw
de Russtau *m*
da branddug
('sodskimmel')
sv sotdagg
es fumagina *f*;
negrón *m*
it fumaggine *f*
i.a. Cladosporium spp.

342
BLACK NIGHTSHADE
fr morelle *f* noire
ne zwarte nachtschade
de schwarzer
Nachtschatten *m*

da sort natskygge
sv nattskata
es hierba *f* mora
it erba *f* morella
Solanum nigrum L.

BLACK OUT, TO,   984

BLACK PEAT,   336

343
(BLACK) PLUM SAWFLY
fr hoplocampe *m* noir du
prunier
ne zwarte pruimezaagwesp
de schwarze Pflaumensäge-
wespe *f*
da blommehveps
sv plommonstekel
es hoplocampa *f* negra del
ciruelo
it oplocampa *f*; piccolo *m*
tentradine delle susine
Hoplocampa minuta
Christ.

344
BLACK RADISH
fr radis *m* noir
ne ramenas
de Rettich *m*
da ræddike *mf*
sv rättika
es rábano *m* negro
it ramolaccio *m*; rafano *m*
Raphanus sativus L.
cv. Rapa L.

345
BLACK ROOT ROT
(LATHYRUS)
fr pourriture *f* noire des
racines
ne wortelrot *n*
de Wurzelbräune *f*
da rodråddenskab
sv rotbrand
es peste *f* de los emilleros;
podredumbre *f* de la raíz
it marciume *m* radicale
Thielaviopsis basicola
(B. et B.) Ferr.

BLACK ROT,   3276

346
BLACK ROT (CARROT)
fr pourriture *f* noire
ne zwarte-plekkenziekte
de Schwarzfäule *f*
da sortråd
sv svartröta
es picado *m* negro
it marciume *m* nero
Stemphylium radicinum
(M., Dr. et E.) Neerg.

347
BLACK ROT
(CRUCIFERS)
fr nervation *f* noire
ne zwartnervigheid
de Adernschwärze *f*
da brunbakteriose
sv brunbakterios
es podredumbre *f* negra
it batteriosi *f* delle
crocifere
Xanthomonas campestris
(Pam.) Dows.

348
BLACK ROT (TOMATO)
fr pourriture *f* noire;
pourriture *f* de la tomate
ne zwartrot
de Schwarzfäule *f*
da sortråd *n*
sv svartröta
es podredumbre negra *f*;
podredumbre *f* del tomate
it marciume *m* nero
Xanthomonas campestris
(P.) Dowson;
Phoma destructiva Plowr.

349
BLACK RUST
fr rouille *f* noire
ne zwarte roest
de Schwarzrost *m*
da sortrust
sv svartrost
es roya *f* negra
it ruggine *f* nera
Puccinia graminis Pers.

350
BLACK SCURF
(POTATO);
BLACK SPECK;
BLACK SCAB
fr rhizoctone *m* noir;
 maladie *f* des collerettes;
 chancre du pied
ne Rhizoctonia-ziekte
de Weisshosigkeit *f*;
 Wurzeltöterkrankheit *f*;
 Triebfäule *f* und
 Fussvermorschung *f*
da rodviltsvamp
sv gråfiltsjuka
es viruela *f*; sarna *f* castrosa;
 rizoctoniosis *f*
it scabbia *f* della patata;
 ipocnosi
 Rhizoctonia solani K.;
 Corticium solani P.

351
BLACK SLIME
fr pourriture *f* noire
ne zwartsnot *n*
de schwarzer Rotz *m*
da hyacinth-knoldbægers-
 vamp
sv svartblötröta
es podredumbre *f* negra
it mucillagine *f* nera
 Sclerotium bulborum
 (Wakk.) Rehm.

352
BLACK SPOT
(DELPHINIUM)
fr maladie *f* des taches
 noires
ne bacterieziekte
de Bakterien-Schwarz-
 fleckenkrankheit *f*
da pletbakteriose
sv fläckbakterios
es enfermedad *f* bacteriana
it mal *m* secco
 Pseudomonas delphinii
 (E. F. Sm.) Stapp.

353
BLACK SPOT (ROSE)
fr taches *f pl* noires
ne sterroetdauw

de Schwarzfleckigkeit *f*;
 Sternrusstau *m*
da stråleplet
sv svartfläckspika
es manchas *f pl* negras;
 negrón *m*
it malattia *f* di macchietta
 negra;
 ticchiolatura *f*
 Diplocarpon rosae Wolf

354
BLACKTHORN; SLOE
fr prunellier *m*
ne sleedoorn
de Schlehe *f*;
 Schwarzdorn *m*
da slåen
sv slån
es endrino *m*
it prugnolo *m*
 Prunus spinosa L.

355
BLADDER SENNA
fr baguenaudier *m*
ne blazenstruik
de Blasenstrauch *m*
da blærebælg
sv blåsärt *n*
es espantalobos *m*
it Colutea *f*
 Colutea L.

356
BLADE (bot.)
fr limbe *m*
ne bladschijf
de Blattspreite *f*
da bladplade
sv bladskiva
es limbo *m*
it limbo *m*

357
BLANCH, TO;
TO BLEACH
fr blanchir
ne bleken
de bleichen
da blege
sv bleka
es blanquear
it imbiancare

358
BLANCHED CELERY
fr céleri *m* à blanchis;
 céleri *m* à côtes
ne bleekselderij
de Bleichsellerie *m*
da blegselleri; bladselleri
sv blekselleri
es apio *m* blanco;
 apio *m* de pencas
it sedano *m*
 Apium graveolens L.
 cv. Dulce

BLASTED, 366

BLEACH, TO, 357

359
BLEACHED SAND
fr sable *m* infertile
ne loodzand *n*
de Bleisand *m*
da blegsand *n*; blysand *n*
sv blekjord
es podsol *m*; plombagina *f*;
 grafito *m*
it sabbia *f* piombifera

360
BLEED
fr robinet *m* purgeur
ne brijnkraan
de Abschäumhahn *m*
da skumhane
sv kran för saltlösning
es grifo *m* de purga
 (de salmuera)
it sfiato *m*

361
BLEEDING HEART
fr coeur *m* de Marie
ne gebroken hartje *n*
de fliegendes Herz *n*
da hjertblomst *mf*
sv blomsterlyra;
 löjtnantshjärta *n*
es Dicentra *f*
it Dicentra *f*
 Dicentra Bernh.

BLEND, TO, 2357

362
BLIGHT
(POTATO, TOMATO)
fr mildiou *m*
ne aardappelziekte
de Kraut- und Knollen-
fäule *f*
da kartoffelskimmel
sv potatisbladmögel *n*
es mildio *m* de la patata
it peronospora *f* della patata
Phytophthora infestans
(Mont.) de By.

BLIGHT, alternaria, 74, 75
—, bacterial, 213, 214, 215,
216, 217
—, cane, 549
—, halo, 1712
—, leaf, 2074, 2075, 2076
—, lilac, 215, 219
—, ray, 3046
—, spur, 3589
see also: blossom blight

363
BLIND FLORETS
(HYACINTH)
fr fleurs *f pl* desséchées
ne stronagels
de Strohnägel *m pl*
da toptørhed
sv döda knoppar
es uñas *f pl* pajosas

364
BLINDNESS
(CAULIFLOWER)
fr chou-borgne *m*
ne draaihartigheid;
belknoppen
de Drehherz *n*
da krusesyge
sv 'krussjuka'
es acalabazado *m* de la col
it cecidomia *f* delle crucifere
Contarinia nasturtii
Kieffer

365
BLINDNESS (IRIS);
THREE-LEAVES (IRIS)
fr avortement *m* de la fleur
ne bloemloosheid

de Blütentaubheit *f*
da blinde
sv blind
es aborto *m* de las floras
it astuzia *f* fiorale

366
BLINDNESS (TULIP);
BLASTED (IRIS)
fr avortement *m* de la fleur
tulipes *f pl* abortées
ne bloemverdroging
de taube, eingetrocknete
Blüten oder Knospen
da blindhed
sv förtorkning av blomman
es abortamiento *m*
it inaridimento *m* fiorale

BLINDNESS, common,
767

367
BLOCK PLANTING
fr plantation *f* en blocs
ne blokbeplanting
de Blockpflanzung *f*
da blokplantning
sv blockplantering
es plantación *m* en bloque
it piantagione *f* specializzata
(a blocchi)

368
BLOOD MEAL
fr sang *m* desséché
ne bloedmeel *n*
de Blutmehl *n*
da blodmel *n*
sv blodmjöl *n*
es harina *f* de sangre
it sangue *m* polverizzata

BLOOM, 1383

369
BLOOM (ON GRAPES)
fr fleur *f*
ne dauw (op druiven)
de Wachsschicht *f*
da vokslag *n*; voksdug
sv dagg; vaxskikt *n*
es pruina *f*; flor *f*
it inceramento *m*

370
BLOOM, TO;
TO FLOWER;
TO BLOSSOM
fr fleurir; être en fleur
ne bloeien
de blühen
da blomstre
sv blomma
es florecer
it fiorire

371
BLOOMING;
FLOWERING;
BLOSSOMING
fr fleuraison *f*; floraison *f*
ne bloei
de Blüte *f*
da blomstring
sv blomning
es floración *f*
it fioritura *f*

372
BLOSSOM
fr fleur *f*
ne bloesem
de Blüte *f*
da blomst
sv blomma
es flor *f*
it fiore *m*

373
BLOSSOM BEETLE
fr méligèthe *m* du colza
ne koolzaadglanskever
de Rapsglanzkäfer *m*
da glansbille; glimmerbøsse
sv rapsbagge
es meligetes *m* de la colza
it meligete *m* della colza
Meligethes aeneus F.

374
BLOSSOM BLIGHT
(PRUNUS);
BLOSSOM WILT
(PRUNUS)
fr moniliose *f*
ne bloesemsterfte
de Monilia-Spitzendürre *f*
da grå monilia

sv grå monilia
es moho *m* de los albari
    coques
it muffa *f* delle albicocche;
    sclerotinia *f*
    Sclerotina laxa Aderh. et
    Ruhl.

BLOSSOM BLIGHT,
bacterial, 218, 219

BLOSSOM-END ROT,
380

BLOSSOM WILT
(PRUNUS), 374

375
BLOTCH (BEAN)
fr anthracnose *f*
ne spikkelziekte
de Ascochyta-Flecken-
    krankheit *f*
da Ascochyta
sv fläcksjuka
es antracnosis *f*
it antracnosi *f*
    Ascochyta phaseolorum
    Sacc.

376
BLOTCH (CUCUMIS)
fr taches *f pl* rouges
ne bladvuur *n*
de Blattbrand *m*
da rudeplet
sv gurkbrand
es abrasado *m* de las hojas
it cernosporiosi *f*
    Corynespora melonis
    (Cke) Lindau

BLOTCH, bacterial, 224
—, brown, 443
—, leaf, 2078
—, purple, 2988
—, sooty, 3510

377
BLOTCHES
fr taches *f pl* d'eau;
    transparence *f*
ne watervlekken

de Wasserflecken *m*
da vandpletter (fugtpletter)
sv vattenfläckar
es manchas *f pl* de agua
it macchie *f pl* d'acqua

378
BLOTCHY;
UNEVEN COULOURED
fr de faux teint
ne wankleurig
de fehlfarbig
da miskulørt
sv missfärgad
es descolorido
it colorito stravagante

379
BLOTCHY RIPENING
fr malade de l'eau
ne waterziek
de wasserkrank
da 'vandsot' (vanddrukkent)
sv vattensjuk
es hidrópico *m*
it ammalato dall'acqua

380
BLOTCHY RIPENING
(TOMATO);
BLOSSOM-END ROT
(TOMATO)
fr pourriture *f* du nez;
    pourriture *f* sèche
nè neusrot
de Blütenendfäule *f*
    Tomatenfruchtfäule *f*
da griffelråd *n*
sv toppröta
es podredumbre *f*
    umbilical;
    podredumbre *f* apical
it marciume *m* apicale

381
BLOW DOWN, TO
fr lâcher
ne afblazen
de abblasen
da udblæse
sv släppa ut ånga
es aflojar
it scaricare

382
BLOW DRAIN, TO
fr vider
ne afspuien
de abblasen
da udlufte
sv rappa av vatten
es vaciar
it svuotare parzialmente

BLOWER, 1291

383
BLOW- OFF COCK
fr robinet *m* d'évacuation
ne spuikraan; afblaaskraan
de Schlammventil *n*;
    Abblashahn *m*
da slamventil
    udblæsningsventil
sv slamavtappningskran;
    bottenventil; ångventil
es grifo *m* de salida
    grifo *m* de descarga
it valvola *f* d'evacuazione

384
BLUE, TO
fr bleuir
ne blauwkleuren
de blaufärben
da blåfarve
sv blåna; blåfärga
es azular
it colorire in azzurro

BLUEBERRY, swamp,
3756

385
'BLUE-GROWING'
(TULIP)
fr bleuissement *m*
ne blauwgroeien (tulp)
de 'blaugewachsene'
    Tulpenzwiebeln
da blomstringsdygtige løg *n*
sv lökarna blir blåaktiga
es azularse
it azzurregiare

386
BLUE SPRUCE;
KOSTER'S SPRUCE
fr Picea *f* pungens 'Glauca'
ne blauwspar

de Blaufichte *f*
da blågran
sv blågran
es abeto *m* azul glauco
it abeto *m* azzuro
   Picea pungens 'Glauca'
   (Regel) Beissw.

BLUE WILLOW BEETLE,
781

387
BLUNT
fr obtus
ne stomp
de stumpf
da stump
sv stump
es obtuso
it ottuso

388
BLUSH
fr rougeur *f*
ne blos
de Röte *f*
da rødmen
sv rodnad
es sonrosado *m*
it rossore *m*

389
BOARD; PLANK
fr planche *f*
ne plank
de Brett *n*
da bræt *n*
sv planka; bräde *n*
es tabla *f*
it asse *f*; tavola *f*

BOG,   2382, 3755

BOG peat,   2652

390
BOG SOIL; PEAT SOIL
fr terre *f* tourbeuse
ne veenaarde
de Moorerde *f*
da mosejord
sv mossjord
es tierra *f* turba
it terra *f* palustre

BOIL, TO,   827

391
BOILER
fr chaudière *f*
ne ketel
de Kessel *m*
da kedel
sv panna
es caldera *f*
it caldaia *f*

392
BOILER MOUNTING
FITTINGS
fr garnitures *f pl* de
   chaudière
ne ketelappendages
de Kesselarmaturen *f pl*
da kedelarmatur
sv pannarmatur
es accesorios *m pl* de caldera
it guarnizioni *f pl* per caldaia

393
BOLTERS
fr plantes *f pl* montant à
   graine
ne schieters
de Schiesser *m pl*
da udløbere
sv stocklöpare
es plantas *f pl* trepadoras
it piante *f pl* evaculazioni

394
BONE FLOUR;
BONE MEAL
fr poudre *f* d'os
ne beendermeel *n*
de Knochenmehl *n*
da benmel *n*
sv benmjöl *n*
es harina *f* de huesos
it farina *f* d'ossi;
   ossa *f* polverizzata

BONE FLOUR,
steamed,   3625

395
BONUS PAYMENT
fr salaire *m* avec primes
ne premieloon *n*
de Prämienlohn *m*
da del i udbytte (bonus)

sv premielön
es salario *m* con primas
it salario *m* con gratifica

396
BOOK VALUE
fr valeur *f* comptable
ne boekwaarde
de Buchwert *m*
da bogført værdi
sv bokfört värde
es valor *m* contable
it valore *m* contabile

397
BORAGE
fr bourrache *f*
ne bernagie
de Boretsch *m*
da hjulkrone; Borago
sv gurkört
es borraja *f*
it borrana *f*
   Borago officinalis L.

398
BORAX
fr borax *m*
ne borax
de Borax *m*
da borax
sv borax
es borax *m*
it borace *m*

399
BORDEAUX MIXTURE
fr bouillie *f* bordelaise
ne Bordeauxse pap
de Kupferkalkbrühe *f*;
   Bordeauxbrühe *f*
da Bordeauxvædske
sv bordeauxvätska
es caldo *m* bordelés
it poltiglia *f* bordolese

400
BORDER; EDGE
fr bord *m*; bordure *f*;
   plate-bande *f*
ne boordbed; border;
   rand; rabat
de Rabatte *f*; Rand *m*;
   Einfassung *f*

da kant; rabat;
randbeplanting
sv kant; rabatt; rand
es platabanda *f*; arriate *m*;
cuartel *m*; borde *m*
it bordo *m* di fiori; aiuola *f*;
bordura *f*

401
BORECOLE
fr chou *m* frisé
ne boerenkool
de Krauskohl *m*;
Blätterkohl *m*
da bladkål
sv bladkål
es col *f* frizada;
col *f* crespa
it cavolo *m* crespo
Brassica oleracea L.
cv. Laciniata Schulz.

402
BORER (CAULIFLOWER)
fr chou *m* feuillé;
chou monté
ne boorder
de Frühblüher *m*
da for tidlig blomstrende
sv förtidigt blomstrande
es floración *f* prematura
it fioritura *f* prematura

BORER, currant, 945
—, peach twig, 2628
—, poplar and willow, 4162
—, red bud, 3057
see also: shot-hole borer

403
BORON
fr bore *m*
ne borium *n*
de Bor *n*
da bor *n*
sv bor *n*
es boro *m*
it boro *m*

404
BORON DEFICIENCY
fr carence *f* en bore
ne boriumgebrek *n*
de Bormangel *m*
da bormangel

sv borbrist
es carencia *f* de boro
it deficienza *f* di boro

405
BORROWED CAPITAL
fr capital *m* extérieur
ne vreemd vermogen *n*
de Fremdkapital *n*
da fremmedkapital *n*
sv främmande kapital *n*;
skulder
es capital *m* ajeno
it capitale *m* preso in
prestito

406
BORROWER
fr demandeur *m* de crédit
ne kredietnemer
de Kreditnehmer *m*;
Schuldner *m*
da (hypothek) debitor
sv låntagare
es prestatario *m*
it debitore *m* ipotecario

407
BOTANICAL
fr botanique
ne botanisch
de botanisch
da botanisk
sv botanisk
es botánico
it botanico

408
BOTANY
fr botanique *f*
ne plantkunde
de Pflanzenkunde *f*;
Botanik *f*
da botanik; plantekendskab
sv botanik
es botánica *f*
it botanica *f*

BOTRYTIS, 1657, 2929

409
BOTRYTIS DISEASE
fr maladie *f* causée par le
Botrytis
ne Botrytisziekte

de Botrytiskrankheit *f*;
Grauschimmelkrankheit *f*
da gråskimmel
sv gråmögelsangrepp
es Botrytis *m*
it malattia *f* botrytis
Botrytis cinerea Pers.

410
BOTRYTIS ROT
(GLADIOLUS);
LETTUCE LEAF SPOT
fr taches *f pl* foliaires;
maladie *f* des taches
ne vuur *n*
de Botrytiskrankheit *f*;
Blattfleckenkrankheit *f*
da gladiolus-gråskimmel
sv gråmögel
es encendido *m*; negrón *m*
(de las plantas de huerta)
it annerimento *m* delle
piante ortensi
Botrytis gladioli (Kleb.)
Pleospora herbarum
(Fr.) Rab.

411
BOTRYTIS SPOT
fr Botrytis
ne Botrytisstip
de Botrytisflecke *m pl*
da gråskimmelplet
sv gråmögelsfläck
es punteado *m* gris
it macchietta *f* da Botrytis

412
BOTTLENECK
fr goulot *m* d'étranglement
ne knelpunt *n*
de Engpass *m*;
begrenzender Faktor *m*;
Flaschenhals *m*
da flaskehals
sv trång sektor; flaskhals
es punto *m* de estrangulamiento
it strozzatura *f*

BOTTOM HEAT, 3483

BOUGH, 421

413
BOULDER CLAY
fr argile *f* caillouteuse
ne keileem *n*
de Geschiebelehm *m*
da bassinler *n*
sv moränlera
es arcilla *f* de silex
it argilla *f*

414
BOUQUET; NOSEGAY;
BUNCH OF FLOWERS
fr bouquet *m*
ne boeket *n*
de Strauss *m*; Bukett *n*
da buket
sv bukett
es ramo *m* de flores;
ramillete *m*
it mazzo *m* di fiori

BOX, 3914

415
BOX (SHRUB)
fr buis *m*
ne palmboompje
de Buchsbaum *m*
da buksbom
sv buxbom
es boj *m*
it palma *f*; Buxus *m*
Buxus L.

416
BOX, TO
fr mettre en jauge
ne opkuilen
de auflagern (in Mieten)
da nedkule
sv nedgräva
es encajonar
it infossare

417
BOX SUCKER;
BOXWOOD PSYLLID
(US)
fr psylle *m* du buis
ne palmbladvlo
de Buchsbaumblattfloh *m*
da buxbom-bladloppe
sv buxbomsbladloppa

es psila *f* del boj
it psilla *f* del bosso
Psylla buxi L.

BOX UP, TO, 1368

BOXWOOD PSYLLID,
417

BRACKEN CLOCK, 1530

418
BRACT
fr bractée *f*
ne schutblad
de Deckblatt *n*; Hochblatt *n*
da højblad *n*; dækblad *n*;
støtteblad *n*
sv skärmblad *n*; högblad *n*
es bráctea *f*
it brattea *f*

419
BRACT CURDS
fr buplèvre *m*
ne doorwas
de Durchwuchs *m*
da genvækst
sv genomväxt
es picarse *m*
it spigazione *f*

420
BRAN
fr son *m*
ne zemelen
de Kleie *f pl*
da klid *n pl*
sv kli *n pl*
es salvado *m*; afrecho *m*
it crusca *f*

421
BRANCH; BOUGH
fr branche *f*
ne tak
de Zweig *m*; Ast *m*
da gren
sv gren
es rama *f*
it ramo *m*

BRANCH, death of, 997
—, fruit, 1471
—, lateral, 2043
—, main, 2242

422
BRANCHED; RAMIFIED
fr ramifié
ne vertakt
de verästelt; verzweigt
da grenet
sv förgrenad
es ramificado
it ramificato

BRANCH OUT, TO, 3028

423
BRANCH TENDRIL
fr sarment *m*
ne takrank
de Zweigranke *f*
da grenranke; årsskud *n*
sv ranka
es sarmiento *m*
it sarmento *m*

424
BRASSICA POD MIDGE
fr cécidomyie *f* du chou
ne koolzaadgalmug
de Kohlgallmücke *f*;
Kohlschotenmücke *f*
da skulpegalmyg
sv skidgallmygga
es mosquito *m* de la col
it cecidomia *f* dei cavoli
Dasyneura brassicae
Winn.

425
BRASSY WILLOW
BEETLE
fr chrysomèle *f* de l'osier
ne bronzen wilgehaantje *n*
de Weidenblattkäfer *m*
da pilebladbille
sv pilglansbagge
es crisomelido *m* del sauce
it crisomela *f* del salice
Phyllodecta vitellinae L.

BREAK, 1411

426
BREAKDOWN;
DECOMPOSITION
fr décomposition *f*
ne ontbinding

de Zersetzung *f*;
Verwesung *f*
da nedbrydning
sv (upp)lösning;
nedbrytning
es descomposición *f*
it decomposizione *f*

BREAKDOWN, internal,
1941

427
BREATHING;
RESPIRATION
fr respiration *f*
ne ademhaling
de Atmung *f*
da ånding
sv andning; respiration
es respiración *f*
it respirazione *f*

428
BREATHING ROOT
fr 'voile' *f* de racines
ne ademwortel
de Atemwurzel *f*
da ånderod
sv andningsrot
es neumatóforos;
raíz *f* respiradora
it radice *f* di respirazione

BREED, TO, 80

429
BREEDER
fr cultivateur *m*
ne kweker
de Züchter *m*
da tiltrækker; dyrker;
avler
sv uppdragare
es cultivador *m*
it coltivatore *m*

430
BREWERS GRAIN
fr drèche *f* de brasserie
ne bostel
de Treber *m*
da mask
sv mäsk
es bagazo *m*
it trebia *f*

431
BRIDAL BOUQUET
fr bouquet *m* de mariée
ne bruidsboeket *n*
de Brautstrauss *m*
da brudbuket
sv brudbukett
es ramo *m* de novia;
it fiori *m pl* nuziali

432
BRIGHT
fr clair
ne licht
de hell
da lyst
sv ljus
es claro
it chiaro; luminoso

433
BRINE
fr saumure *f*
ne pekel
de Sole *f*
da brine
sv saltlösning
es salmuera *f*
it salamoia *f*

434
BROAD BEAN
fr fève *f* de marais
ne tuinboon
de Puffbohne *f*;
Saubohne *f*
da hestebønne
sv bondböna
es haba *f*
it fava *f*
Vicia faba L.

BROAD BEAN
WEEVIL, 272

435
BROADCAST
fr à la volée
ne breedwerpig
de breitwürfig
da bredså
sv bredsådd
es al voleo
it a spaglio

436
BROAD-LEAVED DOCK
fr patience *f* sauvage
ne ridderzuring
de stumpfblättriger
Ampfer *m*
da butbladet skræppe
sv tomtskräppa
es alfileres; aguja *f* de pastor
it romice *m* comune
Rumex obtusifolius L.

437
BROAD-LEAVED TREE
fr feuillu *m*
ne loofboom
de Laubgehölz *n*
da løvtræ *n*
sv lövträd *n*
es árbol *m* de fronda;
árbol *m* frondoso
it albero *m* frondoso

438
BROAD MITE
fr acarien *m* jaune
ne begoniamijt
de Breitmilbe *f*
da topskudmide
sv begoniakvalster *n*
es ácaro *m* de la begonia
it acaro *m*
Hemitarsonemus latus
Banks.

439
BROCCOLI
fr chou-brocoli *m*
ne broccoli
de Brokkoli *m*
da broccoli; aspargeskål
sv broccoli
es bróculi *m*;
brécoles *m pl*
it broccolo *m*

440
BROME GRASS
fr brome *m* doux
ne zachte dravik
de weiche Trespe *f*
da blød hejre
sv luddlosta

es bromo *m* dulce
it Bromus *m*
    Bromus hordeaceus L.

441
BROOK SILT SOIL
fr dépôt *m* de ruisseau
ne beekbezinking
de Bachablagerung *f*
da bækaflejring
sv bäckavlagring
es sedimento *m* de arroyo
it sedimento *m* del ruscello

442
BROOM
fr cytise *m*
ne brem
de Geissklee *m*; Ginster *m*
da gyvel
sv ginst
es retama *f*
it ginestro *m*
    Cytisus L.

BROOM, common,   1556

443
BROWN BLOTCH
(MUSHROOM
GROWING)
fr taches *f pl* brunes
ne bruine vlekken
de Braunfleckigkeit *f*
da brunplettethed
sv brunfläckighet
es manchas *f pl* oscuras
it macchie *f pl* brunastre
    Verticillium malthousei
    Ware

444
BROWN HEART;
CARBON DIOXYDE
INJURY
fr 'coeur brun' *m* dû au CO_2
ne koolzuurbederf *n*
de CO_2 Markbräune *f*;
    Kernhausbräune *f*
da brunt kærneråd *n*
sv kolsyreskada
es deterioro *m* per la;
    acción del ácido
    carbónico
it cuore *m* bruno

445
BROWNISH RED
fr rouge brun;
    couleur de rouille
ne bruinrood
de braunrot
da brunrød
sv brunröd
es pardo rojizo
it rosso brunetto

446
BROWNISH YELLOW
fr jaune brunâtre
ne bruingeel
ne braungelb
da brungul
sv brungul
es pardo amarillento
it giallo brunetto

447
BROWN LEAF WEEVIL;
LEAF EATING WEEVIL
fr phyllobie *f* oblongue
ne bladsnuitkever;
    behaarde snuitkever
de Schmalbauch *m*;
    Spitzmausrüssler *m*
da løvsnudebille
sv avlång lövvivel
es gorgojo *m* de las yemas;
    filobio *m* roedor
it fillobio *m* oblungo;
    tagliagemme *m*
    Phyllobius oblongus L.

448
BROWN PLASTER
MOULD
fr plâtre *n* brun,
    plâtre *m* rouge
ne bruine kalkschimmel
de brauner Gips *m*
da brun gipssvamp
sv brun gipssvamp
es yeso *m* marrón;
    yeso *m* rojo
it muffa *f* marrone
    Papulaspora byssina
    Hotson

449
BROWN ROT (APPLE)
fr moniliose *f*;
    rot *m* brun
ne monilia-rot *n*
de Moniliakrankheit *f*
da monilia
sv gul monilia
es podredumbre *f* negra;
    moniliosis *f*
it marciume *m* bruno;
    monilia *f*
    Sclerotinia fructigena
    A. & R.

450
BROWN SPOT
(LABURNUM)
fr taches *f pl* foliaires
ne bladvlekkenziekte
de Blattfleckenkrankheit *f*
da brunpletsyge
sv bladfläcksjuka
es punteado *m* de las hojas
it macchie *f pl* fogliari e
    seccume
    Pleiochaeta setosa

451
BROWN TAIL MOTH
fr bombyx *m*; chrysorrhée *f*;
    cul-brun *m*
ne bastaardsatijnvlinder
de Goldafter *m*
da guldhale
sv äppelrödgump;
    guldsvansspinnare
es mariposa *f* blanca de cola
    dorada;
    oruga *f* de zurrón;
    polilla *f*;
it bruco peloso degli alberi
    da frutto; bombice *m* dal
    ventre bruno
    Euproctis chrysorrhoea L.;
    Nygmia phaeorrhoea
    (US)

452
BRUISE
fr meurtrissure *f*
ne kneuzing
de Druckstelle *f*
da trykplet

sv stötning
es magulladura f
it lesione f

453
BRUISE, TO
fr meurtrir
ne kneuzen; beschadigen
de quetschen
da trykke
sv krossa; klämma; stöta
es magullar
it ammaccare

454
BRUSH, TO
fr brosser
ne borstelen
de bürsten
da børste
sv borsta
es cepillar; limpiar
it spazzolare

455
BRUSSELS SPROUTS
fr chou m de Bruxelles
ne spruitkool
de Rosenkohl m;
    Sprossenkohl m;
    Brüsseler Kohl m
da rosenkål m
sv brysselkål
es col f de Bruselas
it cavolo m a germogli
    Brassica oleracea L.
    cv. Gemmifera (DC.)
    Schulz.

456
BRUSSELS SPROUTS
PICKING MACHINE
fr récolteuse f de choux de
    Bruxelles
ne spruitenplukker
de Rosenkohlpflück-
    maschine f
da rosenkål-plukke-
    maskine
sv brysselkålsplockare
es cosechadora f de col de
    Bruselas
it raccoglitrice m cavolo
    di Bruxelles

BRYOBIA, apple and
pear, 127
—, gooseberry, 1610
—, ivy, 1965

457
BUCKTHORN
fr nerprun m commun
ne wegedoorn
de Kreuzdorn m
da vrietorn mf
sv getapel
es alaterno m; aladierna f
it ramno m
    Rhamnus cathartica L.

BUCKTHORN, alder, 65
—, sea, 3293

458
BUD
fr bouton m; bourgeon m;
    oeil m
ne knop
de Knospe f
da knop
sv knopp
es nudo m; botoni m;
    yema f
it gemma f

BUD, adventitious, 32
—, axillary, 205
—, dormant, 1103
—, flower, 1389
—, leaf, 2079
—, loose, 2214
—, main, 2243
—, mixed, 2358
—, root, 3172
—, terminal, 3812

459
BUD, TO
fr écussonner
ne oculeren
de äugeln; okulieren
da okulére
sv okulera
es injertar
it innestare

460
BUD AND LEAF
EELWORM
fr anguillule f des feuilles
ne bladaaltje n
de Blattälchen n
da bladnematod
sv bladål
es anguilula f de hojas
it nematode m delle gemme
    e delle foglie
    Aphelenchoides spp.

BUD CUTTING, 1277

461
BUDDING; GRAFTING
fr écussonnage m
ne veredeling
de Veredlung f
da forædling
sv förädling
es injerto m; mejora f
it innesto m

462
BUDDING KNIFE
fr écussonnoir m
ne oculeermes n
de Okuliermesser n
da okulérkniv
sv okuleringskniv
es cuchillo m de injertar
it coltello m da inoculare

463
BUDDLEIA
fr lilas m d'été
ne herfstsering
de Buddleia f
da summerfuglebusk
sv buddleja
es buddleya f
it Buddleia f
    Buddleia L.

464
BUD FORMATION
fr formation f de boutons
ne knopvorming
de Knospenbildung f
da knopdannelse
sv knopplik
es formación f de yemas
it gemmazione f

465
BUDGET
fr budget *m*;
   prévisions *f pl* budgétaires
ne begroting
de Voranschlag *m*;
   Vorschätzung *f*
da budget *n*
sv budget *n*
es previsiones *f pl*
it bilancio *m* preventivo

466
BUD-LIKE;
BUD-SHAPED
fr en forme de bouton
ne knopvormig
de knospig
da knopformet
sv knopplik
es en forma de botón;
   en forma de yema
it gemmiforme

467
BUD MOTH;
EYE-SPOTTED BUD
MOTH (US)
fr tordeuse *f* rouge des
   bourbons
ne rode knopbladroller
de roter Knospenwickler *m*
da røde knopvikler
sv mindre knoppvecklare
es polilla *f* de las yemas;
   tortrix *m* de las yemas
it tortricide *m* fulginae delle
   latifoglie;
   tortrice *f* dei germogli
   Spilonota ocellana Schiff.;
   Tortrix o. (UK)

468
BUD MUTATION; SPORT
fr mutation *f* gemmaire;
   sport *m*
ne knopmutatie; sport
de somatische Mutation;
   Sport *m*
da knopmutation
sv knoppmutation
es mutación *f* en los brotes
it mutazione *f* gemmaria

469
BUD NECROSIS (TULIP)
fr pourriture *f* du coeur
ne kernrot
de 'Innenfäule' *f*
da kenneråd
sv kärnröta
es podredumbre *f* del;
   corazón
it putrefazione *f* cardiaca

470
BUD SCALE
fr écaille *f* de bourgeon
ne knopschub
de Knospenschuppe
da knopskæl *n*
sv knoppfjäll *n*
es escama *f* de la yema
it scaglia *f* gemmaria

BUD-SHAPED, 466

471
BUFFER
fr tampon *m*;
   action-tampon *f*
ne buffer
de Puffer *m*;
   Pufferung *f*
da slødpudde *mf*
sv buffert
es tampón *m*;
   solución *f* tampón
it respingente *m*

472
BUG; TRUE BUG (US)
fr punaise *f*
ne wants; wandluis
de Wanze *f*
da tæge
sv stinkfly; bärfis
es chinche *f*
it cimice *f*; eterottero *m*
   Heteroptera

BUG, meadow spittle,   769
—, mealy,   2314
—, rhododendron,   3123

473
BUGBANE
fr cimicifuge *f*
ne Cimicifuga

de Silberkerze *f*
da sølvlys *n*
sv silverax *n*
es raíz *f* de culebra
it Cimicifuga *f*
   Cimicifuga L.

474
BUGLE
fr bugle *f*
ne zenegroen *n*
de Günsel *m*
da læbeløs
sv suga
es pinillo *m* bugula
it Ajuga *f*
   Ajuga L.

BUILD, TO,   3594

475
BUILT-ON AREA
fr espace *m* bâti
ne beteelde oppervlakte
de bestellte Fläche *f*
da bebygget areal *n*
sv bebyggt område *n*
es superficie *f* cultivada;
   área *f* de cultivo
it area *f* coperta

476
BULB
fr oignon *m*; bulbe *m*;
   oignon *m* à fleurs
ne bol; bloembol
de Zwiebel *f*;
   Blumenzwiebel *f*
da løg *n*; blomsterløg
sv lök: blomsterlök
es bulbo *m*;
   bulbo *m* de flores
it bulbo *m*;
   bulbo *m* da fiore

BULB, mother,   2392
—, offset,   2515

477
BULB CELLAR
fr cave *f* à bulbes
ne bollenkelder
de Zwiebelkeller *m*

da løgkælder
sv lökkällare
es sótano *m* para bulbos
it celliere *m* a bulbi

478
BULB DISTRICT
fr région *f* bulbicole
ne bloembollenstreek
de Blumenzwiebelanbauge-
    biet *n*
da blomsterløgdistrikt *n*
sv blomsterlöksdistrikt
es región *f* bulbicola
it distretto *m* di bulbi

479
BULB FIELD
fr champ *m* de bulbes à
    fleurs
ne bloembollenveld *n*
de Blumenzwiebelfeld *n*
da blomsterløgmark
sv blomsterlöksfält
es campo *m* de bulbos
it campo *m* di bulbi

480
BULB GRADING
MACHINE
fr calibreur *m* de bulbes
ne bloembollensorteer-
    machine
de Blumenzwiebelsortier-
    maschine *f*
da blomsterløgsorter-
    maskine
sv sorteringsmaskin för
    blomsterlökar
es calibrador *m* de bulbos
it calibratrice *f* per bulbi

481
BULBGROWER
fr producteur *m* de bulbes
ne bloembollenkweker
de Blumenzwiebelzüchter *m*
da blomsterløgdyrker
sv blomsterlöksodlare
es cultivador *m* de bulbos
it coltivatore *m* di bulbi

482
BULB GROWING
fr culture *f* des bulbes
ne bollenteelt;
    bloembollencultuur
de Blumenzwiebelkultur *f*
da blomsterløgdyrkning
sv blomsterlökodling
es cultivo *m* de bulbos
it coltivazione *f* di bulbi

483
BULBLET; OFFSET
fr caïeu *m*; bulbillon *m*
ne klister
de Brutzwiebel *f*
da sideløg *n*
sv sidolök
es bulbillo *m*
it bulbillo *m*

484
BULB LIFTER
fr récolteuse *f* de bulbes
    arracheuse
ne bollenrooier
de Blumenzwiebelernte-
    maschine *f*
    Rodemaschine *f*
da blomsterløgoptager
sv upptagare för blomster-
    lökar
es cosecadora *f* de bulbos
    arrancadora
it raccoglitrice *f* per bulbi

485
BULB MITE
fr rhizoglyphe *m* commun
ne bollemijt
de Wurzelmilbe *f*
da løgmide
sv lökkvalster *n*
es ácaro *m* de bulbos
it acaro *m* di bulbo
    Rhizoglyphus echinopus
    Fum. et Rob.

486
BULBOUS
fr bulbiforme
ne bolvormig
de zwiebelförmig,
    zwiebelartig

da løgformet
sv lökformig
es bulbiforme
it sferico; bulbiforme

487
BULBOUS PLANT
fr plante *f* bulbeuse
ne bolgewas *n*
de Zwiebelgewächs *n*
da løgvækst
sv lökväxt
es planta *f* bulbosa
it pianta *f* bulbosa

BULB ROT, grey,   1656

488
BULB SCALE
fr écaille *f* de bulbe;
    tunique *f* de bulbe
ne bolschub; bolrok
de Zwiebelschuppe *f*
da løgskæl
sv lökfjäll
es escama *f* de bulbo;
    túnica *f* de bulbo
it involucro *m* del bulbo

489
BULB SCALE MITE
fr tarsonème *m*
ne narcismijt
de Zwiebelschalenmilbe *f*
da narcis mide
sv narciss-kvalster *n*
es acarido *m* del narciso
it acaro *m* di narciso
    Taeneotarsonemus
    laticeps

490
BULB SHED
fr séchoir *m* à bulbes
ne bollenschuur
de Blumenlagerhaus *n*;
    Blumenzwiebelscheune *f*
da blomsterløgopbevarings-
    hus *n*
sv lagerhus *n* för blomsterlök
es almacén *m* de bulbos
it tettoia *f* a bulbi

491
BULK BIN
fr caisse-palette f
ne stapelkist
de Grosskiste f
da storkasse; palletkasse
sv stapellåda
es caja-paleta f
it cassetta f

492
BUMBLE BEE;
HUMBLE BEE
fr bourdon m
ne hommel
de Hummel m
da humlebi
sv humla
es moscardón m
it bombo m
    Bombus spp.

493
BUNCH
fr botte f
ne bos (bundel)
de Bund n; Bündel n
da bundt n
sv bunt
es manojo m
it fascio m

494
BUNCH, TO
fr bouqueter
ne bossen
de bündeln
da bundte
sv bunta
es amanojar
it affastelare

495
BUNCHED CARROTS
fr carottes f pl bottelées
ne bospeen
de Bundmöhren f pl
da bundtegulerod
sv buntade morötter
es zanahoria f de manojo
it carotas f pl affastellatas

BUNCH OF FLOWERS,
414

496
BUNDLE, TO
fr fardeler
ne bundelen
de bündeln
da bundte
sv bunta
es atar
it legare

497
BURGUNDY MIXTURE
fr bouillie f cuprique
ne bourgondische pap
de Burgunder Brühe f
da kobbersodavædske
sv kopparsodavätska
es caldo m cúprico
it poltiglia f rameica

BURN, 3279

BURN, dry-tip, 1157

498
BURNER
fr brûleur m
ne brander
de Brenner m
da brænder
sv brännare
es quemador m
it bruciatore m

BURNER, emulsifying,
1220
—, pressure atomizing,
2919
—, rotary atomizing, 3196

499
BURNET
fr pimprenelle f
ne pimpernel
de Pimpernell m;
    Bibernelle f
da bibernelle
sv pimpinell
es pimpinela f
it pimpinella f
    Sanguisorba minor Scop.

500
BURST, TO; TO CRACK
fr éclore
ne openbreken
de aufbrechen
da friste
sv bryta upp; bända upp
es romper
it spaccarsi

BUSH, 3393

501
BUSH TREE
fr buisson m
ne struikvorm
de Strauchform f
da buskform
sv buskform
es tallo m bajo
it di forma cespugliosa

502
BUSINESS
ECONOMICAL
BOOKKEEPING
fr tenue f d'un carnet
    d'exploitation
ne bedrijfseconomische
    boekhouding
de betriebswirtschaftliche
    Buchführung f
da driftsøkonomisk
    regnskabsføring
sv kostnadsbokföring
es carnet m de gestión
it contabilità f di gestione

503
BUTCHER'S BROOM
fr fragon m
ne Ruscus
de Mäusedorn m
da musetorn
sv stickmyrten
es rusco m
it Ruscus m
    Ruscus L.

504
BUTTERCUP
fr renoncule f;
    bouton d'or m
ne boterbloem

de Hahnenfuss *m*
da smørblomst; ranunkel
sv smörblomma
es ranúnculo *m*; pomposa *f*;
   francesilla *f*
it ranuncolo *m*
   Ranunculus spp.

505
BUTTERFLY
fr papillon *m*
ne vlinder
de Schmetterling *m*
da sommerfugl
sv fjäril
es mariposa *f*
it farfalla *f*

BUTTERFLY, cabbage
white, 520

—, green-veined white,
1655
—, large white, 2038
—, small white, 3445

506
BUTTERFLY WEED
fr asclépiade *f*
ne zijdeplant
de Seidenpflanze *f*
da silkeplante
sv sidenört
es yerba *f* de la seda
it asclepiade *f*
   Asclepias L.

507
BUTTON (MUSHROOM
GROWING)
fr champignon *m* fermé
ne gesloten champignon

de geschlossener
   Champignon *m*
da lukket champignon
sv sluten champinjon
es champiñon *m* cerrado
it fungo *m* chiuso

BUTTON SNAKE ROOT,
2147

508
BY-PRODUCT
fr sous-produit *m*
ne bijprodukt *m*
de Nebenprodukt *n*
da biprodukt *n*
sv biprodukt *n*
es subproducto *m*
it sottoprodotto *m*

C

509
CABBAGE
fr chou *m*
ne kool
de Kohl *m*
da kål
sv kål
es col *f*; berza *f*
it cavolo *m*
   Brassica spp.

CABBAGE, Chinese, 663
—, headed, 1746
—, oxheart, 2577
—, red, 3058
—, Savoy, 3255
—, white, 4141

510
CABBAGE APHID
fr puceron *m* cendré du
   chou
ne melige koolluis
de mehlige Kohlblattlaus *f*
da kålbladlus; kållus
sv kålbladlus
es pulgón *m* ceroso de la col
it afide *m* ceroso del cavolo
   Brevicoryne brassicae L.

511
CABBAGE CURCULIO
(US)
fr charançon *m* du chou
ne hartboorsnuitkever
de Kohlgallenrüssler *m*
da kålgallesnudebille
sv kålvivel
es gorgojo *m* de la col
it punteruolo *m* galligeno
   della rape
   Ceutho(r)rhynchus rapae
   Gyll.

512
CABBAGE CYST
NEMATODE
fr anguillule *f* des
   crucifères
ne koolcysteaaltje *n*
de Kohlzystenälchen *n*

da kålål
sv kålcystnematod
es nematodo *m* de la col
it nematode *m* delle
   crucifere
   Heterodera cruciferae
   Franklin

513
CABBAGE LETTUCE;
HEADING LETTUCE
fr laitue *f* pommée
ne kropsla
de Kopfsalat *m*
da hovedsalat
sv huvudsallad
es lechuga *f* flamenga
it lattuga *f* cappuccia

CABBAGE MAGGOT,
515

514
CABBAGE MOTH
fr noctuelle *f* du chou
ne kooluil
de Kohleule *f*
da kålugle
sv kålfly
es nóctua *f* de la col
it nottua *f* dei cavoli
   Barathra brassicae L.

515
CABBAGE ROOT FLY;
CABBAGE MAGGOT
(US)
fr mouche *f* du chou
ne koolvlieg
de Kohlfliege *f*
da kålflue
sv kålfluga
es mosca *f* de la col
it mosca *f* del cavolo
   Chortophila spp.;
   Erioischia brassicae
   (UK);
   Hylemya brassicae (US)

CABBAGE SEEDSTALK
CURCULIO, 517

516
CABBAGE SEED
WEEVIL;
CABBAGE SEEDPOD
WEEVIL (US)
fr ceutorrhynque *f* des
   siliques;
   charançon *m* des silices;
   charançon *m* des tiges du
   chou
ne boorsnuitkever;
   koolzaadsnuitkever
de Kohlschotenrüssler *m*;
   Kohlgallenrüssler *m*
da skulpesnudebille
sv blygrå rapsvivel
es hernia *f* de la colza;
   gorgojo *m* de las coles
it punteruolo *m* delle silique
   delle crucifere;
   ceutorrinco *m* delle rape
   *Ceutho(r)rynchus assimi-
   lis* Payk.

517
CABBAGE STEM
WEEVIL;
CABBAGE SEEDSTALK
CURCULIO (US)
fr charançon *m* de la tige du
   chou
ne stengelboorsnuitkever
de gefleckter
   Kohltriebrüssler *m*
da bladribbesnudebille
sv vivel
es potra *f* de la col
it ceutorrinco *m* del cavolo;
   punteruolo *m* della
   nervature delle foglie
   delle rape
   *Ceuto(r)rhynchus
   quadridens* Panz.

518
CABBAGE STUMP
fr trognon *m* de chou
ne koolstronk
de Kohlstrunk *m*

da kålstok
sv kålstock
es troncho *m* de col
it torso *m* di cavolo

519
CABBAGE THRIPS
fr thrips *m* du lin et des
céréales
ne vroege akkertrips
de Kohlrübenblasenfuss *m*
da kålthrips
sv åkertrips
es 'thrips' *m* de los guisantes
it tripide *m*
Thrips angusticeps Uzel.

520
CABBAGE WHITE
BUTTERFLY
fr piéride *f* du chou
ne koolwitje *n*
de Kohlweissling *m*
da kålsommerfugl
sv kålfjäril
es mariposa *f* de la col
it cavolaia *f*
Pieris spp.

521
CACTUS CYST
NEMATODE
fr nématode *m* kystique du
cactus
ne cactusaaltje *n*
de Kakteenzystenälchen *n*
da Heterodera cacti;
kaktusål
sv Heterodera cacti
es Heterodera *f* cacti
it nematode *m* dei cactus
Heterodera cacti
Filipjev & Schuurmans
Stekhoven

522
CAGE WHEEL
fr roue *f* squelette-cage
ne kooiwiel *n*
de Gitterrad *n*
da gitterhjul *n*
sv gallerhjul *n*;
skeletthjul *n*
es rueda *f* de jaula
it rullo *m* a gabbia

523
CALCAREOUS
fr calcarifère
ne kalkhoudend
de kalkhaltig
da kalkholdig
sv kalkhaltig
es calcáreo
it calcareo

524
CALCAREOUS SOIL
fr terre *f* calcaire
ne kalkbodem
de Kalkboden *m*
da kalkjord
sv kalkjord
es tierra *f* calcárea
it terra *f* calcarea

525
CALCEOLARIA
fr calcéolaire *f*
de pantoffelplant
de Pantoffelblume *f*
da taffelblomst
sv toffelblomma
es Calceolaria *f*
it Calceolaria *f*
Calceolaria L.

526
CALCICOLOUS
fr calcicole
ne kalkminnend
de kalkliebend
da kalkelskende
sv kalkälskande
es calcículo
it calcicole

527
CALCIFUGOUS
fr calcifuge
ne kalkschuwend;
kalkmijdend
de kalkmeidend
da kalkskyende
sv kalkskyende
es calcífugo
it calcifuggente

528
CALCIUM
fr calcium *m*
ne calcium *n*
de Kalzium *n*
da kalcium *n*
sv kalcium *n*
es calcio *m*
it calcio *m*

529
CALCIUM AMMONIUM
NITRATE
fr ammonitrate *m*
ne kalkammonsalpeter
de Kalkammonsalpeter *m*
da kalkammonsalpeter
sv kalkammonsalpeter
es amonitrato *m*
it nitrato *m* ammoniacato

530
CALCIUM CARBONATE
fr carbonate *m* de chaux
ne koolzure kalk
de kohlensaurer Kalk *m*
da kulsur kalk
sv kalciumkarbonat *n*
es carbonato *m* cálcico
it carbonato *m* di calce

531
CALCIUM POTASSIUM
SILICATE
fr chaux *f* sili-potassique
ne kalikiezelkalk
de Kalikieselkalk *m*
da kalikiselkalk
sv kalikiselkalk
es silicato *m* potásico-
cálcico
it calce *f* silicica di potassa

532
CALIBRATE, TO
fr calibrer
ne kalibreren
de kalibrieren
da kaliber
sv kalibrera
es calibrar
it calibrare

533
CALICO BUSH;
MOUNTAIN LAUREL
fr Kalmia *m*;
ne kalmia
de Kalmie *f*; Berglorbeer *m*
da Kalmia
sv Kalmia
es calicó *m*; calmia *f*
it Kalmia *f*
Kalmia L.

534
CALIFORNIAN POPPY
fr Eschscholtzia *m*
ne slaapmutsje *m*
de Goldmohn *m*;
Schlafmützchen *n*
da guldvalmue
sv sömntuta
es amapola *f* de California
it Eschscholtzia *f*
Eschscholtzia Cham.

535
CALLUS
fr callosité *f*; cal *m*
ne callus
de Kallus *m*
da kallus
sv kallus
es callosidad *f*
it callo *m*

536
CALORIFIER
fr échangeur *m* de chaleur
ne warmtewisselaar
de Wärmeaustauscher *m*
da varmeveksler
sv värmeväxlare
es intercambiador *m* de
calo
it scambiatore *m* di calore

537
CALYCLE
fr calicule *m*
ne bijkelk
de Nebenkelch *m*
da bibæger *n*
sv ytterfoder
es calícula *f*
it calice *m* accessorio

538
CALYX
fr calice *m*
ne kelk
de Kelch *m*
da bæger *n*
sv foder
es cáliz *m*
it calice *m*

539
CALYX SPLITTING
(CARNATIONS)
fr se crevasser (du calice)
ne scheuren
de platzen
da bægersprækning
sv spricka; rämna
es reventar
it screpolarsi

540
CAMASSIA
fr Camassia *f*
ne Camassia
de Kamassia *f*
da kamassia
sv kamassia
es Camassia *f*
it Camassia *f*
Camassia Lindl.

541
CAMBIUM
fr cambium *m*
ne cambium *n*
de Kambium *n*
da kambium *n*
sv kambium *n*
es cambium *m*
it strato *m* cambiale

542
CAMELLIA
fr camélia *m*
ne camelia
de Kamelie *f*
da kamelia
sv kamelia
es camelia *f*
it camelia *f*
Camellia japonica L.

543
CAMPANULA
ISOPHYLLA
fr étoile *f* de Bethlehem
ne ster van Bethlehem
de Stern *m* von Bethlehem
da Bethlehemsstjerne
sv bethlehemsstjärna
es Campanula *f*
it Campanula *f* di
Betlemme
Campanula isophylla
Morr.

CAMPSHED, TO, 2727

CAMPSHEDDING, 2729

544
CAN; TIN
fr boîte *f* en fer
ne blik *n*
de Blechbüchse *f*
da blikdåse
sv blecklåda; bleckburk
es lata *f*
it scatola *f* di latta

CAN, TO, 3860

545
CANAL
fr canal *m*
ne kanaal *n*
de Kanal *m*
da kanal
sv kanal
es canal *m*
it canale *m*

546
CANDIED FRUITS
fr fruits *mpl* confits
ne geconfijte vruchten
de kandierte Früchte *fpl*
da kandiserede frugter
sv kanderad frukt
es frutas *fpl* confitadas;
frutas *fpl* cristalizadas
it candito *m*

547
CANDY, TO;
TO PRESERVE
fr confire
ne confijten

de einzuckern
da kandisere
sv kandera
es confitar
it candire

548
CANDYTUFT
fr thlaspi *m*
ne scheefbloem
de Schleifenblume *f*
da flipkrave
sv Iberis; flipkrage
es carraspique *m*
it Iberis *m*
  Iberis L.

549
CANE BLIGHT
(RASPBERRY)
fr dessèchement *m* des
  tiges
ne stengelsterfte
de Himbeer-Rutensterben *n*;
  Rutenkrankheit *f*
da grensvamp
sv Leptosphæria coniothy-
  rium
es mal *m* del tallo
it cancro *m* del fusto
  *Leptosphaeria coniothy-
  rium* (Fuck) Sacc.

550
CANE SPOT
(RASPBERRY)
fr anthracnose *f*
ne stengelvlekkenziekte
de Brennfleckenkrankheit *f*
da pletskurv
sv fläckskorv
es salpicón *m* del tallo
it antracnosi *f*
  Elsinoë veneta Jenk.

551
CANKER; BARK
NECROSIS
fr chancre *m*;
  nécrose *f* corticale
ne kanker; bastnecrose
de Krebs *m*
  Rindenbrand *m*
da kræft; barknekrose
sv kräfta; barknekros

es cáncer *m*; chancro *m*;
  necrosis *f* cortical
it cancro *m*;
  necrosi *f* corticale

552
CANKER (APPLE)
fr chancre *m*
ne kanker
de Obstbaumkrebs *m*
da æble-kræft
sv fruktträdskräfta
es cancro *m*
it cancro *m*
  Nectria galligena Bres.

553
CANKER (CABBAGE)
fr taches *f pl* brunes
ne valler
de Umfallkrankheit *f*
da tørforrådnelse
sv torröta
es leptosferiosis *f*
it cancro *m*
  Lepthosphaeria maculans
  (Desm.) Ces. & de Not.

554
CANKER (LARIX)
fr chancre *m*
ne larikskanker
de Lärchenkrebs *n*
da lærkekræft
sv lärkkräfta
es cáncer *m*
it cancro *m*
  *Trychoscyphella
  willkommii*
  (Hart.) Nannf.

CANKER, bacterial, 220,
221
—, black, 331
—, stem, 3637

555
CANKER AND
DIE-BACK (POPULUS)
fr chancre *m* du peuplier
de schorsbrand
de Rindenbrand *m*
da poppelkræft
sv poppelkräfta
es cáncer *m*

it cancro *m*
  Cryptodiaporthe populea
  (Sacc.) Butris

556
CANKER AND
DIE-BACK (ROSA)
fr taches *f pl* sur tiges
ne takziekte
de Zweigkrankheit *f*
da stængelpletter
sv Griphosphæria corticola
es enfermedad *f* de las
  ramas
it tacche *f pl* sui fusti
  Griphosphaeria corticola
  (Fuck) v. Höhnel

557
CANNA
fr balisier *m*
ne Indisch bloemriet *n*
de indisches Blumenrohr *n*
da kanna
sv kanna
es caña *f* de las Indias
it Canna *f*
  Canna indica L.

558
CANNED FRUIT;
TINNED FRUIT
fr conserves *fpl* de fruits en
  boîtes
ne blikvruchten
de Dosenfrüchte *fpl*
da frugtkonserves
sv konserverad frukt
es frutas *mpl* conservadas
  en lata
it conserve *fpl* di frutti

559
CANNED VEGETABLES;
TINNED VEGETABLES
fr conserves *fpl* de légumes
  en boîtes
ne blikgroente
de Dosengemüse *npl*
  Büchsengemüse *npl*
da grøntkonserves
sv konserverade grönsaker
es hortalizas *fpl* conservadas
  en lata
it verdura *f* in scatola

560
CANOPY
fr porte *f* à faux
ne overstek *n*
de Ausladung *f*;
   Überstand *m*
da udhæng
sv takskägg
es dosel *m*; pabellón *m*
it sporgenza *f* a guisa di
   tetto

561
CAP; PILEUS
(MUSHROOM)
fr chapeau *m*
ne hoed
de Hut *m*
da hat
sv hatt
es cabeza *f*
it capello *m*

562
CAPE GOOSEBERRY
fr coqueret *m*; alkékenge *m*
ne lampionplant
de Judenkirsche *f*;
   Lampionpflanze *f*
da jødekirsebær *n*
sv judekörs
es alquequenje *m*;
it fisalis *f*
   Physalis L.

563
CAPE MARIGOLD
fr Dimorphoteca *f*
ne bekergoudsbloem;
   satijnbloem
de Kapkörbchen *n*
da guldblomst; solviser
sv afrikansk ringblomma
es caléndula *f* del cabo
it dimorfoteca *f*
   Dimorphoteca
   aurantiaca DC

564
CAPILLARITY
fr capillarité *f*
ne capillariteit
de Kapillarität *f*
da hårrørskraft

sv kapillarität
es capilaridad *f*
it capillarità *f*

565
CAPILLARY (subst.)
fr capillaire *f*
ne capillair
de Kapillarrohr *n*
da hårrør *n*
sv kapillär
es capilar *m*
it capillare *m*

566
CAPILLARY ACTION
fr action *f* capillaire
ne capillaire werking
de Kapillarwirkung *f*
da kapillær virkning
sv kapillärverkan
es acción *f* capilar
it azione *f* capillare

567
CAPILLARY LEVEL
fr hauteur *f* capillaire
ne capillaire stijghoogte
de kapillare Steighöhe *f*
da kapillær stigehøjde
sv kapillär stighöjd
es altura *f* capilar
it altezza *f* capillare

568
CAPILLARY REGION
fr zone *f* capillaire
ne capillaire zone
de Kapillarzone *f*;
   Kapillarraum *m*
da kapillær zone
sv kapillärzon
es región *f* capilar
it zona *f* capillare

569
CAPILLARY RISE
fr ascension *f* capillaire
ne capillaire opstijging
de kapillares Aufsteigen *n*
da kapillær stigning
sv kapillär stigning
es ascensión *f* capilar
it ascensione *f* capillare

570
CAPITAL ASSETS
fr capital *m*
ne kapitaalgoederen *n pl*
de Aktivkapital
da formeværdier
sv kapitaltillgång
es bienes *m pl* de capital
it capitali *m pl* fissi

571
CAPITAL
REQUIREMENT
fr besoin *m* de capital
ne vermogensbehoefte
de Kapitalbedarf *m*
da kapitalbehov *n*
sv kapitalbehov *n*
es necesidades *f pl* de bienes
it fabbisogno *m* di capitale

572
CAPITULUM (bot.)
fr capitule *m*
ne hoofdje *n*
de Köpfchen *n*; Körbchen *n*
da hoved *n*
sv huvud
es capítulo *m*
it capolino *m*

573
CAPSICUM;
CHRISTMAS CHERRY
fr oranger *m* de savetier
   morelle *f*
ne oranjeappelboompje *n*
de Korallenbäumchen *n*
da koralkirsebær *n*
sv korallbär *n*
es tomatillo *m* de
   Jerusalén
it solano *m* arancione
   Solanum capsicastrum
   Link.

574
CAPSULE (bot.)
fr capsule *f*
ne doosvrucht
de Kapsel *f*
da kapsel
sv kapsel
es cápsula *f*
it capsula *f*

575
CARAWAY
fr carvi *m*;
cumin *m* des prés
ne karwij
de Kümmel *m*
da kommen
sv kummin
es hinojo *m* de prado;
alcaravea *f*
it carvi *m*
Carum carvi L.

576
CARBOHYDRATE
fr hydrate *m* de carbone
ne koolhydraat *n*
de Kohlenhydrat *n*
da kulhydrat *n*
sv kolhydrat *n*
es hidrato *m* de carbono
it idrato *m* di carbonio

577
CARBONATE
fr carbonate *m*
ne carbonaat *n*
de Karbonat *n*
da karbonat *n*
sv karbonat *n*
es carbonato *m*
it carbonato *m*

578
CARBONATE OF
MAGNESIA
fr chaux *f* magnesienne
carbonatée
ne koolzure magnesiumkalk
de Magnesiummergel *m*
da kulsur magniumkalk
sv (kalkstens) dolomitmjöl
es carbonato *m* cálcico-
magnésico
it carbonato *m* di calce
magnesico

579
CARBON BLACK
fr carbone *m*
ne koolstof
de Kohlenstoff *m*
da kulstof *n*
sv kol *n*

es carbono *m*
it carbone *m*

580
CARBON DIOXIDE
f gaz *m* carbonique
ne koolzuurgas
de Kohlensäuregas *n*
da kulsyre
sv kolsyra
es ácido *m* carbónico
it gas *m* d'acido carbonico

581
CARBON DIOXIDE
APPLICATION
fr amendement *m* en CO_2;
enrichement *m* en CO_2;
application *f* de CO_2
ne CO_2-bemesting;
koolzuurgasbemesting;
koolzuurgastoediening
de Kohlensäuregas-
düngung *f*;
Kohlensäuregas-
verabreichung *f*
da kulsyregødning;
kulsyretilførsel
sv kolsyregödsling;
kolsyretillförsel
es enriquecimiento *m* con
ácido carbonico
it concimazione *f* con acido
carbonico;
applicazione *f* d'acido
carbonico

582
CARBON DIOXIDE
ASSIMILATION
fr assimilation *f* du gaz
carbonique
ne koolzuurassimilatie
de Kohlensäureassimila-
tion *f*
da kuldioxydassimilation
sv kolsyreassimilation
es asimilación *f* de gas
carbónico
it assimilazione *f* d'anidride
carbonica

CARBON DIOXYDE
INJURY, 444

583
CARBONIC ACID
fr acide *m* carbonique
ne koolzuur *n*
de Kohlensäure *f*
da kulsyre
sv kolsyra; koldioxid
es ácido *m* carbónico
it acido *m* carbonico

584
CARDBOARD
fr carton *m*
ne karton *n*
de Pappe *f*
da pap *n*; karton
sv kartong
es cartón *m*
it cartone *m*

CARDBOARD,
corrugated, 854

585
CARDBOARD BOX
fr caisse-carton *f*
ne kartonnen doos
de Pappschachtel *f*
da papæske
sv pappask; kartong
es caja *f* de cartón
it scatola *f* di cartone

586
CARDOON
fr cardon *m*
ne kardoen
de Cardy *m*
da kardon
sv kardon
es cardo *m*
it cardone *m*
Cynara cardunculus L.

CAREX PEAT SOIL, 3307

587
CARLINE THISTLE
fr chardon *m* acaule
ne zilverdistel
de Eberwurz *f*
da sølvtidsel
sv silvertistel

es cardo *m* plateado
it Carlina acaulis *f*
 Carlina acaulis

588
CARNATION
fr oeillet *m*
ne anjer
de Nelke *f*
da nellike
sv (trädgårds) nejlika
es clavel *m*
it garofano *m*
 Dianthus caryophyllus L.

589
CARPEL
fr carpelle *m*
ne vruchtblad *n*
de Fruchtblatt *n*
da frugtblad *n*
sv fruktblad *n*
es carpelo *m*
it carpello *m*

590
CARRIER
fr véhicule *m*; porteur *m*
ne draagstof; carrier
 smetstofdrager
de Trägerstoff *m*;
 latenter Träger *m*
da bærestof *n*; smittebærer
sv fyllnadsämne *n*;
 smittbärare
es sustancia *f* portadora;
 portador *m*
it portatore *m*

591
CARRION BEETLE
fr nécrophore *m*
ne aaskever; doodgraver
de Aaskäfer *m*
da ådselbille *m*
sv asbagge
es escarabajo *m*;
 necróforo *m*
it scarfaggio *m*;
 scarabeo *m*
 necroforo *m*
 Silphidae

592
CARRION FLOWER
fr stapelie *f*
ne aasbloem
de Aasblume *f*
da ådselbloms;
 ordensstjerne
sv asblomma
es estapelia *f*;
 flor *f* de lagarto
it Stapelia *f*
 Stapelia L.

593
CARROT
fr carotte *f*
ne peen
de Karotte *f*; Möhre *f*
da karot; gulerod
sv morot
es zanahoria *f*
it carota *f*
 Daucus carota L.

CARROT, topped, 3888
—, wild, 4157

594
CARROT CYST
NEMATODE
fr anguillule *f* de la carotte
ne peencystenaaltje *n*
de Möhrenzystenälchen *n*
da Heterodera carotæ;
 gulerodsål
sv morotcystnematod
es nematodo *m* de la
 zanahoria
it nematode *m* della carota
 Heterodera carotae Jones

595
CARROT FLY;
CARROT RUST FLY (US)
fr mouche *f* de la carotte
ne wortelvlieg;
 peenvlieg
de Möhrenfliege *f*
da gulerodsflue
sv morotsfluga
es mosca *f* de la zanahoria
it mosca *f* della pastinaca;
 mosca *f* della carota
 Psila rosae F.

596
CARROT HARVESTER
fr arracheuse *f* de carottes
ne wortelrooier
de Möhrenroder *m*
da gulerodsoptager
sv upptagare för morötter;
 morotsupptagare
es cosechadora *f* de zana-
 horias
it raccoglitrice *f* di carote

CARROT RUST FLY,
595

597
CARRYING CAPACITY;
STRESS
fr charge *f*
ne belasting
de Belastungsfähigkeit *f*
da bærevne
sv belastningskapacitet
es carga *f*
it caricamento *m*

598
CARYOPSIS; GRAIN
fr caryopse *m*
ne graanvrucht
de Karyopse *f*; Kornfrucht *f*
da skalfrugt; græsfrugt
sv hinnfrukt (hos gräs)
es cariopside *f*
it caryopsis *f*

599
CASE, TO
(MUSHROOM
GROWING)
fr gobeter
ne afdekken
de abdecken
da dække
sv täcka
es cubrir
it mettere terra

600
CASH REALISATION
fr taux *m* d'actualisation
ne contante waarde
de Barwert *m*
da kontant værdi

sv realisationsvärde *n*
es valor *m* en metálico
it valore *m* in contanti

601
CASING SOIL
(MUSHROOM
CULTIVATION)
fr gobetage *m*;
   couche *f* de couverture
ne dekaarde; deklaag
de Deckerde *f*;
   Deckschicht *f*
da dækjord
sv täckjord; täckskikt *n*
es tierra *f* de cubrir;
   cobertura
   casing *f*
it terra *f* di copertura

602
CASK
fr fût *m*; tonneau *m*
ne fust
de Fass *n*
da tonde
sv tunna; fat *n*
es tonel *m*
it barile *m*; fusto *m*

603
CATCHFLY
fr silène *m*
ne Silene
de Leimkraut *n*
da limurt
sv glim
es Silene *f*
it Silene *m*
   Silene L.

604
CATCH PIT
fr puisard *m*
ne zinkput
de Senkgrube *f*
da kloak; sivebrønd;
   rensebrønd
sv kloakbrunn
es sumidero *m*;
   pozo *m* negro
it smaltitoio *m*

605
CATCH-WATER DITCH
fr fossé *m* d'écoulement
ne boezemsloot
de Wasserauffanggraben *m*;
   Vorfluter *m*
da drængrøft
sv grendike;
   dikessystem
   (i en 'polder')
es canal *m* madre
it canale *m* di scolo

606
CAT CLAY
fr argile *f* à chats
ne katteklei
de Pulvererde *f*;
   Maibolt *m*
da katteklæg
sv järnsulfidhaltig lera
es sapropel *m*;
   arcilla *f* ácida

607
CATERPILLAR
fr chenille *f*
ne rups
de Raupe *f*
da sommerfuglelarve;
   kålorm
sv larv (fjäril)
es oruga *f*
it bruco *m*
   larva

CATERPILLAR, alder
sawfly, 67
—, looper, 2212

608
CATKIN
fr chaton *m*
ne katje *n*
de Kätzchen *n*
da rakle
sv hänge
es amento *m*
it amento *m*

CATKIN BEARER, 81

609
CATMINT
fr herbe-aux-chats *f*
ne kattekruid *n*

de Katzenminze *f*
da storblomstret
sv kattmynta; Nepeta
es menta *f* de gato;
   hierba *f* gatera
it gattaia *f*; erba gatta *f*
   Nepeta L.

610
CAT'S TAIL
fr amaranthe *f*
ne kattestaart
de Fuchsschwanz *m*
da rævehale; amarant
sv amarant
es amaranto *m*
it amaranto *m*
   Amaranthus L.

611
CAULIFLOWER
fr chou-fleur *m*
ne bloemkool
de Blumenkohl *m*
da blomkål
sv blomkål
es coliflor *f*
it cavolfiore *m*

CAULIFLOWER,
winter, 4176

CAULIFLOWER
DISEASE OF
STRAWBERRY, 2083

612
CAULIFLOWER MOSAIC
fr mosaïque *f* du chou-fleur
ne bloemkoolmozaïk *n*
de Blumenkohlmosaik *n*
da blomkål-mosaik
sv blomkålsmosaik
es mosaico *m* de la coliflor
it mosaico *m* del cavolfiore

613
CAUSTIC POTASH
SOLUTION
fr potasse *f* caustique
ne kaliloog *n*
de Kaliumlauge *f*
da kalilud
sv kalilut

es solución *f* de potasa
cáustica
it lisciva *f* di potassa

**614
CAUSTIC SODA
SOLUTION**
fr soude *f* caustique
ne natronloog *n*
de Natronlauge *f*
da natronlud
sv natronlut
es solución *f* de sosa
cáustica
it ransio *m* di sodio

CAVE, 2345

**615
CEANOTHUS**
fr céanothe *m*
ne Amerikaanse sering
de Säckelblume *f*
da ceanothus
sv säckbuske
es ceanoto *m*
it ceanota *f*
Ceanothus L.

**616
CEDAR**
fr cèdre *m*
ne ceder
de Zeder *f*
da ceder
sv ceder
es cedro *m*
it cedro *m*
Cedrus Link.

CEDAR, incense, 1886

**617
CEILING FAN**
fr ventilateur *m* du plafond
ne plafondventilator
de Deckenlüfter *m*
da loft ventilator
sv takfläkt
es ventilador *m* de techo
it ventilatore *m* a soffito

**618
CELERIAC**
fr céleri-rave *m*
ne knolselderij
de Knollensellerie *m*
da knoldselleri
sv rotselleri
es apio *m* nabo
it sedano *m* tuberoso
Apium graveolens L.
cv. Rapaceum (Mill.)
DC.

**619
CELERY**
fr céleri *m*;
céleri *m* à couper
ne selderij; snijselderij
de Sellerie *m*;
Schnittsellerie *m*
da selleri; snitselleri
sv selleri; snittselleri
es apio *m*;
Apio *m* para cortar
it sedano *m*
Apium graveolens L.

CELERY, blanched, 358

**620
CELERY FLY**
fr mouche *f* du céleri
ne selderijvlieg
de Selleriefliege *f*
da sellerieflue
sv sellerifluga
es mosca *f* del apio
it mosca *f* dei sedani
Philophylla heraclei L.

**621
CELL**
fr cellule *f*
ne cel
de Zelle *f*
da celle
sv cell
es célula *f*
it cellula *f*

**622
CELL
(PART OF ANTHER)**
fr loge *f*
ne helmhokje *n*

de Fach *n*
da støvsæk
sv pollensäck
es saco *m* polínico
it casella *f*

**623
CELL DIVISION**
fr division *f* cellulaire
ne celdeling
de Zellteilung *f*
da celledeling
sv celldelning
es división *f* de célula
it scissione *f* delle cellule

**624
CELL FORMATION**
fr formation *f* des cellules
ne celvorming
de Zellbildung *f*
da celledannelse
sv cellbildning
es formación *f* de células
it formazione *f* della cellula

**625
CELL SAP**
fr suc *m* cellulaire
ne celvocht *n*
de Zellsaft *m*
da cellesaft
sv cellsaft
es protoplasma *m*
it succhio *m* cellulare

**626
CELL TISSUE;
CELLULAR TISSUE**
fr tissu *m* cellulaire
ne celweefsel *n*
de Zellgewebe
da cellevæv *n*
sv cellvävnad
es tejido *m* cellular
it tessuto *m* della cellula

**627
CELLULOSE**
fr cellulose *f*
ne cellulose; celstof
de Zelluloze *f*; Zellstoff *m*
da cellestof *n*; cellulose
sv cellulosa; cellstoff *n*

es celulosa *f*
it cellulosa *f*

628
CELLULOSE ACETATE
fr acétate *f* de cellulose
ne cellulose-acetaat *n*
de Zellulose-acetat *n*
da cellulose-acetat *n*
sv cellulosaacetat
es acetato *m* de celulosa
it acetate *m* di cellulosa

629
CELL WALL
fr cloison *f* cellulaire;
   membrane *f* cellulaire
ne celwand
de Zellwand *f*
da cellevæg
sv cellvägg
es membrana *f* de la célula
it membrana *f* della cellula

CELOSIA, 733

630
CEMENT, TO
fr mastiquer
ne kitten
de kitten
da kitte
sv kitta
es masticar
it attaccare

631
CENTIPEDE
fr mille-pattes *m*
ne duizendpoot
de Tausendfüßler *m*
da tusindben *n*
sv tusenfoting
es ciempiés *m*
it millepiedi *m*
   Chilopoda

632
CENTRAL LEADER
fr branche *f* axiale
ne harttak
de Mitteltrieb *m*
da midtstamme
sv mittstamm; mittaxel

es tronco *m* guía
it astone *m* centrale;
   branca *f* centrale

633
CENTRE OF
HORTICULTURE
fr centre *m* horticole
ne tuinbouwcentrum
de Gartenbauzentrum *n*
da gartnericenter *n*;
   havebrugscenter *n*
sv trädgårdsodlingscentrum
es centro *m* hortícola
it centro *m* orticole

634
CENTRIFUGAL FAN
fr ventilateur *m* centrifuge
ne centrifugaal ventilator
de Zentrifugalgebläse *n*
da centrifugal ventilator
sv centrifugalfläkt
es ventilador *m* centrífuga
it ventilatore *m* centrifugo

635
CENTRIFUGAL PUMP
fr pompe *f* centrifuge
ne centrifugaalpomp
de Kreiselpumpe *f*;
   Zentrifugalpumpe *f*
da centrifugalpumpe
sv centrifugalpump
es bomba *f* centrífuga
it pompa *f* centrifuga

636
CEP; EDIBLE BOLETUS
fr cèpe *m*; bolet *m*
ne eekhoorntjesbrood *n*
de Steinpilz *m*
da spiselig rörhat
sv Karl Johans-svamp
es cep *m*
it porcino *m*
   Boletus edulis
   Bull. ex Fr.

637
CEPHALOTHORAX
fr céphalothorax *m*
ne kopborststuk *n*
de Kopfbruststück *n*

da forkrop
sv huvud-bröststycke;
   cephalothorax
es cefalotorax *m*
it cefalotorace *m*
   cephalotorax
   (Arachnoidea)

638
CESPITOSE
fr gazonnant
ne zodenvormend
de rasenbildend
da græstorvdannende
sv torvabildande
es cespitoso
it piotante

639
CESSPOOL
fr fosse *f* d'aisance
ne beerput
de Senkgrube *f*
da ajlebeholder
sv latrinbrunn
es pozo *m* negro;
   fosa *f* de letrina
it pozzo *m* nero

640
CHAFER;
WHITE GRUB;
SCARAB (US)
fr ver *m* blanc
ne engerling
de Engerling *m*
da oldenborrelarve
sv ollonborrlarv
es larva *f* del abejorro
it scarabeide *m*;
   larva *f* melolontoide
   Scarabaeidae (larva);
   i.e. Melolontha spp.

CHAFER, garden, 1530
—, summer, 3728

641
CHAIN SAW
fr scie *f* à chaine
ne kettingzaag
de Kettensäge *f*
da båndsav
sv såg

es sierra f de cadena
it sega f

642
CHALKING
fr calcination f;
　calcification f
ne verkalking
de Verkalken n; Verhärten n
da forkalkning
sv förkalkning
es fusariosis f
it calzinazione f
　Fusarium oxysporum
　Schl.

643
CHANGING OF COLOUR
fr changement m de couleur
ne verkleuring
de Verfärbung f
da farveforandring
sv färgförändring
es cambio m de color
it scolorimento m

644
CHANTARELLE
fr girolle f; chanterelle m
ne hanekam, cantharel
de Pfifferling m;
　Eierschwamm m
da kantarel
sv kantarell
es miscalo m; niscalo m
it gallinaccio m
　Cantharellus cibarius
　Fr.

645
CHARLOCK; KEDLOCK
fr sanve f; moutarde f;
　sénevé m; ravenelle f
ne herik
de Ackersenf m
da ager sennep
sv åkersenap
es mostaza f silvestre
it senape
　Sinapis arvensis L.

646
CHAT FRUIT
fr fruit m atrophié
ne kleinvruchtigheid (appel)

de Kleinfrüchtigkeit f des
　Apfels
da småfrugt
sv småfruktighet
es pequeñez f del fruto
　(manzano)
it atrofia f dei frutti
　(del melo)

647
CHECKERBOARD
STACKING
(MUSHROOM
GROWING)
fr empilage f en quinconce
ne dambordstapeling
de schachbrettartiges
　Stapeln n
da skakbrætstabling
sv schackbrädsstapling
sv disposición f en forma de
　tablero de damas
it impilazione m a spazi;
　vuoti m pl alterni

648
CHECKING; CURBING
fr freinant; retardant
ne remmend
de hemmend
da bremsende; hammende
sv hämmande
es refrenante
it rallentante

649
CHEMICAL COMPOUND
fr composé m chimique
ne chemische verbinding
de chemische Verbindung f
da kemisk forbindelse
sv kemisk förening
es compuesto m químico
it composizione f chimica

CHERRY, choke,　675
—, Christmas,　573
—, Japanese,　1972
—, sour,　3517
—, St. Lucie,　3657
see also: bird-cherry

650
CHERRY-BARK
TORTRIX MOTH
fr Carpocapsa f
ne schorsboorder
de Rindenwickler m
da barkvikler
sv vecklare
es taladro m de la corteza
it Carpocapsa f
　Enarmonia formosana
　Scop.; *Carpocapsa;*
　Laspeyresia

651
CHERRY FRUIT FLY
fr mouche f des cerises
ne kersevlieg
de Kirschfruchtfliege f
da kirsebærflue
sv körsbärsfluga
es mosca f de las cerezas
it mosca f delle ciliege
　Rhagoletis cerasi L.

652
CHERRY LAUREL
fr laurier-cerise m
ne laurierkers
de Kirschlorbeer m
da laurbærkirsebær n
sv lagerhägg
es laurel m cerezo;
　laurel m real
it lauro m ceraso
　Prunus laurocerasus L.

653
CHERRY PLUM
fr myrobolan m
ne kerspruim
de Kirschpflaume f
da kirsebærblomme
sv körsbärsplommon
es mirobalano m
it mirobalano m
　Prunus cerasifera Ehrh.

654
CHERVIL
fr cerfeuil m
ne kervel
de Kerbel m
da kørvel; havekørvel

sv körvel
es perifollo *m*
it cerfoglio *m*
    Anthriscus cerefolium
    Hoffm.

655
CHESTNUT
fr châtaignier *m*
ne tamme kastanje
de Edelkastanie *f*
da ægte kastanie
sv äkta kastanj
es castaño *m*
it castagno *m*
    Castanea Mill.

CHESTNUT, horse, 1819

656
CHICKEN MANURE
fr fumier *m* de poule;
    fiente *f*
ne kippemest
de Hühnermist *m*
da hønsegødning;
    fjerkrægödning
sv hönsgödsel
es gallinaza *f*
it letame *m* di pollo

657
CHICKWEED
fr mouron *m* des oiseaux;
    stellaire *f*
ne muur
de Vogelmiere *f*; Hühner-
    darm *m*
da fuglegræs *n*
sv våtarv
es pamplina *f*, regojo *m*
it centocchio *m*
    Stellaria media L. Vill.

CHICKWEED, mouse-ear,
2410

658
CHICORY
fr chicorée *f*
ne cichorei
de Zichorienwurzel *f*
da cikorie
sv cikoria

es achicoria *f* de raíz
it cicoria *f*
    Cichorium intybus L.

CHICORY, witloof, 4193

659
CHILEAN NITRATE
fr nitrate *m* de soude du
    Chili
ne chilisalpeter
de Chilesalpeter *m*
da chilesalpeter *n*
sv chilesalpeter
es nitrato *m* de Chili
it salnitro *m* del Cile

660
CHILE PINE
fr Araucaria *m* du Chile
ne apeboom
de Araukarie *f*;
    Chilenische
    Schmucktanne *f*
da abetræ *n*
sv brödgran; apträd
es Araucaria *f*
it Araucaria *f*
    Araucaria araucana Koch

661
CHIMNEY
fr cheminée *f*
ne schoorsteen
de Schornstein *m*
da skorsten
sv skorsten
es chimenea *f*
it camino *m*

662
CHINA ASTER
fr reine-marguerite *f*
ne Chinese aster
de Sommeraster *f*
da sommeraster; aster
sv trädgårdsaster
es reina margarita *f*;
    maravilla *f*
it astro *m* della Cina
    Callistephus sinensis Nees

CHINA ROSE, 3189

663
CHINESE CABBAGE
fr chou *m* de Chine;
    Pe-Tsai *m*
ne Chinese kool
de Chinakohl *m*
da kinesisk kål
sv kinesisk kål
es col *f* de China
it cavolo *m* cinese
    Brassica cernua (Thbg.)
    Forb. et Hemsl.

664
CHINESE RHUBARB
fr rhubarbe *f* de Tartarie
ne Chinese rabarber
de Medizinalrhabarber *m*
da prydrabarber
sv flikrabarber
es ruibarbo *m* chino
it rabarbero *m* chinese
    Rheum palmatum L.

665
CHIP;
SMALL OFFSET
(BULB GROWING)
fr caïeu *m*
ne spaan (klein)
de platte nichtblühende
    Nebenzwiebel *f*
da små yngelløg *n*
sv sidolök (liten)
es bulbo *m* lateral
it bulbo *m* laterale

666
CHIP (BASKET);
PUNNET
fr cageot-billot *m*;
ne slof (chip)
de Spankorb *m*;
    Span-Flacksteige *f*
da spånkurv
sv spånkorg
es canasta *f*; basquet *m*
it canestro *m* di vimini

667
CHIVES
fr ciboulette *f*
ne bieslook
de Schnittlauch *m*

da purløg
sv gräslök
es ceboletta *f*
it cipolletta *f*
Allium schoenoprasum L.

668
CHLORIDE
fr chlorure *f*
ne chloride *n*
de Chlorid *n*
da klorid
sv klorid
es cloruro *m*
it cloruro *m*

669
CHLORINE
fr chlore *m*
ne chloor *n*
de Chlor *n*
da klor *n*
sv klor *n*
es cloro *m*
it cloro *m*

CHLORINE, low in, 2221

670
CHLOROPHYLL
fr chlorophylle *f*
ne bladgroen *n*
de Blattgrün *n*;
Chlorophyll *n*
da bladgrønt *n*;
klorofyl *n*
sv bladgrönt *n*; klorofyll *n*
es clorofila *f*
it clorofilla *f*

671
CHLOROPICRINE
fr chloropicrine *f*
ne chloorpikrine
de Chlorpikrin; Larvazide *n*
da klorpikrin
sv klorpikring
es cloropicrina *f*
it cloruro *m* di picrico

672
CHLOROSIS
fr chlorose *f*
ne chlorose
de Chlorose *f*

da blegsot; klorose
sv kloros
es clorosis *f*
it clorosi *f*

CHLOROSIS, grafting,
1624

673
CHOCOLATE SPOT
(VICIA)
fr Botrytis *m*
ne chocoladevlekkenziekte
de Schokoladenflecken-
krankheit *f*
da gråskimmel
sv gråmögel
es mal *m* del esclerocio
it vaiolatura *f* bruna
Botrytis spp.

CHOGGING, 3913

674
CHOKEBERRY
fr aronie *f*
ne appelbes
de Apfelbeere *f*;
Zwergvogelbeere *f*
da Aronia
sv Aronia
es cornijuelo *m*
it Aronia *f*
Aronia Pers.

675
CHOKE CHERRY
fr cerisier *m* de Virginie
ne Virginiaanse vogelkers
de rotfrüchtige
Traubenkirsche *f*
da Virginsk hæk
sv virginiahägg
es cerezo *m* de Viriginia
it Prunus *f* virginiana
Prunus Virginiana L.

CHOP OFF, TO, 957

676
CHOPPER; BILLHOOK
fr couperet *m*; hachoir *m*
ne kapmes *n*
de Hackmesser *n*;
Kappmesser *n*

da hakkekniv
sv hackkniv
es machete *m*
it coltellaccio *m*

677
CHRISTMAS CACTUS
fr Zygocactus *m*;
Epiphyllum *m*
ne lidcactus; kerstcactus
de Weinachtskaktus *m*
da julekaktus
sv julkaktus
es cacto *m*
it cacto *m* articolato
Zygocactus bridgesii

CHRISTMAS CHERRY,
573

678
CHRISTMAS ROSE
fr rose *f* de Noël
ne kerstroos
de Christrose *f*
da julerose
sv julros
es eléboro *m*
it rosa *f* di Natale
Helleborus niger L.

CHRISTMAS STAR, 2811

679
CHRIST'S THORN
fr euphorbe *f*
ne Christusdoorn
de Christusdorn *m*
da kristitornekrone
sv kristi törnekrona
es euforbia *f* brillante
it euforbia *f*
Euphorbia milii Desm.

CHRYSALIS, 2982

680
CHRYSANTHEMUM
fr chrysanthème *m*
ne chrysant
de Chrysantheme *f*
da krysanthemum;
vinteraster
sv krysantemum

es crisantemo *m*
it crisantemo *m*
   Chrysanthemum L.

681
CHRYSANTHEMUM
LEAF MINER
fr mouche *f* des
   chrysanthèmes
ne chrysantevlieg
de Erbsenminierfliege *f*
da chrysanthemumflue
sv mindre krysantemum-
   fluga
es mosca *f* del crisantemo
it mosca *f* dei crisantemi e
   dei piselli
   Phytomyza atricornis Mg.

682
CICADA; CICALA;
LEAFHOPPER
fr cicadelle *f*
ne cicade
de Schwarzpunktzikade *f*;
   Singzikade *f*;
   Zwergzikade;
   Kleinzikade
da cikade
sv strit (cikada)
es cicádela *f*
it cicala *f*; cicalina *f*
   Cicadellidae

683
CIDER
fr cidre *m*
ne cider
de Zider *m*
da cider
sv cider; äppelvin
es sidra *f*
it sidro *m*

684
CILIATE
fr cilié
ne gewimperd
de gewimpert
da håret
sv kanthårig
es ciliado
it cigliato

685
CINERARIA
fr cinéraire *f* hybride
ne cineraria
de Cinerarie *f*
da stjerneblomst; cineraria
sv cineraria
es cineraria *f*
it cineraria *f*
   Senecio cruentus DC

686
CINQUEFOIL;
POTENTILLA
fr potentille *f*
ne ganzerik
de Fingerkraut *n*
da potentil
sv fingerört
es Potentilla *f*
it Potentilla *f*
   Potentilla L.

687
CIRCUIT
fr conduite *f*
ne leiding
de Leitung *f*
da ledning
sv ledning
es conducto *m*
it conduttura *f*

688
CIRCULAR
fr circulaire
ne cirkelrond
de kreisförmig
da cirkelrund
sv cirkelrund
es circular
it circolare

689
CIRCULATION
fr circulation *f*
ne circulatie
de Zirkulation *f*
da cirkulation
sv cirkulation; kretslopp *n*
es circulación *f*
it circolazione *f*

690
CITRIC ACID
fr acide *m* citrique
ne citroenzuur *n*
de Zitronensäure *f*
da citronsyre
sv citronsyra
es ácido *m* cítrico
it acido *m* citrico

691
CLADONIA
fr cladonie *f* crénelée
ne bekermos *n*
de Becherflechte *f*
da bægerlav
sv bägarlav
es musgos *m*
it Cladonia *f*
   Cladonia

692
CLAMP LOADER;
CLAMP TRUCK
fr diable *m*
ne klemsteekwagen
de Klemmkarre *f*
da gaffeltruck
   klemmetruck
sv lagerkärra
es diablo *m*
it carello *m* sollevatore

693
CLAW (bot.)
fr onglet *m*
ne nagel
de Nagel *m*
da negl
sv klo (på kronblad)
es uña *f*
it unghia *f*

694
CLAY
fr argile *f*
ne klei
de Ton *m*
da klæg
sv lera
es arcilla *f*
it terra *f* grassa;
   argilla *f*

CLAY, sandy, 3245
—, sea, 3294

695
CLAY-COLOURED
WEEVIL
fr otiorrhynque *m*
ne lapsnuittor
de Lappenrüssler *m*;
   Dickmaulrüssler *m*
da øresnudebille
sv öronvivel
es curculiónido *m* de la vid
it otiorrinco *m*
   Otiorrhynchus singularis
   L.

696
CLAYEY; CLAYISH
fr argileux
ne kleiachtig
de tonig; lehmig
da klægholdig
sv lerig
es arcilloso
it argilloso

697
CLAY SOIL
fr sol *m* argileux
ne kleigrond
de Tonboden *m*;
   Lehmboden *m*
da klægjord
sv lerjord
es suelo *m* arcilloso
it terreno *m* argilloso

698
CLEAN, TO;
TO PURIFY
fr épurer; nettoyer; rincer
ne reinigen
de reinigen
da rengøre; rense
sv rensa
es limpiar
it pulire; nettare

699
CLEAN OFF, TO
fr éplucher
ne pellen
de putzen
da pille

sv rensa lök
es limpiar bulbos
it pulire i bulbi

700
CLEAVERS;
GOOSEGRASS; HERRIF
fr gaillet *m*; rièble *m*;
   gratteron *m*
ne kleefkruid *n*
de Klebkraut *n*;
   klebendes Labkraut *n*
da burre-snerre
sv snärjmåra
es amor *m* de hortelano
it aparine *f*; atlaccamani *f*
   Galium aparine L.

701
CLEFT; SPLIT
fr fendu; fissuré; 'ouvert'
ne gespleten
de gespalten
da delt; kløftet
sv kluven
es hendido
it scisso; fesso

702
CLEFT GRAFTING;
WEDGE GRAFTING
fr greffer en fente
ne spleetenten
de Spaltpfropfung *f*
da spaltepodning
sv klyvimpning
es injerto *m* de hendidura;
   injerto *m* de para
it innestare a spacco

703
CLEMATIS;
OLD MAN'S BEARD
fr clematite *f*
ne Clematis
de Waldrebe *f*
da skovranke; Clematis
sv Clematis
es clemátide *f*
it Clematis
   Clematis L.

704
CLICK BEETLE
fr taupin *m*
ne kniptor
de Schnellkäfer *m*
da smelder
sv (sädes) knäppare
es escarabajo *m*
it elateride *m*
   Elateridae

705
CLIMATE
fr climat *m*
ne klimaat
de Klima *n*
da klima *n*
sv klimat *n*
es clima *m*
it clima *m*

706
CLIMATE CONTROL
fr contrôle *m* des tempéra-
   res; climatisation *f*
ne klimaatbeheersing
de Klimabeherrschung *f*
da klimakontrol
sv klimatkontroll
es control *m* climático
it controllo *m* ambientale

707
CLIMATIC CONDITIONS
fr état *m* de l'atmosphère
ne weersomstandigheden
de Witterungsumstände *m pl*
da vejrforhold
sv väderleksförhållanden
es condición *f* climatológica
it condizioni *f pl* atmos-
   ferice

708
CLIMB, TO
fr grimper
ne ranken
de ranken
da ranke
sv klättra; klänga
es trepar
it avvinghiarsi

709
CLIMBER; RUNNER
fr plante *f* grimpante
ne klimplant; rankplant
de Kletterpflanze *f*
   Schlingpflanze *f*
da klatreplante;
   slyngplante
sv klätterväxt
es planta *f* trepadora
   planta *f* sarmentosa
it pianta *f* rampicante

710
CLIMBING
fr grimpant
ne klimmend
de kletternd
da klatrende
sv klättrande
es remontante
it rampante

711
CLIMBING FRENCH
BEAN
fr haricot *m* à rames
ne stokboon
de Stangenbohne *f*
da stangbønne
sv störböna
es judía *f* alta (de enrame)
it fagiolo *m* rampicante
   Phaseolus vulgaris L.

712
CLIP, TO;
TO TRIM (HEDGE)
fr tondre
ne knippen
de schneiden
da klippe (hæk)
sv klippa (häck)
es cortar; rasurar (setos)
it tagliare

713
CLOCHE
fr cloche *f*
ne teeltklok (glas of kunststof)
de Haube *f*
da dyrkningsklokke
sv miniaturväxthus *n*

es campana *f*
it cloche *f*

CLOCK AUCTION,   192

714
CLOD; LUMP
fr motte *f* de terre
ne kluit
de Scholle *f*; Erdklumpen *m*
da jordklump
sv jordkoka
es terrón *m* de tierra
it motta *f* di terra

715
CLONE
fr clone *m*; lignée *f*
ne kloon
de Klon *m*
da klon
sv klon
es clon *m*
it clone *m*

716
CLOTHES MOTH
fr teigne *f*; mite *f*
ne mot
de echte Motte *f*
da møl *n*
sv mal *n*
es polilla *f*
it tignola *f*
   Tineidae

CLOVE,   3363

717
CLOVER
fr trèfle *m*
ne klaver
de Klee *m*
da kløver
sv klöver
es trébol *m*
it trifoglio *m*
   Trifolium L.

CLOVER, crimson,   891
—, red,   3059
—, white,   4142

718
CLOVER SEED WEEVIL
fr apion *m* du trèfle
ne snuitkever
de Rüsselkäfer *m*
da snudebille
sv spetsvivel
es curculiónido *m* de la
   endivia
it apione *m* dei capolini del
   trifoglio
   Apion assimile Kirby

719
CLUB ROOT (CABBAGE);
FINGER-AND-TOE
DISEASE
fr hernie *f* du chou;
   maladie *f* digitoire
ne knolvoet
de Kohlhernie *f*;
   Kropfkrankheit *f* der
   Kohlgewächse
da kålbrok
sv klumprotssjuka
es hernia *f* de las coles
it ernia *f* del cavolo;
   tubercolosi *f* dei cavoli
   Plasmodiophora brassicae
   Woron.

720
CLUB-SHAPED
fr claviforme
ne knotsvormig
de keulenförmig
da køleformet
sv klubbformig
es claviforme
it claviforme

721
CLUMP; CLUSTER
(MUSHROOM)
fr groupe *f*
ne tros
de Klumpen *m*; Büschel *m*
da klynge
sv klunga
es grupo *m*; roca *f*
it ammasso *m*

CLUSTER,   721

722
COALESCENCE
fr coalescence *f*; union *f*
ne vergroeiing
de Verwachsung *f*
da sammenvoksning;
sammensmeltning
sv sammanväxning;
förening
es cicatrización *f*
it coalescenza *f*;
concrescimento *m*

723
CO_2-APPLICATION;
CO_2-DRESSING
fr amendement *m* en CO_2;
dosage *m* en CO_2
ne CO_2-bemesting
de CO_2-Düngung *f*;
Begasung *f*
da CO_2-gødning;
CO_2-tilførsel
sv CO_2-gödsling
es nutrición *f* por CO_2
concimazione *f* con acido
carbonico

724
COARSE
fr grossier
ne grof
de grob
da grov
sv grov
es grueso
it grosso

725
COARSE GRAINED
fr à gros grains
ne grofkorrelig
de grobkörnig
da grovkornet
sv grovkornig
es de granos gruesos
it a granuli *m pl* grossolani

726
COATING
fr enduction *f*
ne coating
de Beschichtung *f*
da overtræk *m*

sv beläggning; överdrag
es enlucido *m*
it copertura *f*

727
COBALT
fr cobalt *m*
ne kobalt *n*
de Kobalt *m*
da kobolt
sv kobolt
es cobalto *n*
it cobalto *m*

728
COBWEB
fr toile *f* d'arraignée
ne spinneweb
de Spinngewebe *n*
da spindelvæv *n*
sv spindelväv;
spindelnät *n*
es tela *f* de araña
it ragnatela *f*

729
COBWEB DISEASE
fr toile *f*
ne spinnewebziekte :
de Spinnwebkrankheit *f*
da Dactylium
sv spindelvävssjuka;
Dactylium
es velo *m*; dactilio *m*
it mal *m* della tela
Dactylium dendroides
Bull. Fr.

730
COCK
fr robinet *m*
ne kraan
de Hahn *m*
da hane
sv kran; tappventil
es grifo *m*
it rubinetto *m*

731
COCKCHAFER
(LARVA = WHITE
GRUB)
fr hanneton *m*
ne meikever

de gemeiner Maikäfer *m*;
Feldmaikäfer *m*
da almindelig oldenborre
sv vanlig ollonborre
es melolonta *f*;
escarabajo *m* de San Juan,
cochorro *m*
it maggiolino *m*
Melolontha melolontha L.

732
COCK HANDLE
fr clé *f* de robinet
ne kraansleutel
de Hahnschlüssel *m*
da hanenøgle
sv ventilnyckel
es llave *f* de grifo
it chiave *f* del rubinetto

733
COCKSCOMB; CELOSIA
fr crête *f* de coq; célosie *f*
ne hanekam
de Hahnenkamm *m*
da hanekam
sv tuppkam
es cresta *f* de gallo;
amaranto *m*
it cresta *f* di gallo
Celosia cristata L.

734
COCKSFOOT
fr dactyle *m* pelotonné
ne kropaar
de Knaulgras *n*
da hundegræs *n*
sv hundäxing
es dactilo *m* aglomerado
it Dactylis
Dactylis glomerata L.

735
COCOON
fr cocon *m*
ne cocon
de Kokon *m*
da kokon
sv kokong
es capullo *m*
it bozzolo *m*

736
CODLING MOTH
fr carpocapse *m* des
pommes et des poires
ne fruitmot
de Apfelwickler *m*;
Obstmade *f*
da æblevikler
sv äppelvecklare
es polilla *f* de las manzanas;
gusano *m* de las manzanas
y peras;
gusano *m* carpocapsa
it Carpocapsa *f*;
bruco *m*; verme *m* delle
pere e delle mele
Enarmonia pomonella L.;
Laspeyresia p. (UK);
Carpocapsa p. (US)

CO_2-DRESSING, 723

737
COEFFICIENT OF
EXPANSION
fr module d'élasticité;
coefficient *m* de dilatation
ne uitzettingscoëfficiënt
de Ausdehnungs-
koeffizient *m*
da udvidelseskoefficient
sv utvidgningskoefficient
es módulo *m* de elasticidad;
coeficiente *m* de dilata-
ción
it coefficiente *m* di dilata-
zione

738
CO_2-ENRICHMENT
fr fumure *f* carbonique
ne CO_2-toevoer
de CO_2-Düngung *f*
da CO_2-tilförsel
sv CO_2 (koldioxid)-gödsling
es enriquecimiento *m* en
carbonico
it concimazione *f* carbonica

739
COLCHICUM;
AUTUMN CROCUS
fr colchique *m*
ne herfsttijloos
de Herbstzeitlose *f*

da tidløs
sv tidlösa
es colquico *m*
it colchico *m*
Colchicum L.

740
COLD CHAIN
fr chaîne *f* frigorifique
ne koelketen
de Kühlkette *f*
da kølekæde
sv kylkedja
es cadena *f* frigorífica
cadena *f* de frío
it catema *f* del freddo

741
COLD FRAME
fr couche *f* froide
ne koude bak
de kalter Kasten
da kold bænk
sv kallbänk
es cajonera *f* fría
it letto *m* freddo

742
COLD HOUSE
fr serre *f* froide
ne koude kas
de Kalthaus *n*
da kold hus *n*
sv kallhus *n*
es invernadero *m* sin
calefacción
it serra *f* fredda

743
COLD (RADIATION)
STERILISATION
fr (radio) stérilisation *f* à
froid
ne koude (radio) sterilisatie
de kalte (Radio) Sterili-
sierung *f*
da sterilisation
sv sterilisation
es (radio) esterilización *f*
en frío
it sterilizzazione *f* fredda

744
COLD STORAGE
fr entreposage *m* frigorifique;
conservation *f* à froid
ne bewaring in koelruimten;
koelhuisopslag
de Kaltlagerung *f*;
Kühllagerung *f*
da køleopbevaring;
kølelagring
sv kyllagring
es conservación *f* frigorífico
it conservazione *f* frigorifica

745
COLD STORE;
REFRIGERATED ROOM
fr entrepôt *m* frigorifique
ne koelcel; koelhuis
de Kühlhaus *n*; Kühlzelle *f*
da kølehus *n*
sv kylhus *n*
es cámara *f* frigorífica
it cella *f* refrigerante;
frigorifero *m*

746
COLEUS
fr coléus *m*
ne Coleus
de Blumennessel *f*
da koleus; paletblad *n*
sv praktnässla; palettblad *n*
es coléo *m*
it ortica *f* di fiori
Coleus Lour.

747
COLLAR; PIPE COLLAR
fr collet *m*
ne kraag
de Muffe *f*
da rørmuffe
sv rörkrage *n*
es collar *m*
it colletto *m*

748
COLLAR BEAM
fr entrait *m*
ne hanebalk
de Hahnenbalken *m*

da hanebjælke
sv hanbjälke
es tirante *m*
it tirante *m*

749
COLLAR ROT (APPLE)
fr pourriture *f* du collet
ne stambasisrot *n*
de Kragenfäule *f*
da stammebasisråd
sv stambasröta
es gangrena *f* del pie
it marciume *m* del colletto
   Phytophthora cactorum
   Schroet.

COLLAR ROT, fomes,
1418

750
COLLECTIVE
PUBLICITY
fr publicité *f* de masse;
   publicité *f* collective
ne collectieve reclame
de Gemeinschaftswerbung *f*
da kollektiv propaganda
sv gemensam reklam;
   massreklam
es publicidad *f* en masa
it pubblicità *f* di massa

751
COLLOID
fr colloïde *f*
ne colloïd
de Kolloide *f*
da kolloid
sv kolloid
es coloide *m*
it collodio *m*

752
COLLOIDAL
fr colloïdal
ne kolloidaal
de kolloidal
da kolloidal
sv kolloidal
es coloidal
it colloidale

COLONISATION, 2443

753
COLORADO BEETLE;
COLORADO POTATO
BEETLE (US)
fr doryphore *m* de la
   pomme de terre
ne coloradokever
de Kartoffelkäfer *m*
da koloradobille
sv koloradoskalbagge
es dorifora *f*;
   escarabajo *m* de la patata
it dorifora *f* della patata
   Leptinotarsa decemlineata
   Say

754
COLOUR
fr couleur *f*
ne kleur
de Farbe *f*
da farve
sv färg; kulör
es color *m*
it colore *m*

755
COLOURED LIGHT
fr lumière *f* colorée
ne gekleurd licht
de farbiges Licht *n*
da farvet lys *n*
sv färgat ljus *n*
es luz *f* coloreada
it luce *f* colorata

756
COLOURING MATTER
(PROCESSING)
fr colorant *m*;
   matière colorante *f*
ne kleurstof
de Farbstoff *m*
da farvestof
sv färgämne
es materia *f* colorante
it materia *f* colorante

757
COLTSFOOT
fr tussilage *m*;
   pas d'âne *m*
ne klein hoefblad *n*

de Huflattich *m*
da følfod
sv hästhov
es tusilago;
   uña *f* de caballo
it farfara *f*
   Tussilago farfara L.

578
COLUMBINE
fr ancolie *f*
ne akelei
de Akelei *f*
da akeleje
sv akleja
es aguileña *f*;
   pajarilla *f*
it Aquilegia *f*
   Aquilegia L.

759
COLUMN
fr colonne *f*
ne kolom
de Säule *f*
da søjle
sv pelare; stolpe
es columna *f*
it colonna *f*

760
COMBUSTION ARCH
fr autel *m* de foyer
ne vuurbrug
de Feuerbrücke *f*
da forbrændingsbue
sv tändvalv *n*
es arco *m* de combustión
it luogo *m* di combustione

761
COMBUSTION
CHAMBER
fr boîte *f* à feu
ne vlamkast
de Feuerraum *m*;
   Verbrennungsraum *m*
da forbrændingskammer
sv brännkammare
es cámara *f* de combustión
it camera *f* di combustione

762
COMMERCIAL GROWER
fr cultivateur *m*
ne handelskweker
de Erwerbsgärtner *m*
da producent
sv handelsträdgårdsmästare
es cultivador *m* comercial
it agricoltore *m* industriale
   (per il mercato)

763
COMMERCIAL
HORTICULTURE
fr horticulture *f*
   professionnelle
ne beroepstuinbouw
de Erwerbsgartenbau *m*
da erhvervshavebrug *n*
sv yrkesmässig
   trädgårdsodling
es horticultura *f* comercial
it orticoltura *f* professio-
   nale

764
COMMERCIAL POLICY
fr politique *f* commerciale
ne handelspolitiek
de Handelspolitik *f*
da handelspolitik
sv handelspolitik
es política *f* commercial
it politica *f* commerciale

765
COMMISSION AGENT
fr commissionnaire *m*
ne commissionair
de Kommissionär *m*
da kommissionær
sv kommissionär
es comisionista *m*
it commissionario *m*

766
COMMON BEAN
MOSAIC
fr mosaïque *f* commune
   (haricot)
ne rolmozaïek *n* (boon)
de gewöhnliches Bohnen-
   mosaik *n*
da bønne-mosaik

sv vanlig bönmosaik
es mosaico *m* común
   (judías)
it mosaico *m* comune
   (fagiolo)

767
COMMON BLINDNESS
fr sans-coeur *m*
ne hartloosheid
de Herzlosigkeit *f*
da hjerteløshed
sv felslagen huvudbildning
es descogollado *m*
   (coliflor)
it senza-cuore *m*

COMMON BROOM, 1556

768
COMMON EARWIG;
EUROPEAN EARWIG
(US)
fr forficule *m*;
   perce-oreille *m*
ne oorworm
de gemeiner Ohrwurm *m*
da ørentvist
sv tvestjärt
es tijereta *f*
it forfecchia *f*; forbicina *f*
   Forficula auricularia L.

769
COMMON
FROGHOPPER
(NYMPH = CUCKOO
SPIT INSECT);
MEADOW SPITTLE BUG
(US)
fr aphrophore *f* écumeuse;
   crachat *m* de coucou
ne schuimbeestje
de Schaumzirpe *f*;
   Schaumzikade *f*
da skumcikade
sv (vanlig) spottstrit
es Cicádela espumosa;
   espumadora *f*
it sputacchina *f*
   Philaenus spumarius Fall.;
   Philaenus leucophthalmus
   L.

770
COMMON GOOSEBERRY
SAWFLY;
IMPORTED CURRANT
WORM (US)
fr tenthrède *f* jaune du
   groseillier
ne bessebladwesp
de gelbe Stachelbeer-
   blattwespe *f*
da store stikkelsbærhveps
sv krusbärstekel
es tentredino *m* amarillo
   del grosellero
it tentredine *f* del ribes
   Pteronidea ribessi Scop.;
   Nematus r. (US)

771
COMMON HORSETAIL
fr prêle *f* des champs
ne heermoes
de Zinnkraut *n*;
   Ackerschachtelhalm *m*
da ager-padderokke
sv åkerfräken
es cola *f* de rata;
   cola *f* de caballo;
   equiseto *m* menor
it coda *f* cavallina
   Equisetum arvense L.

772
COMMON OAK
fr chêne *m* pédonculé
ne zomereik
de Stieleiche *f*;
   Sommereiche *f*
da alm eg
sv skogsek; vanlig ek
es roble *m* común
it quercia *f* d'estivo
   Quercus robur L.

773
COMMON ORACHE
fr arroche *f* des champs
ne uitstaande melde
de gemeine Melde *f*
da svine-mælde
sv gårdmålla
es armuelle *m*
it erba *f* coreggiola
   Atriplex patula L.

774
COMMON PLANTAIN
fr grand plantain *m*
ne breedbladige weegbree
de grosser Wegerich *m*
da bredbladet vejbred
sv groblad
es ilantén *m*
it piantaggine *f* maggiore
Plantago major L.

775
COMMON SCAB
(POTATO)
fr gale *f*
ne aardappelschurft
de Kartoffelschorf *m*
da kartoffelskurv
sv potatisskorv
es sarna *f*
it scabbia *f* comune
Streptomyces spp.

776
COMMON SPRUCE;
NORWAY SPRUCE
fr épicéa *m* commun
ne fijnspar
de Rotfichte *f*; Rottanne *f*
da rødgran
sv (vanlig) gran
es Picea *f*
it Picea *f* abies
Picea abies Karst.

777
COMMON STORKSBILL
fr bec-de-grue *m*
ne reigersbek
de Reiherschnabel *m*
da hejrenæb *n*
sv skatnäva
es pico *m* de cigüeña
it erba *f* cicutaria
Erodium cicutarium L.
l'Hérit. ex. Hait.

COMMON SWIFT, 1538

778
COMMON THYME
fr thym *m*
ne tijm
de Thymian *m*
da Timian

sv timjan
es tomillo *m*; salsero *m*
it timo *m*
Thymus vulgaris L.

779
COMMON VETCH
fr vesce *f* cultivée
ne voederwikke
de Futterwicke *f*
da foder-vikke
sv fodervicker
es veza *f*
it veccia *f*
Vicia sativa L.

780
COMMON WALNUT
fr noyer *m* commun
ne okkernoot
de Walnuss *f*
da valnød (alm.)
sv valnöt
es nogal *m*
it noce *f*
Juglans regia L.

781
COMMON WASP;
BLUE WILLOW BEETLE
fr chrysomèle *f* de l'osier
ne wilgehaan
de blauer Weidenblatt-
käfer *m*
da pilebladbille
sv pilglansbagge
es crisomelido *m* del sauce;
taladro *m* rojo de los
troncos
it rodilegno *m* rosso
Phyllodecta vulgatissima
L.;
Phyll. vulgaris (UK)

782
COMPACTED LAYER
fr couche *f* d'argile alcaline
ne kniklaag
de Knickschicht
da knikklæglag
sv fortätat skikt i
marskjordar
es capa *f* de arcilla alcalina
it strato *m* d'argilla alcalina

COMPACTING, 793

783
COMPARISON OF FARM
RESULTS
fr comparaison *f* des résultats des exploitations
ne bedrijfsvergelijking
de Betriebsvergleich *m*
da driftssammenligning
sv företagsjämförelse;
resultatjämförelse
es comparación *f* de explo-
taciones
it confronto *m* dei risultati
aziendali

784
COMPATIBILITY
fr compatibilité *f*
ne verenigbaarheid
de Verträglichkeit *f*
da forenelighed
sv sammanväxning
es compatibilidad *f*
it affinità *f*

785
COMPLETE FERTILIZER
fr engrais *m* complet
ne volledige meststof
de Volldünger *m*
da fuldgødning
sv fullgödselmedel *n*
es abono *m* completo
it concime *m* completo

786
COMPOST
fr compost *m*
ne compost
de Kompost *m*
da kompost
sv kompost
es compost *m* de basura
it concime *m*

787
COMPOST, TO
fr composter
ne composteren
de kompostieren
da kompostere
sv kompostera
es fermentar;
   preparación *f* de basuras
it concimare

788
COMPOST HEAP
fr tas *m* de compost
ne composthoop
de Komposthaufen *m*
da kompostdynge
sv komposthög
es montón *m* de basura
it mucchio *m* di concime

COMPOSTING AREA,
790

789
COMPOSTING
ENTERPRISE
(MUSHROOM
GROWING)
fr compostage *m* à façon;
   fermentation *f* à façon
ne compostbedrijf *n*
de Kompostierungs-
   betrieb *m*
da kompostcentral
sv kompostcentral
es venta *f* de estiércol
   preparado
it centrale di preparazione
   del letame

COMPOSTING
MACHINE, 2267

790
COMPOSTING YARD;
COMPOSTING AREA;
FERMENTING WHARF;
BATCH GROUND (US)
fr aire *f* de fermentation
ne fermenteerplaats
de Kompostierungsplatz *m*
da komposteringsplads
sv komposteringsplats

es pista *f* de preparación de
   basuras
it area *f* di fermentazione

791
COMPOUND (bot.)
fr composé
ne samengesteld
de zusammengesetzt
da sammensat
sv sammansatt
es compuesto
it composito

792
COMPOUND
FERTILIZER
fr engrais *m* composé
ne mengmeststof
de Mischdünger *m*
da blandingsgødning
sv blandgödselmedel
es abono *m* compuesto
it concime *m* mescolato

793
COMPRESSION;
COMPACTING
fr compression *f*
ne verdichting
de Verdichtung *f*
da fortætning
sv sammantryckning;
   förtätning
es apelmazarse *m*
it condensamento *m*

COMPRESSION
SPRAYER, 2353

794
COMPRESSOR
fr compresseur *m*
ne compressor
de Verdichter *m*;
   Kompressor *m*
da kompressor
sv kompressor
es compresor *m*
it elemento *m* compressore

795
COMPULSORY
DELIVERY
fr livraison *f* obligatoire
ne veilplicht

de Andienungspflicht *f*
da leveringspligt
sv leveransplikt
   (till auktion)
es entrega *f* obligatoria
it consegna *m* obbligatoria

796
CONCAVE
fr concave
ne verdiept (bot.)
de vertieft
da indhvælvet; konkav
sv konkav
es profundo; cóncavo
it concavo

797
CONCENTRATE
fr concentré *m*
ne concentraat *n*
de Konzentrat *n*
da koncentrat *n*
sv koncentrat *n*
es concentrado *m*
it concentrazione *f*;
   liquore *m* concentrato

798
CONCENTRATION
fr concentration *f*
ne concentratie
de Konzentration *f*
da koncentration
sv koncentration
es concentración *f*
it concentrazione

799
CONCRETE BEAM
fr poutre *f* en béton
ne balk (beton-)
de Betonbalken *m*
da betonbjælke
sv betongbalk
es viga *f* de hormigón
   armado
it trave *f* in cemento
   prefabbricata

800
CONDENSATION
WATER
fr eau *f* de condensation
ne condenswater *n*

de Schweisswasser *n*;
Kondenswasser *n*
da kondensvand *n*
sv kondensationsvatten *n*
es agua *m* de condensación
it acqua *f* di condensazione

801
CONDENSE, TO
fr condenser
ne condenseren
de kondensieren
da kondensere; fortætte
sv kondensera
es condensar
it condensare

802
CONDENSER
fr condenseur *m*
ne condensor
de Verflüssiger *m*
Kondensator *m*
da kondensator
sv kondensator
es condensador *m*
it condensatore *m*

803
CONEFLOWER
fr Rudbeckia *m*
ne Rudbeckia
de Sonnenhut *m*
da solhat; Rudbeckia
sv Rudbeckia; gullboll
es Rudbeckia *f*
it Rudbeckia *f*
Rudbeckia L.

804
CONICAL
fr conique
ne kegelvormig
de kegelförmig
da kegleformet
sv konisk; kägeiformig
es cónico
it conico; coniforme

805
CONIFER
fr conifère *m*
ne naaldboom; conifeer
de Nadelbaum *m*;
Konifere *f*

da nåletræ *n*
sv barrträd
es conífera *f*
it conifero *m*

806
CONIFEROUS LITTER;
FOREST FLOOR
fr litière *f* forestière;
litière d'aiguille de
résineux
ne naaldengrond
de Nadelstreu *f*;
Nadelerde *f*
da grannålejord
sv barrströ *n*; förna
es tierra *f* de agujas;
tierra *f* de pino;
mantillo *m*
it lettiera *f* di conifere

807
CONIFER SPINNING
MITE; SPRUCE SPIDER
MITE (US)
fr araignée *f* rouge de
l'épicéa
ne sparrespintmijt
de Fichtenspinnmilbe *f*;
Nadelholzspinnmilbe *f*
da nåletræspindemide
sv barrträds(spinn)kvalster
es ácaro *m* del abeto
it ragnetto *m* rosso delle
conifere
Paratetranychus ununguis
Jacobi;
Oligonychus u. (UK)

808
CONNECTIVE
fr connectif *m*
ne helmbindsel *n*
de Konnektiv *n*
da støvknapbånd *n*
sv konnektiv
es conectivo *m*
it connettivo *m*

809
CONSISTENCY;
CONSISTENCE
fr consistance *f*
ne consistentie

de Konsistenz *f*
da konsistens
sv konsistens
es consistencia *f*
it consistenza *f*; densata *f*

CONSOLIDATE, TO,
3354

810
CONSUMER
fr consommateur *m*
ne verbruiker
de Verbraucher *m*;
Konsument *m*; Käufer *m*
da forbruger; konsument
sv förbrukare; konsument
es consumidor *m*
it consumatore *m*

811
CONSUMERS' CREDIT
fr crédit *m* à la consommation
ne afnemerskrediet *n*
de Abnehmerkredit *m*;
Kundenkredit *m*
da kundekredit
sv kundkredit
es crédito *m* al cliente
it credito *m* del consumatore

812
CONSUMER UNIT
fr empaquetage *m* au
détail
ne kleinverpakking
de Kleinverpackung *f*
da forbrugerpakning
sv kundförpackning;
minutförpackning
es empaquetado *m* espicial
it imballo *m* minuto

813
CONTACT ACTION
fr action *f* de contact
ne contactwerking
de Kontaktwirkung *f*
da kontaktvirkning
sv kontaktverkan
es acción *f* de contacto
it azione di contatto

814
CONTACT POISON
fr insecticide *m* de contact
ne contactgif *n*
de Kontaktgif *n*
da kontaktgift
sv kontaktgift
es insecticida *f* de contacto
it insetticida *f* di contatto

815
CONTAGIOUSNESS
fr contagiosité *f*
ne besmettelijkheid
de Ansteckbarkeit *f*
da smitsomhed
sv smittsamhet
es contagiosidad *f*
it contagiosità *f*

816
CONTAINERS
fr emballage *m*
ne fust (emballage)
de Verpackung *f*
da pakning
sv förpackning; emballage
es embalaje *m*
it recipiente *m*;
   imballaggio *m*

CONTAMINATE, TO,
1898

CONTAMINATED,   1899

CONTAMINATION,
1900

817
CONTENT
fr teneur *f*
ne gehalte *n*
de Gehalt *m*
da indhold *n*
sv innehåll *n*
es tenor *m*
it contenuto *m*

818
CONTENTS
fr contenu *m*
ne inhoud
de Inhalt *m*

da indhold *n*
sv innehåll *n*
es contenido *m*
it contenuto *m*

819
CONTRACTILE ROOT
fr racine *f* tractive
ne trekwortel
de Zugwurzel *f*
da sugerod
sv sugrot
es raíz *f* contráctil
it radice *f* di trazione

820
CONTRACTING FIRM;
MACHINERY
CONTRACTOR
fr entreprise *f* forfaitaire;
   station *f* de machines
ne loonbedrijf *n*
de Maschinenstation *n*;
   Lohnunternehmen *n*;
   Lohnbetrieb *m*
da lønbedrift
sv maskinstation
es casa *f* contratante
it azienda *f* di ricompensa

821
CONTRACT WORK
fr travaux *m pl* par entreprise *f*
ne loonwerk *n*
de Lohnarbeit *f*
da lønarbejde *n*
sv kontraktsarbete *n*;
   lönearbete *n*
es trabajo *m* contratado
it lavoro *m* salariato

822
CONTROL
fr lutte *f*
ne bestrijding
de Bekämpfung *f*
da bekæmpelse
sv bekämpning
es lucha *f*; control *m*
it lotta *f*

823
CONTROLLED ATMOS-
PHERE STORAGE;
GAS STORAGE
fr entreposage *m* en
   atmosphère artificielle
ne gasbewaring;
   bewaring in gascellen
de Gaslagerung *f*;
   Lagerung *f* in gesteuerter
   Luftzusammensetzung *f*
da kulsyreopbevaring *mf*
   kulsyrelagring
sv kolsyrelagring
es conservación *f* en atmósfera controlada
it conservazione *f* in atmosfera controllata

824
CONTROL MEASURE
fr mesure *f* de lutte
ne bestrijdingsmaatregel
de Bekämpfungs-
   massnahme *f*
da bekæmpelses-
   foranstaltning
sv bekämpningsåtgärd
es medida *f* de control
it misura *f* di lotta

825
CONVEX (bot.)
fr convexe; bombé
ne gewelfd
de gewölbt
da hvælvet
sv välvd; konvex
es arqueado; convexo
it arcuato

826
CONVEYER BELT
fr tapis roulant *m*
ne transportband
de Förderband *n*
da transportbånd *n*
sv transportband *n*;
   bandtransportör
es cinta *f* de transporte
it nastro *m* trasportatore

827
COOK, TO; TO BOIL
fr cuire; bouillir
ne koken
de kochen
da koge
sv koka
es hervir; cocer
it cuocere

828
COOK OUT, TO;
(MUSHROOM
GROWING)
fr recuire;
passer à la vapeur
ne doodstomen
de ausdämpfen
da udkoge; koge ud
sv koka ur; utkoka
es esterilizar con vapor
it sterilizzare

829
COOL, TO
fr refroidir
ne koelen
de kühlen
da køle
sv kyla
es enfriar; refrigerar
it raffreddare

830
COOLER
fr refroidisseur m
ne koeler
de Kühler m
da køler
sv kylare
es refrigerador m
it raffreddatore m

COOLEY SPRUCE GALL,
2733

COOL GLASSHOUSE,
3802

831
COOLING
fr réfrigération f
ne koeling
de Kühlung f

da køling
sv avkylning
es refrigeración f
it raffreddamento m

832
COPPER
fr cuivre m
ne koper
de Kupfer n
da kobber n
sv koppar
es cobre m
it rame m

833
COPPER DEFICIENCY
fr carence f en cuivre
ne kopergebrek n
de Kupfermangel m
da kobbermangel
sv kopparbrist
es carencia f en cobre
it deficienza f di rame

834
COPPER SULPHATE
fr sulfate m de cuivre
ne kopervitriool n
de Kupfervitriol n
da kobbervitriol n
sv kopparvitriol
es sulfato m de cobre;
caparrosa f
it vetriolo m di rame

COPPING, 3747

835
CORAL SPOT
(FOLIAGE TREES)
fr nécrose f du bois
ne vuur
de Rotpustelkrankheit f
da cinnobersvamp
sv rödvårtsjuka
es pústulas f pl rojas,
nectria f
it necrosi f degli organi
legnosi
Nectria cinnabarina
(Tode) Fr.

836
CORAL SPOT (RIBES)
fr maladie f du corail
ne vuur n
de Rotpustelkrankheit f
da zinnobersvamp
sv rödvårtsjuka
es encendido m
it necrosi f da Nectria
Nectria cinnabarina
(Tode) Fr.

837
CORAL TREE
fr érytrine f
ne koraalstruik
de Korallenstrauch m
da koralbusk
sv korallbuske
es árbol m del coral
it albero m del corallo
Erythrina L.

838
CORDATE;
HEART SHAPED
fr cordé; cordiforme
ne hartvormig
de herzförmig
da hjerteformet
sv hjärtformig
es cordiforme
it cordiforme

839
CORDON
fr cordon m
ne snoer n; cordon n
de Schnur f
da snortræ n; cordon
sv kordong (träd)
es cordón m
it cordone m

840
CORE
fr trognon m
ne klokhuis n
de Kerngehäuse n
da kernehus n
sv kärnhus n
es corazón m; antro m
it torsolo m

841
CORE, TO (HYACINTH)
fr percer
ne boren
de ausbohren
da bore
sv urholka
es taladrar
it scavare

842
CORE FLUSH
fr brunissement *m* du coeur
ne inwendig bruin *n*
de Markbräune *f*; Kernhaus-
   bräune *f*
da centerråd *n*
sv kärnhusbrunt
es corazón *m* pardo
it cuore *m* bruno

843
COREOPSIS
fr Coreopsis *m*
ne Coreopsis
de Mädchenauge *n*;
   Wanzenblume *f*
da skønhedsøje *n*
sv skönhetsöga *n*
es Coreopsis *m*
it Coreopsis *m*
   Coreopsis L.

CORING, 3913

844
CORK
fr liège *m*
ne kurk
de Kork *m*
da kork
sv kork
es corcho *m*
it sughero *m*

CORK, TO FORM, 1439

845
CORKY ROOT
fr déformation *f* liégeuse
   des racines
ne kurkwortel
de Korkwurzel *m*;
   Korkwurzelpilz *m*

da korkrod
sv bruna rötter; korkrot
es acorchado *m* de las
   raíces
it radice *f* sugherosa
   Mycelia sterilia spp.

846
CORM
fr tubercule *m*
ne stengelknol
de Stengelknollen *m*
da stængelknold
sv stamknöl
es tubérculo *m* caulino
it tubero *m* di fuste

847
CORMLETS;
CORMELS
(GLADIOLUS)
fr 'kralen' *m pl*
ne kralen
de Brutknollen *f pl*
da gladiolusyngel
sv yngellök
es bulbillos *m pl*
it bulbicine *m pl*

848
CORN COBS
fr rafles *f pl* de maïs
ne spillen van maïskolven
de Maiskolbenachsen *f pl*
da majskolber
sv majskolvsavfall
es panochas *f pl* de maíz
it bottoli *m pl*

CORNEL, 1098

849
CORNFLOWER
fr bleuet *m*
ne korenbloem
de Kornblume *f*
da kornblomst
sv blåklint
es aciano *m*; azulejo *m*;
   liebrecilla *f*
it fiordaliso *m*
   Centaurea cyanus L.

CORN MINT, 1331

850
CORN POPPY
fr coquelicot *m*
ne klaproos
de Klatschmohm *m*
da kornvalmue
sv kornvallmo
es amapola *f*
it rosolaccio *m*
   Papaver Rhoeas L.

851
CORN SALAD; LAMB'S
LETTUCE
fr mâche *f*; doucette *f*
ne veldsla
de Rapünzchen *n*;
   Feldsalat *m*
da vårsalat
sv vårsallad
es lechuga *f* de campo
it raperonzolo *m*;
   dolcetta *f*
   Valerianella locusta L.
   Betcke cv. Oleracea
   Schlecht L.

852
COROLLA
fr corolle *f*
ne kroon
de Blumenkrone *j*
da krone
sv blomkrona
es corola *f*
it corolla *f*

853
CORONA
fr couronne *f*
ne bijkroon
de Nebenkrone *f*
da bikrone; svælgskæl *n*
sv bikrona
es corona *f*
it corolla *f* accessoria

854
CORRUGATED
CARDBOARD
fr carton *m* ondulé
ne golfkarton *n*
de Wellpappe *f*

da bølgepap *n*
sv wellpapp *n*
es cartón *m* ondulado
it cartone *m* ondulato

855
CORRUGATED SHEET
fr plaque *f* ondulée
ne golfplaat
de Wellplatte *f*
da bølgeplade
sv korrugerad platta
es placa *f* ondulata
it latta *f* ondulata

856
CORYMB
fr corymbe *m*
ne schermvormige tros
de Doldentraube *f*
da halvskærm
sv kvast
es corimbo *m*
it corimbo *m*

857
COS LETTUCE
fr laitue *f* romaine
ne bindsla
de Bindesalat *m*;
   Sommerendivie *f*
da bindsalat
sv bindsallad;
   romersk sallad
es lechuga *f* romana
it lattuga *f* romana
   Lactuca sativa L. cv.
   Romana Gart.

858
COSMOS
fr Cosmos *m*
ne Cosmos
de Cosmee *f*
da stolt kavaler
sv skära
es Cosmos *m*
it Cosmos *m*
   Cosmos Cav.

859
COST ALLOCATION
fr imputation *f* des coûts
ne kostensplitsing
de Kostenverteilung *f*

da omkostningsfordeling
sv kostnadsfördelning
es adjudicación *f* de costes
it imputazione *f* dei costi

860
COST CALCULATION
fr calcul *m* des coûts
ne kostencalculatie
de Kostenkalkulation *f*
da omkostningskalkulation
sv kostnadskalkylering;
   kostnadsberäkning
es cálculo *m* de costes
it calcolo *m* dei costi

861
COST OF PRODUCTION
fr frais *m pl* de production
ne produktiekosten
de Produktionskosten *m pl*
da produktions-
   omkostninger
sv produktionskostnader
es costo *m* de producción
it spese *f pl* di produzione

862
COSTS
fr coûts *m pl*; frais *m pl*;
   dépenses *f pl*; charges *f pl*
ne kosten
de Kosten *f pl*; Aufwand *m*
da omkostninger
sv kostnad
es costes *m pl*
it costi *m pl*; spese *f pl*

863
COSTS CATEGORY
fr catégorie *f* de frais
ne kostensoort (-groep)
de Kostenart *f*
da omkostningsart
sv kostnadsslag
es clases *f pl* de costes
it categoria *f* dei costi

864
COSTS OF
ESTABLISHMENT
fr coûts *m pl* de création
ne aanlegkosten
de Anlagekosten *f pl*
da anlægsomkostninger

sv anläggningskostnader
es coste *m* de instalación
it costi *m pl* di constituzione

865
COTTONY MAPLE
SCALE
fr cochenille *f*
   floconneuse
ne woldopluis
de wollige Napfschildlaus *f*
da vinskjoldlus
sv vinsköldlus
es cochinilla *f* lanuginosa;
   cochinilla *f* algodonosa
it cocciniglia *f*
   Pulvinaria vitis L.

866
COTYLEDON;
SEED LEAF
fr cotylédon *m*
ne kiemblad *n*
de Keimblatt *n*
da kimblad *n*
sv hjärtblad *n*
es cotiledón *m*
it cotiledone *m*

867
COUCH GRASS;
TWITCH GRASS;
QUACK GRASS
fr chiendent *m* ordinaire
ne kweekgras *n*
de Queckengras *n*;
   Quecke *f*; kriechende
   Quecke *f*
da kvikgræs *n*
sv kvickrot
es grama *f*
it gramigna *f*
   Agropyrum repens L.
   Pal.

868
COVER
fr couverture *f*
ne bedekking
de Bedeckung *f*;
   Abdeckung *f*
da dække; dækning
sv täckning
es cobertura *f*
it copertura *f*

869
COVER, TO
(BULB GROWING)
fr couvrir; recouvrir;
    paillassonner
ne dekken; afdekken
de zudecken; abdecken
da afdække
sv täcka av
es tapar; abrigar; cubrir
it coprire; proteggere

870
COVERING LIQUID
(PROCESSING)
fr infusion f
ne opgiet
de Aufguss m
da påskænkning
sv lake
es líquido m complementa-
    rio
it infusione f

871
COVERING MATERIAL
fr matériel m de couverture
ne dekmateriaal n
de Deckmaterial n
da dækkemateriale n
sv täckningsmaterial n
es material m de cobertura
it materia f da copertura

872
COVERING SHEET
fr feuille f d'emballage
ne pakblad n
de Einlagepapier n
da dækblad n
sv tråg n
es Capa f
it carta f di rivestimento

873
COVERING WITH SAND
fr couverture f de sable
ne bezanding
de Besandung f
da sanddækning
sv sandtäckning
es cubierta f de arena
it rincalzamento m da
    sabbia

874
COVER SAND
fr sable m de couverture
ne dekzand
de Decksand m
da dæksand
sv täcksand
es arena f para cubrir
it sabbia f a coprire

875
COVER WITH TURF, TO;
TO LAY TURF;
TO TURF
fr gazonner (par placage)
ne bezoden
de mit Rasen (Soden)
    abdecken
da dække med græstørv
sv täcka med grästorv
es cubrir con césped
it piotare

876
COWBERRY;
MOUNTAIN
CRANBERRY
fr airelle f rouge
ne rode bosbes (vossebes)
de Preiselbeere f;
    Kronsbeere f
da tyttebær n
sv lingon n
es arándano m encarnado
it mirtillo m rosso
    Vaccinium vitis-idaea L.

877
COW MANURE
fr fumier m de vache
ne koemest
de Kuhmist m;
    Rindermist m
da kogødning
sv kogödsel
es estiércol m de vaca
it bovina f; buina f

CRABAPPLE,
flowering, 1399

878
CRACK
fr crevasse f
ne barst; scheur

de Riss m
da revne; spalte
sv spricka; rämna
es grieta f; quierba f
it fessura f

CRACK, TO, 500

CRANBERRY,
American, 83
—, European, 1244
—, mountain, 876
—, Virginian, 4040

879
CRANBERRY TREE
LEAF BEETLE
fr galéruque de la boule de
    neige
ne sneeuwbalhaan
de Schneeballblattkäfer m
da syrebladbille
sv olvonbagge
es galerucela f del durillo
it galerucella del viburno
    Galerucella viburni Payk.

880
CRANE FLY
(LARVA = LEATHER
JACKET)
fr tipule f
ne langpootmug
    (larve = emelt)
de Schnake f
    (larve = Erdschnacke f)
da stankelben
sv harkrank
es tipula f
it zanzarone m
    Tipulidae spp.

881
CRANESBILL
fr géranium m;
    bec-de-grue m
ne ooievaarsbek
de Storchschnabel m
da storkenæb
sv näva
es geranio m
it geranio m
    Geranium L.

882
CREDIT
fr crédit *m*
ne krediet *n*
de Kredit *m*
da kredit
sv kredit; lån *n*
es crédito *m*
it credito *m*

883
CREDIT GUARANTEE
fr garantie *f* d'un crédit
ne borgstelling
de Bürgschaft *f*
da sikkerhedsstillelse
sv ställa borgen
es garantía *f* de un crédito
it garanzia *f* di un prestito

884
CREDIT VOLUME
fr volume *m* de crédit
ne kredietruimte
de Kreditvolumen *n*
da kreditomfang *n*
sv kreditvolym
es volumen *m* de crédito
it volume *m* del credito

CREEPING, 3029

885
CREEPING
BENT-GRASS
fr agrostide *m* stolonifère
ne fioringras
de Fioringras *n*;
  Staussgras *n*
da krybende hvene;
  fioringræs *n*
sv krypven
es agróstide *m* de renuevos
it agrostide *f* stolonifera
  Agrostis *f*
  Agrostis stolonifera L.

886
CREEPING THISTLE
fr chardon *m* ordinaire
ne akkerdistel
de Ackerdistel *f*
da ager-tidsel
sv åkertistel

es cardo *m* cundidor
it stoppione *m*;
  cardo *m* campestre
  Cirsium arvense L. Scop.

887
CREEPING WILLOW
fr saule *m* rampant
ne kruipwilg
de Kriechweide *f*
da krybende pil
sv krypvide
es sauce *m* rastrero
it salice *m* strisciante
  Salix repens L.

888
CRENATE
fr crénelé
ne gekarteld
de gekerbt
da takket; rundtakket
sv naggad
es dentado
it dentellato

889
CRESS
fr cresson *m* alénois
ne tuinkers
de Gartenkresse *f*
da karse
sv kryddkrasse
es lepidio *m*
it crescione *m*
  Lepidium sativum L.

CRESS,
marsh yellow, 2300
—, swine, 3774
—, thale, 3820
—, wart, 3774
—, water, 4076

890
CRESTED DOG'S TAIL
fr crételle *f* des prés
ne kamgras *n*
de Kammgras *n*
da kamgræs *n*
sv kamäxing
es cola *f* de perro;
  cinura *f*
it Cynosurus *m* cristatus
  Cynosurus cristatus L.

891
CRIMSON CLOVER
fr trèfle *m* incarnat
ne incarnaatklaver
de Inkarnatklee *m*
da blodkløver
sv blodklöver
es trébol *m* colorado
it trifoglio *m* incarnato
  Trifolium incarnatum L.

892
CRINKLE
fr frisolée *f*
ne krinkel
  (aardappel, aardbei)
de Kräuselkrankheit *f*
da krusesyge
sv krussjuka
es rizadura *f*
it arricciamento *m*

893
CRISP; CURLY
fr crépu; frisé
ne gekroesd
de kraus
da kruset
sv krusig
es crespo
it crespo

894
CROCUS
fr Crocus *m*
ne krokus
de Krokus *m*
da krokus
sv krokus
es Crocus *m*
it croco *m*
  Crocus L.

CROCUS,
autumn, 2312

895
CROP
fr récolte *f*; production *f*
ne opbrengst (oogst)
de Ernte-Ertrag *m*
da høstudbytte *n*
sv skördeavkastning

es producción *f*
it raccolto *m*; produzione *f*

CROP, field, 2559
—, first, 1360
—, glasshouse, 1581
—, green manure, 1652
—, outdoor, 2559
—, second, 3303

896
CROP ANALYSIS
fr analyse *f* foliaire
ne gewasanalyse
de Pflanzenanalyse *f*
da planteanalyse
sv växtanalys;
   vävnadsanalys
es análisis *m* de plantas
it analisi *f* vegetale

897
CROP CONTROL
fr contrôle *m* de la culture
ne gewascontrole
de Gewächskontrolle *f*
da vækstkontrol
sv växtkontroll
es control *m* de plantas
it controllo *m* sulla pianta

898
CROP DETAILS
fr données *f pl* de récolte
ne oogstgegevens
de Ernte-Ergebnisse *n pl*
da høstoplysninger;
   udbyttedata
sv skörderapport
es datos *m pl* de cosecha
it dati *m pl* della raccolta

899
CROP FAILURE
fr récolte *f* déficitaire;
   mauvaise récolte *f*
ne misoogst
de Missernte *f*; Fehlernte *f*
da misvækst
sv missväxt
es perdida *f* de cosecha
it cattivo raccolto *m*

900
CROPPING
ADVANCEMENT;
HARVEST
ADVANCEMENT
fr accélération *f* de récolte
ne oogstvervroeging
de Ernteverfrühung *f*
da høstfremmende;
   tidlighedsgørende
   foranstaltning
sv tvångsmognad
es adelanto *m* de la cosecha
it anticipazione *f* della
   raccolta

901
CROPPING DELAY;
HARVEST DELAY
fr retardation *f* de récolte
ne oogstverlating
de Ernteverspätung *f*
da høstforsinkelse
sv skördeförsening
es retraso *m* de la cosecha
it ritardamento *m* della
   raccolta

902
CROPPING PLAN
fr plan *m* de culture
ne teelttabel
de Kulturplan *m*;
   Anbauplan *m*
da dyrkningsplan
sv odlingsplan; kulturplan
es plan *m* de cultivo
it ordinamento *m* colturale

903
CROPPING ROOM
(MUSHROOM
GROWING)
fr chambre *f* de récolte
ne oogstcel
de Ernteraum *m*
da afdrivningsrum *n*
sv odlingsrum *n*
es sala *f* (local) de cosecha
it stanza *f* di raccolta

CROP PROTECTION
CHEMICALS, 2690

904
CROP ROTATION
fr assolement *m*
ne vruchtwisseling
de Fruchtwechsel *m*
da sædskifte *n*
sv växtföljd
es alternativas *f pl* de
   cosecha
it cambiamento *m* di
   coltura

905
CROP SPRAYER
fr pulvérisateur *m* pour
   cultures
ne veldspuit
de Feldspritze *f*
da marksprøjte
sv fältspruta
es pulverizador *m* de
   cultivo
it irroratrice *f* per colture

906
CROSS;
CROSSBREEDING
fr croisement *m*
ne kruising
de Kreuzung *f*
da krydsning
sv korsning
es cruzamiento *m*
it incrocio *m*

907
CROSS BACK, TO
fr croisement *m* de retour
ne terugkruisen
de zurückkreuzen
da tilbagekrydsning
sv återkorsning
es cruce *m* recurrente
it reincrociare

908
CROSS-OUT, TO
(HYACINTH)
fr inciser
ne snijden
de Kreuzschnitt machen
da krydssnitte
sv skära
es cortar en cruz
it incidere

909
CROSS POLLINATION
fr fécondation f croisée
ne kruisbestuiving
de Fremdbestäubung f
da krydsbetøvning;
   fremdbestøvning
sv korsbefruktning
es fecundación f cruzada;
   polinización f cruzada
it fecondazione f incrociata

910
CROW
fr corneille f
ne kraai
de Krähe f
da krage
sv kråka
es corneja f
it corvo m; cornacchia f
   Corvus

911
CROWN (TREE)
fr couronne f
ne kroon
de Krone f
da trækrone
sv krona
es copa f
it corona f

912
CROWN GALL
fr tumeurs f pl bactériennes
   du collet et des racines
ne wortelknobbel
de Wurzelkropf m
   Kronengalle f
da rodhalsgalle; krongalle
sv rotkräfta
es tumor m bacteriano de la
   raíz
it tumor m batterico
   Agrobacterium
   tumefaciens
   (Sm. et Town.) Conn.

913
CROWN GRAFTING
fr greffer en couronne
ne kroonenten
de pfropfen zwischen Holz
   und Rinde

da fligbarkpodning;
   barkpodning
sv barkympning
es injerto m de corona
it innestare a corona

CROWN IMPERIAL, 1462

914
CROWN OF THORNS
fr épine f du Christ
ne echte Christusdoorn
de Stechdorn m;
   Christusdorn m
da Paliurus
sv äkta Kristi törnekrona
es espina f de Cristo;
   espina f vera
it Paliurus m
   Paliurus spina-Christi
   Mill.

915
CROWN ROT
(DELPHINIUM)
fr sclerotiniose f du pied
ne kroonrot n
de Sklerotienkrankheit f
da Sclerotium delphinii
sv Sclerotium delphinii
es podredumbre f de la
   corona
it mal m dello sclerozio
   Sclerotium delphinii
   Welch.

916
CRUMBLE, TO;
PULVERIZE, TO
fr émietter
ne verkruimelen
de zerkrümeln
da smuldre
sv söndersmula
es desmigar
it sbriciolare

917
CRUMB STRUCTURE
fr structure f grumeleuse
ne kruimelstructuur
de Krümelstruktur f
da krummestruktur
sv aggregatstruktur

es estructura f migajosa
it struttura f in minuzzolo

918
CRYPTOGAM
fr cryptogame f
ne cryptogaam
de Kryptogame f
da kryptogam; lønbo
sv kryptogam
es criptógama f
it crittogame
   Cryptogamae

919
CRYSTALLIZE, TO
fr cristalliser
ne kristalliseren
de kristallisieren
da krystallisere
sv kristallisera
es cristalizar
it cristallizzare

CUCKOO SPIT INSECT,
769

920
CUCUMBER
fr concombre m
ne komkommer
de Gurke f
da agurk
sv gurka
es pepino m
it cetriolino m; cetriolo m
   Cucumus sativus L.

CUCUMBER, ridge, 3133

921
CUCUMBER MOSAIC
fr mosaïque f du
   concombre
ne komkommermozaiek n
de Gurkenmosaik n
da agurk-mosaik
sv gurkmosaik
es mosaico m del pepino
it mosaico m del cetriolo

922
CUCUMBER NECROSIS
fr nécrose f du concombre
ne komkommernecrose
de Gurkennekrose f

da agurk-nekrose
sv gurknekros
es necrosis *m* del pepino
it necrosi *f* del cetriolo

923
CUCUMBER VIRUS
fr virus *m* du concombre
ne komkommervirus
de Gurkenvirus *m*
da agurkmosaikvirus *n*
sv gurkvirus *n*
es virus *m* del pepino
it virus *m* del cetriuolo

924
CULINARY HERBS
fr fines herbes *f pl*;
 plantes *f pl* condimen-
 taires
ne keukenkruiden *n pl*
de Küchenkräuter *n pl*
da krydderurter
sv kryddväxter
es hierbas *f pl* de cocina;
 hierbas *f pl* culinarias
it erbe *f pl* da cucina

925
CULM
fr chaume *m*
ne halm
de Halm *m*
da strå *n*
sv strå *n*
es caña *f*; paja *f*
it culmo *m*

926
CULTIVATE, TO
fr cultiver
ne betelen
de anbauen; kultivieren
da kultivere; dyrke
sv odla; kultivera
es cultivar
it coltivare

927
CULTIVATED
MUSHROOM
fr campignon *m* de couche
ne gekweekte champignon
de (Zucht)Champignon *m*;
 Edelpilz

da champignon
sv champinjon
es champiñon *m* de Paris
it fungo *m* coltivato
 Agaricus bisporus Sing.

928
CULTIVATED PLANT
fr plante *f* cultivée
ne cultuurgewas *n*
de Kulturpflanze *f*
da kulturplante
sv kulturväxt
es planta *f* cultivada
it pianta *f* di coltura

929
CULTIVATION;
CULTURE
fr culture *f*
ne teelt; verbouw
de Anbau *m*; Kultur *f*
da dyrkning; kultur
sv kultur; odling
es cultivo *m*
it coltura *f*; coltivazione *f*

930
CULTIVATOR
fr cultivateur *m*
ne cultivator
de Kultivator *m*
da kultivator
sv kultivator
es cultivador *m*
it coltivatore *m*

CULTIVATOR,
rigid time, 3136
—, rotary, 3197
—, spring tine, 3577

931
CULTURAL HYGIENE
fr hygiëne *f* culturale
ne teelthygiëne
de Anbauhygiene *f*
da kulturhygiejne
sv odlingshygien
es higiene *f* del cultivo
it igiene *f* colturale

932
CULTURAL
TECHNIQUE;
GROWING
TECHNIQUE
fr technique *f* culturale
ne cultuurtechniek
de Kulturtechnik *f*
da dyrkningsteknik
sv kulturteknik;
 odlingsteknik
es ingenieria *f* rural
it tecnica *f* colturale

CULTURE, 929
—, gravel, 1648
—, ridge, 3134
—, soilless, 1648
—, water, 4076

933
CUMIN
fr cumin *m*
ne komijn
de Kreuzkümmel *m*;
 römischer Kümmel *m*
da spidskommen
sv spiskummin
es comino *m*
it comino *m*
 Cuminum cyminum L.

934
CUNEATE;
WEDGE SHAPED
fr cunéiforme
ne wigvormig
de keilförmig
da kileformet
sv kilformig
es cuneiforme
it cuneiforme

935
CUP (bot.)
fr cupule *f*
ne napje *n*
de Näpfchen *n*
da skål
sv skål
es cúpula *f*
it cupola *f*

936
CUP (MUSHROOM
GROWING)
fr champignon *m* à voile
  tendue
ne gevliesde (halfopen)
  champignon
de halboffener
  Champignon *m*
da halvåben champignon
sv halvöppen champinjon
es champiñon *m* a punto de
  abrirse
it fungo *m* semichiuso

937
CUP SHAPED
fr cupuliforme
ne bekervormig
de becherförmig
da bægerformet
sv bågformig
es vasiforme; cupuliforme
it caliciforme

CURBING, 648

CURCULIO,
cabbage, 511
—, cabbage seedstalk, 517

938
CURCULIONID
WEEVIL
fr charançon *m*
ne snuitkever
de Rüsselkäfer *m*
da snudebille
sv vivel
es curculiónido *m*
it curcolionido *m*
  Curculionidae

939
CURL, TO
fr friser; boucler
ne krullen
de kräuseln
da krølle; kruse
sv krusa
es encrespar
it increspare

940
CURLED DOCK
fr rumex crépu;
  oseille *f* crépue; parelle *f*
ne krulzuring
de krauser Ampfer *m*
da kruset skræppe
sv krusskräppa
es lengua *f* de vaca;
  hidrolápato *m* menor
it romice *m*
  Rumex crispus L.

941
CURLED MALLOW
fr mauve *f* frisée
ne dessertblad *n*
de krause Malve *f*
da katost
sv krusmalva
es malva *f* crespa
it malva *f* crispa
  Malva crispa L.

CURLED ROSE SAWFLY,
243

942
CURLY
fr crépu; frisé
ne gekroesd
de kraus
da kruset
sv krusig
es crespo
it crespo

943
CURRANT
(RED, WHITE)
fr groseille *f* (rouge, blanc)
ne bes (rode, witte)
de Beere *f* (rote, weisse)
da bær *n* (rød; hvid)
sv vinbär *n* (röd; vit)
es grosella *f* (roja, blanca)
it ribes *m* (rosso; bianco)

CURRANT,
black, 332
—, golden, 1605
—, mountain, 2405

CURRANT APHID, 3060

CURRANT BORER, 945

944
CURRANT BUSH
fr groseillier *m* à grappes
ne bessestruik
de Beerenstrauch *m*;
  Johannisbeerstrauch *m*
da bærbusk; ribsbusk
sv vinbärsbuske
es grosellero *m*
it arbusto *m* di ribes

945
CURRANT CLEARWING
MOTH;
CURRANT BORER (US)
fr sésie *f* du groseillier
ne besseglasvlinder
de Johannisbeerglas-
  flügler *m*
da ribs-glassværmer
sv vinbärsglasvinge
es mariposa *f* del grosellero
it sesia *f* del ribes
  Synanthedon tipuliformis
  Clerck;
  Aegeria t. (UK);
  Ramosia t. (US)

946
CURRANT JUICE
fr jus *m* de groseilles
ne bessesap *n*
de Johannisbeersaft *m*
da ribssaft; solbærsaft
sv vinbärssaft
es jugo *m* de grosellas
it succo *m* di sultamina

CURRANT MOTH, 2237

947
CURRANT ROOT APHID
fr puceron *m* de l'orme et du
  groseillier
ne bessebloedluis
de Ulmenblattlaus *f*
da ribs-blodlus
sv almbladlus
es pulgón *m* lanigero
it pidocchio *m* delle galle
  piriformi dell' olmo
  Eriosoma ulmi L.;
  Schizonera u. (UK)

948
CURRANT SOWTHISTLE
APHID
fr pou *m* du laiteron
ne groene melkdistelluis
de grünliche
   Gänsedistellaus *f*
da solbærbladlus
sv mjölktistelbladlus;
   svart vinbärsbladlus
es pulgón *m* de la lechuga
it afide *m* verde della
   lattuga
   Hyperomyzus lactucae L.

949
CURRANT WINE
fr vin *m* de groseilles
ne bessenwijn
de Johannisbeerwein *m*
da ribsvin; solbærvin
sv vinbärsvin
es vino *m* de grosellas
it vino *m* di sulamina

950
CURVED SCALPEL;
ROUND SCOOP;
SCOOPING KNIFE
fr couteau *m* à creuser
ne holmesje *n*
de gebogenes
   Spezialmesser *n*
da hulingskniv
sv urholkningskniv
es cuchillo *m* para ahuecar
it coltello *m* a scavare

951
CURVE VEINED
fr à nervures courbelés
ne kromnervig
de krummnervig
da buenervet
sv bågnervig
es curvinervado
it curvinervato

952
CUSTOMS UNION
fr union *f* douanière
ne douane-unie
de Zollunion *f*
da toldunion

sv tullunion
es unión *f* aduanera
it unione *f* doganale

953
CUT; INCISION
fr coupe *f*; incision *f*
ne snede
de Schnitt *m*; Einschnitt *m*
da snit *n*; indsnit *n*
sv snitt *n*; insnitt *n*
es incisión
it incisione *f*

CUT, TO,   2411, 2963

954
CUT BACK, TO;
TO SHORTEN;
TO PRUNE
fr recéper; rabattre
ne insnoeien; terugsnoeien
de einstutzen; zurückschnei-
   den
da skære tilbage
sv skära tillbaka
es podar de rebaja
it potare; tagliare di ritorno

CUT DOWN, TO,   1304

955
CUT FLOWER
fr fleur *f* coupée
ne snijbloem
de Schnittblume *f*
da snitblomst
sv snittblomma
es flor *f* cortada
it fiore *m* d'intaglio

956
CUT FOLIAGE;
CUT GREEN;
ORNAMENTAL
FOLIAGE
fr feuillage *m* ornamental;
   verdure *f*
ne snijgroen
de Schnittgrün
da snitgrønt *n*
sv snittgrönt
es hojas *f pl* ornamentales
it verdura *f* ornamentale

957
CUT OFF, TO;
TO CHOP OFF
fr élaguer; couper; tailler
ne afhakken
de abhacken
da afhugge; afskare
sv avhugga
es cortar
it staccare

958
CUTTER BAR
fr barre *f* de coupe
ne maaibalk
de Mähbalken *m*
da knivbjælke; knivstang
sv knivbalk
es barra *f* guadanadora
it trave *f* mietitrice

959
CUTTER LOADER
fr faucheuse *f* auto-
   chargeuse
ne maailader
de Mähselbstlader *m*
da høste-læsseredskab *n*
sv slåtterlastare
es segadora *f* auto-
   cargadora
it falciatrice *f* auto-
   caricatrice

960
CUTTING
fr bouture *f*
ne stek
de Steckling *m*
da stikling
sv stickling
es estaca *f*; esqueje *m*
it talea *f*

CUTTING, bud,   1277
—, eye,   1277
—, hardwood,   1727
—, leaf,   2080

961
CUTTING BED
fr table pour bouturer
ne stektablet *n*
de Vermehrungsbeet *n*

da stikkebord
sv sticklingsbord *n*;
    arbetsbord *n*
es mesa *f* de esqueja
it tavoletta *f* da moltiplica-
    zione

962
CUTTING EDGE
fr tranchant *m*
ne snijkant
de Schneide *f*
da æg *n*
sv egg *n*
es filo *m*
it parte *f* tagliente

963
CUTTING-LETTUCE
fr laitue *f* à couper
ne snijsla
de Schnittsalat *m*
da snitsalat
sv bladsallad
es lechuga *f* de cortar
it lattuga *f* da taglio
    Lactuca sativa L.
    cv. Longifolia Lam. L.

964
CUTTING WITH HEEL
fr bouture *f* à talon
ne stek met een hieltje
de Steckling *m* mit Astring;
    Rissling *m* mit Astring
da hælstikling *mf*
sv stickling med klack
es estaca *f* de ramo calzado
it talea *f* a zampa di cavallo

CUT TURF, TO, 2608

CUTWORM, 2486

965
CYANAMIDE
(OF CALCIUM)
fr cyanamide *f* calcique
ne kalkstikstof
de Kalkstickstoff *m*

da kalkkvælstof *n*
sv kalkkväve *n*
es cianamida *f* en calcio
it azoto *m* calcareo

966
CYANIC ACID GAS
fr acide *m* prussique;
    acide *m* cyanhydrique
ne blauwzuurgas *n*
de Blausäuregas *n*
da blåsyregas
sv blåsyregas
es ácido *m* cianhídrico
it acido *m* cianidrico

967
CYCLAMEN
fr Cyclamen *m*
ne Cyclamen
de Alpenveilchen *n*
da cyclamen; alpeviol
sv Cyclamen
es ciclamen *m*;
    violeta *f* de las alpes
it ciclamen *m*
    Cyclamen L.

CYCLAMEN MITE, 3682

968
CYCLIC LIGHTING
fr éclairage *m* cyclique
ne cyclische belichting
de zyklische Beleuchtung *f*
da cyclisk belysning
sv intermittent belysning
es iluminación *f* cíclica
it illuminazione *f* ciclica

969
CYME
fr cyme *f*
ne gevorkt bijscherm *n*
de Trugdolde *f*
da gaffelkvast
sv tvåsidigt knippe
es cima *f*
it umbella *f* accessoria

CYME, helicoid, 1774
—, scorpioid, 3281

CYNIPID, 1521

CYPRESS, false, 1285
—, swamp, 3757

970
CYPRIPEDIUM
fr sabot *m* de Venus
ne venusschoentje
de Frauenschuh *m*
da Venussko (stukket)
sv guckusko
es zapatillo *m* de Venus
it cipripedio *m*
    Cypripedium L.

971
CYST
fr kyste *f*
ne cyste
de Zyste *f*
da cyste
sv cysta
es quiste *m*
it cisti *f*

CYSTERSHELL SCALE,
2434

972
CYST NEMATODE
fr nématode *m* kystique
ne cysteaaltje *n*
de Zystenälchen *n*
da cystenematod
sv cystenematod
es nematodo *m* ustógeno
it nematode *m* cisticolo
    Heterodera spp.

CYST NEMATODE,
beet, 293
—, cabbage, 512
—, cactus, 521
—, carrot, 594
—, pea, 2629
—, potato, 2866

D

973
DAFFODIL; NARCISSUS
fr narcisse *m*
ne narcis
de Narzisse *f*
da narcis
sv narciss
es narciso *m*
it narciso *m*
 Narcissus L.

974
DAHLIA
fr Dahlia *m*
ne Dahlia
de Dahlie *f*
da dahlie; georgine
sv Dahlia
es dalia *f*
it dalia *f*
 Dahlia Cav.

975
DAISY
fr pâquerette *f*
ne madeliefe *n*
de Massliebchen *n*
da tusindfryd
sv tusensköna
es chirivita *f*; margarita *f*
it margheritina *f*;
 pratolina *f*
 Bellis perennis L.

DAISY, Michaelmas, 179,
2334
—, sharta, 3367

DAMAGE, insect, 1921
—, leaf, 2081

976
DAMAGE BY BIRDS
fr dégât *m* causé par les
 oiseaux
ne vogelschade
de Schaden durch Vögel
da fugleskade
sv fågelskada

es daño *m* causado por
 pájaros
it danno *m* causato dagli
 uccelli

977
DAMAGE BY GAME
fr dégât *m* causé par le
 gibier
ne wildschade
de Wildschaden *m*
da vildtskade *mf*
sv viltskada
es daño *m* causado por
 animales de caza
it danno *m* causati dai
 selvatici

978
DAMPER
fr registre *m* de cheminée
ne schoorsteenschuif
de Rauchschieber *m*
da røglem
sv spjäll *n*; rökspjäll *n*
es registro *m*; ventanilla *f*;
 pestillo *m* de chimenea
it valvola *f* di tiraggio

979
DAMPING OFF
fr fonte *f* des semis
ne kiemschimmel;
 omvalziekte; smeul
de Keimlingskrankheit *f*;
 Wurzelbrand
da kimskimmel; rodbrand
sv förökningssvamp;
 groddbrand; rotbrand;
 svartrot
es caída *f* de almácigo;
 fusario *m*;
 hongos *m pl* del semillero
it marciume *f* delle pinatine
 dei semenzai;
 muffa *f* germinativa
 *Pythium-;Thanatephorus-
 and Fusarium*-spp.

980
DAMPING OFF
(LETTUCE)
fr pourriture *f* noire
ne zwartrot *n*
de Schwarzfäule *f*
da rodfiltsvamp
sv rotfiltsjuka;
 föröknïngssvamp
es podredumbre *f* negra
it marciume *m* nero;
 scabbia *f*; ipocnosi *f*
 Thanatophorus cucumeris
 (Fr.) Donk.

DAMPING OFF, foot rot
and, 1424

981
DAMSON
fr prune *f* de Damas
ne kwets
de Zwetsche *f*
da svedskeblomme
sv plommon *n*
es ciruela *f* damascena
it prugna *f* di Damasco
 Prunus domestica L.

982
DANDELION
fr pissenlit *m*; dent *f* de lion
ne paardebloem; molsla
de Löwenzahn *m*;
 Kuhblume *f*
da mælkebøtte;
 løvetandssalat
sv maskros
es diente *m* de león
it dente *m* di leone;
 soffione *m*
 Taraxacum officinale
 Web.

983
DAPHNE
fr daphné *n*
ne peperboompje *n*
de Seidelbast *m*

da pebertræ *n*
sv tibast; pepparbuske
es dafne *m*
it dafne *f*
Daphne mezereum L.

984
DARKEN TO;
TO BLACK OUT
fr obscurcir
ne verduisteren
de verdunkeln
da formørke
sv mörklägga
es obscurecer
it oscurare

985
DARKENING
fr obturation *f*
ne verduistering
de Verdunklung *f*
da mørklægning
sv mörkläggning
es obturación *f*
it oscurità *f*; oscurimento *m*

986
DARK LEAF SPOT
(CABBAGE)
fr alternariose *f*
ne spikkelziekte
de Kohlschwärze *f*
da skulpesvamp
sv svartfläcksjuka
es antracnosis *f*
it alternariosi *f*
Alternaria brassicae
(Beck.) Sacc.

987
DART MOTH
(LARVA = CUTWORM)
fr noctuelle *f* des moissons;
ver *m* gris
ne aardrups
de Erdraupe *f*
da knoporm
sv sädesbroddflylarv;
jordflylarv
es nóctua *f* común de las
mieses; gusanos grises;
gusano *m* de tierra

it nottua *f*; agrotide *f*
Agrotis spp.

988
DATE
fr datte *f*
ne dadel
de Dattel *f*
da daddel
sv dadel
es dátil *m*
it dattero *m*

989
DATE PALM
fr dattier *m*
ne dadelpalm
de Dattelpalme *f*
da daddelpalme
sv dadelpalm
es palma *f* datilera
it dattero *m*
Phoenix dactylifera L.

990
DAYLIGHT
fr lumière *f* diurne
ne daglicht *n*
de Tageslicht *n*
da dagslys *n*
sv dagsljus *n*
es luz *f* diurna
it luce *f* diurna

991
DAY LILY
fr hémérocalle *f*
ne daglelie
de Taglilie *f*
da daglilie
sv daglilja
es lirio *m* de San Juan
it giglio *m* diurno
Hemerocallis L.

DAYNETTLE,   1779

992
DEADLY NIGHTSHADE;
DWALE
fr belladonne *f*
ne doodskruid *n*
de Tollkirsche *f*
da galnebær *n*

sv belladonna
es belladona *f*
it erba *f* mortuaria
Atropa belladonna L.

993
DEADNETTLE
fr lamier *m*
ne dovenetel
de Taubnessel *f*
da døvnælde
sv rödplister; vitplister
es ortiga *f* muerta
it ortica *f* falsa; ortica *f*
morta
Lamium

994
DEAD WEIGHT
fr poids *m* propre
ne eigen gewicht *n*
de Eigengewicht *n*
da egenvægt
sv egenvikt
es peso *m* limpio
it peso *m* morto

995
DE-AERATE, TO
fr purger
ne ontluchten
de entlüften
da udlufte
sv utlufta
es ventilar
it ventilare

996
DEATH OF BRANCH
fr tavelure *f* des branches
ne taksterfte
de Aststerben *n*
da toptørhed *mf*
sv grentorka
es tristeza *f*
it tristezza *f*

997
DEATH'S HEAD HAWK
MOTH
fr sphinx *m* tête de mort
ne doodshoofdvlinder
de Totenkopf *m*
da dødningehoved *n*

sv dödskallefjäril
es mariposa *f* de la calavera
it sfinge *f* testa di morto
   Acherontia atropos L.

DEBLOSSOMING, 1747

DEBRIS, 1853

998
DECALCIFY, TO
fr décalcifier
ne ontkalken
de entkalken
da afkalke
sv avkalka
es descalcificar
it decalcifiare

999
DECAY;
PUTREFACTION;
DECOMPOSITION
fr pourriture *f*;
   détérioration *f*
ne rotting; bederf
de Fäulnis *f*; Verderb *m*;
   Verfall *m*
da rådne; fordærve
sv förruttnelse
es putrefacción *f*;
   deterioro *m*
it putrefazione *f*

1000
DECAYED
fr vermoulu; pourri
ne vermolmd
de morsch; vermodert
da formuldet
sv förmultnad; murken
es carcomido
it intarlato

1001
DECIDUOUS
fr à feuilles caduques
ne bladverliezend
de laubabwerfend
da løvfældende
sv bladfällande
es de hoja caduca
it a foglia caduca

1002
DECOLORIZE, TO
fr décolorer
ne ontkleuren
de entfärben
da affarve
sv avfärga
es decolorar
it scolorare

1003
DECOMPOSE, TO
fr se décomposer
ne vergaan
de verwesen
da forgå
sv multna; murkna
es descomponerse
it putrefarsi

1004
DECOMPOSITION;
BREAKDOWN
fr décomposition *f*
ne afbraak
de Abbau *m*; Zersetzung *f*
da nedbrydning
sv nedbrytning
es descomposición *f*
it decomposizione *f*

DECORATIVE GARDEN,
2554

1005
DECORATIVE PLANT;
ORNAMENTAL PLANT
fr plante *f* d'ornement
ne sierteeltgewas *n*
de Zierpflanze *f*
da prydplante
sv prydnadsväxt
es planta *f* de adorno;
   planta *f* ornamental
it pianta *f* ornamentale

DECORATIVE SHRUB,
2556

1006
DECUSSATE
fr décussé
ne kruisgewijs tegenover-
   staand

de kreuzständig
da korsvis modsat
sv korsvis motsatta;
   korsställda
es decusado
it decussato

1007
DEEP FREEZING;
QUICK FREEZING
fr congélation *f* à basse température
ne diepvriezen
de tiefgefrieren; tiefkühlen
da dybfryse
sv djupfrysa
es congelación *f* a baja temperatura
it congelazione *f* a basse
   temperatura

1008
(DEEP)FROZEN
VEGETABLES
fr légumes *mpl* surgelés
ne diepgevroren groenten
de Tiefkühlgemüse *mpl*
da dybfrosne grønsager
sv djupfrysta grönsaker
es hortalizas *fpl* congeladas
it verdura *f* congelata

1009
DEEP PLOUGHING;
SUBSOIL PLOUGHING
fr labourer en profondeur
ne diepploegen
de tiefpflügen
da dybpløje
sv djupplöja
es arar profundamente
it arare a fondo

1010
DEEP ROOTING
fr enracinement *m* profond
ne diepwortelend
de tiefwurzelnd
da dybrodende
sv djuprotande
es de enraizamiento
   profundo
it radicazione *f* profonda

1011
DEFICIENCY; DEFICIT
fr déficit *m*
ne tekort *n*
de Mangel *m*; Defizit *n*
da mangel
sv brist; underskott *n*
es déficit *m*
it deficit *m*

DEFICIENCY, boron,   404
—, iron,   1953
—, light,   2152
—, magnesium,   2232
—, manganese,   2257
—, moisture,   2369
—, nitrogen,   2483
—, potassium,   2862

1012
DEFICIENCY DISEASE
fr maladie *f* de carence
ne gebreksziekte;
     voedingsziekte
de Mangelkrankheit *f*
da mangelsygdom
sv bristsjukdom
es enfermedad *f* carencial
it malattia *f* per carenza;
     malattia *f* di deficienza

1013
DEFICIENY OF
PHOSPHORUS
fr carence *f* en phosphore
ne fosforgebrek
de Phosphormangel *m*
da fosformangel
sv fosforbrist
es carencia *f* en fósforo
it deficienza *f* di fosforo

1014
DEFICIENCY OF RAIN
fr manque *m* de pluie
ne neerslagtekort *n*
de Niederschlagsmangel *m*
da regnmangel
sv nederbördsunderskott;
     regnbrist
es falta *f* de lluvia
it deficienza *f* di pioggia

1015
DEFICIENCY SYMPTOM
fr symptôme *m* de carence
ne gebreksverschijnsel
de Mangelerscheinung *f*
da mangelsymptom
sv bristsymtom *n*
es síntoma *m* de carencia
it sintomo *m* di deficienza

1016
DEFICIENT IN LIME
fr pauvre en chaux
ne kalkarm
de kalkarm
da kalktrængende
sv kalkfattig
es pobre en cal
it privo di calce

DEFICIT,   1011

1017
DEFOLIATE, TO
fr effeuiller
ne ontbladeren
de entblättern
da afblade
sv avblada
es deshojar
it sfogliare

1018
DEFOLIATING
MACHINE
fr machine *f* à effeuiller
ne ontbladermachine
de Entblätterungsmaschine *f*
da afbladningsmaskine
sv avbladningsmaskin
es máquina *f* deshojadora
it macchina *f* a sfogliare

1019
DEFOLIATION;
LEAF DROP
fr défoliation *f*;
     chute *f* des feuilles
ne bladval
de Laubfall *m*; Blattfall *m*
da bladfald *n*; løvfald *n*
sv lövfall *n*; bladfall *n*
es defoliación *f*
     caída *f* de las hojas
it cascata *f* fogliale

DEFORMATION,   2252

1020
DEFORMED
fr défiguré
ne misvormig
de missförmig *f*
da misformet; misdannelse
sv missformad
es deforme
it deforme

1021
DEGENERATE, TO
fr dégénérer
ne verbasteren
de ausarten; entarten
da udarte; degenerere
sv degenerera; urarta
es degenerar; bastardear
it degenerare; imbastardire

1022
DEGENERATION
fr dégénerescence *f*
ne degeneratie
de Degeneration *f*
     Entartung *f*
da degeneration
sv degeneration
es degeneración *f*
it degenerazione *f*

1023
DEGREE; RATE
fr degré *m*
ne graad
de Grad *m*
da grad
sv grad
es grado *m*
it grado *m*

1024
DEGREE OF
DISSOCIATION
fr degré *m* de dissociation
ne dissociatiegraad
de Dissoziazionsgrad *m*
da dissociationsgrad
sv dissociationsgrad
es grado *m* de disociación
it grado *m* di dissociazione

1025
DEGREE OF MOISTURE
fr degré *m* d'humidité
ne vochtigheidsgraad
de Feuchtigkeitsgrad *m*
da fugtighedsgrad
sv fuktighedsgrad
es grado *m* de humedad
it grado *m* d'umidità

1026
DEGREE OF
SATURATION
fr degré *m* de saturation
ne verzadigingsgraad
de Sättigungsgrad *m*
da mætningsgrad
sv mättningsgrad
es grado *m* de saturación
it grado *m* di saturazione

1027
DEHISCENT FRUIT
fr fruit *m* déhiscent
ne splitvrucht
de Spaltfrucht *f*
da spaltefrugt
sv klyvfrukt
es fruto *m* deshicente
it frutto *m* spaccato

1028
DEHYDRATE, TO
fr sécher
ne drogen
de trocknen
da tørre
sv torka
es secar; deshidratar
it disidratare

1029
DEHYDRATED
VEGETABLES
fr légumes *m pl* déshydratés
ne gedroogde groenten
de Trockengemüse *n pl*
da tørrede grønsager
sv torkade grönsaker
es hortalizas *f pl* deshidrata-
das; hortalizas *f pl* secas
it ortaggi *m pl* secchi

1030
DEHYDRATION
fr déshydration *f*
ne wateronttrekking
de Wasserentziehung *f*;
Wasserentzug *m*
da dehydrering
sv vattenfrånskiljning
es deshidratación *f*
it sottrazione *f* d'acqua

1031
DEHYDRATION-PLANT
fr usine *f* de déshydration
de légumes
ne groentedrogerij
de Gemüsetrockenanstalt *f*
da grønsagstørreri *n*
sv torkinrättning för
grönsaker
es secadero *m* de verduras
it seccagione *f* di verdura

1032
DELIVERY
fr livraison *f*
ne aflevering
de Lieferung *f*
da levering
sv leverans
es entrega *f*
it consegna *f*

DELPHINIUM,   2039

1033
DELTOID
fr deltoïde
ne deltavormig
de deltaförmig
da deltaformet
sv deltaformig
es deltoideo
it deltoide

1034
DEMINERALISATION
fr déminéralisation *f*
ne demineralisatie
de Entmineralisation *f*;
Entdüngung *f*;
Entsalzung *f*
da demineralisation
sv demineralisation

es desmineralización *f*
it levamento *m* dei minerali

1035
DENITRIFY, TO
fr dénitrifier
ne denitrificeren
de denitrifizieren
da denitrificere
sv denitrificera
es desnitrificar
it denitrificare

1036
DENTATE; TOOTHED
fr denté; dentelé
ne getand
de gezähnt
da tandet
sv tandad
es dentado
it dentato

1037
DEPARTMENT STORE
fr grand magasin *m*
ne grootwinkelbedrijf *n*
de Grossraumladen *m*
da engros-forretning
sv supermarket; varuhus *n*
es tienda *f* al por mayor
it grande magazzino *m*

1038
DEPOSITED, TO BE
fr se déposer
ne afzetten (zich)
de sich ablagern;
sich absetzen
da afsætte sig
sv avsätta sig
es depositarse
it depositarsi

1039
DEPRECIATION
fr amortissement *m*;
dépréciation *f*
ne afschrijving
de Abschreibung *f*
da afskrivning
sv avskrivning
es depreciación *f*
it deprezzamento *m*

1040
DEPRESSION; HOLLOW
fr dépression f de terrain;
    bas-fond m; endroit-bas m
ne laagte
de Niederung f
da lavning; sænkning
sv sänka; dalgång
es depresión f del terreno;
    valle m
it depressione f;
    bassopiano m

1041
DEPTH OF DRAINAGE
fr profondeur f de drainage
ne drainagediepte
de Dräntiefe f
da drænrørsdybde
sv dräneringsdjup n
es profundidad f del drenaje
it profondità f di drenaggio

1042
DE-SALT, TO
fr dessaler
ne ontzilten
de entsalzen
da afsalte (jord)
sv urlaka (jord);
    avsalta
es desalar
it dissalare

1043
DESCENT
fr origine f; descendance f
ne afstamming
de Abstammung f
da afstamning
sv ursprung; härstamning
es descendencia f
it discendenza f; origine f

1044
DESICCATE, TO;
TO DRY OUT
fr dessécher
ne uitdrogen; indrogen;
    verdrogen
de austrocknen; eintrocknen
da udtørre; indtørre
sv förtorka; uttorka
es desecar
it disseccare

1045
DESICCATION
fr dessèchement m
ne verdroging
de Vertrocknung f;
    Austrocknung f
da udtørring
sv uttorkning
es secarse; desecación f
it disseccamento m

1046
DETERIORATION OF
SOIL STRUCTURE
fr structure f dégradée
ne structuurbederf n
de Strukturverfall m
da strukturødelæggelse
sv strukturförstöring
es deterioro m de la
    estructura
it corruzione f della
    struttura

1047
DEUTZIA
fr Deutzia f
ne Deutzia; bruidsbloem
de Deutzia f
da stjernetop; Deutzia
sv stjärnbuske
es Deutzia f
it Deutzia f
    Deutzia Thunbg.

1048
DIADELPHOUS
fr diadelphe
ne tweebroederig
de zweibrüderig
da tveknippet
sv diadelphia
es diadelfo
it didelfo

1049
DIAL
fr cadran m
ne wijzerplaat
de Zifferblatt n
da urskive
sv urtavla
es esfera f
it quadrante m

1050
DIAMOND-BACK
MOTH
fr teigne f des crucifères
ne koolmot
de Kohlschabe f
da kålmøl n
sv kålmal n
es polilla f de la col
it plutella f delle crucifere
    Plutella maculipennis
    Curt.

1051
DIAPHRAGM PUMP
fr pompe f à membrane
ne membraanpomp
de Membranpumpe f
da membranpumpe
sv membranpump
es bomba f de membrana
it pompa f a membrana

1052
DIBASIC
fr dibasique
ne tweebasisch
de zweibasisch
da tobasisk
sv tvåbasisk
es dibásico
it dibasico

1053
DICALCIUM
PHOSPHATE
fr biphosphate m de
    calcium
ne dubbelkalkfosfaat n
de Dicalciumphosphat n
da dobbelkalkfosfat n
sv dubbelkalkfosfat n
es bisuperfosfato m cálcico
it bifosfato m calcario

1054
DIDYMELLA STEM ROT
(TOMATO)
fr chancre m
ne kanker
de Tomatenstengelfäule f
da tomatsyge
sv tomatkräfta

es chancro *m*
it marciume *m* del fusto
  Didymella lycopersici
  Kleb.

1055
DIE
fr dé *m*
ne poer
de Würfel *m*
da terning
sv tärning
es pilastra *f*
it dado *m* per colonna

1056
DIE, TO; TO PERISH
fr dépérir
ne afsterven
de absterben
da dø; visne
sv dö; vissne
es morir; secarse *m*
it disseccare; inaridire

1057
DIE-BACK
fr dépérissement *f*;
  dégénérescence *f*
ne afstervingsziekte
de Absterbekrankheit *f*;
  Degenerationskrankheit *f*;
  'Dieback' *f*
da degenerationssyge;
  'die-back'
sv degenerationssjuka;
  'die-back'
es languidez *f*
it moria *f*

1058
DIE-BACK
(CHAMAECYPARIS);
STRANGLING DISEASE
OF CONIFERS
fr maladie *f* du collet des
  plantes d'épicéa et de
  sapin
ne insnoeringsziekte
de 'Einschnürungskrank-
  heit' *f*
da 'indsnøringssygdom'
sv 'insnöringssjuka'

es mal *m* pestifero;
  mal *m* del cuello de las
  plantas forestales
it pestalozzia;
  mal *m* del colletto delle
  piantine forestali
  Pestalotia funerea Desm.;
  Pestalotia hartigii Tub.

DIE-BACK,
canker and, 555, 556
—, mushroom, 2431

1059
DIFFERENCE IN
TEMPERATURE
fr différence *f* de température
ne temperatuurverschil *n*
de Temperaturunterschied *m*
da temperaturforskel
sv temperaturskillnad
es diferencia *f* de temperatura
it differenza *f* di temperatura

DIFFERENTIAL, 3818

1060
DIFFERENTIATION
fr différenciation *f*
ne differentiatie
de Differenzierung *f*
da differentiering
sv differentiering
es diferenciación *f*
it differenziazione *f*

1061
DIFFERING
fr taré; présentant des dé-
  fauts; non conforme
ne afwijkend
de abweichend
da afvigende
sv avvikande
es diferente
it stravagante

1062
DIFFUSING FITTING
fr appareil *m* dispensif
ne breedstraler
de Breitstrahler *m*

da diffus-stråler
sv bredstrålare
es aparato *m* dispersivo
it apparecchio *m* di diffusione

1063
DIG, TO
fr bêcher
ne spitten
de graben; umspaten
da grave
sv gräva
es cavar
it vangare

1064
DIGITATE
fr digité
ne handvormig samengesteld
de handförmig
da håndnervet sammensat;
  koblet fingret
sv fingrad
es digitada; palmeado compuesto
it digitato

1065
DIG OFF, TO
fr déterrer
ne afgraven
de abgraben
da afgrave
sv gräva bort
es escavar; desfondar;
  desmontar
it scavare

1066
DIG OFF SAND, TO
fr enlever le sable
ne afzanden
de Sand abgraben
da afgrave sand
sv avgräva sand
es quitar la arena;
  desarenar
it dissabbiare

1067
DIG UP, TO; TO LIFT
fr déterrer; arracher
ne rooien; opgraven
de ausgraben; roden

da rydde; grave op
sv taga upp; röja
es arrancar
it sarchiare; scavare

DIKE, 1090

1068
DIKE, TO; TO EMBANK
fr endiguer
ne bedijken
de eindeichen
da inddæmme
sv invalla
es cercar con diques
it cingere di dighe

1069
DILL
fr aneth m
ne dille
de Dill m
da dild
sv dill
es eneldo m; aneldo m
it aneto m
   Anethum graveolens L.

1070
DILUTOR
fr mètre m de concentration
ne concentratiemeter
de Konzentrationsmess-
   gerät n
da koncentrationsmåler;
   osmometer n
sv koncentrationsmätare
es diluidor m
it misuratore m di
   concentrazione

1071
DIOECIOUS
fr dioïque
ne tweehuizig
de zweihäusig
da tvebo; særbo
sv tvåbyggare
es dióico
it dioico

1072
DIP, TO
fr tremper
ne dompelen; dopen

de tauchen;
   ins Wasser hängen
da vådafsvampe
sv doppa
es sumergir; remojar
it tuffare

1073
DIRECT COSTS
fr coût m direct
ne directe kosten
de direkte Kosten pl;
   Einzelkosten pl;
   Spezialkosten pl
da direkte omkostninger
sv direkta kostnader
es costes m pl directos
it costi m pl diretti

1074
DIRECTION
fr instruction f;
   prescription f
ne voorschrift n
de Vorschrift f
da vejledning; instruktion
sv föreskrift; instruktion
es instrucción f
it istruzione f; ordine m

1075
DIRECTIONS FOR USE
fr mode m d'emploi
ne gebruiksaanwijzing
de Gebrauchsanweisung f
da brugsanvisning
sv bruksanvisning
es modo m de empleo
it istruzione f per l'uso

1076
DIRT DEPOSIT
fr dépôt m d'impuretés
ne vuilneerslag
de Schmutzniederschlag m
da smudsnedfald n
sv smutsnederbörd
es precipitación f sucia
it deposito m di sudiciume

1077
DISBUD, TO;
(e.g. CARNATIONS)
fr éboutonner
ne pluizen

de entknospen
da udknoppe; knoppe
sv utknoppa; knoppa
es desbotonar; desyemar
it allontamento m dei
   bottoni

1078
DISC COULTER
fr coutre m circulaire
ne schijfkouter n
de Scheibensech n
da skiveforplov
sv skivrist
es reja f (del arado) circular
it coltro m a disco

1079
DISC HARROW
fr herse f à disques
ne schijveneg
de Scheibenegge f
da tallerkenharve
sv tallriksharv
es grada f de discos
it erpice m a dischi

1080
DISCOLORATION;
FADING
fr décoloration f
ne verkleuring
de Entfärbung f
da affarvning
sv färgförlust
es decoloración f
it decolorazione f

DISCOLOURED, 229

1081
DISEASE
fr maladie f
ne ziekte
de Krankheit
da sygdom
sv sjukdom
es enfermedad f
it malattia f

—, deficiency, 1012
—, Dutch elm, 1177
—, Eckelrade, 1195
—, finger-and-toe, 719
—, fungal, 1499
—, fungus, 1502
—, gum spot, 3387
—, ink, 1914
—, inspection for, 1926
—, leaf, 2082
—, lily, 2168
—, mosaic, 2390
—, mould, 2403
—, mummy, 2428
—, M-virus, 2440
—, plant, 2768
—, red fire, 3063
—, shot-hole, 3387
—, skin, 3433
—, soil, 3479
—, storage, 3666
—, strangling, 1058
—, S-virus, 3754
—, vascular, 4005
—, virus, 4043
—, wart, 4063
—, weeping, 4117
—, wilt, 4025
—, witches' broom, 4189
—, Y-virus, 4235

1082
DISEASE CONTROL;
PEST CONTROL
fr lutte *f* contre les maladies
ne ziektebestrijding
de Krankheitsbekämpfung *f*
da sygdomsbekæmpelse
sv sjukdomsbekämping
es lucha *f* contra enferme-
    dades de plantas
it lotta *f* contro le malattie

1083
DISEASED; ILL
fr malade
ne ziek
de krank
da syg
sv sjuk
es enfermo
it ammalato

1084
DISEASE SYMPTOM
fr symptôme *m* morbide
ne ziekteverschijnsel *n*
de Krankheitserscheinung *f*
da sygdomssymptom *n*
sv sjukdomssymtom
es síntoma *m* de enfermedad
it sintomo *m* morboso

DISINFECT, TO, 3647

DISINFECTANT,
liquid, 2190
—, seed, 3315

1085
DISINFECTION
fr désinfection *f*
ne ontsmetting
de Desinfektion *f*;
    Entseuchung *f*
da desinfektion
sv desinfektion
es desinfección *f*
it disinfezione *f*

DISINFECTION,
dry, 1150
—, soil, 3480

1086
DISSOLVE, TO
fr dissoudre
ne oplossen
de auflösen; lösen
da opløse
sv upplösa
es disolver
it dissolvere

1087
DISTILLATE
fr produit *m* de distillation
ne destillaat *n*
de Destillat *n*
da destillat *n*
sv destillat *n*
es destilado *m*
it distillato *m*

1088
DISTRIBUTION
fr distribution *f*
ne distributie

de Vermarktung *f*;
    Distribution *f*
da fordeling
sv distribution; fördelning
es distribución *f*
it distribuzione *f*

1089
DISTRIBUTION
CHANNEL
fr circuit *m* de distribution
ne distributiekanaal *n*
de Absatzkanal *m*;
    Marktkanal *m*
da afsætningskanal
sv distributionskanal;
    avsättningskanal
es canal *m* de distribución
it circuit *m* distributivo

1090
DITCH; DIKE;
SURFACE DRAIN
fr fossé *m*
ne greppel; sloot
de Graben *m*
da grøft; fure
sv dike *n*; fåra
es zanja *f*; cuneta *f*
it fossa *f*; fossatello *m*

DITCH, catch-water, 605

DITCHING, 3746

1091
DITCHING PLOUGH;
TRENCHING PLOUGH
fr charrue *f* rigoleuse
ne greppelploeg
de Grabenräumer *m*;
    Grabenpflug *m*
da grøfteplov
sv dikesplog; dräneringsplog
es arado *m* abrezanjas
it aratro *m* di fognatura

1092
DITCH WATER LEVEL
fr niveau *m* d'eau de fossé
ne slootwaterpeil *n*
de Grabenwasserstand *m*
da vandspejl *i* grøft
sv dikesvattenstånd

es nivel *m* del agua en un
    canal
it livello *m* d'acqua di fossato

1093
DIVIDE, TO (bot).
fr diviser (multiplication)
ne scheuren
de teilen
da dele
sv dela
es multiplicar por división
it moltiplicare da divisione

1094
DIVISION OF LABOUR
fr division *f* du travail
ne arbeidsverdeling
de Arbeitsteilung *f*
da arbejdsdeling
sv arbetsfördelning
es división *f* del trabajo
it divisione *f* del lavoro

DOCK, broad leaved,   436
—, curled,   940

1095
DOCK APHID
fr puceron *f* de l'oseille
ne zuringbladluis
de Ampferblattlaus *f*
da syrebladlus
sv syrebladlus
es pulgón *f* de la acedera
it afide *m* dell'erba acetosa
    Aphis rumicis L.;
    Doralis rumicis Leach

1096
DOCK SAWFLY
fr hoplocampe *m* du
    pommier
ne zuringbladwesp
de Ampferblattwespe *f*
da syrehveps
sv syrastekel
es tentredo *m* de las acederas
it tentredine *f* del meline
    Ametastegia glabrata
    Fall.

1097
DOG ROSE
fr églantier *m*
ne hondsroos

de Hundsrose *f*
da hunderose
sv nyponros
es rosal *m* perruno
    escaramujo *m*
it Rosa *f* canina;
    rosa *f* di macchia
    Rosa canina L.

1098
DOGWOOD; CORNEL
fr cornouiller *m*
ne kornoelje
de Hartriegel *m*
da kornel
sv kornell
es cornejo *m*
it corniola *f*
    Cornus L.

1099
DOLOMITE
fr dolomie *f*; dolomite *f*
ne dolomiet *n*
de Dolomit *m*
da dolomit
sv dolomit
es dolomía *f*; dolomita *f*
it dolomite *m*

1100
DOMESTIC COMPOST
fr compost *m* d'ordures
ne huisvuilcompost
de Hausabfallkompost *m*
da byaffald *n*;
    dagrenovation; skrald *m*
sv avfallskompost
es basura *f* casera;
    compost *m*
it concime *m* della
    immondezza

1101
DOMINANCE
fr dominance *f*
ne dominantie
de Dominanz *f*
da dominans
sv dominans
es dominancia *f*
it dominanza *f*

1102
DOOR POST
fr montant *m*
ne deurstijl
de Türpfosten *m*
da dørstolpe
sv dörrpost
es montante *m*
it montante *m*

1103
DORMANT BUD
fr oeil *m* dormant
ne slapend oog *n*
de schlafendes Auge *n*
da sovende øje *n*;
    proventivknop
sv sovande öga *n*
es ojo *m* durmiente;
    yema *f* dormida
it occhio *m* dormiente

1104
DORMANT CORMS
(FREESIA)
fr dormeurs *mpl*
ne slapers
de nichtwachsende
    Freesienknollen *mpl*
da sovende knolde
sv sovande knölar
es tubérculos *mpl* falto de
    fuerza vegetativa
it tuberi *m* dormenti

1105
DOSAGE
fr dosage *m*
ne dosering
de Dosierung *f*
da dosering
sv dosering
es dosificación *f*
it dosatura *f*

1106
DOTTED
fr ponctué; pointillé
ne gestippeld
de punktiert
da punkteret
sv punkterad
es punteado; moteado
it punteggiato

1107
DOUBLE
fr double
ne gevuld
de gefüllt
da fyldt
sv dubbel; fylld
es doble
it pieno; riempito

DOUBLE CROPPING,
1937

DOUBLE DIG, TO, 3927

1108
DOUBLE NOSE
(BULB GROWING)
fr double-nez m
ne dubbelneus
de Doppelnase f
da dobbeltnæse
sv dubbelnäs
es nariz f doble
it bulbo m a doppio
    rampollo

1109
DOUBLESPAN FRAME
fr châssis m double
ne dubbele bak
de Doppelkasten m
da dobbelt bænk
sv drivbänk i sadeltaksform
es cajonera f doble;
    cajonera f de dos
it letto m gemello

1110
DOUBLE (OR TRIPLE)
SUPERPHOSPHATE
fr bisuperphosphate m
ne dubbelsuperfosfaat
de Hyperphos n
da dubbelsuperfosfat n
sv Aubbelsuperfosfat n
es bisuperfosfato m
it bisuperfosfato m

DOUBLE WORKING,
1939

1111
DOUGLAS FIR
fr sapin m de Douglas
ne Douglaspar
de Douglaise f;
    Douglastanne f
da douglasgran
sv douglasgran
es pino m de Douglas
it Pseudotsuga f
    Pseudotsuga Carr.

DOUGLAS FIR
ADELGES, 2733

1112
DOWN PIPE
fr tube f de retour d'eau
ne zakbuis
de Fallrohr n
da faldrør n
sv returrör n
es tubo m de retorno de agua
it tubo m di ritorno

1113
DOWN-THE-ROW
THINNER
fr pré-démarieuse f et
    bineuse f
ne rijendunner
de Reihenhack- und
    Ausdünnungsmaschine f
da uttynde maskine
sv uttunnare
es binadora f; aclaradora f
it diradatrice f

1114
DOWNY PUBESCENT
(bot).
fr duveté; pubescent
ne donzig
de flaumig behaart
da dunet
sv dunig; fjunig
es pubescente
it piumoso

1115
DOWNY MILDEW
fr mildiou m
ne valse meeldauw
de falscher Mehltau m

da skimmel
sv bladmögel
es mildio m falso
it peronospora f falsa
    Peronosporaceae

1116
DOWNY MILDEW
(SPINACH); LEAF MOLD
fr mildiou m
ne wolf
de falscher Mehltau m;
    Spinatschimmel m
da spinatskimmel
sv bladmögel
es mildio m
it Peronospora f
    Peronospora spinaciae
    (Mont) de By.

DRAIN, main, 2244
—, spacing of, 3525
—, suction, 3717

1117
DRAIN, TO
fr drainer; assécher
ne draineren; afwateren;
    droogleggen
de dränieren; entwässern;
    trockenlegen
da dræne; afvande; tørlægge
sv dränera; avvattna;
    torrlägga
es drenar; desecar
it fognare; bonificare

1118
DRAINAGE
fr drainage m;
    découlement m
ne drainage; ontwatering;
    waterafvoer
de Dränierung f;
    Entwässerung f;
    Abfluss m
da dræn n; afløb
sv dränering; avlopp
es drenaje m
it drenaggio m; fognatura f

DRAINAGE,
depth of, 1041
—, mole, 2373

—, open, 2530
—, polder, 2818
—, pump, 2977
—, supplementary, 3742
—, tile, 3849

1119
DRAINAGE DISTRICT;
POLDER DISTRICT
fr administration f des eaux
ne waterschap n
de Wasserverband m;
   Wasser- und Bodenver-
   band m
da vandstandsbræt
sv Vattenadministration
es administración f de aguas
it amministrazione f delle
   acqua

1120
DRAINAGE DITCH
fr fossé m de drainage
ne afvoergreppel;
   poldersloot
de Abflussgraben m;
   Ableiter m;
   Poldergraben m
da afløbsrende;
   poldergrøft
sv avloppsdike n;
   'polder'dike n
es cuneta f;
   zanja f de drenaje
it fossatello m di scolo

1121
DRAINAGE TILE
fr drain;
   tuyau m de drainage
ne draineerbuis
de Dränrohr n
da drænrør n
sv dräneringsrör n
es tubo m de drenaje
it tubo m di fognatura

1122
DRAINAGE USING
BRUSHWOOD
fr drainage m en bois
ne houtdrainage
de Kastendränung f
da trædræning

sv träddränering
es drenaje m a base de
   madera
it drenaggio m da legno

1123
DRAIN COCK
fr robinet m de vidange
ne aftapkraan
de Auslaufhahn m
da aftapningshane
sv bottenventil;
   tömningskran
es grifo m de vaciado
it valvola f di scarico

1124
DRAINING MACHINE
fr machine f draineuse
ne draineermachine
de Dränmaschine f
da drænemaskine
sv dräneringsmaskin
es máquina m de drenaje
it macchina f di fognatura

1125
DRAIN OUT, TO;
TO RECLAIM
fr assécher
ne droogleggen
de trockenlegen
da tørlægge
sv torrlägga
es desecar
it bonificare

1126
DRAIN PIPE;
DRAINAGE TILE
fr drain m;
   tuyau m de drainage
ne draineerbuis
de Dränröhre f
da drænrør n
sv dräneringsrör n
es tubo m de drenaje
it tubo m di fognatura

1127
DRAIN TRENCHER
fr draineuse f
ne draineermachine
de Dränmaschine f

da dræningsmaskine
sv täckdikningsmaskin;
   grävmaskin för täck-
   dikning
es drenadora f
it scavafossi f

1128
DRAUGHT (AIR)
fr courant m d'air
ne tocht
de Zugluft f
da gennemtræk
sv luftdrag
es corriente f de aire
it corrente f (d'aria)

1129
DRAUGHT STABILIZER
fr stabilisateur m de tirage
ne trekregelaar
de Zugstabilisator m;
   Zugregler m
da trækregulator
sv dragregulator
es estabilizador m de tiraje
it stabilizzatore m di
   tiraggio

1130
DREDGE; SLUDGE
fr boue f; curages m pl
ne bagger
de Modder m; Schlamm m
da mudder n
sv mudder; slam
es cieno m de dragado
it fango m; melma f

DREDGE, hand, 1716

1131
DREDGE, TO
fr draguer; curer
ne uitbaggeren; baggeren
de ausbaggern; baggern
da udgrave med maskine;
   opmudre
sv muddra upp; rensa upp
es dragar;
it dragare; dissabbiare

1132
DREDGER
fr drague *f*; dragueur *m*
ne baggermolen
de Baggermaschine *f*
da muddermaskine
sv mudderverk *n*
es draga *f*
it draga *f*

1133
DREDGING BARGE
fr bateau-drague *m*
ne baggerschuit
de Baggerboot *n*
da mudderpram
sv mudderpråm
es lancha *f* dragadora:
it cavafango *m*

1134
DRIED BEAN BEETLE;
BEAN WEEVIL (US)
fr bruche *f* des haricots
ne bonekever
de Speisebohnenkäfer *m*
da bønnebille
sv bönbagge; bönsmyg
es gorgojo *m* de las judías
it tonchio *m* dei fagioli
Acanthoscelides obtectus
Say

1135
DRILL COULTER
fr soc *m* de semoir
ne vorentrekker
de Drillschar *f*
da rellemaskine
sv bill (för radsånings-
maskin)
es reja *f* de sembradora
it assolcatore *m*

1136
DRILL HOE
fr rayonneur *m*; traceur *m*
ne rijentrekker
de Rillenzieher *m*
da rillemaskine; markør
sv markör
es marcador *m*
it tiratore *m* di file

1137
DRILLING;
SOWING IN ROWS
fr semis *m* en ligne;
semaille *f* en ligne
ne rijenzaai
de Reihensaat *f*
da rækkeudsæd
sv radsådd
es siembra *f* en líneas
it semina *f* a rigo

1138
DRIP-PROOF FITTING
fr appareil *m* d'éclairage
protégé des gouttes
ne druipwaterdichte
armatuur
de tropfwassergeschützte
Leuchte *f*
da stænktæt armatur
sv droppskyddad armatur
es aparato *m* de iluminación
protegido de las gotas
it apparecchio *m* d'illuminazione a tenuta stagna

1139
'DRIVE-IN FLOOR'
fr installation *f* de séchage
par parois ventilées
ne inrijvloer
de Trocknungsanlage *f* für
Zwiebeln
da løgrammerne stables
på pallets
sv systemet för torkning av
lök
es secado *m* especial para
bulbos
it seccatoio *m* speciale per
bulbi

1140
DRONE
fr abeille-mâle *m*;
faux-bourdon *m*
ne dar
de Drohne *f*
da drone
sv drönare
es zángano *m*
it fuco *m*

1141
DROOPING; HANGING;
PENDULOUS
fr retombant
ne overhangend
de überhängend
da overhængende
sv överhängande
es inclinado
it sporgente; pendente

1142
DROP FORMATION
fr formation *f* des gouttes
ne druppelvorming
de Tropfenbildung *f*
da dråbedannelse
sv droppbildning
es formación *f* de gotas
it formazione *f* di goccia

1143
DROP IN TEMPERATURE
fr refroidissement *m*;
réfrigération *f*
ne afkoeling
de Abkühlung *f*
da afkøling
sv avkylning
es enfriamiento *m*
it raffreddamento *m*

1144
DROPPER
(BULB GROWING)
fr tulipe *f* qui s'enfonce en
terre
ne zinker
de 'Sinker' *m*
da sænkere
sv utlöpare
(som växer ner i jorden)
es rosario *m* subbulboso

1145
DROPPING OF BUDS
fr chute *f* des boutons
ne knopval
de Knospenfall *m*
da knopfald *n*
sv knoppfall *n*
es caída *f* de botones;
caída de yemas
it cascata *f* delle gemme

1146
DRUPE; STONE FRUIT
fr drupe *f*
ne steenvrucht
de Steinfrucht *f*
da stenfrugt
sv stenfrukt
es fruto *m* de hueso
it frutto *m* a nocciolo

1147
DRY, TO;
TO DEHYDRATE
fr sécher
ne drogen
de trocknen
da tørre
sv torka
es secar; deshidratar
it seccare; disidratare

DRYBERRY MITE,
3038

1148
DRY BRUSHING
MACHINE
fr brosseur *m*
ne borstel(poets)machine
de Bürstenmaschine *f*;
   Putzmaschine *f*
da pudsemaskine
sv putsningsmaskin
es cepillo *m* mecánico
it spazzolatrice *f*

1149
DRY BUBBLE
fr môle *f* sèche;
   Verticillium *m*
ne droge mol
de trockene Molle *f*;
   Trockenfäule *f*;
   'Bovist' *m*
da Verticillium
sv Verticillium
es mole *f* seca; gota *f* seca;
   Verticillium *m*
it Verticilliosi *f*
   Verticillium malthousei
   Ware

1150
DRY DISINFECTION;
SEED DRESSING
fr traitement *m* à sec
ne droogontsmetting
de Trockenbeize *f*
da tørafsvampning
sv torrbetning
es desinfección *f* en seco
it disinfezione *f* secca

1151
DRYING-FLOOR
(BULB GROWING)
fr plancher *m* de séchage
ne droogvloer
de Trocknungsraum *m*;
   Trocknungsboden *m* mit
   Gebläse
da tørrekasse
sv torkgolv
es suelo *f* secadero
it seccatoio *m*

1152
DRY MATTER
fr matière *f* sèche;
   extrait *m* sec
ne droge stof
de Trockensubstanz *f*
da tørstof *n*
sv torrsubstans
es materia *f* seca
it materia *f* secca

DRY OUT, TO,   1044

1153
DRY ROT
fr fusariose *f*
ne fusarium *n*
de Fusariumfäule *f*
da fusarium
sv fusarium
rs podredumbre *f* seca
it marcimento *m* secco

1154
DRY ROT (FREESIA
AND GLADIOLUS)
fr pourriture *f* molle
ne droogrot *n*
de Stromatinia- Knollen-
   trockenfäule *f*

da tørforrådnelse
sv torröta
es gangrena *f* seca
it cancrena *f* secca
   Stromatinia gladioli
   (Drayt.) Whetz.

1155
DRY ROT (POTATO);
GANGRENE
fr pourriture *f* séche
ne droogrot *n*
de Trockenfäule *f*
da tørforrådnelse
sv torröta
es gangrena *f* seca
it marciume *m* secco dei
   tuberi; seccume *m*
   Fusarium spp.;
   Phoma solanicola

1156
DRY SALE (BULBS)
fr vente *f* au grainier
ne droogverkoop
de Trockenzwiebelverkauf *m*
da tørsalg
sv försäljning av lökar i
   konsumentförpakning
es venta *f* de bulbos secos
it vendita *f* di bulbi secci

1157
DRY TIPBURN
(LETTUCE)
fr déshydratation *f*
ne droogrand
de Trockenrand *m*
da brune bladrande;
   'tipburn'
sv vissna bladkanter
es margen *f* seca
it margine *m* disseccato

1158
DRY UP, TO; TO WITHER
fr sécher; flétrir
ne verdorren
de vertrocknen
da hentørre
sv förtorka; torka ut
es marchitarse
it seccarsi

1159
DRY WALL
fr mur *m* de pierres sèches;
   murette *f*
ne stapelmuur
de Trockenmauer *f*
da plantemur; tørmur
sv kallmur
es muro *m* en seco
it muro *m* di pietra secca

1160
DRY WEIGHT;
TOTAL SOLUBLE SALT
fr teneur *m* totale en sels
   (cendres)
ne gloeirest
de Glührest *m*
da inddampningsrest
sv glödrest
es cenizas *f pl*
it resto *m* dell'incandescen-
   za

1161
DUCKWEED
fr lentille *f* d'eau
ne eendekroos *n*
de Wasserlinse *f*
da andemad
sv andmat; lämna
es lenteja *f* de agua
it lemna *f*
   Lemna L.

DUMPED, 3982

1162
DUMPER
fr déchargeur *m*
ne kistenlediger
de Aufkippvorrichtung *f*
da tippeanordning
sv tippanordning
es descargador *m*
it autocarro *m* con cussons
   ribaltabile

1163
DUNESAND
fr sable *m* des dunes;
   sable *m* dunal
ne duinzand *n*
de Dünensand *m*

da klitsand *n*
sv flygsand *n*
es arena *f* de dunas
it sabbia *f* delle dune

DUNG, 2261
—, well decayed, 4125

DUNG, TO, 2262

1164
DUNG HILL;
MANURE HEAP
fr tas *m* de fumier
ne mesthoop; mestvaalt
de Dunghaufen *m*;
   Misthaufen *m*
da gødningsbunke
sv gödselhög
es montón *m* de estiércol
it concimaia *f*

1165
DURABILITY;
KEEPING QUALITY
fr durabilité *f*
ne houdbaarheid
de Haltbarkeit *f*
da holdbarhed
sv hållbarhet
es durabilidad *f*
it durevolezza *f*

1166
DURABILITY;
RESISTANCE
fr durée *f* d'usage
ne duurzaamheid
   (kunststof)
de Dauerhaftigkeit *f*
da holdbarhed
sv varaktighet; beständighet
es duración *f* de uso
it durata *f* d'uso

DURABLE, 2042

1167
DURABLE PRODUCTION
MEANS
fr moyens *m pl* de production durables
ne duurzame produktie-
   middelen *n pl*

de dauerhafte Produktionsmittel *n pl*;
   Anlagekapital *n*
da varige produktionsmidler
   n pl
sv varaktiga produktions-
   medel *n pl*
es medios *m pl* durables de
   producción
it mezzi *m pl* di produzione
   durevoli

1168
DURATION OF
ILLUMINATION
fr durée *f* d'exposition;
   temps *m* de pose
ne belichtingstijd
de Belichtungszeit *f*
da belysningstid
sv belysningstid
es duración *f* de exposición
it durata *f* della delucidazione

1169
DURATION OF LIFE;
LIFE
fr durée *f* de vie
ne levensduur
de Lebensdauer *f*;
   Nutzungsdauer *f*
da levetid
sv livslängd
es vida *f* útil
it vita *f*; durata *f* della vita

1170
DURATION OF
STORAGE
fr durée *f* de conservation
ne bewaarduur
de Lagerdauer *f*
da lagringsperiode;
   opbevaringstid
sv lagringstid
es duración *f* de la conservación
it durata *f* della conservazione

1171
DURATION OF USE
fr durée *f* d'emploi
ne gebruiksduur

de Nutzungsdauer *f*
da levetid
sv varaktighet;
　　användningstid
es duración *f* de utilización
it durata *f* d'impiego

1172
DURMAST OAK
fr chêne *m* rouvre
ne wintereik
de Traubeneiche *f*
da vintereg
sv bergek; vinterek
es roble *m* albar
it quercia *f* petraea
　　quercia *f* d'inverno
　　Quercus petraea
　　(Lieblein)

1173
DUSKY CLEARWING
MOTH
fr petite sésie *f* du peuplier
ne populiereglasvlinder
de kleiner Apfelglasflügler *m*
da Sciapteron tabaniformis
sv svart poppelglasvinge
es mariposa *f* transparante
　　del chopo
it sesia *f* del pioppo
　　Paranthrene tabaniformis
　　Rott.; *Sciapteron t.* (UK)

1174
DUST, TO
(DISEASE CONTROL)
fr poudrer
ne bestuiven
de bestäuben
da pudre
sv bepudra
es pulverizar
it polverizzare

1175
DUSTER
fr poudreuse *f*
ne poederverstuiver
de Stäubegerät *n*
da pudderblæser
sv pudra; puderspridare
es espolvoreadora *f*
it impolveratrice *f*

DUSTER, power-driven
knapsack, 2888
—, rotary, 3198

1176
DUSTING
PREPARATION
fr produit *m* de poudrage
ne stuifmiddei *n*
de Stäubemittel *n*
da puddermiddel *n*
sv pudermedel *n*
es producto *m* en polvo
it prodotto *m* per
　　polverizzazione

1177
DUTCH ELM DISEASE
fr maladie *f* de l'orme;
　　dépérissement *m* de
　　l'orme
ne iepziekte
de Ulmensterben *n*
da elmesyge
sv almsjuka;
　　almdödaren
es enfermedad *f* holandesa
　　del olmo
it moria *f* degli olmi;
　　grafiosi *f* dell'olmo
　　Ceratocystis ulmi
　　(Buism.) C. Moreau

1178
DUTCH HOE
fr ratissoire *f*
ne schoffel
de Häufelhacke *f*
　　Schaufel *f*
da skuffejern *n*
sv skyffeljärn
es escardillo *m*; almocafre *m*
it sarchio *m*; zappa *f*

1179
DUTCH LIGHT
fr châssis *m* hollandais
ne eenruiter
de Holländerfenster *n*
da hollandsk bænkevindue *n*;
　　enrudevindue *n*
sv holländska drivbänks-
　　fönster *n*
es ventana *m* de un vidrio
it vetro *m* olandese

DUTCH LIGHTS, 1449

1180
DUTCH LIGHT
STRUCTURE
fr serre *f* démontable;
　　serre *f* à châssis mobiles
ne warenhuis *n*
de Holländerblock *m*
da 'warenhuis' *n*;
　　småhusblok;
　　væksthusblok
sv växthus *n*
es invernadero *m* desmon-
　　table
it serra *f* smontabile

DWALE, 992

1181
DWARF FLOWERING
ALMOND
fr prunier *m* à trois lobes
ne drielobbige amandel
de Mandelbäumchen *n*
da rosenmandel
sv rosenmandel
es Prunus *m* triloba
it Prunus *f* triloba
　　Prunus triloba Lindl.

1182
DWARF FRENCH BEAN
fr haricot *m* nain
ne stamboon
de Buschbohne *f*
da buskbønne; krybbønne
sv krypböna
es judías *f pl* bajas;
　　frejoles *m pl* enanos
it fagiolo *m* nano;
　　mangiattuto
　　Phaseolus vulgaris L.
　　cv. Nonscandens Bail

1183
DWARFING; STUNTING
fr nanisme *m*;
　　rabougrissement *m*
ne dwerggroei
de Zwergwuchs *m*; Kümmer-
　　wuchs *m*; Nanismus *m*;
　　Stauchung *f*;
　　Verzwergung *f*

da dværgvækst
sv dvärgväxt
es enanismo *m*
it nanismo *m*; rachitismo *m*

1184
DWARF SLICING BEAN
fr haricot *m* nain sabre
ne stamsnijboon
de Buschschwertbohne *f*
da lav snittebønne
sv låg skärböna
es judía *f* verde para cortar de pie bajo

it fagiolino *m* nano
Phaseolus sp.

1185
DWARF SNAP BEAN
fr haricot *m* nain mange-tout
ne stamslaboon
de Buschbrechbohne *f*
da lav brydbønne
sv låg brytböna
es judía *f* verde de pie bajo
it fagiolo *m* comune nano
Phaseolus sp.

1186
DYING OFF PROCESS
fr processus *m* de dépérissement
ne afstervingsproces *n*
de Absterbungsprozess *m*
da dødsforløb *n*; visning
sv vissneförlopp *n*
es marchitamiento *m*; necrosis *f*
it processo *m* di disseccazione

E

1187
EARLINESS; PRECOCITY
fr précocité *f*
ne vroegheid
de Frühe *f*
da tidlighed
sv tidighet
es precocidad *f*
it precocità *f*

1188
EARLY BROWNING (PEA)
fr brunissement *m* précoce
ne vroege verbruining
de frühe Bräune *f*
da tidlig brunfarvning
sv tidig brunfärgning
es pardeamiento *m* precoz
it imbrunimento *m* precoce

1189
EARLY FORCING
fr forçage *m* hâtif
ne vroegbroei
de Frühtreiberei *f*
da tidlig drivning
sv tidig drivning
es forzado *m* temperano
it forzatura *f* precoce

1190
**EARLY LETTUCE;
YOUNG LETTUCE;
CUTTING LETTUCE**
fr laitue *f* à couper;
laitue *f* jeune
ne dunsel
de junger Lattich *m*
da pluksalat
sv sallad (ung)
es lechuga *f* tierna
it lattuga *f* primaticcia

1191
EARLY POTATOES
fr pommes *fpl* de terre
hâtives
ne vroege aardappelen
de Frühkartoffeln *f pl*
da tidlige kartofler
sv tidig potatis
es patatas *fpl* tempranas
it patate *fpl* precoci

1192
EARTH; SOIL
fr terre *f*; sol *m*
ne grond
de Erde *f*; Boden *m*
da jord
sv jord; jordmån
es suelo *m*; tierra *f*
it suolo *m*

EARTH, black, 1535

1193
**EARTH UP, TO;
TO RIDGE UP**
fr butter
ne aanaarden
de häufeln; anhäufeln
da hyppe
sv kupa
es aporcar
it rincalzare; rinterrare

1194
EARTHWORM
fr ver *m* de terre; lombric *m*
ne regenworm
de Regenwurm *m*
da regnorm
sv daggmask
es gusano *m* de tierra;
lombriz *f*
it lombrico *m*

EARWIG, common, 768
—, European, 768

1195
ECKELRADE-DISEASE (CHERRY)
fr maladie *f* de Pfeffingen
ne Eckelraderziekte
de Pfeffinger Krankheit *f*
da Eckelrader-syge
sv raspblad *n*;
'Pfeffingersjuka'
es hojas *f pl* rasposas
it deperimento *m*;
malattia *f* di Pfeffing

1196
EDELWEISS
fr edelweiss *m*;
immortelle *f* des neiges
ne edelweis
de Edelweiss *n*
da edelweiss
sv edelweiss (äkta)
es pie *m* de léon;
flor *f* de nieve
it stella *f* alpina
Leontopodium alpinium
Cass.

EDGE, 400

1197
EDGING KNIFE
fr bèche *f* à découper;
couteau *m* à découper
ne graskantensteker
de Rasenkantenstecher *m*
da kantskærer; kantklipper
sv kantskärare; kanthuggare
es tijera *f* cortadora de
bordes
it coltello *m* da bordos

1198
EDIBLE
fr comestible
ne eetbaar
de essbar
da spiselig
sv ätlig; ätbar
es comestible
it mangiabile

EDIBLE BOLETUS, 636

1199
EEC;
EUROPEAN ECONOMIC
COMMUNITY
fr CEE *f*;
   Communauté *f* Economi-
   que Européenne
ne EEG;
   Europese Economische
   Gemeenschap
de EWG *f*;
   Europäische Wirtschafts-
   gemeinschaft
da EEC;
   Fællesmarkedet
sv EEC;
   Europeiska Ekonomiska
   Gemenskapen
es CEE;
   Comunidad *f* Económica
   Europea
it CEE;
   Comunità *f* Economica
   Europea

1200
EELWORM; NEMATODE
fr anguillule *f*; nématode *m*
ne aaltje *n*
de Aelchen *n*
da nematod
sv trådmask
es anguilula *f*
it anguillula *f*
   Nematodes

EELWORM, bud and
leaf, 460

1201
EELWORM
INFESTATION
fr anguillulose *f*
ne aaltjesziekte
de Aelchenkrankheit *f*
da åleangreb *n*
sv nemetodangrepp
es enfermedad *f* por
   anguilulas
it Nematosis *f*
   Nematosis

EELWORMS,
free living, 3176

1202
EFFERVESCE, TO
fr mousser
ne mousseren
de moussieren
da bruse
sv mussera; skumma
es espumar
it spumeggiare

1203
EGG
fr oeuf *m*
ne ei *n*
de Ei *n*
da æg *n*
sv ägg *n*
es huevo *m*
it uovo *m*

1204
EGG DISPOSITION
fr depôt *m* d'oeufs
ne eiafzetting
de Eiabsatz *m*
da æganbringelse;
   æglægning
sv äggläggning
es puesta *f* de huevos
it ovulazione *f*

1205
EGG PLANT
fr aubergine *f*
ne eierplant
de Eierfrucht *f*
da ægplante; ægfrugt;
   aubergine
sv äggplanta; aubergine
es berengena *f*
it melanzana *f*
   Solanum melongena L.

1206
ELDER
fr sureau *m*
ne vlier
de Holunder *m*
da hyld
sv fläder
es saúco *m*
it sambuco *m*
   Sambucus L.

ELDER, ground, 1668

1207
ELECTRICAL SOIL
HEATING
fr chauffage *m* électrique du
   sol
ne elektrische grondverwarming
de elektrische
   Bodenheizung *f*
da elektrisk jordopvarmning
sv elektrisk jordupp-
   värmning
es calentamiento *m* eléctrico
   del suelo
it riscaldamento *m* elettrico
   del terreno

1208
ELECTRONIC LEAF
fr feuille *f* électronique
ne elektronisch blad *n*
de elektronischer Sprüh-
   nebelregler *m*
da elektronisk blad *n*
sv elektroniskt blad *n*
es hoja *f* electrónica
it foglia *f* elettronica

1209
ELECTRON
MISCROSCOPY
fr microscopie *f* électronique
ne elektronenmicroscopie *f*
de Elektronenmikroskopie *f*
da elektronmikroskopi
sv elektronmikroskopi
es microscopía *f* electrónica
it microscopia *f* elettronica

1210
ELEMENT
fr élément *m*
ne element *n*
de Element *n*
da element *n*
sv element *n*
es elemento *m*
it elemento *m*

ELEMENT, active, 22

1211
ELEVATOR DIGGER
fr arracheuse *f* avec éleva-
   teur
ne zeefkettingrooier
de Siebkettenroder *m*
da optager med transport-
   bånd
sv upptagare med sålmatta
es cosechadora *f* con elevador
it cavatuberi *f* ad elevatore

1212
ELLIPTICAL
fr elliptique
ne elliptisch
de elliptisch
da elliptisk
sv elliptisk
es elíptico
it ellittico

1213
ELM
fr orme *m*
ne iep
de Ulme *f*; Rüster *f*
da elm
sv alm
es olmo *m*
it olmo *m*
   Ulmus L.

ELM, scotch, 4222
—, wych, 4222

1214
ELM BARK BEETLE
fr scolyte *m* de l'orme
ne grote iepespintkever
de grosser Ulmensplint-
   käfer *m*
da elmebarkbille
sv almsplintborre
es barrenillo *m* del olmo
it scolitide *m* dell'ormo
   Scolytus scolytus F.

1215
ELONGATION
fr élongation *f*
ne strekking
de Streckung *f*

da strækning
sv sträckning
es extensión *f*
it allungamento *m*

EMBANK, TO, 1068

1216
EMPLOYEE
fr salarié *m*; employé *m*
ne werknemer
de Arbeitsnehmer *m*
da arbejdstager; arbejder
sv arbetstagare; löntagare
es obrero *m*; empleado *m*
it lavoratore *m*;
   prenditore *m* di lavoro

1217
EMPLOYER
fr patron *m*; employeur *m*
ne werkgever
de Arbeitgeber *m*
da arbejdsgiver
sv arbetsgivare
es patrón *m*; empresario *m*
it padrone *m*; principale *m*

1218
EMPOLDER, TO
fr endiguer
ne inpolderen
de einpoldern; eindeichen
da 'inddigning'
sv indämma; invalla
es construir un polder *m*
it bonificare

EMPOLDERING, 2817

1219
EMULSIFIABLE
fr émulsionnable
ne emulgeerbaar
de emulgierbar
da emulgerbar
sv emulgerbar
es emulsionable
it emulsionabile

1220
EMULSIFYING BURNER
fr brûleur *m* à émulsion
ne emulsiebrander
de Injektorbrenner *m*

da emulsionsfyr *n*
sv emulsionsbrännare
es quemador *m* de emulsión
it bruciatore *m* ad
   emulsione

1221
EMULSION
fr émulsion *f*
ne emulsie
de Emulsion *f*
da emulsion
sv emulsion
es emulsión *f*
it emulsione *f*

1222
ENAMELLED COPPER
   WIRE
fr fil *m* de cuivre émaillé
ne geëmailleerd koperdraad
   n
de emaillierter
   Kupferdraht *m*
da emailleret kobbertråd
es emaljerad koppartråd
es hilo *m* de cobre esmaltado
it filo *m* di rame smaltato

1223
ENCLOSURE; FENCE
fr enceinte *f*; clôture *f*
ne omheining
de Einfriedigung *f*
da indhegning
sv inhägnad
es cerca *f*
it siepe *f*; recinto *m*

1224
ENDIVE
fr chicorée frisée *f*;
   scarole *f*
ne andijvie
de Endivie *f*
da endivie
sv endiviesallat
es escarola *f*
it indivia *f*

1225
ENDLESS BELT
fr tapis *m* roulant
ne lopende band
de Förderband *n*

da endeløst bånd *n*
sv löpande band *n*
es banda *f*;
cinta *f* continua
it trasportatore a nastro

1226
ENDOSPERM
fr endosperme *m*
ne kiemwit *n*
de Endosperm *n*
da frøvide
sv frövita
es endospermo *m*
it albume *m*

1227
END-PIPE;
LONG OUTFALL PIPE
fr bouche *f* de drainage
ne eindbuis
de Ausmündungsrohr *n*
da enderør *n*
sv slutrör; dikesöga
es tubo *m* terminal
it tubo *m* finale

ENSILAGE, 2991

1228
ENTIRE (bot.)
fr indivis; entier
ne ongedeeld; gaafrandig
de ungeteilt; ganzrandig
da udelt; helrandet
sv odelad; helbräddat
es indiviso; entero; total
it intero

1229
ENTOMOLOGIST
fr entomologiste *m*
ne entomoloog
de Insektologe *m*;
Entomologe *m*
da entomolog
sv entomolog
es entomólogo *m*
it entomologo *m*

1230
ENTREPRENEUR'S NET
PROFIT
fr bénéfice *m* net de l'entrepreneur
ne ondernemerswinst

de Unternehmergewinn *m*
da driftsherregevinst
sv företagarvinst
es beneficio *m* del empresario
it profitto *m* dell'imprendi-
tore

1231
ENVIRONMENT
fr milieu *m*
ne milieu *n*
de Milieu *n*
da omgivelser
sv miljö
es ambiente *m*
it ambiente *m*

1232
EPIDEMIOLOGY
fr épidémiologie *f*
ne epidemiologie
de Epidemiologie *f*
da epidemiologi
sv epidemiologi
es empidemiología *f*
it epidemiologia *f*

1233
EPIDERMIS
fr épiderme *m*
ne opperhuid
de Oberhaut *f*
da overhud
sv epidermis; överhud
es epidermis *f*
it epidermide *f*

1234
EPIPHYLLUM;
LEAF CACTUS
fr épiphyllum *m*
ne bladcactus
de Blattkaktus *m*
da bladkaktus
sv bladkaktus
es epifilo *m*
it cacto *m* foglioso
Epiphyllum Haw.

1235
EPOXY RESIN
fr résine *f* époxyde
ne epoxyhars
de Epoxyharze *f*

da epoxyharpiks
sv epoxiplast
es resina *f* 'epoxido'
it resina *f* epossidica

1236
EQUIPMENT
fr équipement *m*
ne inventaris (uitrusting)
de Inventar *n*;
Geschäftsanlagen *fpl*
da inventar *n*; udrustning
sv utrustning
es inventario *m*; equipo *m*
it equipaggiamento *m*;
attrezzatura *f*

1237
ERADICATE, TO;
TO EXTIRPATE
fr extirper; déraciner;
détruire
ne verdelgen
de vertilgen
da udrydde; tilintetgøre
sv utrota
es exterminar
it sterminare; distruggere

1238
ERECT
fr dressé
ne rechtopstaand
de aufrechtstehend
da opretstående
sv upprätt (stående)
es erguido; erecto
it diritto

ERICA, 1756

1239
EROSION
fr érosion *f*
ne erosie
de Erosion *f*; Abspülung *f*
da erosion
sv erosion
es erosión *f*
it erosione *f*

1240
ESCAPE CLAUSE
fr clause *f* de sauvegarde
ne vrijwaringsclausule

de Sicherungsklausel *f*
da sikkerhedsklausul
sv skyddsklausul
es cláusula *f* de seguridad
it clausola *f* di salvaguardia

1241
ESPALIER
fr arbre *m* en espalier
ne leiboom
de Spalierbaum *m*
da espaliertræ *n*
sv spaljéträd *n*
es árbol *m* en espaldera
it spalliera *f*

1242
ESSENTIAL OILS
fr huiles *f pl* essentielles
ne etherische oliën
de ätherische Oele *n*
da æteriske olier
sv eteriska oljor
es aceites *m pl* essenciales
it oli *m pl* eterei

EUONYMUS, 3554

EUROPEAN APPLE
SAWFLY, 134

1243
EUROPEAN BIRD
CHERRY
fr cerisier *m* à grappes
ne vogelkers
de Traubenkirsche *f*
da hæg
sv hägg
es ciruelo *m* de Bahama
it ciliegio *m* selvatico
   Prunus padus L.

1244
EUROPEAN
CRANBERRY
fr canneberge *f*
ne veenbes
de Moosbeere *f*
da tranebær *n*
sv tranbär *n*
es baya *f* de turbera
it ossicocco *m* palustre
   Vaccinium uliginosum L.

EUROPEAN EARWIG,
768

EUROPEAN ECONOMIC
COMMUNITY, 1199

EUROPEAN FRUIT
SCALE, 2578

EUROPEAN PINE
SHOOT MOTH, 2736

EUROPEAN RED MITE,
1487

1245
EUROPEAN WALNUT
APHID
fr petit puceron *m* du noyer
ne gele okkernootluis
de Walnussblattlaus *f*
da valnødsbladlus
sv valnötsbladlus
es pulgón *m* amarillo del
   nogal
it afide *m* inferiore del
   noce
   Chromaphis juglandicola
   Kltb.

1246
EURYDEMA OLERACEA
fr punaise *f* du chou;
   punaise *f* potagère
ne koolwants
de Kohlwanze *f*
   Gemüsewanze *f*
da kåltæge
sv rapssugare
es chinche *f* de la col
it cimice *f* dei cavoli
   Eurydema oleraceum L.;
   E. oleracea (UK)

1247
EVAPORATE, TO
fr évaporer
ne verdampen
de verdunsten; verdampfen
da fordampe
sv avdunsta
es evaporar
it svaporare

1248
EVAPORATION
fr évaporation *f*
ne verdamping
de Verdunstung *f*;
   Verdampfung *f*
da fordampning
sv avdunstning
es evaporación *f*
it evaporazione *f*

1249
EVAPORATOR
fr évaporateur *m*
ne verdamper
de Verdampfer *m*
da fordamper
sv avdunstare
es evaporador *m*
it vaporizzatore *m*

1250
EVENING PRIMROSE
fr onagre *f*; oenothère *f*
ne teunisbloem
de Nachtkerze *f*
da natlys *n*
sv nattljus *n*
es enotera *f*;
   hierba *f* del asno;
   onagra *f*
it Oenothera *f*
   Oenothera L.

1251
EVERGREEN
fr à feuilles persistantes
ne groenblijvend
de immergrün
da stedsegrøn
sv städsegrönt
es de hojas perennes
it a foglia persistente

1252
EVERLASTING FLOWER
fr immortelle *f*
ne strobloem
de Strohblume *f*
da evighedsblomst
sv evighetsblomma;
   eternell
es siempreviva *f*
it elicriso *m*
   Helichrysum Gaertn.

1253
EXCESS
fr excès
ne overmaat
de Uebermass *n*
da overmål *n*
sv övermått
es exceso *m*
it soprappiú *m*

1254
EXCESS OF
MANGANESE
fr excès *m* de manganèse
ne mangaanovermaat
de Manganüberschuss *m*
da manganoverskud
sv manganöverskott
es exceso *m* de manganeso
it soprappiú *m* di manganese

1255
EXCESS OF NITROGEN
fr excès *m* en azote
ne stikstofovermaat
de Stickstoffübermass *n*
da kvælstofovermål *n*
sv kväveöverskott
es exceso *m* de nitrógeno
it soprappiú *m* d'azoto

1256
EXOTIC
fr exotique; étranger
ne uitheems
de ausländisch
da udenlandsk
sv utländsk
es exótico; extranjero
it straniero; estraneo

1257
EXPANSION PIPE
fr tube *m* compensateur
ne expansiepijp
de Ausdehnungsrohr *n*
da expansionsrør *n*
sv expansionsrör *n*
es tubo *m* compensador
it tubo *m* espanione

1258
EXPANSION TANK
fr réservoir *m* de détente
ne expansievat *n*
de Ausdehnungsgefäss *n*
da expansionsbeholder
sv expansionskärl *n*
es tanque *m* de expansión
it scambiatore *m* di calore

1259
EXPERIMENT; TRIAL;
TEST
fr essai *m*
ne proef
de Versuch *m*
da forsøg *n*
sv försök *n*
es prueba *f*; ensayo *m*;
experimento *m*
it esperimento *m*

EXPERIMENTAL CROP,
3817

1260
EXPERIMENTAL
DESIGN
fr procédé *n* d'essai
ne proefopzet
de Versuchsanordnung *f*
da forsøgsplan
sv försöksplan
es disposición *f* del ensayo
it disegno *m* da esperimento

1261
EXPERIMENTAL FIELD;
TRIAL PLOT
fr champ *m* d'essai
ne proefveld *n*
de Versuchsfeld *n*
da forsøgsmark
sv försöksfält *n*
es campo *m* experimental
it terreno *m* sperimentale

1262
EXPERIMENTAL
GARDEN;
TRIAL GARDEN
fr jardin *m* d'essai
ne proeftuin
de Versuchsgarten *m*

da forsøgshave
sv försöksträdgård
es jardín *m* de ensayos
it campo *m* sperimentale

1263
EXPERIMENTAL
NURSERY;
EXPERIMENTAL
HOLDING
fr ferme *f* expérimentale;
culture *f* d'essais
ne proefbedrijf *n*
de Versuchsbetrieb *m*
da demonstrationsgartneri
sv försöksföretag
es granja *f* experimental
it azienda *f* sperimentale

1264
EXPERIMENTAL
SCHEDULE;
EXPERIMENTAL
PROGRAMME
fr protocole *m* d'essai
ne proefopzet
de Versuchsanordnung *f*
da forsøgsplan
sv försöksplan
es disposición *f* del ensayo
it disegno *m* da esperimento

EXPERIMENTAL
STATION, 3105

1265
EXPORT;
EXPORTATION
fr exportation *f*
ne uitvoer
de Ausfuhr *f*
da udførsel (eksport)
sv export; utförsel
es exportación *f*
it esportazione *f*

1266
EXPORT CONTROL
fr contrôle *m* des exportations
ne uitvoercontrole
de Ausfuhrkontrolle *f*
da eksportkontrol

sv export-kontroll;
    utförsel-kontroll
es control *m* de la
    exportación
it controllo *m* sull'esporta-
    zione

1267
EXPORT DUTY (TAX)
fr droit *m* à l'exportation
ne uitvoerheffing
de Ausfuhrabgabe *f*
da eksportafgift
sv exportavgift
es derecho *m* de exportación
it tassa *f* sull'esportazione

1268
EXPORTER
fr exportateur *m*
ne exporteur
de Exporteur *m*
da eksportør
sv exportör
es exportador
it esportatore *m*

1269
EX POST-CALCULATION
fr calcul *m* ex-post
ne na-calculatie
de Nachkalkulation*f*
da efterkalkulation
sv efterkalkylering
es cálculo *m* ex-post
it calcolo *m* es-post

1270
EXPROPRIATION
fr expropriation *f*
ne onteigening

de Enteignung *f*
da expropriation
sv expropriering
es expropiación *f*
it espropriazione *f*

EXTENSION SERVICE,
35

1271
EXTENSIVE
fr extensif
ne extensief
de extensiv
da extensiv
sv extensiv
es extensivo
it estensivo

1272
EXTERNAL TARIFF
fr tarif *m* extérieur
ne buitentarief *n*
de Aussentarif *m*
da ydre tarif
sv extern taxa
es tarifa *f* exterior
it tariffa *f* esterna

EXTIRPATE, TO,   1237

1273
EXTRACT
fr estrait *m*
ne extract *n*
de Extrakt *m*; Auszug *m*
da ekstrakt
sv extrakt
es extracto *m*
it estratto *m*

1274
EXTRACT, TO
fr extraire
ne extraheren
de extrahieren
da ekstrahere
sv extrahera
es extraer
it estrarre

1275
EXTRACTION
fr extraction *f*
ne extractie
de Extraktion *f*
da ekstraktion
sv extrahering
es extracción *f*
it estrazione *f*

1276
EYE
fr oeil *m*
ne oog *n*
de Auge *n*
da øje *n*
sv öga *n*
es ojo *m*
it occhio *m*

1277
EYE CUTTING;
BUD CUTTING
fr bouture *f* à 'un oeil'
ne oogstek
de Augensteckling *m*
da knopstikling
sv ögonstickling
es esqueje *m* de yema
it germoglio *m* ad occhio

EYE-SPOTTED BUD
MOTH,   467

F

1278
FADE, TO; TO FLAG;
TO WILT
fr se faner; se flétrir
ne verwelken
de verblühen; verwelken
da visne; afblomstre
sv vissna; förtvina
es marchitarse
it appassire

FADING, 1080

1279
FAGGOT
fr fagot *m*
ne takkebos
de Reisigbündel *n*
da risknippe *n*
sv risknippa *n*; faskin
es haz *m* de leña
it fascina *f*

1280
FALL (OF FRUITS)
fr chute *f*
ne val
de Fall *m*
da fald *n*
sv fall *n*
es caída *f*
it cascata *f*;
cascola *f* dei frutti

1281
FALL-OFF
fr fonte *f*
ne wegval
de Ausfall *m*
da bortfald *n*
sv bortfall *n*
es caída *f* prematura
it scatto *m*

1282
FALLOW
fr en jachère; en friche
ne braak
de brach

da brak
sv träda
es baldío; barbecho
it incolto

FALLOW, TO, 2149

1283
FALLOW LAND
fr friche *f*; jachére *f*
ne braakland *n*
de Brachland *n*
da brakjord
sv träde; trädesjord
es barbecho *m*
it maggese *m*

1284
FALSE ACACIA
fr robinier *m*; faux-acacia *m*
ne acacia
de gemeine Akazie *f*;
Robinie *f*
da robinie; uægte akacie
sv vanlig robinia
es acacia *f* falsa
it acacia *f*
Robinia pseudo-acacia L.

1285
FALSE CYPRESS
fr faux-cyprès *m*
ne Chamaecyparis
de Scheinzypresse
da dværgcypres
sv falsk cypress
es Chamecyparis *m*
it cipresso *m* falso
Chamaecyparis spp.

1286
FALSE TRUFFLE
fr fausse truffe *f*
ne valse truffel
de falscher Trüffel *m*
da trøffelsygdom
sv tryffelsjuka

es trufa *f* falsa
it mal *m* del tartufo
Diehlioyces microspora
(Diehl et Lamgert)
Gilkey

1287
FALSE WHORL;
PSEUDO WHORL
fr pseudo-verticille *m*
ne schijnkrans
de Scheinquirl *m*
da falsk årring
sv skenkrans; falsk årsring
es pseudo verticilo *m*
it pseudo cerchio *m*

1288
FAMILY FARM;
FAMILY HOLDING
fr exploitation *f* familiale
ne gezinsbedrijf *n*
de Familienbetrieb *m*
da familiebedrift
sv familjeföretag
es explotación *f* familiar
it azienda *f* familiare

1289
FAMILY LABOUR
fr travail *m* familial
ne gezinsarbeid
de Familienarbeit *f*
da familiearbejde
sv familjens arbete
es mano *f* de obra familiar
it lavoro *m* familiare

1290
FAMILY WORKERS
fr main d'oeuvre *f* familiale
ne eigen arbeidskrachten
de familieneigene Arbeits-
kräfte *f pl*
da familiens arbejdskraft
sv familjens arbetskraft;
egen arbetskraft
es mano *f* de obra familiar
propia
it mano *f* d'opera familiare

1291
FAN; BLOWER
fr ventilateur m
ne ventilator
de Ventilator m; Lüfter m
da blæser; ventilator
sv ventilator; fläkt
es ventilador m
it ventilatore m

FAN SHAPED,   1365

1292
FARMER'S BANK
fr banque f de crédit
   agricole
ne boerenleenbank
de Raiffeisenbank f;
   Spar- und Darlehnskasse f
da landmandsbank
sv jordbrukskassa
es banca m de crédito
   agrícola
it banca f di credito
   agricola

1293
FARMERS' CREDIT
BANK
fr banque f de crédit agricole
de Raiffeisenbank f;
   landwirtschaftliche
   Rentenbank f
da jordbrugsbank
sv jordbrukskreditkassa
es banco m de crédito
   agrícola
it banca f di credito
   agricolo

1294
FARM MANAGEMENT
fr gestion f des exploitations
ne bedrijfseconomie
de Betriebswirtschaft f
da driftsøkonomie
sv företagsekonomie
es economía f de la
   explotación
it economia f aziendale

1295
FARM STRUCTURE
fr structure f d'exploitation
ne bedrijfsstructuur

de Betriebsstruktur f
da driftsstruktur
sv företagsstruktur
es estructura f de la
   explotación
it struttura f aziendale

1296
FARM TYPE
fr type f d'exploitation
ne bedrijfstype n
de Betriebstyp m
da driftstype
sv företagstyp
es tipo m de la explotación
it tipo m d'azienda

1297
FARMYARD MANURE
fr fumier m de ferme
ne stalmest
de Stallmist m
da staldgødning
sv stallgödsel
es abono m de granja
it stallatico m

1298
FASCIATION
fr fasciation f
ne bandvorming
de Verbänderung f;
   Fasziation f
da bånddannelse
sv fasciation; förbandning
es fasciación f
it fasciatura f

FASCIATION,
leaf gall and,   2083

1299
FATHEN
fr chénopode m
ne ganzevoet
de Gänsefuss m
da gäsefod
sv svinmålla
es anserina f; pie m de anade
it chenopodio m
   Chenopodium

1300
FATIGUE
fr fatigue f
ne vermoeidheid

de Ermüdung f
da træthed
sv trötthet
es fatiga f
it fatica f

FEATHER VEINED,
2746

FEED, TO,   2262

FEEDING,   2269, 2502

1301
FEED PIPE
fr tuyau m d'alimentation
ne voedingspijp
de Speiserohr n;
   Speiseleitung f
da føderør n; tilledning
sv matarrör n
es tubo m de alimentación
it tubo m d'alimentazione

1302
FEED PUMP
fr pompe f alimentaire
ne voedingspomp
de Speisepumpe f
da fødepumpe
sv matarpump
es bomba f de alimentación
it pompa f d'alimentazione

1303
FEELER ARM
fr tâteur m
ne tastarm
de Taster m
da følerarm
sv skämmare; kännararm
es llave f falsa
it organo m tastatore

1304
FELL, TO;
TO CUT DOWN
fr abattre
ne vellen
de fällen
da fælde
sv fälla; avverka
es talar; abatir
it abbattere

1305
FEMALE
fr femelle
ne vrouwelijk
de weiblich
da hunlig
sv honlig
es femenino
it femminile

1306
FENCE; SCREEN
fr clôture f
ne schutting
de Zaun m
da hegn n
sv stängsel
es valla f
it steccato m

1307
FENCE, TO
fr clôturer
ne omheinen
de einfriedigen
da indhegne
sv inhägna
es cercar
it chiudere con siepe

1308
FENNEL
fr fenouil m
ne venkel
de Fenchel m
da fennikel
sv fänkål
es hinojo m
it finocchio m
Foeniculum vulgare Mill.

1309
FERMENT, TO
fr fermenter
ne gisten
de gären
da gære
sv jäsa
es fermentar
it fermentare

1310
FERMENTATION
fr fermentation f
ne gisting

de Gärung f
da fermentering; gæring
sv jäsning
es fermentación f
it fermentazione f

1311
FERMENTATION
PROCESS
fr processus m de fermentation
ne gistingsproces n
de Gärungsvorgang m
da gæringsproces
sv jäsningsprocess
es proceso m de fermentación
it processo m di fermentazione

FERMENTING WHARF,
790

1312
FERN
fr fougère f
ne varen
de Farn m
da bregne
sv ormbunke
es helecho m
it felce f; asplenio m
Filices

FERN, maidenhair, 2239
—, royal, 3213
—, staghorn, 3598

FERN LEAF, 3875

1313
FERRIFEROUS;
FERRUGINOUS
fr ferrigineux; ferrifère
ne ijzerhoudend
de eisenhaltig
da jernholdig
sv järnhaltig
es ferruginoso
it ferruginoso; ferrifero

1314
FERROUS SULPHATE
fr sulfate m de fer
ne ijzervitriool n

de Eisenvitriol n
da jernsulfat n
sv järnsulfat n
es sulfato m de hierro
it solfato m di ferro;
vetriolo m di ferro

1315
FERTILE
fr fertile
ne fertiel; vruchtbaar
de fruchtbar
da fertil
sv bördig; fruktbar (jord)
es fértil; fecundo
it fertile

1316
FERTILITY;
FRUITFULNESS
fr fertilité f
ne vruchtbaarheid
de Fruchtbarkeit f
da frugtbarhed
sv bördighet; fruktbarhet
es fertilidad f
it fertilità f

1317
FERTILITY GRADIENT
fr gradient m de fertilité
ne vruchtbaarheidsverloop n
de Fruchtbarkeitsverlauf m
da frugtbarhedsændring
sv fruktbarhetsförlopp n
es gradiente m de fertilidad
it decorso f della fertilità

1318
FERTILIZATION
fr fécondation f
ne bevruchting
de Befruchtung f
da befrugtning
sv befruktning
es fecundación f
it fecondazione f

FERTILIZE, TO, 2262

1319
FERTILIZE
ADDITIONALLY, TO
fr fertiliser additionellement
ne bijmesten
de beidüngen

da overgøde; tilskudsgøde
sv övergödsla
es recebar (abonado)
it concimare in aggiunta

1320
FERTILIZER
fr engrais *m* chimique;
   engrais *m* artificiel
ne kunstmest; meststof;
   hulpmeststof
de Kunstdünger *m*
   Hilfsdünger *m*
da kunstgødning *n*;
   hjælpegødning
sv gödselmedel för över-
   gödsling
es abono *m* químico;
   abono *m* artificial
it concime *m*;
   concime *m* ausiliario

FERTILIZER, complete,
785
—, compound, 792
—, nitrogenous, 2485
—, potash, 2858

1321
FERTILIZER AND LIME
BROADCASTER
fr épandeur *m* à la volée
ne centrifugaalstrooier
de Schleuderdüngerstreuer *m*
da kunstgødningsspreder
sv centrifugalspridare
   (för handelsgödsel)
es distribudor *m* de abonos
   a voles
it spondiconcime *m*

1322
FERTILIZER DILUTER
fr doseur *m* d'engrais
   chimiques
ne concentratiemeter
de Konzentrationsmesser *n*
da koncentrationsmåler
sv koncentrationsmätare
es dosimetro *m*
it contatore *m* di
   concentrazione

1323
FERTILIZER
PLACEMENT
fr fumure *f* en ligne;
   fumure *f* en sillon
ne rijenbemesting
de Reihendüngung *f*
da rækkegødning
sv radgödsling
es abono *m* en línea
it concimazione *f*
   filatamente

1324
FERTILIZER
RECOMMENDATION
fr avis *m* de fumure
ne bemestingsadvies *n*
de Düngungsempfehlung *f*
da gødningsvejledning
sv gödslingsråd
es consejo *m* de abonado
it avviso *m* per la concimazione

1325
FERTILIZER
REQUIREMENT
fr besoin *m* en engrais
ne mestbehoefte
de Düngerbedürfnis *n*
da gødningsbehov *n*
sv gödslingsbehov *n*
es requerimiento *m* de
   abono
it fabbisogno *m* di concime

FERTILIZING, 2269

1326
FETCH IN, TO
(INTO GLASSHOUSE);
TO SHIFT
fr enserrer
ne inhalen
de hereinholen
da indtagning
sv intagning
es almacenar
it portare dentro

1327
FIBROUS
fr fibreux
ne vezelig

de faserig
da trævlet
sv fibrös; trådig
es fibroso
it filamentoso

FICUS, 1341

1328
FIELD
fr champ *m*
ne akker
de Acker *m*
da mark; ager
sv åker; fält
es campo *m*
it campo *m*

FIELD, experimental, 1261

1329
FIELD BINDWEED
fr liseron *m* des champs
ne akkerwinde
de Ackerwinde *f*
da ager-snerle
sv åkervinda
es enredadera *f*; corregüela *f*;
   correhuela *f*
it vilucchio *m*
   Concolvulus arvensis L.

1330
FIELD CAPACITY
fr capacité *f* de rétention au
   champ minimum
ne veldcapaciteit
de Feldkapazität *f*;
   Wasserkapazität *f*
da markkapacitet
sv fältkapacitet
es capacidad *f* de campo
it capacità *f* del campo

FIELD CROP, 2559

1331
FIELD MINT;
CORN MINT
fr menthe *f* des champs
ne akkermunt
de Acker-Minze *f*
da ager-mynte
sv åkermynta

es menta *f* silvestre
it menta *f* selvatica
Mentha arvensis L.

1332
FIELD PANSY;
HEART'S EASE
fr pensée *f* sauvage
ne akkerviooltje *n*
de Stiefmütterchen *n*;
   Ackerveilchen *n*
da ager-stedmoderblomst
sv åkerviol
es pensamiento *m* silvestre
it viola *f* del pensiero
   selvatica
   Viola arvensis Murr.

1333
FIELD PEAS
fr pois *mpl* gris
ne capucijners
de graue Erbsen *f pl*
da markært
sv grå ärt
es arvejas *f pl* grises;
   guisantes *m pl* grises
it ceci *mpl* grigi
   Pisum sativum L. cv.
   Arvense L. Poir.

FIELD PENNYCRESS,
3654

1334
FIELD RESISTANCE
fr résistance *f* au champ
ne veldresistentie
de Feldresistenz *f*
da mark-resistens
sv fältresistens
es resistencia *f* a campo
it resistenza *f* in campo

1335
FIELD SLUG;
GREY GARDEN SLUG
(US)
fr limace *f* rouge
ne naakte slak
de Ackerschnecke *f*
da agersnegl
sv åkersnigel

es limaco *m*; babosa *f*
it limaccia *f* grigia
   Arion rufus L.;
   Deroceras reticulatum
   Müll.

1336
FIELD SPEEDWELL
fr véronique *f* des champs
ne akkererereprijs
de Ackerehrenpreis *m*
da flerfarvet ærenpris
sv åkerärenpris;
   åkerveronika
es verónica *f* agreste
it pavarina *f*;
   veronica *f* delle messi
   Veronica agrestis oeder

1337
FIELD STOCK
fr stock *m* de champs
ne veldinventaris
de Feldinventar *n*;
   Feldvorräte *m pl*
da jordforråd *n*
sv växande gröda
es inventario *m* de cosechas
   pendientes
it inventario *m* dei terreni

1338
FIELD SURVEYING
fr arpentage *m*
ne landmeting
de Feldmessung *f*
da landmåling
sv fältmätning
es agrimensura *f*
it agrimensura *f*

1339
FIELD WOUNDWORT
fr épiaire *f* des champs
ne akkerandoorn
de Ackerziest *m*
da ager galtetand
sv åkersyska
es ortiga *f* hedienda
it erba *f* strega
   Stachys arvensis L.

1340
FIG
fr figue *f*
ne vijg
de Feige *f*
da figen
sv fikon
es higo *m*
it fico *m*

1341
FIG TREE; FICUS
fr figuier *m*
ne vijgeboom
de Feigenbaum *m*
da figentræ *n*
sv fikonträd *n*
es higuera *f*
it fico *m*
   Ficus L.

1342
FILAMENT (bot.)
fr filet *m* (étamine)
ne helmdraad
de Staubfaden *m*
da støvtråd *n*
sv ståndarsträng
es fiiamento *m*
it filamento *m* dello stame

FILBERT, 1742

1343
FILIFORM
fr filiforme
ne draadvormig
de fadenförmig
da trådformet
sv trådlik
es ahilado; filiforme
it filiforme

1344
FILLER (IN ORCHARD)
fr (arbre) temporaire *m*
ne wijker
de Füller *m*
da mellemplantningstræ *n*
sv fyllnadsträd *n*
es árbol *m* temporale
it albero *m* temporale;
   pianta *f* della temporanea

1345
FILL IN, TO (GAPS)
fr combler; compléter
ne inboeten
de nachpflanzen
da efterplante
sv komplettera
es rellenar
it rimettere

1346
FILM
fr film *m*
ne folie
de Folie *f*
da folie
sv folie; film
es película *f*
it foglio *m*

1347
FILTER
fr filtre *m*
ne filter *n*
de Filter *m*
da filter *n*
sv filter *n*
es filtro *m*
it filtro *m*

1348
FILTER, TO
fr filtrer
ne filtreren
de filtrieren
da filtrere
sv filtrera
es filtrar
it filtrare

1349
FILTER PAPER
fr papier-filtre *m*
ne filtreerpapier *n*
de Filterpapier *n*
da filterpapir *n*
sv filterpapper *n*
es papel *m* filtro
it carta *f* da filtrare

1350
FINE GRAINED
fr à grain fin
ne fijnkorrelig
de feinkörnig

da finkornet
sv finkornig
es de grano fino
it di grana fine

1351
FINELY GROUND
COPPER SLAG
fr scories *f pl* de cuivre
    moulues
ne koperslakkenbloem
de Kupferschlackenmehl *n*
da kobberslaggemel *n*
sv pulveriserad kopparslagg
es escorias *f pl* de cobre
    molidas
it scorias *f pl* di rame
    polverizzata

FINGER-AND-TOE
DISEASE, 719

1352
FIRE (HYACINTHUS)
fr Botrytis *m*
ne vuur *n*
de Grauschimmelkrank-
    heit *f*
da hyacinth-gråskimmel
sv gråmögel
es encendido *m*
it Botrytis *m*
    Botrytis hyacinthi
    Westerd. et v. Beyma

1353
FIRE (TULIPA)
fr Botrytis *m*
ne vuur *n*
de Grauschimmelkrank-
    heit *f*
da tulipan-gråskimmel
sv tulpangråmögel
es encendido *m*
it Botrytis *m*
    Botrytis tulipae
    (Lib.) Lind

1354
FIRE BRICK;
REFRACTORY
fr brique *f* réfractaire
ne vuurvaste steen
de Flintenstein *m*

da ildfaste sten
sv eldfast sten;
    eldfast tegel *n*
es piedra *f* refractaria
it mattone *m* refrattario

1355
FIRE DOOR
fr porte *f* de foyer
ne vuurdeur
de Feuerungstür *f*
da fyrelem
sv eldningslucka
es puerta *f* de la fogaina de
    la cámara de combustión
it porta *f* a fuoco

1356
FIREFANG
(MUSHROOM
GROWING)
fr zone *f* blanche
    d'actinomycètes
ne brandplekken
de Actinomyceten *f pl*;
    Weissbrandflecken *m pl*
da actinomyceter; firefang
sv firefang
es zona *f* blanca de
    actinomicetos
it attinomiceti *m pl*

1357
'FIRE' SPOTTING
(TULIP)
fr tulipes *fpl* piquées
ne pokken
de 'Stippen';
    Flecken *fpl* auf Blüten
    und Blättern
da gråskimmelpletter
sv gråmögelsfläckar
es 'pokken' (enfermedad de
    tulipán)
it 'pokken'
    (macchie vaiolosa)
    Botrytis

1358
FIRETHORN;
PYRACANTHA
fr buisson-ardent *m*
ne vuurdoorn
de Feuerdorn *m*

da ildtorn
sv eldtorn
es piracanta *m*
it piracanta *f*
  Pyracantha Roem.

1359
FIRE TUBE;
SMOKE FLUE TUBE
fr tube *f* de chaudière
ne vlampijp
de Flammrohr *n*; Heizrohr *n*
da flammerør *n*
sv flamrör *n*
es tubo *m* de llama;
  tubo *m* de humo
it tubo *m* di fiamma

FIRMNESS, 1725

1360
FIRST CROP
fr pré-culture *f*; culture *f*
  antérieure; culture précé-
  dente
ne voorteelt
de Vorkultur *f*
da forkultur
sv förkultur
es precultivo *m*
it coltura *f* precorrente

1361
FISCAL BOOKKEEPING
fr tenue *f* de comptabilité
ne fiscale boekhouding
de Fiskalbuchführung *f*
da skattemæssig;
  regnskabsføring
sv affärsbokföring
es contabilidad *f* fiscal
it contabilità *f* fiscale

1362
FIX, TO
fr attacher; fixer
ne vastleggen
de festlegen
da fastlægge
sv fastlägga; fixera
es atar; fijar
it fissare

1363
FIXATION
fr fixation *f*
ne fixatie
de Fixierung *f*
da fiksering
sv fixering
es fijación *f*
it fissazione *f*

FIXATION,
nitrogen, 2484

1364
FIXED COSTS
fr coûts *m pl* fixes
ne constante kosten
de Festkosten *f pl*;
  Fixkosten *f pl*
da faste omkostninger
sv fasta kostnader
es costes *m pl* fijos
it costi *m pl* fissi

1365
FLABELLIFORM;
FAN SHAPED
fr flabelliforme; en éventail
ne waaiervormig
de fächerförmig
da vifteformet
sv solfjäderformig
es flabeliforme
it flabelliforme

FLAG, TO, 1278

1366
FLAME GUN
fr lance-flammes *m*
ne vlammenwerper
de Flammenwerfer *m*
da flammekaster
sv eldkastare
es lanza-llamas *f*
it lanciafiamma *m*

1367
FLAT
(BULB GROWING)
fr clayette *f*
ne broeikist (broeibak)
de Treibkiste *f*
da drivkasse

sv drivlåda
es caja *f* para forzar
it stufa *f*

1368
FLAT, TO
(BULB GROWING);
TO BOX UP
fr planter pour le forçage
ne opplanten
de aufpflanzen
da lægge løg
sv plantera
es plantar
it piantare in cassette

1369
'FLAT' MUSHROOM
fr champignon *m* ouvert;
  champignon *m* gallipède
ne open champignon
de offener Champignon *m*
da åben champignon
sv öppen champinjon
es champiñon *m* abierto
it fungo *m* aperto

1370
FLAT ROLLER
fr rouleau *m* plombeur
ne gladde rol
de Glattwalze *f*
da glattromle
sv slätvält
es rulo *m* apisonador
it rullo *m* compressore

1371
FLATTEN, TO
fr aplatir
ne afplatten
de abplatten
da aflade
sv avplatta
es aplanar
it appiattire

1372
FLAVOUR
fr arome *m*
ne aroma *n*
de Arom *n*; Aroma *n*
da aroma
sv arom

es aroma *f*
it aroma *m*

1373
FLAX
fr lin *m*
ne vlas *n*
de Flachs *m*
da hør
sv lin
es lino *m*
it lino *m*
  Linum L.

1374
FLEABANE
fr érigéron *m*
ne fijnstraal
de Berufskraut *n*;
  Beschreikraut *n*
da bakkestjerne
sv binka
es erigerón *m*
it Erigeron *m*
  Erigeron L.

1375
FLEA BEETLE
fr altise *f*; puce de terre *f*
ne aardvlo
de Erdfloh *m*
da jordloppe
sv jordloppa
es altisa *f*
it altica *f*
  Haltica spp.

1376
FLESHY
fr charnu
ne vlezig
de fleischig
da kødet
sv köttig
es carnoso
it carnoso

1377
FLEXIBLE
fr flexible; souple
ne buigzaam
de biegsam
da bøjelig
sv böjlig; smidig

es flexible
it flessibile

1378
FLOODING
fr rincer
ne doorspoelen
de durchwässern;
  durchspülen
da gennemspule
sv genomspola
es enjuagar
it pulire

1379
FLORAL
ARRANGEMENT
fr bouquet *m*
ne bloemstuk *n*
de Strauss *m*
da buket
sv blombukett;
  blomsteruppsats
es ramo *m* de flores;
  ramillete *m*
it mazzo *m* di fiori

1380
FLORICULTURE
fr floriculture *f*
ne bloemisterij
de Zierpflanzenbau *m*
da blomstergartneri *n*
sv blomsterodling
es floricultura *f*
it fioricultura *f*

FLORICULTURIST, 1395

FLORIFEROUS, 1450

1381
FLORIST
fr fleuriste *m & f*
ne binder; bindster
de Blumenbinder *m*;
  Florist *m*; Floristin *f*
da binder; binderske
sv bindare; binderska
es florista *m & f*
it fioraio *m*; fioraia *f*

1382
FLORISTRY
fr art *m* floral
ne binderij
de Blumenbinderei *f*
da binderi *n*
sv binderi
es arte *m* floral;
  confección *m* de ramilletes
it arte *f* fiorale

FLORIST'S, 1403

FLOW, 60

1383
FLOWER; BLOOM
fr fleur *f*
ne bloem
de Blume *f*; Blüte *f*
da blomst
sv blomma
es flor *f*
it fiore *m*

FLOWER, cut, 955
—, globe, 1598
—, everlasting, 1252
—, ligulate ray, 2163

1384
FLOWER, TO;
TO BLOSSOM;
TO BLOOM
fr fleurir
ne bloeien
de blühen
da blomstre
sv blomma
es florecer
it fiorire

1385
FLOWER ARRANGING
fr arranger des fleurs
ne bloemschikken
de Blumen binden
da blomsterbinderi *n*
sv blomsterbinderi
  och dekoration
es confección *f* de ramilletes;
  disposición *f* de flores
it assettare dei fiori

1386
FLOWER BASKET
fr corbeille *f* à fleurs
ne bloemenmand
de Blumenkorb *m*
da blomsterkurv
sv blomsterkorg
es cestillo *m* de las flores
it cestino *m* di fiori

1387
FLOWER BEARING
fr fleuri; à fleurs
ne bloemdragend
de blütentragend
da blomsterbærende
sv blombärande
es floreciente
it fiorifero

1388
FLOWER BED;
FLOWER BORDER
fr plate-bande *f*; corbeille *f*
ne bloembed *n*; bloemperk *n*
de Blumenbeet *n*
da blomsterbed *n*
sv blomsterbädd; rabatt
es cuadro *m* de flores;
    arriate *m*
it aiuola *f*

FLOWER BORDER, 1388

1389
FLOWER BUD
fr bouton *m* floral
ne bloemknop
de Blütenknospe *f*
da blomsterknop
sv blomknopp
es botón *m* floral;
    yema *f* floral
it boccia *f*

1390
FLOWER CARPET
fr tapis *m* de fleurs
ne bloemtapijt *n*
de Blumenteppich *m*
da blomstertæppe *n*
sv blomstermatta
es tapiz *m* de flores
it tappeto *m* di fiori

1391
FLOWERCLUSTER
fr grappe *f* de fleurs
ne bloemtros
de Blütentraube *f*
da blomsterklynge
sv blomklase
es racimo *m*
it grappolo *m* di fiori

1392
FLOWER CORSO;
FLOWER PARADE
fr bataille *f* de fleurs;
    corso *m* fleuri
ne bloemencorso *n*
de Blumenkorso *m*
da blomsterkors
sv blomstertåg *n*
es batalla *f* de flores
it corso *m* di fiori

1393
FLOWER FORMATION
fr formation *f* des fleurs
ne bloemvorming
de Blütenbildung *f*
da blomsterdannelse
sv blombildning
es formación *f* de las flores
it formazione *f* dei fiori

1394
FLOWER GARDEN
fr jardin *n* fleuriste
ne bloementuin
de Blumengarten *m*
da prydhave; blomsterhave
sv blomsterträdgård
es jardín *m* de flores
it giardino *m* di fiori

1395
FLOWER GROWER;
FLORICULTURIST
fr jardinier-fleuriste *m*;
    producteur *m* de fleurs
ne bloemist; bloemkweker
de Zierpflanzengärtner *m*
da blomstergartner
sv blomsterodlare
es floricultor *m*
it fioricultore *m*

1396
FLOWER GROWING;
FLORICULTURE
fr floriculture *f*
ne bloementeelt
de Zierpflanzenbau *m*
da blomsterdyrkning
sv blomsterodling
es floricultura *f*
it fioricultura *f*

1397
FLOWERING;
BLOSSOMING
fr floraison *f*
ne bloei
de Blüte *f*
da blomst; blomstring *f*
sv blomning
es floración *f*
it fioritura *f*

1398
FLOWERING ALMOND
fr Prunus *m* triloba
ne drielobbige amandel
de Mandelbäumchen *n*;
    dreilappige Mandel *f*
da rosenmandel
sv rosenmandel
es Prunus *m* triloba
it mandorlo *m* triloba
    Prunus triloba Lindl.

FLOWERING ALMOND,
dwarf, 1181

1399
FLOWERING
CRABAPPLE
fr pommier *m* d'ornement
ne sierappel
de Zierapfel *m*
da bæræble *n*; paradisæble *n*
sv prydnadsapel
es manzano *m* de adorno
it Malus *m*
    Malus spp.

1400
FLOWERING TIME
fr (époque *f* de) floraison *f*
ne bloeitijd
de Blütezeit *f*

da blomstringstid
sv blomningstid
es época f de floración
it stagione f della fioritura

1401
FLOWER INITIATION
fr initiation f florale;
   formation f des fleurs
ne bloemaanleg
de Blumenanlage f
da blomsteranlæg n
sv blomanlag n
es primordio m de flor;
   iniciación f floral
it iniziazione f fiorale

FLOWER PARADE,   1392

1402
FLOWER POT
fr pot m à fleur
ne bloempot
de Blumentopf m
da blomsterpotte; potte
sv blomkruka
es maceta f; tiesto m
it vaso m di fiori; testo m

FLOWER REMOVAL,
1747

1403
FLOWER SHOP;
FLORIST'S
fr boutique f de fleuriste
ne bloemenwinkel
de Blumengeschäft n
da blomsterbutik
sv blomsteraffär
es tienda f de flores
it bottega f di fiori

1404
FLOWERS OF SULPHUR
fr fleur f de soufre
ne bloem van zwavel
de Schwefelblüte f
da svovlblomme
sv svavelblomma
es azufre m en flor
it solfo m polverizzato

1405
FLOWER STALK;
PEDICLE
fr pédoncule n
ne bloemsteel
de Blütenstiel m
da blomsterstilk
sv blomstjälk
es pedúnculo m
it peduncolo m

1406
FLOWER VASE
fr vase m à fleurs
ne bloemenvaas
de Blumenvase f
da blomstervase; vase
sv blomvas
es jarrón m; vaso m para
   flores
it vaso m di fiori

1407
FLOW OP SAP;
SAP STREAM
fr circulation f de la sève
ne sapstroom
de Saftstrom m
da saftstrøm
sv saftström
es circulación f de la savia
it circolazione f del succhio

1408
FLOW PATTERN
fr programme m permanent
   de base; programme m de
   routière; programme m
   usuel
ne route-schema n
de Verfahrensschema n
da rutineskema n
sv flödesplan
es programa m permanente
   de base; programa m de
   rutina
it programma m di lavoro

1409
FLUE GAS
fr gaz m fumigène
ne rookgas n
de Rauchgas n
da røggas n

sv rökgas n
es gas m fumigeno
it gas m da fumo

1410
FLUORESCENT LAMP
fr tube f fluorescente
ne fluorescentielamp
de Leuchtstofflampe f
da lysstofrør n
sv lysrör n; lysämnesrör n
es lámpara f fluorescente
it tubo m di luca fluorescente

1411
FLUSH; BREAK
(MUSHROOM
HARVESTING)
fr volée f
ne vlucht
de Welle f; Trieb m; Flug m
da bræk
sv bräck
es florada f
it volata f

FLUSH, core,   842

1412
FLY
fr mouche f
ne vlieg
de Fliege f
da flue
sv fluga
es mosca f
it mosca f

FLY, asparagus,   170
—, bean seed,   275
—, cabbage root,   515
—, carrot,   595
—, carrot rust,   595
—, celery,   620
—, cherry fruit,   651
—, crane,   880
—, glasshouse white,   1590
—, humpbacked,   2701
—, large narcissus,   2445
—, lesser bulb,   3444
—, march,   308
—, mediterranean fruit,
2320

—, narcissus bulb, 2445
—, onion, 2524
—, phorid, 2701

1413
FOAMED CONCRETE
fr béton cellulaire *m*
ne gasbeton *n*
de Schaumbeton *m*
da gasbeton
sv gasbetong
es hormigón *m* celular
it calcestruzzo *m* cellulare

1414
FOLIAGE
fr feuillage *m*
de loof *n*; gebladerte *n*
de Laub *n*
na løvblad *n*
sv löv *n*
es follaje *f*
it fogliame *m*

FOLIAGE, cut, 956
—, ornamental, 956

1415
FOLIAGE PLANT
fr plante *f* à feuillage ornamental
ne bladplant
de Blattpflanze *f*
da bladplante
sv bladväxt
es planta *f* de hojas
it pianta *f* fogliosa

1416
FOLIATE; LEAFY
fr feuillé
ne bebladerd
de beblättert
da bladrig
sv med blad; lövad
es foliado
it fogliato

1417
FOLLICLE
fr follicule *m*
ne kokervrucht
de Balgkapsel *m*
da bælgkapsel

sv baljkapsel
es folículo *m*
it follicolo *m*

1418
FOMES COLLAR ROT (RIBES)
fr Polypore *m*
ne kraagrot *n*
de Holzschwamm *m*
da ribs-porresvamp
sv krusbärsticka
es podredumbre *f* del cuello de la raíz
it marciume *m* da fomes
Fomes ribes Fr.

1419
FOMES ROT
fr pourridié *f* des vergers; pourriture *f* blanche
ne boomgaardzwam
de Weissfäule *f*
da rodfordærver
sv rötticka
es caries *f* del tronco; podredumbre *f* blanca de la madera
it carie *f* del tronco
Fomes pomaceus
(Pers. ex Gray) Lloyd.

1420
FOOD
fr nourriture *f*
ne voedsel *n*
de Nahrung *f*
da føde; næring
sv föda
es alimento *m*
it nutrimento *m*

1421
FOOD ADDITIVES
fr additives *fpl* alimentaires
ne toevoegingen
de Lebensmittelzusätze *mpl*
da additiver
sv (livsmedels) tillsats
es aditamentos *m pl* a productos alimentacios
it addizioni *m pl* d'alimentazione

1422
FOOTING; FOUNDATION
fr fondation *f*
ne fundering
de Gründung *f*;
Fundament *n*
da fundament *n*; sokkel
sv grund; fundament *n*
es fundamento *m*
it fondazione *f*

1423
FOOTROT
fr pourriture *f* du pied
ne voetrot *n*; rotpoot
de Kragenfäule *f*; Umfallkrankheit *f*;
Fusskrankheit *f*
da rodhalsråd; tomatkræft
sv rothalsröta
es podredumbre *f* del pie
it marcitura *f* al piede
Didymella lycopersici Kleb.;
Botrytis cinerea Fr.;
Tanatephorus cucumeris

1424
FOOT ROT AND DAMPING OFF (TOMATO)
fr mildiou *m*
ne osseogenziekte
de 'Ochsenaugenkrankheit' *f*
da bukkeøje
sv bocköga
es mildio *m*
it marciume *m* del colletto
Phytophthora parasitica Dast.

1425
FORCE, TO; TO ADVANCE
fr forcer; hâter
ne broeien; trekken vervroegen
de treiben; verfrühen
da drive
sv driva
es forzar
it forzare; anticipare

1426
FORCED DRAUGHT
fr tirage *m* forcé
ne geforceerde trek
de Druckzug *m*
da overtryks-træk
sv drag *n* genom övertryck
es tiraje *m* forzado
it tiraggio *m* forzato

FORCING, indoor, 1896

1427
FORCING COMPOST
fr compost *m* à fermenter;
  compost *m* de couches
ne broeicompost
de Gärkompost *m*
da varmebedsmateriale *n*
sv (driv-)bänkströ *m*
es compost *m* para cama-
  caliente
it concime *m* riscaldante

1428
FORCING HOUSE
fr serre *f* de forçage
ne trekkas
de Treibhaus *n*
da drivhus*n*;formeringshus*n*
sv drivningshus *n*
es invernadero*m*deforzado;
  estufa *f* de forzado
it serra *f* per forzatura

1429
FORCING MATERIAL
fr matériel *m* de couches
ne broeimateriaal *n*
de Gärmaterial *n*;
  Treibmaterial *n*
da drivmateriale *n*
sv drivmaterial *n*
es material *m* para cama-
  caliente
it materia *f* riscaldante

1430
FORCING TRENCH
fr tranchée *f* de couche
ne broeivoor
de Gurkentreibfurche *f*
da grube; rende
sv drivfåra; drivränna

es zanja *f* para cama-
  caliente
it solco *m* riscaldante

FOREST,   4194

FOREST FLOOR,   806

1431
FOREST LITTER;
FOREST FLOOR
fr litierè *f* forestière
ne bosstrooisel *n*
de Waldstreu *f*
da skovbunds-afrivning
sv skogsförna
es cama *f* de hojas
it lettiera *f* forestale

1432
FOREST SOIL
fr terre *f* de forêt
ne bosgrond
de Waldboden *m*
da skovjord
sv skogsjord
es tierra *f* de bosque
it terra *f* forestale

1433
FOREST TREE
fr arbre *m* forestier
ne woudboom
de Waldbaum *m*
da skovtræ
sv skogsträd *n*
es árbol *m* forestal
it albero *m* silvestre

1434
FOREST TREES AND
HEDGING PLANTS
fr plants *mpl* forestiers et
  ornementaux
ne bos- en haagplantsoen *n*
de Waldsträucher *m pl* und
  Heckengehölze *n pl*
da skov- og hækplanter
sv skogs- och häckplantor
es plantas *f pl* forestales
it piante *f pl* forestali e da
  siepe

1435
FORGET-ME-NOT
fr Myosotis *m*
ne vergeet-mij-niet
de Vergissmeinnicht *n*
da forglemmigej
sv förgätmigej
es miosota *f*; nomeolvides *f*
it miosotide *f*
  Myosotis L.

1436
FORK
fr fourche *f* à fumier
ne greep; vork
de Dunggabel *f*;
  Mistgabel *f*; Forke *f*
da greb
sv grep
es horquilla *f*
it forca *f*

1437
FORK LIFT
fr élévateur *m* à fourche
ne hefmast
de Heckstapler *m*
da lift
sv gaffelstaplare
es elevador *m* de horquilla
it elevatore *m* a forca

1438
FORK LIFT TRUCK
fr fourche *f* élévatrice
ne vorkheftruck
de Gabelstapler *m*;
  Heckstapler *m*
da gaffeltruck
sv gaffeltruck
es carro *m* elevador
it elevatore *m* a forca

1439
FORM CORK, TO
fr durcir comme liège
ne verkurken
de verkorken
da forkorkning
sv förvedas
es acorchar
it divenire sugheroso

1440
FORM PRUNING
fr taille *f* de formation
ne vormsnoei
de Erziehungsschnitt *m*
da formbeskæring
sv formbeskärning
es poda *f* de formación
it potatura *f* di formazione;
  potatura *f* di modella-
  mento

1441
FORMULATION
fr formulation *f*
ne formulering
de Formulierung *f*
da formulering
sv formulering
es formulación *f*
it formulazione *f*

1442
FORSYTHIA
fr Forsythia *f*
ne Chinees klokje *n*
de Forsythie *f*;
  Goldglöckchen *n*
da vårguld *n*; Forsythia
sv Forsythia; gullbuske
es forsitia *f*
it Forsythia *f*
  Forsythia L.

FOUNDATION, 1422

1443
FOUR-WHEELED
TRACTOR
fr tracteur *m* à 4 roues
ne vierwielige trekker
de Schlepper *m*
da traktor
sv fyrhjulig traktor
es tractor *m* de 4 ruedas
it trattrice *f* a quattro ruote

1444
FOXGLOVE
fr digitale *f*
ne vingerhoedskruid *n*
de Fingerhut *m*
da fingerbøl *n*
sv fIngerborgsblomma

es digital *f*
it digitale *f*
  Digitalis L.

1445
FRAGRANCE;
SCENT (OF FLOWER)
fr parfum *m* des fleurs
ne bloemengeur
de Blumenduft *m*
da blomsterduft
sv blomdoft
es fragancia *f* de las flores
it odore *m* di fiori

1446
FRAGRANT; SCENTED
fr odorant
ne welriekend
de wohlriechend
da vellugtende
sv välluktande; doftande
es oloroso
it odoroso

1447
FRAISE HOOD
fr carter *m* de protection
ne freeskap
de Schutzhaube *f*
da fræserkappe
sv fräshuv; skyddshuv över
  fräsvals
es carter *m* de protección
it cofano *m*

1448
FRAME
fr bâche *f* à chassis
ne bak
de Treibbeet *n*; Mistbeet *n*
da bænk
sv bänk
es cajonera *f*
it letto *m*

FRAME, doublespan, 3533
—, spanroof, 3533

1449
FRAMES;
DUTCH LIGHTS
fr couches *f pl*; châssis *m pl*
ne plat glas *n*
de Treibbeete *n pl*

da drivbænke
sv drivbänkar
es cajoneras *f pl*
it vetri *m pl* olandese

1450
FREE FLOWERING;
FLORIFEROUS
fr florifère
ne rijkbloeiend
de reichblühend
da rigtblomstrende
sv rikblommig;
  rikblomstrande
es de abundante floración;
  florido
it fiorente abondantemente

1451
FREE FROM CHLORINE
fr sans chlore
ne chloorvrij
de chlorfrei
da klorfri
sv klorfri
es libre de cloro
it esente di cloro

FREE LIVING
EELWORMS, 3176

1452
FREESIA
fr Freesia *f*
ne fresia
de Freesie *f*
da Freesia
sv Freesia
es fresia *f*
it fresia *f*
  Freesia Klett.

1453
FREEZE, TO
fr geler
ne bevriezen
de gefrieren
da fryse
sv frysa
es helar; congelar
it gelare

1454
FREEZE DRYING
fr lyophilisation *f*;
   cryo-dessication *f*
ne vriesdroging
de Gefriertrocknung *f*
da frysetørring
sv frystorkning
es criodesecación *f*;
   liofilización *f*
it concentrazione *f* per
   sublimazione;
   liofilizzazione *f*

1455
FRENCH BEAN
fr haricot *m* vert;
   haricot *m* mange-tout
ne sperzieboon;
   slaboon
de Brechbohne *f*;
   Salatbohne *f*
da brækbønne; brydbønne
sv sparrisböna; brytböna
es judía *f* verde
it fagiolino *m*

FRENCH BEAN,
climbing, 711
—, dwarf, 1182

FRENCH MARIGOLD,
44

1456
FRESH AIR HOOD
fr prise *f* d'air frais
ne verseluchtkap
de Lüftungsklappe *f*
da friskluftshætte
sv friskluftkappa
es toldo *m* de aire fresco
it cappa *f* d'aria fresca

1457
FRESH FRUIT
fr fruits *m pl* frais
ne vers fruit *n*
de Frischobst *n*
da frisk frugt
sv färsk frukt
es fruta *f* fresca
it frutta *f* fresca

1458
FRESH VEGETABLES
fr légumes *m pl* frais
ne verse groente
de Frischgemüse *n*
da friske grønsager
sv färska grönsaker
es verduras *f pl* frescas
it ortaggio *m* fresco

1459
FRESH WATER
fr eau *f* douce
ne zoet water *n*
de Süsswasser *n*
da ferskvand *n*
sv sötvatten *n*
es agua *f* dulce
it acqua *f* dolce

1460
FRIABLE
fr meuble; friable
ne mul
de locker
da løs
sv lös
es mueble; friable
it arenoso

1461
FRINGED
fr frangé
ne gefranjerd
de gefranst
da frynset
sv fransad
es franjeado
it frangiato

1462
FRITILLARY;
CROWN IMPERIAL
fr fritillaire *f*;
   couronne *f* impériale
ne keizerskroon
de Kaiserkrone *f*
da kejserkrone
sv kejsarkrona
es corona *f* imperial;
   Fritillaria *f*
it corona *f* imperiale
   Fritillaria imperialis L.

FROGHOPPER,
common, 769

1463
FROST DAMAGE
fr dégât *m* causé par la gelée
ne vorstschade
de Frostschaden *m*
da frostskade
sv frostskada
es daño *m* de heladas
it danno *m* causato dal gelo

1464
FROSTPROOF
fr à l'abri de la gelée;
   'antigel' (emballage)
ne vorstvrij
de frostfrei
da frostfri
sv frostfri
es incongelable
it al riparo del gelo

1465
FROST RINGS
fr anneaux *m pl* de gel
ne vorstringen
de Frostringe *m pl*
da frostringe
sv frostringar
es estrangulación *f* de frutos
   por las heladas
it anelli *m pl* da gelo

1466
'FROZEN BLACK PEAT
AMENDMENT'
fr tourbe *f* noire soumise
   à l'action de la geleé
ne doorgevrorenzwartveen*n*
de durchgefrorener
   Schwarztorf *m*
da gennemfrosset tørv
sv genomfrusen torv;
   starkt humifierad
   vitmosstorv
es turba *f* negra decongelada
it torba *f* nera gelata per
   lungo e per largo

1467
FROZEN FOOD
fr produit *m* congelé;
   produit *m* surgelé
ne vriesprodukt
de Tiefkühlkost *f*

da fryseprodukt *n*
sv frysprodukt
es producto *m* congelado
it derrata *f* congelata

1468
FROZEN PEAT
fr tourbe *f*
ne tuinturf
de Gartentorf *m*
da frostbehandlet tørvejord
sv trädgårdstorv
es tierra *f* turba
it torba *f* da giardino

FROZEN VEGETABLES,
1008

FRUCTIFICATION, 2738

1469
FRUIT
fr fruits *m pl*
ne fruit *n*
de Obst *n*
da frugt
sv frukt
es fruta *f*
it frutta *f*

1470
FRUIT
fr fruit *m*
ne vrucht
de Frucht *f*
da frugt
sv frukt
es fruto *m*
it frutto *m*

FRUIT, dehiscent, 1027
—, fresh, 1457
—, hard, 1723
—, pseudo, 2969
—, small, 3443
—, soft, 3467
—, stewed, 3648
—, stone, 1146
—, subtropical, 3709

FRUIT BARK
BEETLE, 2032

1471
FRUIT BRANCH
fr branche *f* fruitière
ne vruchttak
de Fruchtzweig *m*
da frugtgren
sv fruktved *n*
es tallo *m* frutal
it ramo *m* fruttifero

1472
FRUIT CLUSTER
fr grappe *f* de fruits;
régime *m*
ne vruchtkluwen
de Fruchtknäuel *m*
da nøgle; hovede *n*
sv fruktklunga
es racimo *m* de frutas
it grappolo *m* di frutti

1473
FRUIT CONSERVES
(THICK)
fr marmelade *f*
ne vruchtenmoes *n*
de Obstmus *n*
da frugtmos
sv fruktmos *n*
es pulpa *f* de frutas;
pasta *f* de frutas
it passata *f* di frutti

1474
FRUIT FLESH
fr chair *f*; mésocarpe *m*
ne vruchtvlees *n*
de Fruchtfleisch *n*
da frugtkød *n*
sv fruktkött *n*
es pulpa *f*; mesocarpio *m*
it polpa *f* di frutta

1475
FRUIT FORMATION
fr fructification *f*
ne vruchtvorming
de Fruchtbildung *f*
da frugtdannelse
sv fruktbildning
es fructificación *f*
it fruttificazione *f*

FRUITFULNESS, 1316

1476
FRUIT-GROWER
fr arboriculteur *m*; producteur *m* des fruits
ne fruitteler
de Obstbauer *m*
da frugtavler
sv fruktodlare
es fruticultor *m*
it frutticoltore *m*

1477
FRUIT GROWING
fr culture *f* fruitière
ne fruitteelt
de Obstbau *m*
da frugtavl
sv fruktodling
es fruticultura *f*
it frutticoltura *f*

1478
FRUIT JUICE
fr jus *m* de fruits
ne vruchtesap *n*
de Fruchtsaft *m*
da frugtsaft
sv fruktsaft
es jugo *m* de frutas
it succo *f* di frutta

1479
FRUIT PROCESSING
fr transformation *f* de fruits;
technologie *f* des fruits
ne fruitverwerking
de industrielle Obstverarbeitung *f*; Konservierung *f*
da forarbejdning af frugt
sv fruktberedning
es transformación *f* de frutas
it trasformazione *f* di frutti

1480
FRUIT PULP
fr pulpe *f* de fruit
ne vruchtenpulp
de Fruchtpulpe *f*
da frugtpulp
sv fruktpulpa
es pulpa *f*
it polpa *f* di frutta

1481
FRUIT SETTING
fr mise f à fruit
ne vruchtzetting
de Fruchtansatz m
da frugtsætning
sv fruktsättning
es fructificación f
it allegazione f

1482
FRUITSHED;
PACKING SHED
fr fruitier m; fruiterie f
ne fruitschuur
de Obstlager n
da frugtlager n
sv fruktlagerhus n
es almacén m de frutas
it deposito m di frutta

1483
FRUIT STALK
fr pédoncule n
ne vruchtsteel
ne Fruchtstiel m
da frugtstilk
sv fruktskaft
es rabillo m; pedúnculo m
it peduncolo m

1484
FRUIT THINNING
fr ciselage m;
éclaircissage m des fruits
ne vruchtdunning
de ausdünnen der Früchte
da frugtudtynding
sv fruktgallring
es aclarar frutos
it diramazione f;
diramento m

1485
FRUIT TRAILER
fr remorque f
ne fruitwagen
de Plattformwagen m
da frugtplukkevogn
sv flåkvagn
es remolque m
it rimorchio m

1486
FRUIT TREE
fr arbre m fruitier
ne vruchtboom
de Obstbaum m
da frugttræ n
sv fruktträd n
es árbol m frutal
it albero m da frutto

1487
FRUIT TREE RED
SPIDER (MITE);
EUROPEAN RED MITE
(US)
fr araignée f rouge
ne fruitspint
de Obstbaumspinnmilbe f;
rote Spinne f
da frugttræspindemide
sv fruktträdsspinnkvalster n
es arañuela f roja; ácaro m
rojo; araña f de los frutales
it ragno m rosso delle piante
da frutto
Metatetranychus ulmi
C.L. Koch;
Panoychus u. (UK/US)

1488
FRUIT WINE
fr vin m de fruit
ne vruchtenwijn
de Obstwein m
da frugtvin
sv fruktvin
es vino m de frutas
it vino m di frutti

1489
FUEL
fr combustible m
ne brandstof
de Brennstoff m
da brændstof n; brændsel n
sv bränsle n
es combustible m
it combustibile m

1490
FULL (FLOWERS)
fr double
ne gevuld

de gefüllt
da fyldt (fyldtblomstret);
dobbelt
sv dubbel; fylld
es doble; lleno; pleno
it pieno

1491
FULLY AUTOMATIC
fr entièrement automatique
ne volautomatisch
de vollautomatisch
da fuldautomatisk
sv helautomatisk
es automático
it interamente automatico

1492
FUMIGANT
fr fumigant m
ne rookmiddel n
de Ausräucherungsmittel n
da rygemiddel n
sv rökmedel n
es fumigante m
it fumigante m

1493
FUMIGATE, TO
fr faire une fumigation
ne roken
de räuchern
da ryge
sv röka
es fumigar
it fumigare

1494
FUMIGATING CANDLE;
SMOKEBOMB
fr bombe f fumigène
ne rookkaars
de Raücherkerze f
da røgbombe; røgkegle
sv rökkägla
es cartucho m fumigante
it candela f da fumo

1495
FUMIGATING POWDER
fr poudre f fumigène
ne rookpoeder n
de Räucherpulver n
da rygepulver n

sv rökpulver *n*
es polvo *m* fumigante
it polvere *f* fumigante

1496
**FUMIGATION WITH
SULPHUR;
TO SULPHURIZE**
fr soufrer
ne zwavelen
de schwefeln
da svovle
sv svavla
es azufrar
it solforare

1497
FUMITORY
fr fumeterre *m* officinal
ne duivekervel
de gemeiner Erdrauch *m*
da læge-jordrøg
sv jordrök
es Fumaria *f*;
sangre *f* de cristo
it Fumaria *f*
Fumaria officinalis L.

1498
FUNERAL WREATH
fr couronne *f* funéraire
ne grafkrans
de Grabkranz *m*
da krans
sv (grav) krans
es corona *f* mortuoria
it corona *f* funeraria

1499
FUNGAL DISEASE
fr mycose *f*;
maladie *f* cryptogamique
ne schimmelziekte
de Schimmelpilz *m*
da skimmelsygdom
sv mögel
es enfermedad *f* criptogámica
it malattia *f* di muffa

1500
FUNGICIDAL
fr fongicide
ne zwamdodend
de pilztötend

da svampedræbende
sv svampdödande
es fungicido
it fungicido

1501
FUNGUS
fr champignon *m*
ne zwam
de Pilz *m*
da svamp
sv svamp
es hongo *m*; seta *f*
it fungo *m*
Fungus

1502
FUNGUS DISEASE
fr maladie *f* cryptogamique
ne zwamziekte
de Pilzkrankheit *f*
da svampesygdom
sv svampsjukdom
es enfermedad *f* cripto-
gámica
it malattia *f* crittogamica

1503
**FUNKIA;
PLAINTAIN LILY**
fr Hosta *f*; Funkia *f*
ne funkia
de Funkie *f*
da Funkia
sv Funkia; blålilja
es funquia *f*;
hermosa *f* de día
it Hosta *f*
Hosta Tratt.

1504
FUNNEL SHAPED
fr infundibuliforme
ne trechtervormig
de trichterförmig
da tragtformig
sv trattformig
es infundibuliforme
it imbutiforme

1505
FURNACE
fr foyer *m*
ne vuurgang
de Feuerung *f*

da fyring ild
sv eldstad
es fogaina *f*;
cámara *f* de combustión
it camera *f* di combustione

1506
FURROW
fr sillon *m*
ne voor; veur
de Furche *f*
da fure; plovfure
sv fåra
es surco *m*
it solco *m*

1507
FURROW, TO
fr rigoler
ne begreppelen
de mit Rinnen durchziehen;
entwässern
da grave grøfter; grøftegrave
sv förse med fåror;
göra fåror
es zanjar; abrir zanjas
it sistemare dei fossatelli

1508
FURROWED (bot.)
fr sillonné
ne gegroefd
de gefurcht
da furet
sv fårad
es acanalado
it scanalato

1509
FURZE; GORSE
fr ajonc *m*
ne gaspeldoorn
de Stechginster *m*
da tornblad *n*
sv arttörne *n*
es tojo *m*; aulaga *f*
it ginestra *f* a spina
Ulex L.

1510
**FUSARIUM CORM ROT
(GLADIOLUS)**
fr fusariose *f*
ne fusariumrot *n*

de Fusarium-Krankheit *f*
da slimskimmel
sv slemmögel
es fusariosis *f*
it marciume *m* secco;
　　giallume *m*
　　Fusarium oxysporum
　　Schl.; *f. gladiolo* (Massey)
　　Snyd. & Hans.

1511
FUSARIUM ROT
(CROCUS)
fr fusariose *f*
ne fusariumrot *n*
de Fusarium-Fäule *f*
da slimskimmel
sv fusarium-röta;
　　slemmögel
es fusariosis *f*
it fusariosi *f*
　　Fusarium spp.

1512
FUSARIUM ROT
(FREESIA)
fr fusariose *f*
ne voetrot *n*
de Fusskrankheit *f*
da fusariose
sv fusarium-röta

es podredumbre *f* del pie
it fusariosi *f*
　　Fusarium oxysporum
　　Schl. sensu Snyd. & Hans

1513
FUSARIUM ROT
(TULIPA)
fr fusariose *f*;
　　pourriture *f* acide
ne zuur *n*
de Fusarium-Zwiebel-
　　trockenfäule *f*
da fusariumråd
sv fusarium-röta
es fusariosis *f*
it fusariosi *f*
　　Fusarium oxysporum
　　Schl. sensu Snyd. & Hans

1514
FUSARIUM WILT
fr maladie *f* du flétrissement
ne Amerikaanse vaatziekte
de Fusariumwelke *f*
da fusariumvisnesyge
sv slemmögelvissnesjuka
es fusariosis *f*
it malattia *f* d'appassimento
　　Fusarium oxysporum
　　Schlecht.

1515
FUSARIUM WILT
(BEAN)
fr fusariose *f*
ne vaatziekte
de Welkekrankheit *f*
da slimskimmel; visnesyge
sv vissnesjuka
es grasa *f*
it fusariosi *f*
　　Fusarium oxysporum
　　Schl. sensu Snyd. & Hans.;
　　f. phaseoli
　　Kendr. & Snyd.

1516 FUSARIUM WILT
(PISUM)
fr maladie *f* de flétrissement
ne Amerikaanse vaatziekte
de Fusariumwelke *f*
da slimskimmel;
　　fusariumvisnesyge;
　　rodbrand
sv slemmögelvissnesjuka
es fusariosis *f*
it fusariosi *f*; avizzimento *m*
　　Fusarium oxysporum
　　Schlecht. sensu Snyd. &
　　Hans.; *f. pisi* (Linford)
　　Snyd. & Hans.

G

1517
GAILLARDIA
fr gaillarde *f*
ne Gaillardia
de Kokardenblume *f*
da kokardeblomst
sv kokardblomster *n*
es gallardia *f*
it Gaillardia *f*
   Gaillardia Foug.

1518
GALL (AZALEA,
RHODODENDRON)
fr fausse cloque *f*;
   galles *f pl* foliaires
ne oortjesziekte
de Ohrläppchenkrankheit *f*;
   Klumpenblätterkrank-
   heit *f*; Löffelkrankheit *f*
da klumpblad
sv azeleasvulst
es agalla *f* foliar
it galle *f pl* delle foglie;
   galle *f pl* fogliari
   Exobasidium japonicum
   Shir.; *Exobasidium*
   rhododendri Cram.

GALL, azalea leaf, 209
—, leaf, 2083
—, leafy, 2116, 2117

GALLANT SOLDIER,
1995

1519
GALL APPLE
fr noix *f* de galle
ne galappel
de Gallapfel *m*
da galæble *n*
sv galläpple *n*; gallbildning
es agalla *f*; abogalla *f*
it galla *f*
   Cecidium

1520
GALL MIDGE; MIDGE;
LEAF-MINING MIDGE
fr cécidomyie *f*
ne galmug
de Gallmücke *f*
da galmyg
sv gallmygga
es mosquito *m* de agalla
it cecidomia *f*
   Dasyneura spp;
   Cecidomyiidae (= Itoni-
   didae)

1521
GALL WASP
CYNIPID (US)
fr cynips *m*
ne galwesp
de Gallwespe *f*
da galhveps
sv gallstekel
es avispa *f* de agallas
it cinipide *m*
   Cynipoidea;
   Cynipidae (US)

GALL WASP,
bedeguar, 282
—, mossy-rose, 282

1522
GALVANIZE, TO
fr galvaniser; zinguer
ne verzinken
de verzinken
da galvanisere
sv galvanisera
es galvanizar
it zincare; galvanizzare

1523
GAMETE
fr gamète *m*
ne gameet
de Gamete *f*
da gamet; kønscelle
sv könscell
es gameto *m*
it gametophyte *m*

1524
GAMOPETALOUS
(COROLLA)
fr gamopétale
ne vergroeidbladig
de verwachsen
da sambladet
sv sambladigt
es gamopétalo
it gamopetalo

1525
GAMOSEPALOUS
(CALYX)
fr gamosépale
ne vergroeidbladig
de verwachsen
da sambladet
sv sambladigt
es gamosépalo
it gamosepalo

GANGRENE, 1155

1526
GAP UP, TO
fr combler; completer
ne inboeten
de nachpflanzen
da efterplante
sv komplettera
es rellenar
it rimettere

1527
GARDEN
fr jardin *m*
ne tuin
de Garten *m*
da have
sv trädgård
es jardín *m*
it giardino *m*

GARDEN,
ornamental, 2554

1528
GARDEN, TO
fr jardiner
ne tuinieren
de gärtnern
da havearbejde
sv arbeta i trädgården
es trabajar la huerta;
   trabajar el jardín
it lavorare nel giardino

1529
GARDEN CENTRE
fr grainier m
ne zaadwinkel
de Samengeschäft n
da frøforretning
sv fröhandel
es tienda f de semillas
it negozio m di semi

1530
GARDEN CHAFER;
BRACKEN CLOCK
fr hanneton m horticole
ne rozekever
de Gartenlaubkäfer m
da gåsebille
sv trädgårdsborre
es abejorro m
it carruga f degli orti
   Phyllopertha horticola L.

1531
GARDEN DESIGN;
GARDEN LAY OUT
fr plantation f
ne tuinaanleg
de Anlage f
da anlæg n
sv anläggning
es plantación f;
   construcción f
it disegno m

1532
GARDENER
fr jardinier m
ne tuinman; hovenier
de Gärtner m
da anlægsgartner ;
   havemand
sv trädgårdsanläggare
es jardinero m; hortelano m
it giardiniere

1533
GARDEN HOSE
fr tuyau m d'arrosage
ne tuinslang
de Gartenschlauch m
da haveslange
sv trädgårdsslang
es manga f de riego
it tubo m per annaffiare

GARDEN LAY OUT,
1531

1534
GARDEN LINE;
PLANTING LINE
fr cordeau m
ne pootlijn
de Pflanzleine f
da plantesnor
sv planteringslina
es cordél m
it corda f da piantare

1535
GARDEN MOULD;
BLACK EARTH
fr terreau m
ne teelaarde; tuinaarde
de Gartenerde f
da agerjord
sv trädgårdsjord; matjord
es tierra f de cultivo
it terra f vegetale

1536
GARDEN PEA;
GREEN PEA
fr pois m à écosser
ne doperwt
de Schalerbse f; Palerbse f
da skalært
sv spritärt
es guisante m verde
it pisello m dolce
   Pisum sativum L.

1537
GARDEN REFUSE
fr détritus m de jardin
ne tuinafval n
de Gartenabfälle m pl
da haveaffald n
sv trädgårdsavfall n

es detritus m de jardín
it cascame m di giardino

1538
GARDEN SWIFT MOTH;
COMMON SWIFT
fr hépialidé m
ne wortelboorder
de Wurzelspinner m
da konvalæder
sv konvaljrotätare;
   lerfärgad rotfjäril
es barreno m de la raíz de la
   peonía
it epialidi f
   Hepialus lupulinus L.;
   H. lupulina (US)

1539
GARDEN TROWEL
fr déplantoir m
ne planteschopje n
de Pflanzkelle f
da planteskovl
sv planteringsspade
es trasplantador m
it zappetta f da piantare

1540
GARLIC
fr ail m
ne knoflook n
de Knoblauch m
da hvidløg n
sv vitlök
es ajo m
it aglio m
   Allium sativum L.

GAS, 1544

1541
GAS BURNER
fr brûleur m à gaz
ne gasbrander
de Gasbrenner m
da gasfyr n
sv gasbrännare
es quemador m de gas
it bruciatore m a gas

1542
GAS DAMAGE
fr dommage m par gaz
ne gasschade

de Gasschäden *m*
da gasskade
sv gasskada
es daño *m* de gas
it danno *m* da gas

1543
GASEOUS DISCHARGE
LAMP
fr lampe *f* à décharge
ne gasontladingslamp
de Gasentladungslampe *f*
da gasudladningslampe
sv urladdningslampa;
gasurladdningslampa
es lámpara *f* de descarga
it lampada *f* a scarica nei gas

1544
GASEOUS FUEL; GAS
fr combustible *m* gazeux
ne gasvormige brandstof
de gasförmiger Brennstoff *m*
da gasformig brændstof *n*
sv gasformigt bränsle *n*
es combustible *m* gaseoso
it combustibile *m* gassoso; gas *m*

1545
GAS MASK
fr masque *m* à gaz
ne gasmasker *n*
de Gasmaske *f*
da gasmaske
sv gasmask
es máscara *f* de gas; careta *f*
it maschera *f* antigas

1546
GAS POISONING
fr intoxication *f* par gaz
ne gasvergiftiging
de Gasvergiftung *f*
da gasforgiftning
sv gasförgiftning
es asfixia *f*
it attossicamento *m* da gas

GAS STORAGE, 823

1547
GAS STORAGE ROOM
fr chambre *f* à atmosphère contrôlée
ne gascel
de Gaslagerraum *m*
da gaslagerrum *n*; kulsyrerum *n*
sv kolsyrelagerrum
es cámara *f* de atmósfera controlada
it cabina *f* da gas

GATE VALVE, 3437

1548
GATHERING; PICKING
fr cueillette *f*
ne pluk
de Ernte *f*; Pflücken *n*
da plukning; høst
sv plockning
es recolección *f*; cosecha *f*
it raccolta *f*

1549
GAUZE; WIRE NETTING
fr grillage *m*; toile *f* métallique
ne gaas *n*
de Drahtgeflecht *n*
da trådvær *n*; hønsetrad
sv ståldrådsnät *n*
es tela *f* métalica
it rete *f* metallica

GEAN, 4158

1550
'GEEST SOIL'
fr lande *f* sablonneuse
ne geestgrond
de Geestboden *m*
da geest
sv 'geest' jord
es mezcla *f* de turbera y arena
it terreno *m* sabbioso a piede dei dune

1551
GENE
fr gène *m*
ne gen

de Gen *n*
da gen *n*
sv gen *n*
es gen *m*; factor *m* interno
it geno *m*

1552
GENERATION
fr génération *f*
ne generatie
de Generation *f*
da generation
sv generation
es generación *f*
it generazione *f*

1553
GENERATIVE
fr par réproduction sexuée
ne generatief
de generativ
da generativ
sv generativ
es generativo
it gamico

1554
GENERATIVE
PROPAGATION
fr multiplication *f* sexuée
ne geslachtelijke vermenigvuldiging
de geschlechtliche Vermehrung *f*
da kønnet formering; frø formering
sv könlig förökning; förökning med frö
es multiplicación *f* sexual
it moltiplicazione *f* sessuale

GENERATOR, 47

1555
GENERIC NAME
fr nom *m* générique
ne geslachtsnaam
de Gattungsname *m*
da slægtsnavn *n*
sv släktnamn *n*
es nombre *m* generífico
it nome *m* generico

1556
GENISTA;
COMMON BROOM;
GORSE
fr genêt *m*
ne heidebrem
de Stechginster *m*;
    Besenginster *m*
da visse
sv ginst
es retama *f* de los montes
it ginestra *f* di brughiera
    Genista L.

1557
GENOTYPE
fr génotype *m*
ne genotype *n*
de Genotypus *m*
da anlægspræg; genotype
sv genotyp; anlagstyp
es genotipo *m*
it genotipo *m*

1558
GENTIAN
fr gentiane *f*
ne gentiaan
de Enzian *m*
da gensian
sv Gentiana
es genciana *f*
it genziana *f*
    Gentiana L.

1559
GENUS
fr genre *m*
ne geslacht *n*
de Gattung *f*
da slægt
sv släkt; släkte
es género *m*
it genere *m*

1560
GEOLOGICAL
fr géoligique
ne geologisch
de geologisch
da geologisk
sv géologisk
es geológico
it geologico

1561
GERANIUM
fr Geranium *m*
ne ooievaarsbek
de Storchschnabel *m*
da storkenæb *n*
sv näva
es geranio *m*;
    hierba *f* de San Roberto
it geranio *m*
    Geranium L.

1562
GERM
fr germe *m*
ne kiem
de Keim *m*
da kim
sv grodd
es germen *m*
it germe *m*

1563
GERMAN WASP
fr guêpe *f*
ne wesp
de gemeine Wespe *f*
da hveps
sv vanlig geting
es avispa *f*
it vespa *f*
    Paravespula spp.

1564
GERMINATE, TO
fr germer
ne ontkiemen
de keimen
da spire
sv gro; spira
es germinar
it germinare

1565
GERMINATING POWER;
VIABILITY
fr énergie *f* germinative;
    capacité germinative
ne kiemkracht
de Keimfähigkeit *f*;
    Keimkraft *f*
da spireevne
sv grobarhet;
    groningsförmåga

es poder *m* germinativo
it forza *f* germinativa

1566
GERMINATION
fr germination *f*
ne kieming
de Keimung *f*
da spiring
sv groning
es germinación *f*
it germinazione *f*

GERMINATION,
polen, 2823
—, rate of, 3041

1567
GERMINATION
APPARATUS
fr germoir *m*
ne kiemtoestel *n*
de Keimapparat *m*
da spireapparat *n*
sv groningsapparat *n*
es aparato *m* de germinación
it apparecchio *m* da
    germinare

1568
GERMINATION
CAPACITY
fr faculté *f* germinative
ne kiemvermogen *n*
de Keimfähigkeit *f*
da spireevne
sv grobarhet
es facultad *f* germinativa
it capacità *f* germinativa

1569
GET SLIMY, TO
fr devenir visqueux;
    devenir gluant
ne verslijmen
de verschleimen
da slime
sv uppslamma
es formación *f* de mucílago
it diventare mucillagginoso

GEUM, 199

GHERKIN, 3133

1570
GIANT REED;
GREAT REED
fr canne f de Provence
ne pijlriet n; reuzenriet n
de Pfahlrohr n; italienisches
   Rohr n
da kæmperør n
sv jättevass; italienskt rör
es caña f común
it canna f palustre;
   canna f sagittata
   Arundo donax L.

1571
GIBBERELLIC ACID
fr gibberelline f
ne gibberelline
de Gibberellin n
da gibberellin
sv gibberellin
es ácido m giberélico
it gibberilline

GILL, 2023
—, hard, 1724

1572
GIPSY MOTH
fr bombyx m disparate;
   spongieuse f
ne plakker
de Schwammspinner m
da løvskovsnonne
sv lövskogsnunna
es lagarta f peluda de las
   encinas
it bombice m dispari;
   limantria f
   Limantria dispar L.;
   Porthetria d. (US)

1573
GLABROUS
fr glabre
ne kaal
de kahl
da nøgen
sv hårlös; glatt
es glabro
it glabro

1574
GLADIOLUS
fr glaïeul m
ne gladiool
de Siegwurz f; Gladiole f
da jomfrufinger; Gladiolus
sv Gladiolus; sabellilja
es gladiolo m
it giagiolo m
   Gladiolus L.

1575
GLAND
fr glande f
ne klier
de Drüse f
da kirtel
sv körtel
es glándula f
it glandola f

GLAND, nectar, 2456

1576
GLANDULAR
fr glanduleux
ne klierachtig
de drüsig
da kirtelagtig
sv körtelaktig
es glanduloso
it glandoloso

1577
GLANDULAR HAIR
fr poil m sécréteur
ne klierhaar n
de Drüsenhaar n
da kirtelhår n
sv körtelhår n
es pelo m glandular
it pelo m glandolare

1578
GLASS FIBRE
fr fibre f de verre
ne glasvezel
de Glasfaser f
da glasfiber
sv glasfiber
es fibra f de vidrio
it fibra f di vetro

1579
GLASSHOUSE
fr serre f
ne kas
de Gewächshaus n
da væksthus n; drivhus n
sv växthus n
es invernadero m
it serra f

GLASSHOUSE, cool, 3802
—, heated, 1835
—, lean-to, 2118
—, mobile, 2361
—, multispan, 2427
—, span roofed, 3532
—, temperate, 3802
—, tower, 3892

1580
GLASSHOUSE BENCH
fr tablette f
ne kasbedding
de Vermehrungsbeet n
   (unter Hochglass)
da væksthusbed n
sv växthusbädd
es mesa f para plantas
it tavoletta f della serra

1581
GLASSHOUSE
CULTIVATION;
GLASSHOUSE CROP
fr culture f de serre
ne kascultuur
de Gewächshauskultur f
da væksthuskultur;
   drivhuskultur
sv växthuskultur
es cultivo m de invernadero
it coltura f da serra

1582
GLASSHOUSE CULTURE
fr culture f sous verre
ne glasteelt
de Unterglaskultur f
da drivhuskultur
sv växthuskultur
es cultivo m en invernadero
it coltura f in serra

1583
GLASSHOUSE
FOREMAN
fr chef *m* de multiplication
    (en pépinière)
ne kasbaas
de Gewächshausvormann *m*
da blokformand
sv växthusförman
es encargado *m* de inverna-
    dero
it capo-mastro *m* di serra

1584
GLASSHOUSE NURSERY
fr exploitation miroitante *f*;
    exploitation de serre
ne glasbedrijf *n*
de Hochglasbetrieb *m*
da væksthusgartneri *n*
sv växthusodling
es establecimiento (*m*)
    hortícola bajo cristal
it azienda *f* con colture
    sotto vetro

1585
GLASSHOUSE PLANT
fr plante *f* de serre
ne kasplant
de Gewächshauspflanze *f*
da drivhusplante
sv växthusväxt
es planta *f* de invernadero
it pianta *f* di serra

1586
GLASSHOUSE POTATO
APHID
fr puceron *m* de la pomme
    de terre
ne boterbloemluis
de Fingerhutblattlaus *f*
da kartoffelbladlus
sv potatisbladlus
es pulgón *m* verde de la
    patata
it afide *m* della patata
    Aulacorthum solani Kltb.

1587
GLASSHOUSE SOIL
fr terres *f pl* de serres
ne kasgrond
de Gewächshausboden *m*

da væksthusjord
sv växthusjord
es tierra *f* de invernadero
it terra *f* di serra

1588
GLASSHOUSE THRIPS
fr thrips *m* des serres
ne kastrips
de schwarzer Gewächs-
    hausblasenfuss *m*
da væksthusthrips
sv svart växthustrips
es piojillo *m* de los inverna-
    deros
it tripide *m* emorroidale
    della serre
    Heliothrips haemorrhoi-
    dalis Bouché

1589
GLASSHOUSE TYPE
fr type *m* de serre
ne kastype *n*
de Gewächshaustyp *m*
da hustype; væksthustype
sv växthustyp
es tipo *m* de invernadero
it tipo *m* di serra

1590
GLASSHOUSE
WITHEFLY;
GREENHOUSE
WHITEFLY (US)
fr mouche *f* blanche;
    aleurode *f* des serres
ne witte vlieg
de weisse Fliege *f*
da hvid flue
sv vita flygare;
    växthusmjöllus
es mosca *f* blanca;
    aleurode *f* de la col
it aleurode *f* del cavolo;
    mosca *f* bianca delle serre
    Trialeurodes vaporariorum
    Westw.

1591
GLASS POLLUTION
fr salissement *m* du verre;
    salissure *f* du verre
    encrassement *m* du verre
ne glasvervuiling

de Glasverschmutzung *f*
da glasforurening;
    tilsmudsning
sv nedsmutsning av glas
es ensuciado *m* del cristal
it insudiciamento *m* del
    vetro

1592
GLASS REBATE
fr feuillure *f* à verre
ne sponning
de Kittfalz *m*
da kitfals
sv kittfals
es ranura *f* para poner
    mastic
it incastro *m* per mastica

1593
GLASS WOOL
fr laine *f* de verre
ne glaswol
de Glaswolle *f*
da glasuld
sv glasull
es lana *f* de vidrio
it lana *f* di vetro

1594
GLAZE, TO
fr vitrer
ne beglazen
de verglasen
da glasere; sætte ruder i
sv sätta i glas (glaza)
es acristalar
it copriro da vetri

1595
GLAZING
fr vitrage *m*
ne beglazing
de Verglasung *f*
da glasilægning
sv glasning
es poner vidrios
it montaggio *m* di vetri;
    vetratora *f*

1596
GLAZING BAR
fr barre *f* à vitrage;
    petit bois *m*
ne glasroede

de Sprosse *f*
da sprosse
sv spröjs
es barra *f* para poner vidrios
it intelaiatura *f* vetrata

1597
GLOBE ARTICHOKE
fr artischaut *m*
ne artisjok
de Artischocke *f*
da artiskok
sv kronärtskocka
es alcachofa *f*
it carciofo *m*
  Cynara scolymus Pers.

1598
GLOBE FLOWER
fr trolle *m*
ne Trollius; globebloem
de Trollblume *f*
da engblomme
sv daldocka; smörboll
es calderones *m pl*
it trollo *m*
  Trollius L.

1599
GLOBE THISTLE
fr échinops *m*
ne kogeldistel
de Kugeldistel *f*
da kugletidsel
sv bolltistel
es cardo *m*; yesquero *m*
it cardo *m* globoso
  Echinops L.

1600
GLOBE VALVE
fr soupape *f* droite
ne klepafsluiter
de Durchgangsventil *n*
da gennemløbsventil
sv klaffventil
es válvula *f* derecha
it valvola *f* a globo

1601
GLORY-OF-SNOW
fr Chionodoxa *m*
ne sneeuwroem
de Schneeglanz *m*

da snepryd
sv vårstjärna
es gloria *f* de la nieve
it Chionodoxa *f*
  Chionodoxa

1602
GLOXINIA
fr gloxinia *m*
ne gloxinia
de Gloxinie *f*
da Sinningia; gloxinia
sv gloxinia
es gloxinia *f*
it glossinia *f*
  Sinningia speciosa
  Benth. et Hook

1603
GLUME
fr glume *f*
ne kafje *n*
de Spelze *f*
da avne
sv agn
es gluma *f*
it loppa *f*

1604
GOAT MOTH
fr cossus *m* gâtebois
ne wilgehoutrups;
  houtrups
de Weidenbohrer *m*
da træboren; pileborer
sv träfjäril; träddödare
es coso *m*; taladro *m*
it perdilegno *m* rosso;
  rodilegno *m* rosso
  Cossus cossus L.

GOLD BAND LILY, 1606

1605
GOLDEN CURRANT
fr groseillier *m* doré
ne gele alpenbes
de Gold-Johannisbeere *f*
da guldribs
sv gullrips
es grosellero *m* dorado
it Ribes *m* aureum
  Ribes aureum Pursh.

1606
GOLDEN RAYED LILY;
GOLD BAND LILY (US)
fr lis *m* à bandes dorés
ne goudbandlelie
de Goldbandlilie *f*
da guldbåndslilje
sv guldbandslilja
es azucena *f* aúrea
it giglio *m* aureo
  Lilium auratum Lindl.

1607
GOLDEN ROD
fr verge *f* d'or
ne guldenroede
de Goldrute *f*
da gyldenris *n*
sv gullris *n*
es vara *m* de oro; Solidago *m*
it Solidago *m*
  Solidago L.

1608
GOLDEN WILLOW
fr osier *m*
ne bindwilg
de Dotterweide *f*
da guldpil
sv gulpil
es sauce *m* cabruno
it salice *f* bianca
  Salix alba 'Vitellina' L.
  Stokes

GOLD-TAILED MOTH,
1654, 4230

1609
GOOSEBERRY
fr groseille *f* à maquereau
ne kruisbes
de Stachelbeere *f*
da stikkelsbær
sv krusbär *n*
es grosella *f* espinosa
it uva *f* spina

GOOSEBERRY, Cape,
562

1610
GOOSEBERRY BRYOBIA
fr araignée *f* rouge du
  groseillier
ne kruisbessespintmijt

de rote Stachelbeermilbe *f*
da stikkelsbærmide
sv krusbärskvalster *n*
es ácaro *m* de la grosella
   espinosa
it briobia *f* del ribes
   Bryobia ribis Thom.

1611
GOOSEBERRY BUSH
fr groseillier *m* à maquereau
ne kruisbessestruik
de Stachelbeerstrauch *m*
da stikkelsbærbusk
sv krusbärsbuske
es grosellero *m* espinoso
it arbusto *m* d'uva spina
   Ribes uva-crispa L.

GOOSEGRASS, 700

GORSE, 1509, 1556

1612
GOURD; MARROW
fr courge *f*; citrouille *f*;
   patisson *m*
ne pompoen
de Kürbis *m*
da mandelgræskar *n*
sv matpumpa
es calabaza *f* común;
   calabacín *m*
it zucca *f*
   Cucurbita pepo L.

GOUTWEED, 1668

1613
GRADE (QUALITY)
fr catégorie *f*
ne kwaliteitsklasse
de Güteklasse *f*
da kvalitet
sv kvalitetsklass
es categoría *f*
it grado *m* qualitativo

GRADE, TO, 3513

1614
GRADER; SORTER;
SIZER
fr trieuse *f*; calibreuse *f*
ne sorteermachine

de Sortiermaschine *f*;
   Kalibriermaschine *f*
da sorteremaskine
sv sorteringsmaskin
es máquina *f* calibradora;
   máquina *f* clasificadora
it calibratrice *f*;
   selezionatrice *f*

1615
GRADING
fr triage *m*
ne sortering
de Sortierung *f*
da sortering
sv sortering
es elección *f*; clasificación *f*
it graduamento *m*

1616
GRADING
REGULATION
fr règles *fpl* de normalisation
ne sorteringsvoorschrift *n*
de Sortierungsvorschrift *f*
da sorteringsregler
sv sorteringsregler
es reglas *f pl* de normalización
it regole *f pl* di normalizzazione

1617
GRADING ROOM
fr local *m* de triage
ne sorteerruimte
de Sortierraum *m*
da sorteringsrum *n*
sv sorteringsrum *n*
es lugar *m* de clasificación;
   lonja *m* de clasificación
it luogo *m* d'assortimento

1618
GRADING TO QUALITY
fr triage *m* sélectionné
ne kwaliteitssortering
de Gütesortierung *f*
da kvalitetssortering
sv sortering efter kvalitet
es clasificación *f*
it calibratrice *f* in base alla
   qualità

1619
GRADING TO WEIGHT
fr calibrage *m* par le poids
ne gewichtssortering
de Gewichtssortierung *f*
da vægtsortering
sv sortering efter vikt
es calibrado *m* por el peso
it calibratrice *f* in base al
   peso

1620
GRAFT
fr greffon *m*
ne ent; zetling
de Edelreis *n*; Pfropfreis *n*;
   Propfen *n*
da podekvist; ædelris *n*
sv ädelris *n*; ympkvist
es injerto *m*
it innesto *m*

1621
GRAFT, TO
fr greffer; enter
ne enten
de pfropfen
da pode
sv ympa
es injertar
it innestare

1622
GRAFT HYBRID
fr hybride *m* de greffe;
   chimère *f*
ne entbastaard
de Pfropfbastard *m*
da podningsbastard
sv ympbastard
es hibrido *m* de injerto;
   hibrido *m* vegetativo
it ibrido *m* d'innesto

1623
GRAFTING
fr greffage *m*
ne enting
de Propfung *f*
da podning
sv ympning
es injertación *f*
it innesto *m*

GRAFTING, cleft, 702
—, crown, 913
—, wedge, 702

1624
GRAFTING
CHLOROSIS
fr chlorose *f* cicatrique
ne entchlorose
de Pfropfchlorose *f*
da podningsklorose
sv ympkloros
es clorosis *f* de los injertos
it clorosi *f* d'innesto

1625
GRAFTING KNIFE
fr greffoir *m*
ne entmes *n*; zetmes *n*
de Pfropfmesser *n*
da podekniv
sv ympkniv
es cuchillo *m* de injertar;
   injertador *m*
it coltello *m* da innestare

1626
GRAFTING WAX
fr mastic *m* à greffer
ne entwas
de Baumwachs *n*
da podenvoks *n*
sv ympvax *n*
es betún *m* de injertar;
it cera *f* da innesto

1627
GRAFT UNION
fr zone *f* génératrice
ne entplaats
de Pfropfungsstelle *f*
da podested *n*
sv förädlingsställe *n*
es unión *m* del injerto
it luogo *m* d'innesto

1628
GRAIN
fr grain *m*
ne korrel
de Korn *n*
da korn *n*
sv korn
es grano *m*
it granello *m*

GRAIN, pollen, 2824

1629
GRAIN SIZE
fr diamètre *m* du grain
ne korrelgrootte
de Korngrösse *f*
da kornstørrelse
sv kornstorlek
es tamaño *m* del grano
it grossezza *f* del granello

1630
GRAIN SPAWN
(MUSHROOM
GROWING)
fr blanc *m* sur grain
ne korrelbroed *n*
de Körnerbrut *f*
da kornmycelium
sv kornmycelium
es semilla *f* de grano;
   semilla *f* granulada
it bianco *m* su grano

GRAIN STRUCTURE,
single, 3421

1631
GRANULAR
fr granuleux
ne korrelvormig; korrelig
de kornförmig; körnig
da kornet; granuleret
sv granulerad
es granuloso
it granuloso

1632
GRANULATE
fr granule *m*
ne granulaat *n*
de Granulat *n*
da granulat *n*
sv granulat *n*
es granulado *m*
it granulato *m*

1633
GRANULATE, TO
fr grener; granuler
ne korrelen
de körnen; granulieren
da granulere

sv korna
es granular
it granulare

1634
GRANULATED PEAT
fr poussière *f* de tourbe
ne molm
de Torfmull *n*
da tørvesmuld
sv torvmull
es turba *f*; cama *f* de turba
it polvere *f* di torba

1635
GRAPE
fr raisin *m*; grain *m* de raisin
ne druif
de Traube *f*
da drue
sv druva; vindruva
es uva *f*; grano *m*
it uva *f*

1636
GRAPE FRUIT
fr pamplemousse *m*
ne pompelmoes
de Pampelmuse *f*
da pompelmus
sv pompelmus; grapefrukt
es pomelo *m*
it pampelino *m*
   Citrus paradisi Macf.

1637
GRAPE HOUSE
fr serre *f* à vigne
ne druivenkas
de Traubenhaus *n*
da 'vinhus';
   drivhus til vindyrkning
sv vinhus *n*; vinkast
es estufa *f* para vides
it serra *f* da vite

1638
GRAPE HYACINTH
fr jacinthe *f* musquée;
   Muscari *m*
ne druifhyacint;
   blauwe druifjes
de Traubenhyazinthe *f*

da perlehyacinth;
   druehyacinth
sv pärlhyacint
es Muscari *m*;
   jacinto *m* racimoso
it Muscari *m*;
   giacinto *m* a grappolo
   Muscari Mill.

1639
GRAPE JUICE
fr jus *m* de raisin
ne druivesap *n*
de Traubensaft *m*
da druesaft
sv druvsaft
es zumo *m* de uvas
it succo *m* dell' uva

1640
GRAPE PHYLLOXERA
fr phylloxéra *m*
ne druifluis
de Reblaus *f*
da vinlus
sv vinlus
es filoxera *f*
it fillossera *f*
   Viteus vitifolii Fitch

1641
GRASS
fr gramen *m*; herbe *f*
ne gras *n*
de Gras *n*
da græs *n*
sv gräs *n*
es hierba *f*; gramínea *f*
it graminaceo *m*
   Graminae

GRASS, barnyard,   253
—, brome,   440
—, couch,   867
—, creeping bent-,   885
—, pampas,   2593
—, quack,   867
—, sweet scented
vernal,   3769
—, twitch, 867
—, whitlow,   4151
see also: meadowgrass

1642
GRASS STALK
fr brin *m* d'herbe
ne grasspriet
de Grashalm *m*
da græsstrå *n*
sv grässtrå *n*
es tallo *m* herbáceo
it erbicina *f*

1643
GRASS STRIP
fr bande *f* enherbée
ne grasstrook
de Grasstreifen *m*
da græsstribe
sv gräsremsa
es banda *f* de pasto
it striscia *f* inerbita;
   striscia *f* d'erba

1644
GRATE BAR
fr barre *f* de grille
ne roosterijzer *n*
de Roststab *m*
da ristgitter *n*
sv roststav
es barra *f* de grilla
it barra *f*

1645
GRATER; RASP
fr râpe *f*
ne rasp
de Raspe *f*
da rasp
sv rasp
es rallo *m*; rallador *m*
it grattugia *f*

1646
GRATING
fr grille *f*
ne rooster *n*
de Rost *m*
da rist
sv rost
es grilla *f*
it reticolo *m*

1647
GRAVEL
fr gravier *m*
ne grint *n*

de Kies *m*
da grus *n*
sv grus *n*
es grava *f*
it ghiaia *f*

1648
GRAVEL CULTURE;
SOILLESS CULTURE;
HYDROPHONICS
fr culture *f* sans sol;
   culture *f* sur gravier
ne grintcultuur;
   teelt zonder aarde
de Kieskultur *f*;
   Hydrokultur *f*; erdlose
   Kultur *f*
da gruskultur; vandkultur
sv gruskultur; hydrokultur;
   jordfri odling
es cultivo *m* sobre grava;
   cultivo *m* sin suelo;
   hidropónicos *m pl*
it Coltivazione *f* senza
   terra; idrocoltura *f*

1649
GRAVEL SOIL
fr cailloutis *m*
ne grintbodem
de Kiesboden *m*
da grusjord
sv grusjord
es suelo *m* de grava
it suolo *m* ghiaioso

GREAT REED,   1570

1650
GREEN APPLE APHID;
APPLE APHID (US)
fr puceron *m* vert du
   pommier
ne groene appeltakluis
de grüne Apfelblattlaus *f*
da grønne æblebladlus
sv grön äppelbladlus
es pulgón *m* verde de las
   ramas del manzano
it afide *m* verde del melo
   Aphis pomi de G.

1651
GREENBACK
fr de tige vert-foncé
ne groenkraag
de Grünkragen *m*
da grønskjold *n* (-nakke)
sv grönsjuka
es cuello *m* verde
it a collo *m* verde

GREENHOUSE
WHITEFLY, 1590

1652
GREEN MANURE CROP
fr culture *f* à engrais vert
ne groenbemestingsgewas *n*
de Gründüngungspflanze *f*
da grøngødningsafgrøder
sv gröngödslingsväxt
es cultivo *m* de abono en
verde; forraje *f*
it pianta *f* dessovescio

1653
GREEN MANURING
fr engrais *m* vert
ne groenbemesting
de Gründüngung *f*
da grøngødning
sv gröngödsling
es abono *m* verde
it sovescio *m*

1654
GREEN OAK TORTRIX
MOTH;
GOLD-TAILED MOTH
fr tordeuse *f* verte du chêne
ne eikebladroller
de grüner Eichenwickler *m*
da egevikler
sv ekvecklare
es brugo *m*; lagarta *f*
pequeña de la encina
it tortrice *f* verde delle
querce; portesia simile;
bombice dal ventro
dorato
Tortrix viridana L.

GREEN PEA, 1536

GREEN PEACH APHID,
2625

1655
GREEN-VEINED
WHITE BUTTERFLY
fr piéride *f* du navet
ne geaderd *n* koolwitje
de Rapsweissling *m*
da (grønårede)
kålsommerfugl
sv rapsfjäril
es oruga *f* de la col
it navoncella *f*;
pieride *f* del navone
Pieris napi L.

1556
GREY BULB ROT
(TULIPA)
fr pourriture *f* grise;
pourriture à Sclerotium
ne kwadegrondziekte;
wegblijver
de Sklerotium-Krankheit *f*
Zwiebelrotgraufäule *f*
da tulipan rodfiltsvamp
sv blomsterlökröta
es podredumbre *f* gris
esclerocio *m*
it putrefazione *f* grigia
mal *m* dello sclerozio
Sclerotium tuliparum
Kleb.

GREY GARDEN
SLUG, 1335

1657
GREY MOULD;
BOTRYTIS
fr pourriture *f* grise
ne grauwe schimmel
de Grauschimmel *m*
da gråskimmel
sv gråmögel
es podredumbre *f*;
moho *m* gris
it marcio *m* grigio
Botrytis spp.

1658
GREY MOULD;
DAMPING OFF
fr pourriture *f* grise
ne grauwe schimmel;
smeul; kiemschimmel

de Grauschimmel *m*
da gråskimmel
sv gråmögel
es mal *m* del esclerocio
moho *m* gris
it muffa *f* grigia
Botrytis spp.

1659
GREY MOULD;
NECK ROT (ONION)
fr pourriture *f* grise
ne grauwe schimmel
de Halsfäule *f*
da løg-gråskimmel
sv gråmögel
es mal *m* del exclerocio
it marciume *m*
Botrytis allii Munn.

1660
GREY MOULD (ABIES)
fr Botrytis *m*
ne treurziekte
de Grauschimmel *m*
da gråskimmel
sv gråmögel
es moho *m* gris
it marciume *m* del fusto
Botrytis cinerea Fr.

1661
GREY MOULD
(CEDRUS)
fr Botrytis *m*
ne ruiziekte
de Graufäule *f*
da gråskimmel
sv gråmögel
es moho *f* gris
it muffa *f* grigia
Botrytis cinerea Fr.

1662
GREY MOULD (VINE)
fr pourriture *f* grise
ne meiziekte
de 'Maikrankheit' *f*;
Botrytis-Fäule *f*
da majsyge
sv gråmögel
es podredumbre *f* gris
it muffa *f* grigia
Botrytis cinerea Fr.

1663
GREY POPLAR
fr peuplier *m* grisaille
ne grauwe abeel
de Grauappel *f*
da gråpoppel
sv gråpoppel
es álamo *m* negro
it pioppo *m* grigio
   Populus canescens
   (Ait.) Sm.

1664
GREY THYME
fr serpolet *m*
ne gewone tijm
de Thymian *m*
da smalbladet timian
sv backtimjan
es tomillo *m*; serdol *m*
it timo *m* comune
   Thymus serpyllum L.

1665
GROOVE
fr rainure *f*
ne groef
de Nut *f*
da not; fals
sv fals
es ranura *f*; encaje *m*
it scanalatura *f*

1666
GROSS WEIGHT
fr poids *m* brut
ne brutogewicht *n*
de Brutto-Gewicht *m*
da bruttovægt
sv bruttovikt
es peso *m* bruto
it peso *m* lordo

1667
GROSS YIELD
fr rendement *m* brut
ne bruto-opbrengst
de Rohertrag *m*;
   Betriebsertrag *m*
da bruttoudbytte *n*
sv intäkt (bruttoavkastning)
es producción *f* bruta
it produzione *f* lorda

1668
GROUND ELDER;
GOUTWEED;
BISHOP'S WEED
fr herbe-aux-goubleux *f*;
   herbe-de-Saint Gérard *f*
ne zevenblad *n*
de Zipperleinskraut *n*;
   Geissfuss *m*; Giersch *m*
da skvalderkål
sv kirskål
es hierba *f* de S. Gerardo
it castaldina *f*; erba *f*
   girarda; podagraria *f*
   Aegopodium podagraria L.

1669
GROUND IVY
fr lierre *m* terrestre
ne hondsdraf
de Gundelrebe *f*;
   Gundermann *m*
da korsknap
sv jordreva
es hiedra *f* terrestre
it edera *f* terrestre
   Glechoma hederacea L.

GROUND KEEPERS,
4050

1670
GROUND LEATHER
fr farine *f* de cuir
ne ledermeel *n*
de Ledermehl *n*
da lædermel *n*
sv lädermjöl *n*
es harina *f* de cuero
it cuoio *m* polverizzato

1671
GROUND LIMESTONE;
AGRICULTURAL
CHALK
fr chaux *f* carbonatée
   agricole
ne koolzure landbouwkalk
de Kalkmergel *m*
da jordbrugskalk
sv kalkstensmjöl
es carbonato *m* cálcico
it carbonato *m* di calce
   agricolo

1672
GROUND MAGNESIAN
LIME STONE
fr chaux *f* magnésienne en
   poudre
ne magnesiapoederkalk
de Magnesiumbranntkalk *m*
da magnesiumbræntkalk
sv dolomitmjöl *n*
es magnesio *m* de cal en
   polvo
it calce *f* magnesiaca
   polverizzata

1673
GROUNDSEL
fr petit séneçon *m*;
   séneçon *m* des oiseaux
ne kruiskruid *n*
de gemeines Greiskraut *n*;
   Kreuzkraut *n*
da almindelig brandbæger
sv korsört
es hierba *f* cana;
   casaruelos *m*
it senecio *m*;
   erba *f* cardellina
   Senecio vulgaris L.

1674
GROUNDWATER
fr eau *f* souterraine
ne grondwater *n*
de Grundwasser *n*
da grundvand *n*
sv grundvatten *n*
es agua *f* subterránea
it acqua *f* sotterranea

GROUNDWATER
LEVEL, 4099

1675
GROW, TO
fr croître; pousser
ne groeien
de wachsen
da vokse
sv växa; gro
es crecer
it crescere

1676
GROW, TO;
TO CULTIVATE
fr cultiver
ne telen; opkweken
de erzeugen; kultivieren
da tiltrække
sv odla; draga upp
es cultivar
it coltivare

1677
GROWER
fr maraîcher *m*; cultivateur
ne tuinder; teler
de Erzeuger *m*; Züchter *m*
da dyrker
sv odlare
es cultivador *m*
it coltivatore *m*

1678
GROWING LICENSE
fr législation culturale *f*;
légalisation *f* en matière
des emblavements;
droit *m* de culture
ne teeltrecht *n*
de Anbaurecht *n*;
Züchtungsrecht *n*
da dyrkningsrettighed;
dyrkningstilladelse
sv odlingsrättighet
es derecho *m* de cultivo
it autorizzazione *f* per la
coltura;
licenza *f* colturale

1679
GROWING METHOD;
WAY OF GROWING
fr façon *f* culturale;
méthode *f* de culture
ne teeltwijze
de Kulturweise *f*
da dyrkningsmetode
sv odlingsmetod
es modo *m* de cultivo
it metodo *m* di coltura

1680
GROWING PERIOD
fr période *f* de croissance
ne groeiperiode

de Wachstumszeit *f*
da vækstperiode;
vækstsæson
sv växtperiod
es período *m* de crecimiento
it periodo *m* vegetativo

1681
GROWING-POINT;
GROWTH TIP
fr point *m* de végétation
ne groeitop
de Terminalknospe *f*
da vækstpunkt *n*
sv tillväxtpunkt
es ápice *m*
it apice *m*

1682
GROWING POWER;
VITALITY
fr force *f* végétative;
faculté *f* végétative
ne groeikracht
de Wuchskraft *f*
da vækstkraft
sv växtkraft
es poder *m* vegetativo
it vigoria *f*;
forza *f* vegetativa

1683
GROWING ROOM
(MUSHROOM)
fr chambre *f* de culture
ne kweekcel
de Kulturraum *m*
da dyrkningsrum *n*
sv odlingsrum *n*
es sala *f* (local) de cultivo
it stanza *f* di coltura

1684
GROWING TECHNIQUE;
TECHNIQUE OF
CULTIVATION
fr technique *f* culturale
ne teelttechniek
de Kulturtechnik *f*;
Anbautechnik
da dyrkningsteknik
sv odlingsteknik
es técnica *f* del cultivo
it tecnica *f* colturale

1685
GROWTH
fr croissance *f*
ne groei
de Wachstum *n*
da vækst
sv tillväxt
es crecimiento *m*
it crescita *f*

1686
GROWTH CHAMBER
fr phytotron *m*
ne groeikamer
de Klimakammer *f*
da klimakammer *n*
sv klimatkammare
es fitotrón *m*
it fitotrone *m*

1687
GROWTH FACTOR
fr facteur *m* de croissance
ne groeifactor
de Wachstumsfaktor *m*
da vækstfaktor
sv växtfaktor
es factor *m* vegetativo
it fattore *m* della crescita

GROWTH HORMONE,
1690

1688
GROWTH INHIBITION
fr inhibition *f* de croissance;
freinage *f* de la croissance
ne groeiremming
de Wachstumshemmung *f*
da væksthæmning
sv tillväxthämning
es refreno *m* del crecimiento
it impedimento *m* della
crescenza

GROWTH INHIBITOR,
1691

1689
GROWTH REDUCTION
fr réduction *f* de croissance
ne groeiremming
de Wachstumshemmung *f*
da væksthæmning

sv tillväxthämning
es retardación f del
   crecimiento
it riduzione f di sviluppo

1690
GROWTH REGULATING
SUBSTANCE;
GROWTH HORMONE
fr substance f de croissance;
   auxine f
ne groeistof
de Wuchsstoff m; Hormon n
da vækststof n; roddanner
sv tillväxtämne n
es hormona f vegetal;
   regulador m de
   crecimiento
it hormone m

1691
GROWTH RETARDANT;
GROWTH INHIBITOR
fr retardant m
ne remstof
de Hemmstoff m
da vækstretarderende stof n
sv växthämmande medel n
es inhibidor m; retardante m
it hormone m d'impedi-
   mento

1692
GROWTH STIMULATION
fr excitation f de la
   croissance
ne groeibevordering
de Wachstumsforderung f
da vækstfremning
sv växtfrämjande;
   växtuppiggning
es estimulación f del crecimiento
it acceleramento m della
   crescita

GROWTH TIP, 1681

1693
GROW TOGETHER, TO
fr se souder
ne vegroeien
de verwachsen
da sammenvokse

sv sammanväxa; växa över
es cicatrizadar
it unitare

GRUB, 2229

GRUB, white, 640, 731

1694
GRUB, TO
fr fouiller
ne woelen
de wühlen
da kultivere
sv kultivera
es cavar; abrir galerías
it grufolare

1695
GRUBBER
fr cultivateur m
ne woeler
de Grubber m; Kräuel n
da undergrundsplov
sv kultivator
es subsolador m
it macchina f a grufolare

1696
GUANO
fr guano m
ne guano
de Guano m
da guano
sv guano
es guano m
it guano m

1697
GUARANTEE PRICE
fr prix m garanti
ne garantieprijs
de garantierter Mindest-
   preis m
da garanteret (mindste) pris
sv garantipris
es precio m garantizado
it prezzo m garantito

1698
GUMMOSIS (GHERKIN)
fr nuile f; cladosporiose f
ne vruchtvuur n
de Gurkenkrätze f;
   Braunflechigkeit f

da gummiflåd n
sv gurkfläckssjuka
es encendido m; gomosis f
it cladosporiosi f
   Cladosporium
   cucumerinum
   Ell. et Arth.

GUM SPOT DISEASE,
3387

1699
GUSSET PLATE
fr gousset m
ne schetsplaat
de Knotenblech n
da knudeplade
sv knutplåt
es cartela f;
   placa f de unión
it piastra f modale di testa

1700
GUTTER
fr gouttière f
ne goot
de Dachrinne f
da tagrende
sv takränna
es canal m
it grondaia f

1701
GYPSOPHILA
fr gypsophile f
ne gipskruid n
de Schleierkraut n
da gipsurt; brudeslør n
sv slöjblomma
rs gisofila f
it gipsofila f
   Gypsophila L.

1702
GYPSUM
fr plâtre m; gypse m
ne gips n
de Gips m
da gips
sv gips
es yeso m
it gesso m

H

1703
HABIT OF GROWTH;
MANNER OF GROWTH
fr croissance *f*;
type *m* de croissance
ne groeiwijze
de Wuchs *m*
da vækstmåde (vækstform)
sv växtsätt *n*; växtform
es hábito *m* de crecimiento
it modo *m* vegetativo

1704
HACKLING MACHINE
fr machine *f* à sérancer
ne hekelmachine
de Hechelmaschine *f*
da heglemaskine
sv häckla
es máquina *f* de peinar
it scapecchiatoio *m*

1705
HAIR ROOT
fr radicelle *f*; chevelu *m*
ne haarwortel
de Haarwurzel *f*
da hårrod
sv hårrot
es pelo *m* radical
it barbolina *f*

1706
HAIRY; PILOSE;
HIRSUTE
fr poilu; pileux; pubescent
ne behaard
de behaart
da håret
sv hårig
es velloso
it lanuginoso

1707
HAIRY BITTERCRESS
fr cardamine *f* velue;
cardamine *f* hérissée
ne kleine veldkers
de behaartes Schaumkraut *n*

da roset-karse
sv bergbräsma
es mastuerzo *m*
it billeri *f*
Cardamine hirsuta L.

1708
HAIRY TARE
fr vesce *f* hérissée,
vesce *f* hirsutée
ne ringelwikke
de rauhaarige Wicke *f*
da tofrøet vikke
sv duvvicker
es veza *f* hirsuta
it tentennino *m*
Vicia hirsuta L. S.F.
Gray

1709
HAIRY VETCH
fr vesce *f* velue
ne zandwikke
de Sandwicke *f*
da sandvikke
sv luddvicker
es veza *f* vellosa
it veccia *f* villosa
Vicia villosa Roth.

1710
HALF STANDARD
fr arbre *m* demi-tige;
demi-tige *m*
ne halfstam
de Halbstamm *m*
da halvstammet
sv stam
es árbol *m* a medio viento
it albero *m* a mezzo fusto

1711
HALF TURN PLOUGH
fr charrue *f* reversible demi-
tour
ne aanbouwwentelploeg
de Anbaudrehpflug *m*
da ophængt vendeplov
sv påhängsväxelplog för
½ varvs viidning

es arado *m* reversible
½ vuelta
it aratro *m* portato a ½ di giro

1712
HALO BLIGHT
(OF BEAN)
fr graisse *f* du haricot
ne vetvlekkenziekte
de Fettfleckenkrankheit *f*
da fedtpletsyge
sv fettfläcksjuka
es enfermedad *f* de manchas
de la grasa
it batteriosi *f* del fagiolo
Pseudomonas phaseolicola
(Burkh.) Dowson

1713
HAMMERED GLASS
fr verre *m* martelé
ne gehamerd glas *n*
de Klarglas *n*
da råglas *n*
sv råglas *n*
es vidrio *m* amartillado
it vetro *m* picchiato

1714
HAMMERING BENCH
fr outil *m* à battre
ne haarapparaat *n*
de Dengelapparat *m*
da bankeapparat
sv hamringsapparat
es útil *m* para martillar la
hoz
it affilacoltelli *m*

1715
HAMPER
fr corbeille *f* d'osier
ne tenen mand
de Weidenkorb *m*
da kurv
sv korg
es cesto *m* de mimbre
it cesta *f* di vinco

1716
HAND DREDGE
fr drague *f* à bras
ne baggerbeugel
de Baggernetz *n*
da muddernet *n*
sv mudderbygel
es draga *f* de mano
it draga *f* a mano

HAND FORK TRUCK,
2587

1717
HANDLING AND
PACKING
fr conditionnement *m*
ne behandeling en
   verpakking
de Aufbereitung *f* und
   Verpackung *f*
da tilberedning og pakning
sv behandling och packning
es acondicionamiento *m*;
   preparación *f*
it lavorazione *f*

1718
HANDPALLET TRUCK
fr élévateur *m* à palettes
ne handhefwagen
de Gabelstapler *m*
da håndløftevogn
sv gaffelvagn (handdragen)
es elevador *m* de paletas
it elevatore *m* a palette

1719
HAND SEED DRILL
fr semoir *m* à bras
ne handzaaimachine
de Handdrillmaschine *f*
da håndsåmaskine
sv handsåmaskin
es sembrador *m* de mano
it seminatrice *f* a mano

HAND WORK, 2260

HANGING, 1141

1720
HARD COATED
fr testacé
ne hardschalig

de hartschalig
da hårdskallet
sv hårdskalig
es de cáscara dura
it di mallo duro

1721
HARDEN OFF, TO
fr endurcir
ne (af)harden
de abhärten ·
da afhænde; harde
sv avhärda
es endurecer
it indurare

1722
HARD FESCUE
fr fétuque *f* durette
ne hard zwenkgras *n*
de harter Schwingel *m*
da stivbladet svingel
sv hårdsvingel
es cañuela *f* durilla
it Festuca *f*
   Festuca longifolia Thuill.

1723
HARD FRUIT
fr fruits *m pl* à pépins
ne hard fruit *n*
de Kernobst *n*
da kernefrugt
sv kärnfrukt
es fruto *m* pomo
it frutta *f* a granella;
   frutto *m* duro

1724
HARD GILL (MUSH-
ROOM GROWING);
OPEN VEILS (US)
fr champignon *m* ouvert de
   nature
ne open vlies
de velumlos; schleierlos;
   hartlamellig
da 'hård' champignon
sv hård champinjon
es champiñones *m pl*
   abiertos
it spessimento *m* degli
   stipiti

1725
HARDNESS; FIRMNESS
fr dureté *f*
ne hardheid; stevigheid
de Härte *f*
da hårdhed
sv hårdhet; styvhet
es consistencia *f*; dureza *f*
it durezza *f*; compattezza *f*

1726
HARD ROT AND LEAF
SPOT (FREESIA,
GLADIOLUS)
fr septoriose *f*
ne hardrot *n*
de Hartfäule *f*
da hårdforrådnelse
sv hård röta
es podredumbre *f* aguda;
   septoriosis *f*
it septoriosi *f*
   Septoria gladioli Pass.

1727
HARDWOOD CUTTING
fr bouture *f* de rameau
   aoûté
ne twijgstek
de Steckholz *n*
da skudstikling
sv vedartad stickling
es estaca *f* leñosa
it talea *f* legnosa

1728
HARDY
fr résistant au froid;
   rustique
ne winterhard
de winterhart
da vinterfast; hårdfør
sv vinterhärdig
es resistente al frío
it resistente al freddo

1729
HARE (BROWN)
fr lièvre *m*
ne haas
de Hase *m*
da hare
sv tysk hare
es liebre *f*
it lepre *f*

1730
HARROW
fr herse *f*
ne eg
de Egge *f*
da harve
sv harv
es grada *f*
it erpice *m*

HARROW, disc, 1079
—, light spiked chain, 2160
—, zig-zag, 4236

1731
HARROW, TO
fr herser
ne eggen
de eggen
da harve
sv harva
es gradear
it erpicare

1732
HARROW TINE;
HARROW TOOTH
fr dent *f* de herse
ne egtand
de Eggenzinken *m*
da harvetand
sv harvpinne
es diente *m* de grada
it dento *m* d'erpice

1733
HARVEST
fr récolte *f*
ne oogst
de Ernte *f*
da høst
sv skörd
es cosecha
it raccolta *f*

1734
HARVEST, TO; TO REAP;
TO PICK
fr récolter; cueillir
ne oogsten
de ernten
da høste
sv skörda
es recolectar; cosechar
it raccogliere

HARVEST
ADVANCEMENT, 900

HARVEST DELAY, 901

HARVESTER, bean, 274
—, carrot, 596
—, onion, 2525

1735
HARVEST YEAR
fr campagne *f*
ne oogstjaar *n*
de Erntejahr *n*
da høstår *n*
sv skördeår *n*
es campaña *f* agrícola
it annata *f* agraria

1736
HASTATE
fr hasté
ne spiesvormig
de spiessförmig
da spydformet
sv spjutformig
es lanceolado
it lanciforme

1737
HATCH, TO
(EGG, LARVA)
fr éclore
ne uitkomen
de schlüpfen; auskommen
da klække
sv kläcka
es salir del huevo;
    salir del cascarón
it schiudersi

1738
HAULM PULLER
fr arrache-fanes *m*
ne looftrekker
de Krautzieher *m*
da løvsamler
sv blastryckare
es arrancador *m* de hojas
it estirpatrice *f*

1739
HAULM STRIPPER
fr déchiqueteuse *f* de fanes
ne loofklapper
de Krautschläger *m*

da afløver; aftopper
sv blastkrossare
es despedazador *m* de hojas

1740
HAWK MOTH;
SPHINX MOTH (US)
fr sphinx *m*
ne pijlstaartvlinder
de Schwärmer *m*
da sværmer
sv svärmare
es falena *f*; esfinge *f*
it sfinge *f*
    Sphingidae

HAWK MOTH,
death's head, 997
—, privet, 2935

1741
HAWTHORN
fr aubépine *f*
ne meidoorn
de Weissdorn *m*
da tjørn
sv hagtorn
es espino *m* blanco
it biancospino *m*
    Crataegus L.

1742
HAZEL; FILBERT
fr noisetier *m*
ne hazelaar
de Haselnussstrauch *m*
da hassel
sv hassel
es avellano *m*
it nocciuolo *m*
    Corylus avellana L.

HAZEL, witch, 4190

1743
HAZEL LEAF ROLLER
WEEVIL
fr cigarier *m*
ne sigarenmaker
de Trichterwickler *m*
da Rhynchites betulæ
sv björkrullvivel
es cortabrotes *m* del abedul
it rinchite *m* della betulla
    Rhynchites betulae F.

1744
HAZEL NUT
fr noisette *f*
ne hazelnoot
de Haselnuss *f*
da hasselnød
sv hasselnöt
es avellana *f*
it nocciuola *f*

HAZEL SAWFLY, 67

1745
HEAD
fr tête *f*
ne kop
de Kopf *m*
da hoved *n*
sv huvud
es cabeza *f*
it capo *m*; testa *f*
    caput

HEAD DOWN, TO, 3883

1746
HEADED CABBAGE
fr chou *m* cabus
ne sluitkool
de Kopfkohl *m*
da hovedkål
sv huvudkål
es col *f*; repollo *m*
it cavolo *m* cappuccio
    Brassica oleracea
    c.v. Capitata L.

1747
HEADING
(BULB GROWING);
FLOWER REMOVAL;
DEBLOSSOMING
fr écimage *m* des fleurs
ne koppen
de köpfen
da kappe
sv toppa blommor
es descabezar
it decapitare;
    asportare i fiori

HEADING LETTUCE,
513

1748
HEADLAND
fr fourrière *f*
ne wendakker
de Vorgewende *f*
da forager
sv vändteg
es cabecera *f*
it capezzagna *f*

1749
HEALTHY
fr sain
ne gezond
de gesund
da sund
sv frisk; sund
es sano
it sano

1750
HEART FORMATION
fr formation *f* de la pomme
ne kropvorming
de Kopfformung *f*;
    Kopfbildung *f*
da hoveddannelse
sv knytning; huvudbildning
es acogollarse
it formazione *f* del cesto

1751
HEART ROT
fr pourriture *f* du coeur
ne hartrot *n*
de Herzfäule *f*
da hjerteforrådnelse
sv hjärtröte
es podredumbre *f* del
    corazón
it marciume *m* del cuore

1752
HEART ROT (FAGUS)
fr amadouvier *m*
ne tonderzwam
de Weissfäule *f*
da tøndesvamp
sv rötticka
es hongo *m* de la yesca;
    yesquero *m*
it muffa *f* grigia
    Fomes fomentarius (Fr.)
    Kickx

HEART'S EASE, 1332

HEART SHAPED, 838

1753
HEARTWOOD
fr coeur *m* de bois;
    duramen *m*
ne kernhout *n*
de Kernholz *n*
da kerneved *n*
sv kärnved
es duramen *m*
it legno *m* durno

1754
HEAT, TO
fr chauffer
ne verwarmen
de heizen
da varme; opvarme
sv uppvärma
es calentar
it riscaldare

HEATED FRAME, 1833

HEATED GLASSHOUSE,
1835

1755
HEATED HOUSE;
HOTHOUSE
fr serre *f* chaude
ne warme kas
de Warmhaus *n*
da varmthus *n*
sv varmhus *n*
es estufa *f* caliente;
    invernadero *m* calentado
it serra *f* calda

1756
HEATH; ERICA
fr bruyère *f*
ne dopheide
de Glockenheide *f*
da klokkelyng
sv klockljung
es érica
it Erica *f*
    Erica L.

1757
HEATH SOIL
fr terre f de bruyère
ne heidegrond
de Heideerde f
da hedejord
sv ljungjord; hedmark
es tierra f de brezo; brezal m
it brughiera f

HEATHER, 2183

1758
HEATING CABLE
fr câble m de chauffage
ne verwarmingskabel
de Heizkabel m
da varmekabel
sv värmekabel
es cable m de calefacción
it gomena d di riscaldamento

HEATING
INSTALLATION, 1760

1759
HEATING MANURE
fr fumier m de réchaud
ne broeimest
de Treibmist m
da varmegivende gødning
sv bänkgödsel
es estiércol m caliente
it stabbio m riscaldante

1760
HEATING PLANT;
HEATING INSTALLA-
TION
fr installation f de chauffage
ne stookinstallatie
de Heizinstallation f
da varmeinstallation
sv värmeinstallation
es instalación f de cale-
    facción
it impianto m focolare

1761
HEATING TECHNICIAN
fr technicien m de chauffage
ne stooktechnicus
de Heizungstechniker m
da varmetekniker

sv värmetekniker
es técnico m de la cale-
    facción
it tecnico m focolare

1762
HEATING WITH OPEN
VENTILATORS
fr chauffer à sec
ne droogstoken
de trockenheizen
da tørre ved fyring
sv upptorkning genom
    eldning
es desecar
it scaldare a prosciuga-
    mento

1763
HEAT LOSS
fr perte f à feu;
    perte f par cuisson
ne gloeiverlies n
de Glühverlust m
da glødetab n
sv glödförlust
es pérdida f por cocción;
    pérdida f por calcinación
it perdita f al fuoco;
    perdita f dell'incandes-
    cenza

1764
HEAT OF RADIATION
fr chaleur f de rayonnement
ne stralingswarmte
de Strahlungswärme f
da strålesvarme
sv strålningsvärme
es calor m radiante
it calore m d'irraggiamento

1765
HEAT STORED (BULBS)
fr traité par la chaleur
ne heetgestookt
de heissbehandelt
da varmekurered
sv värmebehandlad
es tratado con calor;
it riscaldato a temperatura
    elevata

1766
HEAT TREATMENT;
HEAT THERAPY
fr thermothérapie f
ne warmtebehandeling
de Wärmebehandlung f;
    Wärmetherapie f
da varmebehandling;
    termoterapi
sv värmebehandling;
    värmeterapi
es tratamiento m
    por calor;
    termo-terapía f
it trattamento m
    col calore;
    termoterapia f

1767
HEDGE
fr haie f
ne heg; haag
de Hecke f
da hæk
sv häck
es seto m vivo
it siepe f

1768
HEDGE BINDWEED;
BELLBINE
fr liseron m des haies;
    grand liseron m
ne haagwinde
de Zaunwinde f
da gærde-mergle
sv snårvinda
es correhuela f major
it vilucchione m
    Galystegia sepium L.
    R. Br.

1769
HEDGEHOG
fr hérisson m
ne egel
de Igel m
da pindsvin n
sv igelkott
es erizo m
it riccio m; porcospino m
    Erinaceus europaeus L.

1770
HEDGE MUSTARD
fr herbe *f* au chantre;
   vélar *m* officinal
ne raket
de Wegranke *f*
da rank vejsennep
sv vägsenap
es erismo *m*;
   hierba *f* de los cantores
it erba *f* cornacchia
   Sisymbrium officinale L.
   Scop.

1771
HEDGEROW
fr système *m* de haie fruitière
ne haagsysteem *n*
de Heckensystem *n*
da hæksystem *n*
sv häcksystem *n*
es plantación *f* de frutales
   en forma de selo
it sistema *m* a siepe

1772
HEDGE SHEARS
fr cisailles *f pl* à haies
ne heggeschaar
de Heckenschere *f*
da hæksaks
sv häcksax
es tijeras *f pl* de jardinero
it forbicioni *f pl*

1773
HEIGHT OF ASCENT
fr hauteur *f* ascensionnelle
ne opvoerhoogte
de Steighöhe *f*
da stigehøjde
sv uppfordringshöjd
es altura *f* ascensional
it livello *m* ascencionale

1774
HELICOID CYME (bot.)
fr cyme *f* hélicoïde
ne schroef
de Schraubel *m*
da skrue
sv skruvknippe
es inflorescencia *f* helicoidal
it cime *m* cocleato

1775
HELIOTROPE
fr héliotrope *m*
ne heliotroop
de Heliotrop *n*
da heliotrop
sv heliotrop
es heliótropo *m*
it vaniglia *f*
   Heliotropium L.

1776
HEMLOCK SPRUCE
fr hemlock *m*; pruche *f*
ne hemlockden
de Hemlocktanne *f*; Tsuga *f*
da tsuga
sv hemlock
es tuya *f* de Canadá
it Tsuga *f*
   Tsuga Carr.

1777
HEMP
fr chanvre *m*
ne hennep
de Hanf *m*
da hamp
sv hampa
es cáñamo *m*
it canapa *f*
   Cannabis L.

HEMP, African, 3535

1778
HAMP AGRIMONY
fr eupatoire *f*; chanvrine *f*
ne leverkruid *n*
de Wasserdost *m*
da leverurt
sv flockel
es eupatorio *m*
it agrimonia *f*
   Eupatorium L.

1779
HEMPNETTLE;
DAYNETTLE
fr ortie *f* royale
ne hennepnetel
de stechender Hohlzahn *m*
da almindelig hanekro
sv pipdån

es galeopsis *f*
it canapa *f* selvatica
   Caleopsis tetrahit L.

1780
HENBANE
fr jusquiame *f* noire
ne bilzenkruid *n*
de Bilsenkraut *n*
da bulmeurt
sv bolmört
es beleño *m* negro
it giusquiamo *m*
   Hyoscyamus niger L.

1781
HENBIT
fr lamier *m*; ortie *f* morte
ne hoenderbeet
de stengelumfassende
   Taubnessel *f*
da liden tvetand
sv mjukplister
es ortiga *f* muerta
it erba *f* rotella
   Lamium amplexicaule L.

1782
HEPATICA
fr hépatique *f*;
   herbe *f* de la Trinité;
   anémone *f* hépatique
ne leverbloempje *n*; Hepatica
de Leberblümchen *n*
de blå anemone
sv blåsippa
es hepática *f*
it Hepatica *f*; fegatella *f*
   Hepatica nobilis Mill.

1783
HERBACEOUS
fr herbacé; herbeux
ne kruidachtig
de krautartig; krautig
da urteagtig
sv örtartad
es herbaceo
it erbaceo

1784
HERBACEOUS
PERENNIALS
fr plantes *f pl* vivaces
ne vaste planten

de Stauden *f pl*;
ausdauernde Pflanzen *f pl*
da stauder
sv perenna växter;
fleråriga växter
es plantas *f pl* vivaces
it piante *f pl* perenni

1785
HERB GROWING
fr culture *f* de plantes
aromatiques et médicinales
ne kruidenteelt
de Kräuteranbau *m*
da dyrkning af krydderurter
sv medicinal äxtodling
es cultivo *m* de plantas
medicinales
it coltura *f* d'erbe

1786
HERBICIDE
fr herbicide *m*;
désherbant *m*
ne onkruidbestrijdings-
middel *n*; herbicide
de Unkrautbekämpfungs-
mittel *n*
da ukrudtsbekæmpelses-
middel
sv ogräsbekämpnings-
medel *n*
es herbicida *m*
it erbicida *f*; disherbante *m*

1787
HERB PATIENCE
fr patience *f*;
oseille-épinard *f*
ne patientie
de Gemüseampfer *m*;
englischer Spinat *m*
da Engelsk spinat
sv trädgårdssyra;
engelsk spenat
es hierba *f* de la paciencia
it pazienza *f*
Rumex patientia L.

1788
HERBS
fr herbes *f pl* aromatiques
ne kruiden

de Kräuter *n pl*
da krydderurter
sv örter; medicinalväxter
es hierbas *f pl* aromáticas
it erbe *f pl*

HERBS, medicinal, 2319

1789
HEREDITY
fr hérédité *f*
ne erfelijkheid
de Erblichkeit *f*
da arvelighed
sv ärftlighet
es herencia *f*
it ereditarietà *f*

1790
HERON'S BILL
fr bec-de-grue *m*
ne reigersbek
de Reiherschnabel *n*
da hejrenæb *n*
sv skatnäva; näva
es pico *m* de cigueña;
erodio *m*
it erodia *f*
Erodium l'Hérit.

HERRIF, 700

1791
HETEROSIS;
HYBRID VIGOUR
fr hétérosis *m*
ne heterosis
de Heterosis *m*
da krydsningsfrodighed;
heterosis
sv heterosis
es heterosis *f*;
vigor *m* híbrido
it heterosis *f*

1792
HETEROZYGOTE
fr hétérozygote
ne heterozygoot
de heterozygot
da heterozygot
sv heterozygot
es heterozigoto
it heterozygote

1793
HIBERNATE, TO;
TO WINTER
fr hiverner
ne overwinteren
de überwintern
da overvintre
sv övervintra
es invernar
it svernare

1794
HIGHER PLANTS
fr plantes *f pl* supérieures
ne hogere planten
de höhere Pflanzen *f pl*
da højere planter
sv högre växter
es plantas *f pl* superiores
it piante *f pl* superiori

1795
HIGH PRESSURE AREA
fr zone *f* à haute pression
ne hogedrukgebied
de Hochdruckgebiet *n*
da højtryksområde *n*
sv högtrycksområde *n*
es anticiclón *m*
it settore *m* d'alta pressione
atmosferica

1796
HIGH PRESSURE
BOILER
fr chaudière *f* à haute
pression
ne hogedrukketel
de Hochdruckkessel *m*
da højtrykskedel
sv högtryckspanna;
ångpanna
es caldera *f* de alta presión
it caldaia *f* ad alta pressione

1797
HIGH PRESSURE
MERCURY VAPOUR
LAMP
fr lampe *f* à vapeur de
mercure haute pression
ne hogedrukkwiklamp
de Hochdruckqueck-
silberlampe *f*

da høgtrykskviksølvlampe
sv högtryckskvicksilver-
lampa
es lámpara *f* de vapor de
mercurio de alta presión
it lampada *f* a vapore di
mercurio ad alta
pressione

1798
HILUM
fr ombilic *m*; hile *m*
ne navel (micropyle)
de Nabel *m*
da navle
sv micropyle; fröämnesmun
es ombligo *m*
it omblico *m*; micropile *m*

HIRSUTE,   1706

HOARY,   2962

1799
HODDESDON PIPE
fr tuyau *m* Hoddesdon;
tuyau *m* pour désinfec-
tion à la vapeur
ne graafrek
de Dämpfgabel *f*;
Dämpfrost *m*
da damprist
sv ångspjut *n*
es tubo *m* Hoddesdon;
tubo *m* para desinfección
con vapor
it tubo *m* perforato per
disinfezione con il vapore

1800
HOE
fr houe *f*; binette *f*
ne hak
de Hacke *f*
da hakke
sv hacka
es azada *f*; binadora *f*
it zappa *f*; piccone

HOE, drill,   1136
—, Dutch,   1178
—, motor,   2397
—, roller-type hand,   3162
—, to,   3287
—, wheel hand,   4135

1801
HOLDING; NURSERY
fr exploitation *f*
ne bedrijf *n*
de Betrieb *m*
da virksomhed; bedrift
sv företag
es explotación *f*
it azienda *f*

1802
HOLLOW
fr creux
ne hol (adj.)
de hohl
da hul
sv ihålig
es hueco
it scavato

1803
HOLLY
fr houx *m*
ne hulst
de Stechpalme *f*
da kristtorn
sv järnek
es acebo *m*
it agrifoglio *m*
Ilex L.

HOLLY, pyramid,   2996
—, sea,   3296

1804
HOLLYHOCK
fr rose *f* trémière
ne stokroos
de Stockrose *f*
da stokrose
sv stockros
es malva *f*
it altea *f*
Althaea rosea Cav.

1805
HOLLY LEAF MINER
fr Phytomyza *f* ilicis
ne hulstvlieg
de Ilexminierfliege *f*;
Blattminierfliege *f*
da kristtorn-minérflue
sv järneksminerarfluga

es mosca *m* del acebo
it Phytomyza *f* ilicis
Phytomyza ilicis Curt.

1806
HOME FREEZER
fr congélateur *m*
domestique;
congélateur *m* ménager
ne vrieskist; vrieskast
de Haushaltgefriermöbel *n*;
Gefriertruhe *f*
da hjemmefryser
sv frysbox
es congelador *m* doméstico
it congelatore *m* domestico

1807
HOMOGENIZATION
(SOIL)
fr homogénisation *f*
ne homogenisatie
de Homogenisation *f*
da homogenisering af jord
sv homogenisering av jord
es homogeneización *f*
it omogeneizzazione *f*

1808
HOMOZYGOTE
fr homozygote
ne homozygoot
de homozygot
da homozygot
sv homozygot
es homozigoto
it homozygote

1809
HONESTY
fr lunaire *f*; monnaie *f* de
pape; herbe *f* aux écus
ne judaspenning
de Judassilberling *m*
da judaspenge; måneskulpe
sv Judassilverpenningar;
månviol
es Lunaria *f*
it Lunaria *f*
Lunaria L.

1810
HONEY
fr miel *m*
ne honing

de Honig *m*
da honning
sv honung
es miel *f*
it miele *m*

1811
HONEY DEW
fr miellat *m*; miellée *f*;
   miélaison *f*
ne honingdauw
de Honigtau *m*
da honningdug
sv honungsdagg
es segrecación *f* azucarada;
   ligamaza *f* mielada
it mielata *f*

1812
HONEY LOCUST
fr févier *m*
ne valse Christusdoorn
de Dornkronenbaum *m*;
   Lederhülsenbaum *m*
da tretorn
sv korstörne *n*
es acacia *f* de tres espinas;
   acacia *f* de puñal
it Gleditsia *f*
   Gleditsia triacanthos L.

1813
HONEYSUCKLE
fr chèvrefeuille *m*
ne kamperfoelie
de Heckenkirsche *f*
da gedeblad *n*; kaprifolie
sv kaprifol
es madreselva *f*
it madreselva *f*
   Lonicera

HOOF AND HORN,   1818

1814
HORIZONTALLY
SPREADING
fr étalé; ouvert
ne uitstaand
de abstehend
da udstående
sv utstående
es abierto; separado
it aperto

1815
HORNBEAM
fr charme *m*
de haagbeuk
de Hainbuche *f*
da avnbøg
sv avenbok
es ojaranzo *m*; carpe *m*
it carpina *m*
   Carpinus betulus L.

1816
HORN CHIPS
fr corne *f* râpée
ne hoornspaanders
de Hornspäne *f pl*
da hornspåner
sv hornspån *n pl*
es virutas *f pl* de cuerno
it squame *f pl* di corno

1817
HORNET
fr frelon *m*
ne hoornaar
de Hornisse *f*
da gedehams
sv bålgeting
es sesia *f* apiforme
it calabrone *m*
   Vespa crabro L.

1818
HORN MEAL;
HOOF AND HORN
fr corne *f* torréfiée
ne hoornmeel *n*
de Hornmehl *n*
da hornmel *n*
sv hornmjöl *n*
es harina *f* de cuerno
it corno *m* polverizzato

1819
HORSE CHESTNUT
fr marronnier *m* d'Inde
ne paardekastanje
de Rosskastanie *f*
da hestekastanie
sv hästkastanj
es castaño *m* de Indias
it ippocastano *m*
   Aesculus L.

1820
HORSE DROPPING
fr crottin *m* de cheval
ne paardevijg
de Pferdeapfel *m*
da hestepaere
sv hästspillning
es boñiga *f*
it letame *m* equino

1821
HORSE MANURE
fr fumier *m* de cheval
ne paardemest
de Pferdemist *m*
da hestegødning
sv hästgödsel
es estiércol *m* de caballo
it stallatico *m* cavallino

1822
HORSE RADISH
fr raifort *m*
ne mierikswortel
de Meerrettich *m*
da peberrod *f*
sv pepparrot
es rábano *m* rusticano
it ramolaccio *m*
   Armoracia rusticana
   G.M. Sch.

HORSETAIL, common,
771
—, marsh,   2295

1823
HORTICULTURAL
ADVISER
fr conseiller *m* horticole
ne tuinbouwconsulent
de Gartenbauberater *m*
da havebrugskonsulent
sv trädgårdskonsulent
es asesor *m* hortícolo
it consultore *m* orticolo

1824
HORTICULTURAL
CENTRE;
HORTICULTURAL
DISTRICT
fr région *f* horticole
ne tuinbouwgebied *n*

de Gartenbaugebiet *m*
da havebrugsområde *n*
sv trädgårdsområde *n*
es región *f* hortícola
it regione *f* orticola

1825
HORTICULTURAL
LAMP
fr irradiateur *m* de plantes
ne plantenbestraler
de Pflanzenbestrahler *m*
da plantebestråler;
plantelampe
sv lampa för växtbelysning
es lámpara *f* hortícola
para irradiación de
plantas
it lampada *f* per illuminazione di piante

1826
HORTICULTURAL
SCHOOL
fr école *f* d'horticulture
ne tuinbouwschool
de Gartenbauschule *f*
da havebrugsskole
sv trädgårdsskola
es escuela *f* de horticultura
it scuola *f* d'orticoltura

1827
HORTICULTURE
fr horticulture *f*
ne tuinbouw
de Gartenbau *m*
da gartneri; havebrug *n*
sv trädgård-skötsel
es horticultura *f*
it orticoltura *f*

HORTICULTURE,
centre of, 633
—, commercial, 763

1828
HORTICULTURE
UNDER GLASS
fr horticulture *f* de serre
ne glastuinbouw
de Gartenbau *f* unter Glas
da væksthusgartneri *n*

sv trädgårdsodling under
glas
es horticultura *f* bajo cristal
it orticoltura *f* sotto vetro

1829
HORTICULTURIST
fr technicien *m* horticole
ne tuinbouwkundige
de Gartenbaukundiger *m*
da havebrugskyndig
sv trädgårdssakkunnig
es técnico *m*; horticultor *m*
it orticultore *m*

1830
HOSE
fr tuyau *m*; manche *f* à eau
ne slang
de Schlauch *m*
da slange
sv slang
es tubo *m*; manga *f* (de agua)
it tubo *m*; manica *f*

1831
HOST PLANT
fr plante-hôte *f*
ne gastheer; waardplant
de Wirtspflanze *f*
da værtplante
sv värdväxt
es planta *f* huésped
it pianta *f* ospite

1832
HOT AIR HEATER
fr générateur *m* d'air chaud
ne warmeluchtkachel;
heteluchtkachel
de Warmluftofen *m*;
Lufterhitzer *m*
da varmeluftovn;
kalorifere
sv varmluftblock *n*
es quemador *m* de aire
caliente
it generatore *m* d'aria calda

1833
HOTBED;
HEATED FRAME
fr couche *f* chaude
ne broeibak; warme bak

de Frühbeet *n*; Warmbeet *n*
da drivbænk; varmebænk
sv varmbänk
es cama *f* caliente
it letto *m* caldo

HOTBED, open, 2531

1834
HOT DIP GALVANISED
fr galvanisé à chaud
ne thermisch verzinkt
de feuerverzinkt
da galvaniseret;
varmforsinket
sv varmgalvaniserad;
varmförzinkad
es galvanizado en caliente
it galvanizzato a caldo

1835
HOTHOUSE;
HEATED GLASSHOUSE
fr serre *f* chauffée
ne stookkas
de Warmhaus *n*
da varmthus *n*
sv varmhus *n*
es invernadero *m* caliente
it serra *f* calda; stufa *f*

1836
HOT WATER BOILER
fr chaudière *f* d'eau chaude
ne warmwaterketel
de Warmwasserkessel *m*
da varmtvandskedel
sv varmvattenpanna
es caldera *f* de agua
caliente
it caldaia *f* ad acqua calda

1837
HOT-WATER TANK
fr bac *m* de traitement
ne kookketel
de Warmwasserbehälter *m*
zum Kochen
da varmtvandstank
sv varmvattenbehandlings-
reservoar
es caldera *f* para cocer
it calderone *m*

1838
HOT WATER
TREATMENT
fr traitement *m* à l'eau
   chaude
ne warmwaterbehandeling
de Warmwasserbehandlung*f*
da varmtvandsbehandling
sv varmvattenbehandling
es tratamiento *m* con agua
   caliente
it trattamento *m* con acqua
   calda

1839
HOURLY RATE
fı salaire *m* horaire
ne tijdloon *n*
de Zeitlohn *m*
da timeløn; tidsløn
sv tidlön
es salario *m* según tiempo
it retribuzione *f* oraria

1840
HOUSEHOLD REFUSE
fr ordures *f pl* ménagères
ne huisvuil *n*
de Hauskehricht *m*;
   Hausmüll *m*
da husholdningsaffald *n*
sv hushållsavfall *n*
es basuras *f pl* de ciudad
it immondezze *f pl*

1841
HOUSE LEEK
fr joubarbe *f*
ne huislook *n*
de Hauswurz *f*; Hauslauch *n*
da husløg *n*
sv (äkta) taklök
es siempreviva *f*
it sempreviva *f*
   Sempervivum tectorum L.

1842
HOUSE PLANT;
POTPLANT
fr plante *f* d'appartement
ne kamerplant
de Zimmerpflanze *f*
da stueplante
sv rumsväxt

es planta *f* ornamental
it pianta *f* di camera

1843
HOUSE SPARROW
fr moineau *m* commun
ne huismus
de Haussperling *m*
da gråspurv
sv gråsparv
es gorrión *m*
it passero *m*
   Passer domesticus L.

1864
HULL (bot.); SHELL
fr écaille *f*
ne bolster
de Schale *f*
da hase; hylster *n*; skal
sv skal
es cáscara *f*
it riccio *m*

HUMBLE BEE,   492

1845
HUMIC
fr humique
ne humusachtig; humeus
de humös
da muldrig; humusrig
sv humusartad; myllrik
es mantilloso; humifero
it umoso

1846
HUMIC ACID
fr acide *m* humique
ne humuszuur *n*
de Humussäure *f*
da humussyre
sv humussyra
es ácido *m* húmico
it acido *m* d'umus

1847
HUMIFEROUS LAYER
fr couche *f* humique
ne humusdek *n*
de Humusdecke *f*;
   Humusauflage *f*
da humusdække
sv humustäcke *n*

es cobertura *f* de humus;
   capa *f* de humus
it copertura *f* d'umus

1848
HUMIFICATION
fr humification *f*
ne humificatie
de Humifizierung *f*
da humificiering
sv humifiering
es humuficación *f*
it umoficazione *f*

HUMPBACKED FLY,
2701

1849
HUMUS
fr humus *m*
ne humus
de Humus *m*
da muld; humus
sv mull; mulla; humus
es mantillo *m*; humus *m*
it umus *m*; ferricio *m* vegetale

HUMUS, acid,   18
—, mild,   2341

1850
HUMUS CONTENT
fr taux *f* humique;
   teneur *f* en humus
ne humusgehalte *n*
de Humusgehalt *m*
da humusindhold *n*
sv humushalt
es contenido *m* de humus
it contenuto *m* d'umus

1851
HUMUS FORMATION
fr formation *f* d'humus
ne humusvorming
de Humusbildung *f*
da humusdannelse
sv humusbildning
es formación *f* del humus
it formazione *f* d'umus

1852
HURDLE
fr claie *f*
ne horde

de Hürde f
da gaerde n
sv gärde n
es zarzo m; cañizo m
it canniccio m; graticcio m

1853
HUSKS (BULB); DEBRIS
fr déchets m pl
ne bolhuidresten
de Zwiebelhautreste m pl
da løgskalrester
sv lökhudrester
es residuos mpl de la túnica;
   cáscaras fpl
it rimanenti mpl delle
   tuniche

1854
HYACINTH
fr jacinthe f
ne hyacint
de Hyazinthe f
da hyacinth
sv hyacint
es jacinto m
it giacinto m
   Hyacinthus L.

HYACINTH, grape,   1638
—, summer,   3731

1855
HYBRID
fr hybride m
ne hybride
de Hybride m
da hybrid; krydsning
sv hybrid
es hibrido m
it ibrido m

HYBRID, graft,   1622

HYBRID VIGOUR,   1791

1856
HYDRANGEA
fr hortensia m
ne hortensia

de Hortensie f
da hortensie
sv hortensia
es hortensia f
it ortensia f
   Hydrangea macrophylla
   (Thunb.) Ser.

1857
HYDRAULIC LIFT;
POWER LIFT
fr relevage m hydraulique
ne hydraulische hefinrichting
de Kraftheber m
da løfteanordning
sv hydraulisk lyftanordning
es elevador m hidráulico
it sollevatore m idraulico

1858
HYDROCHLORIC ACID
fr acide m chlorhydrique
ne zoutzuur n
de Salzsäure f
da saltsyre
sv saltsyra
es ácido m clorhídrico
it acido m cloridrico

1859
HYDROCOOLING
fr refroidissement m par eau
   glacée
ne waterkoeling
de Eiswasserkühlung f
da vandkøling
sv vattenkylning
es enfriamiento m del agua
it raffreddamento m ad
   acqua

1860
HYDROGEN
fr hydrogène m
ne waterstof
de Wasserstoff m
da brint
sv väte n
es hidrógeno m
it idrogeno m

1861
HYDROLYSIS
fr hydrolyse f
ne hydrolyse
de Hydrolyse f
da hydrolyse
sv hydrolys
es hidrólisis f
it idrolisi f

HYDROPONICS,   3488

1862
HYGROMETER
fr hygromètre m
ne luchtvochtigheidsmeter
de Luftfeuchtigkeitsmesser
   m
da luftfugtighedsmåler
   hygrometer n
sv luftfuktighetsmätare;
   hygrometer
es higrómetro m
it igrometro m

1863
HYPHA
fr hyphe f
ne hyfe; zwamdraad
de Hyphe f; Pilzfaden n
da hyfe; svampetråd
sv hyf; svamptråd
es hifa f
it filaccia f

HYPOCRATERIFORM,
3239

1864
HYSSOP
fr hysope f
ne hysop
de Ysop m
da isop
sv isop
es hisopo m
it issopo m
   Hyssopus officinalis L.

I

1865
ICE
fr glace *f*
ne ijs *n*
de Eis *n*
da is
sv is
es hielo *m*
it ghiaccio *m*

1866
ICEBERG LETTUCE;
COS-LETTUCE
fr laitue *f* pommée frisée
ne ijssla
de Eissalat *m*
da issalat
sv issallat
es lechuga *f* crespa de hielo

1867
ICELAND POPPY
fr pavot *m* d'Islande
ne Papaver nudicaule
de Islandmohn *m*
da sibirisk valmue
sv siberisk vallmo
es adormidera *f*
it papavero *m* d'Islanda
  Papaver nudicaule L.

1868
ICE PLANT;
MESEMBRYANTHEMUM
fr ficoïde *f* glaciale
ne ijskruid *n*
de Eiskraut *n*
da isplante
sv isört
es hierba *f* de hielo
it cristallina *f*
  Mesembrianthemum
  crystallinum L.

1869
ICHNEUMON
fr ichneumon *m*
ne sluipwesp
de Schlupfwespe *f*

da snyltehveps
sv parasitstekel
es icneumón *m*
it icneumonide *m*
  Ichneumonidae

ILL, 1083

1870
ILLUMINATE, TO
fr éclairer
ne belichten
de beleuchten
da belysne
sv belysa
es iluminar
it illuminare

ILLUMINATION, 160,
2156
—, duration of, 1168

1871
ILLUMINATION PERIOD
fr durée *f* de l'éclairage
ne belichtingstijd
de Belichtungszeit *f*
da belysningstid
sv belysningstid
es duración *f* de la
  iluminación
it durata *f* d'illuminazione

1872
IMAGO; ADULT INSECT
fr imago *f*; insecte *m* parfait
ne volwassen insekt *n*
de Imago *f*
da imago;
  fuldvoksent insekt *n*
sv fullbildad insekt *n*; imago
es insecto *m* adulto;
  imago *m*
it insetto *m* aldulto;
  immagine *f*

IMMATURE, 3981

1873
IMMUNITY
fr immunité *f*
ne onvatbaarheid;
  immuniteit
de Immunität *f*
da immunitet
sv immunitet
es inmunidad *f*
it resistenza *f* assoluta;
  immunità *f*

1874
IMPACT RESISTANCE
fr résistant au choc
ne slagvast
de schlagfest
da slagfast
sv slagfast
es resistente al choque
it resistente all'urte

IMPATIENS, 238

1875
IMPERMEABLE;
IMPERVIOUS
fr imperméable
ne ondoorlatend
de undurchlässig
da ugennemtrængelig
sv ogenomtränglig
es impermeable
it impermeabile

IMPERVIOUS, 1875

IMPORTED CABBAGE
WORM, 3445

IMPORTED CURRANT
WORM, 770

1876
IMPORTER
fr importateur *m*
ne importeur
de Importeur *m*
da importør

sv importör
es importador *m*
it importatore *m*

1877
IMPOVERISHMENT
(SOIL)
fr appauvrissement *m*
ne verarming
de Bodenverhagerung *f*;
  Bodenverarmung *f*
da udpining
sv utarmning; utsugning
es empobrecimiento *m*
it impoverimento *m*

IMPREGNATION, 2443

IMPROVE, TO, 80

1878
IMPROVEMENT
fr amélioration *f*
ne veredeling
de Veredlung *f*
da forædling; podning;
  okulation
sv förädling
es injerto *m*; mejoría *f*;
  mejoramiento *m*
it miglioramento *m*

1879
IMPROVEMENT OF THE
SOIL STRUCTURE
fr amélioration *f* structurale
ne structuurverbetering
de Strukturverbesserung *f*
da struktuforbedring
sv strukturförbättring
es mejoría *f* estructural
it miglioramento *m* della
  struttura

1880
IMPURITY
fr impureté *f*
ne onzuiverheid
de Unreinheit *f*;
  Unsauberkeit
da urenhed
sv orenhet
es impureza *f*
it impurità *f*

1881
IMPUTED COSTS
fr charges *f pl* calculées
ne berekende kosten
de kalkulatorische Kosten
  f pl
da kalkuleredeomkostninger
sv kalkylerade kostnader
es costes *m pl* calculados
it costi *m pl* calcolati

1882
INARCH, TO
fr greffer en approche
ne afzuigen (zoogenten)
de ablaktieren
da afsuge
sv aflaktera; afsuga
es succionar; injerto *m* de
  aproximación
it innestare per approssima-
  zione

1883
INBREEDING
fr inbreeding *f*;
  endogamie *f*
ne inteelt
de Inzucht *f*
da indavl
sv inavel
es endogamía *f*

1884
INCANDESCENT LAMP
fr lampe *f* à incandescence
ne gloeilamp
de Glühlampe *f*
da glødelampe
sv glödlampa
es lámpara *f* de incandescencia
it lampada *f* ad incandas
  cenza

1885
INCASEMENT
fr couverture *f*
ne omhulling
de Umhüllung *f*
da ombinding; beskyttelse
sv omhöljning
es cubierta *f*
it viluppo *m*

1886
INCENSE CEDAR
fr libocèdre *m*
ne riviercypres
de Flusszeder *f*
da Libocedrus
sv cedertuja
es libocedro *m*
it libocedro *m*
  Libocedru. decurrens
  Torr.

1887
INCISION
fr entaille *f*; incision *f*
ne insnijding
de Einschnitt *m*
da indsnit *n*
sv insnitt *n*; inskärning
es incisión *f*
it incisione *f*

1888
INCOMPATIBILITY
fr incompatibilité *f*
ne onverenigbaarheid
de Unvereinbarkeit *f*;
  Unverträglichkeit *f*
da uforligelighed;
  uforenelighed
sv oförenlighet
es incompatibilidad *f*
it incompatibilità *f*

1889
INCREASE (ec.)
fr accroissement *m*
ne aanwas
de Zunahme *f*; Zuwachs *m*
da forøgelse
sv tillväxt
es incremento *m*
it incremento *m*

1890
INCREASE IN
PRODUCTION
fr suppléments *m pl* de
  rendement
ne meeropbrengst
de Mehrertrag *m*
da merudbytte *n*
sv merutbyte *n*
es producción *f* superior
it rendita *f* di piú

1891
INCUBATION PERIOD
fr periode f d'incubation
ne incubatietijd
de Inkubationszeit f
da inkubationstid
sv inkubationstid
es período m de incubacíon
it periodo m d'incubazione

INDICATOR, 2812

1892
INDICATOR PLANT
fr plante f indicatrice
ne indicatorplant
de Indikatorpflanze f
da indikatorplante
sv indikatorväxt
es plante f indicadora
it pianta f indicatrice

1893
INDIGENOUS; NATIVE
fr indigène
ne inheems
de einheimisch
da indenlandsk
sv inhemsk
es indigeno
it indigeno; nativo

1894
INDIGO
fr indigotier m
ne indigostruik
de Indigostrauch m
da indigo
sv indigo
es añil m; índigo m
it anile m
    Indigofera L.

1895
INDIRECT COSTS
fr coûts m pl indirects
ne indirecte kosten
de indirekte Kosten f pl;
    Gemeinkosten f pl
da indirekte omkostninger
sv indirekta kostnader
es costes m pl indirectos
it costi m pl indiretti

INDOOR-COMPOSTING,
2632

1896
INDOOR FORCING
(BULB GROWING)
fr forçage m en appartement
ne huisbroei
de Haustreiberei f
da stuedrivning
sv lökdrivning i hemmet
es forzado m casero
it forzatura f in casa

1897
INERT MATTER;
BALLAST MATERIAL
fr matière f inerte;
    matière f négligeable
ne onschadelijke onzuiver-
    heid
de unschädliche Verunreini-
    gung f
da uskadelig forurening
sv oskadlig förorening
es impureza f anodina
it impurità f innocua

1898
INFECT, TO;
TO CONTAMINATE
fr infecter; contaminer
ne besmetten
de infizieren
da smitte; inficere
sv nedsmitta; infektera
es contaminar; infectar
it infettare; contagiare

1899
INFECTED;
CONTAMINATED
fr infecté; contaminé
ne besmet
de infiziert
da smittet; inficeret
sv nedsmittad; infekterad
es infectado
it infetto

1900
INFECTION;
CONTAMINATION
fr infection f;
    contamination f
ne besmetting
de Ansteckung f

da smitte; infektion
sv smitta; infektion
es infección f
    contaminación f
it infettamento; contagio m

1901
INFECTIVITY
fr pouvoir m infectieux
ne infectievermogen n
de Infektiosität f
da infektivitet
sv infektionsförmåga
es infectividad
it infettività f

1902
INFERIOR (bot.)
fr infère
ne onderstandig
de unterständig
da undersædig
sv undersittande
es infero; hipógino
it infero; ipogino

1903
INFERIOR PRODUCT;
LOW-GRADE PRODUCT
fr piqué m; produit m en-
    tiché
ne minderwaardig produkt;
    stek; kroet
de Ausschussware f pl
da affald n; frasorteret;
    udskudsvare
sv utskott n; avfall n
es fruto m dañado;
    fruto m inferior
it prodotto m stravagante

1904
INFESTATION
fr attaque f parasitaire
ne aantasting
de Befall m
sv angreb n
sv sjukdomsangrepp
es ataque m
it attacco m

INFESTATION,
eelworm, 1201

1905
INFILTRATION
fr infiltration f
ne infiltratie; inzijging
de Versickerung f;
    Infiltration f
da infiltration
sv infiltration
es infiltración f
it infiltrazione f

1906
INFLATABLE HOUSE
fr serre f gonflable;
    serre f pneumatique
ne overdrukkas
de Tragluftgewächshaus n
da opblæst hus n
sv luftburet hus n;
    ballonghus n
es invernadero m neumático
it serra f pneumatica

1907
INFLORESCENCE
fr inflorescence f
ne bloeiwijze
de Blütenstand m
da blomsterstand
sv blomställning
es inflorescencia f
it florescenza f

1908
INFLUENCING OF
FLOWERING
fr influence f sur la floraison
ne bloeibeïnvloeding
de Blühbeeinflussung f
da blomstringspåvirkning
sv blomningspåverka
es influir en la floración
it influenzamento m fiorale

1909
INFRARED
fr infra-rouge
ne infrarood
da infrarot
da infrarød
sv infraröd
es infra-rojo
it infrarosso

INHIBITION, growth,
1688

1910
INHIBITOR
fr inhibiteur m; réducteur m;
    anti-auxine f
ne remstof
de Hemmstoff m
da hamningsstof n
sv växtreglerande medel n
es materia f refrenante
it hormone m d'impedi-
    mento

INHIBITOR, growth, 1691

1911
INJECT, TO
fr injecter
ne injecteren
de einspritzen; injizieren
da injicere; indsprøjte
sv inspruta; injicera
es inyectar
it iniettare

INJURIOUS, 2490

INJURIOUS ANIMALS,
2491

1912
INJURIOUS IMPURITY
fr impureté f nuisible
ne schadelijke onzuiverheid
de schädliche Verunreini-
    gung f
da skadelig forurening
sv skadlig förorening
es impurezas f pl perjudi-
    ciales
it impurità f dannosa

1913
INK CAP
fr coprin m
ne inktzwam
de Tintenpilz m; Tintling f
da blækhat
sv bläcksvamp
es coprino m
it coprino m
    Coprinus spp.

1914
INK DISEASE (IRIS)
fr taches f pl noires
ne inktvlekkenziekte
de 'Tintenfleckenkrankheit' f
da blæksyge
sv bleksot
es tinta f
it mal m dell'inchiostro
    Mystrosporium adustum
    Massee

1915
INNER BARK
fr liber m
ne bast
de Rinde f
da bast
sv bast n
es corteza f
it corteccia f

1916
INOCULATION
fr inoculation f
ne inoculatie
de Inokulation f
da inokulation; smitning
sv inokulering
es inoculación f
it inoculazione f

1917
INORGANIC
fr inorganique; mineral
ne anorganisch
de anorganisch
da uorganisk
sv oorganisk
es inorgánico
it inorganico

1918
INORGANIC MATTER
fr matière f inorganique;
    substance f minérale
ne anorganische stof
de anorganischer Stoff m
da uorganisk stof
sv oorganiskt ämne n
es materia f inorgánica
it elemento m inorganico

INPUT, 2533

1919
INSECT
fr insecte *m*
ne insekt *n*
de Insekt *n*
da insekt *n*
sv insekt
es insecto *m*
it insetto *m*

INSECT, adult, 1872
—, scale, 3258

1920
INSECT CONTROL
fr lutte *f* contre les insectes
ne insektenbestrijding
de Insektenbekämpfung *f*
da insektbekæmpelse
sv insektsbekämpning
es lucha *f* contra los
 insectos; combate *m* de
 insectos
it lotta *f* contro gli insetti

1921
INSECT DAMAGE
fr dégat *m* causé par les
 insectes
ne vreterij; vraat
de Frass *m*
da gnavpletter
sv gnagskada
es daño *m* de insectos
 masticadores
it danno *m* causato dagli
 insetti

INSECT FEEDER, 1923

1922
INSECTICIDE
fr insecticide *f*
ne insekticide
de Insektenvertilgungs-
 mittel *n*
da insektdræbende middel
sv insektsdödande medel
es insecticida
it insetticida *f*

1923
INSECTIVORE;
INSECT FEEDER
fr insectivore *m*
ne insekteneter

de Insektenfresser *m*
da insektæder *mf*
sv insektsätare
es insectivoro *m*
it insettivoro *m*

1924
INSPECTION
fr contrôle *m*
ne keuring
de Prüfung *f*;
 Kontrollbesichtigung *f*
da bedømmelse
sv prövning; undersökning;
 inspektion
es inspección *f*
it esame *m*; verifica *f*

1925
INSPECTION BELT
fr table *f* de visite
ne leesband
de Verleseband *n*
da kontrolbånd *n*
sv sorteringsband *n*
es cinta *f* de control
it cinghia *f* d'ispezione

1926
INSPECTION FOR
DISEASE
fr recherche *f* des bulbes
 malades
ne ziekzoeken *n*
de auf Krankheiten über-
 prüfen
da sygdomskontrol
sv sjukdomskontroll
es buscar plantas enfermas
it cercare pianti ammalati

1927
INSPECTION SERVICE
fr service *m* de contrôle
ne keuringsdienst
de Anerkennungsdienst *m*;
 Anerkennungsstelle *f*
da kåringstjeneste
sv kontrollverksamhet
es servicio *m* de inspección
it servizio *m* di controllo

1928
INSPECTOR
fr inspecteur *m*;
 contrôleur *m*
ne keurmeester
de Besichtiger *m*; Prüfer *m*
da inspektør
sv inspektör; kontrollant
es inspector *m*
it ispettore *m*

1929
INSULATING FILM;
PLASTIC FILM
fr feuille *f* isolante
ne isolatiefolie
de Isolierfolie *f*
da isoleringsfolie
sv isoleringsfolie
es hoja *f* aislante
it foglio *m* isolanto

1930
INSULATING MATERIAL
fr matériau *m* isolant
ne isolatiemateriaal *n*
de Isoliermaterial *n*
da isolationsmateriale
sv isoleringsmaterial *n*
es material *m* aislante
it isolante *m*

1931
INSULATION
fr isolation *f*
ne isolatie
de Isolation *f*
da isolation
sv isolering
es aislamiento *m*
it isolamento *m*

1932
INTEGRAL UNIT COSTS
fr prix *m* de revient complet
ne integrale kostprijs
de Vollkosten *pl*
da integral fremstillings-
 omkostninger
sv total självkostnad
es coste *m* de producción
 total
it costi *m pl* unitari totali
 (integrali)

1933
INTENSIFICATION
fr intensification f
ne intensivering
de Intensivierung f
da intensivering
sv intensifiering
es intensificación f
it intensificazione f

1934
INTENSITY OF
LIGHTING
fr intensité f d'exposition
ne belichtingssterkte
de Beleuchtungsstärke f
da belysningsstyrke
sv belysningsstyrka
es fuerza f de iluminación;
   poder m de iluminación
it intensità f della delucidazione

1935
INTENSITY OF RADIA-
TION (IRRADIANCE)
fr éclairement m énergétique
ne bestralingssterkte
de (Be)strahlungsstärke f
da bestrålingsintensitet
sv strålningsintensitet
es iluminación f energética
it intensità f della radiazione

1936
INTERCHANGEABLE
fr interchangeable
ne uitwisselbaar
de auswechselbar
da ombyttelig
sv utbytbar
es intercambiable
it permutabile

1937
INTERCROPPING;
DOUBLE CROPPING
fr culture f intercalaire
ne tussenteelt
de Zwischenkultur f
da mellemkultur
sv mellankultur
es cultivo m intercalado
it coltura f interposta

1938
INTERFERING LAYER
fr couche f perturbante
ne storende laag
de Störschicht f
da forstyrende lag n
sv störande skikt n
es capa f perturbante
it strato m importuno

1939
INTERGRAFTING;
DOUBLE WORKING
fr intermédiaire m
ne tussenstam
de Zwischenpfropfung f;
   Zwischenveredlung f;
   Stammbildner m
da mellempodning
sv mellanförädling
es injerto m intermedio
it innesto m ad intermediario

1940
INTERMEDIATE STOCK
fr intermédiaire m
ne tussenstam
de Zwischenstamm m
da mellemstamme
sv intermediär
es tronco m intermedio
it intermediario m

1941
INTERNAL
BREAKDOWN
fr brunissure f interne
ne inwendig bederf n
de Fleischbräune f
da møsk
sv invändig förruttnelse
es descomposición f interna
it disfacimento m interno

1942
INTERNAL DEVELOP-
MENT INSPECTION
(BULBS)
fr vérification f du stade
ne stadiumonderzoek n
de Stadien-Untersuchung f
da stadium-undersøgelse
sv stadiumundersökning

es comprobación f de fases
   en el crecimiento
it examen m della fase

1943
INTERNAL TRANSPORT
fr transport m intérieur
ne intern transport n
de innerbetrieblicher Transport m
da internere transport
sv intern transport
es transporte m interior
it trasporto m interno

1944
INTERNODE
fr entrenoeud m
ne lid n
de Glied n
da led n; internodie n
sv led
es entrenudo m
it internodo m

1945
INTERRUPTION IN
GROWTH
fr arrêt m de croissance;
   perturbation f dans la
   croissance
ne groeistoornis
de Wachstumsstockung f
da vækstafbrydelse
sv tillväxtrubbning
es interrupción f
   del crecimiento
it disturbo m vegetativo

1946
INTERVEINAL MOSAIC
fr mosaïque f internervaire
ne tussennervig mozaïek n
de Zwischennervenmosaik n
da mellemnerve-mosaik
sv internervmosaik
es mosaico m internerval
it mosaico m internervale

1947
INTERVENTION
fr intervention f
ne interventie

de Intervention *f*
da intervention
sv intervention
es intervención *f*
it intervento *m*

1948
INVEST, TO
fr placer; investir
ne investeren
de investieren
da investere
sv investera
es invertir
it investire

1949
INVOLUCRE
fr involucre *m*
ne omwindsel *n*
de Hülle *f*
da svøb *n*
sv svepe *n*
es envoltura *f*
it involucro *m*

1950
IONIZING RADIATION
fr rayonnement *m* ionisant;
    radiation *f* ionisante
ne ioniserende straling
de ionisierende Strahlung *f*
da ioniserende stråling
sv joniserande strålning
es radiación *f* ionizante
it irradiazione *f* ionisante

1951
IRIS
fr Iris *m*
ne Iris
de Schwertlilie *f*; Iris *f*
da Iris
sv Iris
es Iris *f*
it Iris *f*
    Iris L.

1952
IRON
fr fer *m*
ne ijzer *n*
de Eisen *n*
da jern *n*

sv järn *n*
es hierro *m*
it ferro *m*

1953
IRON DEFICIENCY
fr carence *f* en fer
ne ijzergebrek *n*
de Eisenmangel *m*
da jernmangel
sv järnbrist
es carencia *f* en hierro
it deficienza *f* di ferro

1954
IRON PAN (SOIL)
fr alios *m*
ne oerbank
de Ortstein *m*
da al
sv ortsten
es arenisca *f* ferruginosa;
    capa *f* de hierro
it limonite *f*

IRONWEED, 2001

1955
IRRADIATE, TO
fr rayonner sur; irradier
ne bestralen
de bestrahlen
da bestråle
sv bestråla
es irradiar
it rischiarare

1956
IRRADIATION
fr irradiation *f*
ne bestraling
de (Be)strahlung *f*
da bestråling
sv bestrålning
es irradiación *f*
it irradiazione *f*

1957
IRREGULAR
fr irrégulier
ne onregelmatig
de unregelmässig
da uregelmæssig
sv oregelbunden

es irregular
it irregolare

1958
IRRIGATE, TO
fr irriguer
ne bevloeien; irrigeren
de bewässern
da vande; overrisle
sv bevattna
es regar; irrigar
it irrigare

1959
IRRIGATION;
WATERING
fr irrigation *f*; arrosage *n*
ne bevloeiing; irrigatie;
    beregening
de Bewässerung *f*;
    Berieselung *f*
da overrisling; irrigation
sv bevattning
es irrigación *f*; riego *m*
it irrigazione *f*; aspersione *f*

IRRIGATION, trickle,
3931

1960
IRRIGATION
CONTROLLER
fr programmateur *m*
    d'irrigation
ne regenautomaat
de Beregnungsautomat *m*
da vandningsautomat
sv bevattningsautomatik
es programador *m* de riego
it regolatore *f* d'irrigazione

IRRIGATION
EQUIPMENT, 3580

1961
IRRIGATION WATER
fr eau *f* d'arrosage
ne gietwater *n*
de Giesswasser *n*
da vandingsvand *n*
sv vattningsvatten *n*
es agua *m* de riego
it acqua *f* annaffiale

1962
ISOLATED (bot.)
fr isolé
ne alleenstaand
de einzelstehend
da enkelt
sv ensamställd
es aislado; solitario
it isolato

1963
ITALIAN RYEGRASS
fr ray-grass *m* d'Italie
ne Italiaans raaigras *n*
de italienisches Raygras *n*
da Italiensk rajgræs *n*
sv italienskt rajgräs *n*
es ballico *m* italiano
it loglio *m* italiano
Lolium multiflorum Lam.

1964
IVY
fr lierre *m*
ne klimop

de Efeu *m*
da vedbend
sv murgröna
es hiedra *f*
it edera *f*
Hedera helix L.

IVY, ground,   1669

1965
IVY BRYOBIA
fr bryobia *f*
ne klimopspintmijt
de Spinnmilbe *f*
da vedbend spindemide
sv murgrönskvalster *n*
es ácaro *m* de la hiedra
it briobia *f* dell'edera
Bryobia kissophila
v. Eyndh.

1966
IVY LEAVED
PELARGONIUM
fr géranium *m* à feuilles de
lierre; géranium-lierre *m*

ne hanggeranium
de Efeupelargonie *f*
da hængepelargonie
sv murgrönspelargon;
hängpelargon
es geranio *m* hiedra
it geranio *m* ederaceo
Pelargonium peltatum
Ait.

1967
IVY-LEAVED
SPEEDWELL
fr veronique *f* à feuille de
lierre
ne klimopbladige ereprijs
de efeublättriger Ehren-
preis *m*
da vedbend-ærenpris
sv murgrönsveronika
es té *m* de Europa
it morso *m* di gallina;
veronica *f* ederella
Veronica hederafolia L.

J

1968
JACKET COOLING
fr chambre f à double parois
ne mantelkoeling
de Mantelkühlung f
da kappekøling
sv mantelkylning
es cámara f de doble pared
it magazzino m a doppia
    parete

1969
JACOB'S LADDER
fr polemonie m
ne jacobsladder
de Himmelsleiter f;
    Sperrkraut n
da Jacobsstige
sv blågull; jakobs stege
es escala f de Jacob
it Polemonium m
    Polemonium coeruleum L.

1970
JAM
fr confiture f
ne jam
de Marmelade f
da syltetøj n
sv sylt n
es mermelada f
it marmellatta f

1971
JAPANESE AZALEA
fr azalée f du Japon
ne Japanse azalea
de japanische Azalee f
da Japansk azalea
sv japansk azalea
es azalea f del japón
it azalea f giapponese
    Rhododendron obtusum
    (Lindl.) Planch.

1972
JAPANESE CHERRY
fr cerisier m du Japon
ne Japanse sierkers

de japanische Zierkirsche f
da Japansk kirsebær n
sv japanskt körsbär n
es ciruelo m japonés;
    cerezo m de adorno
    japonés
it ciliegio m giapponese
    Prunus serrulata Lindl.

1973
JAPANESE QUINCE
fr cognassier m du Japon
ne Japanse kwee
de japanische Scheinquitte f
da lille japankvæde
sv liten rosenkvitten
es membrillero m japonés
it cologno m giapponese
    Chaenomeles japonica
    Schn.

1974
JAPANESE WINEBERRY
fr ronce f du Japon
ne Japanse wijnbes
de japanische Weinbeere f
da vinbrombær n
sv vinhallon n
es zarza f japonesa
it Rubus m phoenicolasius
    Rubus phoenicolasius
    Maxim.

JASMINE, winter, 4177

1975
JAY
fr geai m
ne vlaamse gaai
de Holzhäher m
da skovskade
sv nötskrika
es grajo m
it ghiandaia f
    Garrulus glandarius L.

1976
JELLY
fr gelée f
ne gelei

de Gelee n
da gelé
sv gelé n
es jalea f; gelatina f
it gelatina f

1977
JERUSALEM
ARTICHOKE
fr topinambour m
ne aardpeer
de Topinambur f
da jordskok
sv jordärtskocka
es tupinambo m
it topinambur m
    Helianthus tuberosus L.

JEW'S MALLOW, 1994

1978
JOINT (bot.); **NODE**
fr noeud m
ne knoop
de Knoten m
da knude
sv ledknut
es nudo m
it nodo m

1979
JOINT COSTS
fr coûts m pl liés
ne gemeenschappelijke
    kosten
de Gemeinkosten f pl der
    Kuppelproduktion
da generalomkostninger
sv samkostnad;
    gemensam kostnad
es costes m pl unidos
it costi m pl congiunti

1980
JONATHAN SPOT
fr taches f pl de Jonathan
ne Jonathanvlekken
de Jonathanfleckenkrank-
    heit f

da jonathanplet
sv Jonathanfläck
es manchas *f pl* de Jonathan
it macchie *fpl* di Jonathan

1981
JONQUIL
fr jonquille *f*
ne trosnarcis
de Jonquille *f*; Tazette *f*;
  Doldennarzisse *f*
da sivnarcis
sv flocknarciss
es narciso *m* arracimado
it narciso *m* a grappolo

1982
JUDAS TREE
fr gainier *m*; arbre *m* de
  Judée
ne Judasboom
de Judasbaum *m*
da judastræ
sv judasträd *n*
es árbol *m* de Judea
it albero *m* di Giuda
  Cersis siliquastrum L.

1983
JUICE; SAP
fr suc *m*; jus *m*; sève *f*
ne sap *n*
de Saft *m*
da saft

sv saft; sav
es jugo *m*
it succo L.

1984
JUICY
fr juteux; plein de sève
ne sappig
de saftig
da saftig
sv saftig
es jugoso
it sugoso

JUNE BERRY,   3456

1985
JUNE DROP
fr chute *f* des fruits
ne rui; vruchtrui;
  junival
de Junifall *m*
da frugtfald
sv kartfall
es caída *f* intempesiva
it cascata *f* estivale di frutti

1986
JUNIPER
fr genévrier *m*
ne jeneverbes
de Wacholder *m*
da ene
sv en

es enebro *m*
it ginepro *m*
  Juniperus L.

1987
**JUNIPER WEBBER
MOTH;
JUNIPER WEBWORM
(US)**
ne jeneverbesmot
da enebærmøl *n*
sv enbärsmal *n*
es polilla *f* del enebro
  Dichomeris marginellus F.

1988
JUTE
fr jute *m*
ne jute
de Jute *f*
da jute
sv jute; juteväv
es yute *m*
it iuta *f*

1989
JUVENILE FORM
fr forme *f* juvénile
ne jeugdvorm
de Jugendform *f*
da ungdomsform
sv ungdomsform
es forma *f* juvenil
it periodo *m* giovanile

K

KAIL, 1991

1990
KAINIT
fr kaïnite f
ne kainiet n
de Kainit m
da kainit
sv kainit
es cainita f
it cainito m

1991
KALE; KAIL
fr chou m vert
ne groene kool; boerenkool
de Grünkohl
da grønkål
sv grönkål
es col f enana
it cavolo m verzotto
    Brassica oleracea L.

KEDLOCK, 645

1992
KEEL (bot.)
fr carène f
ne kiel
de Kiel m
da køl
sv köl
es quilla f
it carena f

KEEPING QUALITY,
1165

1993
KENTIA PALM
fr kentia m
ne kentia
de Kentie f
da kentia (palme)
sv kentiapalm;
    förmakspalm
es kentia f
it kentia f
    Howeia Becc.

KERNEL, 2747

1994
KERRIA;
JEW'S MALLOW
fr corète m du Japon
ne Kerria; ranonkelstruik
de Ranunkelstrauch m;
    Kerria f
da ranunkelbusk
sv Kerria
es querria f
it Kerria f
    Kerria DC

1995
KEW WEED;
GALLANT SOLDIER
fr galinsoga m
ne knopkruid n
de kleinblütiges Franzosen-
    kraut n; Knopfkraut n
da håret kortstråle
sv gängel
es galinsoa f
it galinsoga f
    Galinsoga parviflora Car.

1996
KIDNEY BEAN
fr haricot m blanc; haricot
    m rouge
ne slaboon; witte boon;
    bruine boon
de Schwertbohne f; braune
    Bohne f
da snittebønne; brune bønne
sv skärböna; brun böna
es frijole m; alubia f blanca
it fagiolo m
    Phaseolis vulgaris L.

KIDNEY-SHAPED, 3100

1997
KIESERITE
fr kiésérite f
ne kieseriet n
de Kieserit m

da kieseret
sv kieserit
es cieserita
it sulfato m magnesiaco

KITCHEN GARDEN,
4008

1998
(KITCHEN)SALT
CONTENT
fr teneur m en chlorure de
    sodium
ne keukenzoutgehalte;
    NaCl-gehalte
de NaCl-Gehalt m
da kogsaltinhold n
sv (kok)salthalt n
es concentración f salina
it valore m di sale comune

1999
KNAPSACK
MISTBLOWER
fr atomiseur m à dos
ne rugnevelspuit
de Motorrückensprühgerät n
da motordrevet rygtåge-
    sprøjte
sv motorryggsspruta
es atomizador m de espalda
it atomizzatore m a spalla

2000
KNIFE COULTER
fr coutre m droit
ne meskouter
de Sech n
da plovøskær n
sv knivrist
es cuchilla f del arado
it coltro m a coltello

2001
KNOTGRASS;
IRONWEED
fr renouée f des oiseaux
ne varkensgras n
de Vogelknöterich m

da vej-pileurt
sv trampört
es correhuela *f* de los cami-
    nos;
    sanguinaria *f* major
it coreggiola*f*;
    centinodia *m*
    Polygonum aviculare L.

2002
KNOTTY
fr noueux
ne knoestig
de knotig
da knudret
sv kvistig
es nudoso
it nodoso

2003
KNOTWEED
fr renouée *f*
ne duizendknoop
de Knöterich *m*
da pileurt; slangeurt
sv pilört; ormrot
es bistorta *f*
it centinodia *f*
    Polygonum L.

2004
KNUCKLING (TULIP)
fr col *m* de cygne
ne onderzeeër
de Blindwuchspflanze *f*
da ombøjede spirer;
    'blindgænger' *m*

sv omböjd stjälk; 'blind-
    gångare'
es tallo *m* encorvado
it germoglio *m* ripiegato

2005
KOHL RABI
fr chou-rave *m*
ne koolrabi
de Kohlrabi *m*
da knudekål;
    glaskålrabi
sv kålrabbi
es colirábano *m*
it cavolo rapa *m*

KOSTER'S SPRUCE,
386

L

2006
LABEL
fr étiquette *f*
ne etiket *n*
de Etikett *n*
da etiket; mærkepind
sv etikett
es etiqueta *f*
it etichetta *f*

2007
LABIATE; TWO-LIPPED
fr bilabié
ne tweelippig
de zweilippig
da tolæbet
sv tvåläppig
es bilabial
it bilobo

2008
LABIATE
fr labiée *f*
ne lipbloem
de Lippenblume *f*
da læbeblomst
sv läppformig blomma
es labiada *f*
it fiore *m* labiato
Labiatae

2009
LABOUR BUDGET
fr budget *m* de travail
ne bedrijfsbegroting
de Arbeitsvoranschlag *m*
da virksomhedsbudget *n*
sv driftsplanering
es presupuesto *m* de trabajo
it piano *m* di lavoro

2010
LABOUR COSTS
fr coût *m* du travail
ne arbeidskosten
de Arbeitskosten *pl*; Lohn-
   aufwand *m*
da lønomkostninger
sv arbetskostnader

es coste *m* de la mano de
   obra
it costo *m* del lavoro

2011
LABOUR FORCE
fr main d'oeuvre *f* disponible
ne arbeidsbezetting
de Arbeitskräftebesatz *m*
da arbejdsstyrke
sv arbetsstyrka
es mano *f* de obra utilizada
it forza *f* di lavoro

2012
LABOUR INTENSIVE
fr demandant une main-
   d'oeuvre nombreuse
ne arbeidsintensief
de arbeitsintensiv
da arbejdsintensiv
sv arbetsintensiv
es de trabajo intensivo
it esigente molto lavoro

2013
LABOUR
PRODUCTIVITY
fr productivité *f* du travail
ne arbeidsproduktiviteit
de Arbeitsproduktivität *f*
da arbejdsproduktivitet
sv arbetsproduktivitet
es productividad *f* del
   trabajo
it produttività *f* del lavoro

2014
LABOUR RECORDING
sv enregistrement *m* du
   travail
ne tijdschrijving
de Arbeitsbuchführung *f*
   Arbeitsaufzeichnungen *f*
   pl
da arbejdsregnskab
sv arbetsredovisning;
   registrering

es registro *m* del trabajo
it registrazione *f* del lavoro

2015
LABOUR RECORDS
fr enregistrement *m* du
   travail
ne arbeidsboekhouding
de Arbeitsbuchführung *f*
da arbejdsregnskab *n*
sv arbetsbokföring
es registro *m* del trabajo
it registrazione *f* del lavoro

2016
LABOUR REQUIREMENT
fr besoins *m pl* en main
   d'oeuvre
ne arbeidsbehoefte
de Arbeitsbedarf *m*
da arbejdsbehov *n*
sv arbetsbehov *n*
es necesidades *f pl* de mano
   de obra
it fabbisogno *m* di manodo-
   pera

2017
LABOUR SAVING
fr économie *f* de travail
ne arbeidsbesparing
de Arbeitsersparnis *n*
da arbejdsbesparelse
sv arbetsbesparing
es economía *f* de trabajo
it economia *f* del lavoro

2018
LABURNUM
fr cytise *m*; aubour *m*;
   faux-ébénier *m*
ne gouden regen
de Goldregen *m*
da guldregn
sv gullregn *n*
es laburno *m*; citiso *m*;
   lluvia *f* de oro
it pioggia *f* d'oro; citiso *m*
   Laburnum Med.

2019
LABURNUM LEAF
MINER
fr chenille *f* mineuse du
   cytise
ne goudenregendamschijf-
   mineermot
de Goldregenminiermotte *f*
da guldregnmøl
sv guldregns(minerar)mal
es minadora *f* dorada
it minatrice *f* del maggio-
   ciondolo
   Leucoptera laburnella St.

2020
LACE WING
fr névrophère
ne gaasvlieg
de Netzflügler *m*
da årevingede
sv nätvingar
es neuroptero *m*
it neurottero *m*
   Neuroptera

2021
LACKEY MOTH
fr bombyx *m* à livrée
ne ringelrupsvlinder
de Ringelspinner *m*
da ringspinder
sv ringspinnare
es oruga *f* de librea;
   oruga *f* galoneada
it bombice *m* gallonato;
   malacosoma *m*
   Malacosoma neustria L.

LADY-BIRD, 2073

2022
LADY'S MANTLE
fr alchémille *f*; manteau-de
   Notre-Dame *m*
ne vrouwenmantel
de Frauenmantel *m*
da alm-løvefod
sv daggkåpa
es alquimila *f*; pie de léon *m*
it stellaria *f*
   Alchemilla vulgaris L.

LAMB'S LETTUCE, 851

2023
LAMELLA; GILL
(MUSHROOM)
fr lamelle *f*
ne lamel; plaatje *n*
de Lamelle *f*; Blätter *n pl*
da lamel
sv lamell
es laminilla *f*
it lamella *f*

2024
LAMINATED BEAM
fr poutre *f* lamelée collée
ne gelijmde balk
de geleimter Holzbinder *m*
da lamineret træbjælke
sv limmad balk
es viga *f* laminada
it trave *f* composta

2025
LAND; SOIL
fr sol *m*; terrain *n*; terre *f*
ne land *n*; grond; bodem
de Land *n*; Erde *f*; Boden *m*
da land; jord
sv mark; land *n*; jord
es suelo *m*; tierra *f*
it terra *f*; suolo *m*

2026
LAND CHARGES
fr charges *f pl* foncières
ne grondlasten
de Grundbesitzabgaben *f pl*
da grundværskatter
sv fastighetsskatt; egen-
   domsskatt
es impuestos *m pl* de arraigo
it spese *f pl* fondiarie

2027
LAND-CONSOLIDATION;
RE-ALLOCATION
fr remembrement *m*
ne ruilverkaveling
de Flurbereinigung *f*
da jordfordeling
sv strukturrationalisering;
   laga skifte
es concentración *f* parcelaria
it ricomposizione *f*;
   scambio *m* di lotti di terre

2028
LANDSCAPE; SCENERY
fr paysage *m*
ne landschap *n*
de Landschaft *f*
da landskab *n*
sv landskap *n*
es paisaje *m*
it paesaggio *m*

2029
LAP
fr recouvrement *m*
ne overlap
de Ueberdeckung *f*
da overlapning
sv överlappning
es recubrimiento *m*
it sovrapposizione *f*

2030
LARCH
fr mélèze *m*
ne lork
de Lärche *f*
da lærke
sv lärk
es alerce *m*
it larice *m*
   Larix Mill.

LARCH ADELGES, 4203

2031
LARCH CASEBEARER
fr coléophore *m* du mélèze
ne lariksmot
de Lärchenminiermotte *f*
da lærke-sækmøl *n*
sv lärksäckmal *n*
es polilla *f* minadora del
   alerce; minadora *f* de las
   hojas del alerce
it minatrice *f* delle foglie di
   larice; coleofora *f* del
   larice
   Coleophora laricella Hb.

2032
LARGE FRUIT BARK
BEETLE; FRUIT BARK
BEETLE; SHOT-HOLE
BORER (US);
LARGER SHOT-HOLE
BORER (US)
fr scolyte *m* du pommier
ne spintkever

de grosser Obstbaumsplintkäfer *m*
da Scolytus mali
sv större fruktträdssplint-
borre
es barrenillo *m* grande del
manzano
it scolito *m* degli alberi da
frutto
Scolytus rugulosus Rtzb.;
Soclytus mali Bachst.

2033
LARGE LEAVED
fr à grandes feuilles
ne grootbladig
de grossblättrig
da storbladet
sv storbladig
es de grandes hojas
it grandifoglia

LARGE NARCISSUS
FLY, 2445

2034
LARGE OFFSET (BULB
GROWING)
fr caïeu *m*
ne spaan (groot)
de platte; blühfähige
Nebenzwiebel *f*
da store yngleløg *n*
sv sidölök (stor)
es bulbo *m* lateral
it bulbo *m* laterale

2035
LARGE POPLAR
LONGHORN
fr grande saperde *f* du
peuplier
ne populiereboktor
de grosser Pappelbock *m*
da poppelbuk
sv större aspvedbock
es saperda *f* de los chopos
it saperda *f* maggiore del
pioppo
Saperda carcharias L.

2036
LARGE ROSE SAWFLY
(ARGE ROSAE)
fr tenthrède *f* du rosier
ne gele rozebladwesp

de Rosen-Bürstenhorn-
wespe *f*
da Arge rosae
sv gul borsthornsstekel
es falsa oruga *f* amarilla del
rosal; tentredino *m*
común del rosal
it ilatoma *f* delle rosa
Arge ochropus Gm.;
Arge rosae L.

LARGER SHOT-HOLE
BORER, 2032

2037
LARGE WHITE BUTTER-
FLY
fr piéride *f* du chou
ne groot koolwitje *n*
de grosser Kohlweissling *m*
da stor kålsommerfugl
sv kålfjäril
es gran mariposa *f* blanca
de la col
it cavolaia *f* maggiore
Pieris brassica L.

2038
LARGE WILLOW APHID
fr gros puceron *m* du saule
ne dromedarisluis
de grosse Weidenrinden-
laus *f*
da pilbarklus
sv pilbarklus
es pulgón *m* del sauce
it lacno *m* del salice; afide
m del salice
Tuberolachnus salignus
Gmel.

2039
LARKSPUR;
DELPHINIUM
fr pied *m* d'alouette
ne ridderspoor
de Rittersporn *n*
da ridderspore
sv riddarsporre
es espuela *f* de caballero
it cappuccio *m*
Delphinium L.

2040
LARVA
fr larve *f*
ne larve
de Larve *f*
da larve
sv larv
es larva *f*
it larva *f*

LARVA, sawfly, 3257

2041
LARVICIDAL
fr larvicide
ne larvedodend
de larvazide
da larvedræbende
sv larvdödande
es larvicido
it larvicido

2042
LASTING; DURABLE
fr durable
ne houdbaar
de haltbar
da holdbar
sv hållbar
es durable
it durevole

2043
LATERAL BRANCH
fr branche *f* latérale
ne zijtak
de Seitenast *m* ;
da sidegren
sv sidogren
es rama *f* lateral
it ramo *m* laterale

2044
LATERAL ROOT
fr racine *f* secondaire
ne bijwortel; zijwortel
de Nebenwurzel *f*; Seiten-
wurzel *f*
da birod; siderod
sv birot; sidorot
es raíz *f* secundaria
it radice *f* laterale

2045
LATERAL SHOOT
fr pousse *f* latérale
ne zijscheut
de Seitentrieb *m*
da sideskud *n*
sv sidoskott *n*
es vástago *m* lateral; chupón
   m; retoño *m* lateral; tallo
   m lateral
it germoglio *m* laterale

2046
LATEX;
VEGETABLE MILK
fr latex *m*
ne melksap *n*
de Milchsaft *m*
da mælkesaft
sv mjölksaft
es latex *m*; jugo *m* lechoso
it sugo *m* lattiginoso

2047
LATH
fr latte *f*
ne lat
de Latte *f*
da lægte
sv läkt; ribba; spjäla
es listón *m*
it travicello *m*

2048
LATIN SQUARE
fr carré *m* latin
ne latijns vierkant *n*
de lateinisches Quadrat *n*
da kvadrat *n*
sv kvadrat *n*
es cuadrado *m* latino
it quadro *m* latino

2049
LATTICE GIRDER
fr ferme *f* à treillis
ne vakwerkspant
de Fachwerkbinder *m*
da gitterspær *n*
sv fackverkstakstol
es cercha *f* de malla
it travatura *f* a traliccio

LAUREL, 268
—, cherry, 652
—, mountain, 533

2050
LAVENDER
fr lavande *f*
ne lavendel
de Lavendel *m*
da lavendel
sv lavendel
es espliego *m*; lavándula *f*
it lavanda *f*
   Lavandula L.

2051
LAWN
fr gazon *m*; pelouse *f*
ne gazon *n*
de Rasen *m*
da plæne
sv gräsplan; gräsmatta
es césped *m*
it tappeto *m* erboso

2052
LAWN BROOM; BESOM
fr balai *m* à gazon
ne gazonbezem
de Rasenbesen *m*
da plænekost
sv trädgårdskvast
es escoba *f* de retamas
it scopa *f* a tappeto

2053
LAWN MOWER
fr tondeuse *f*
ne gazonmaaier
de Rasenmäher *m*
da græsslåmaskine
sv gräsklippare
es segadora *f* de hierba
it falciatrice *f* da prato

2054
LAWN RAKE
fr râteau *m* à herbe
ne grashark
de Grasharke *f*
da græsrive
sv lövräfsa
es rastrillo *m*
it rastrello *m* ad erba

2055
LAWN SPRINKLER
fr arroseur *m*
ne gazonsproeier
de Rasensprenger *m*
da plænevander
sv bevattningsapparat för
   gräsplan
es regardor *m* de prado
it spruzzatore *m* a tappeto

2056
LAYER
fr marcotte *f*
ne aflegger
de Absenker *m*; Ableger *m*
da aflægger
sv avläggare
es acodo *m*
it margotta *f*

2057
LAYER
fr couche *f*
ne laag
de Schicht *f*
da lag *n*
sv skikt *n*; lag *n*
es capa *f*
it strato *m*

LAYER, compacted, 782

2058
LAYER, TO
fr marcotter
ne afleggen
de ablegen; absenken
da aflægge
sv avlägga
es acodar
it margottare

2059
LAYER OF EARTH
fr couche *f* de terre
ne grondlaag
de Erdschicht *f*
da jordlag *n*
sv jordlager *n*
es capa *f* de tierra
it strato *m* di terra

2060
LAYERS, IN;
STRATIFORM
fr par couches
ne laagsgewijs
de lagenweise
da lagvis
sv i skikt; varvvis
es en capas
it stratiforme

2061
LAY FLAT TUBE
fr tube *m* souple
ne buisfolie
de Schlauchfolie *f*
da slangefolie
sv slangfolie
es tubo *m* flexible
it tubo *m* parete liscio

2062
LAYING-IN GROUND
fr jauge *f*
ne kuilhoek
de Einschlag *m*; Einschlag-
    platz *m*
da indslagsplads
sv plats för att gräva ned
es terreno *m* destinado para
    el almacenamiento de
    plantas
it sito *m* per mettere pianta

LAY TURF, TO, 875

2063
LEACH, TO (SOIL)
fr lessiver
ne uitspoelen
de ausspülen
da udvaskning
sv tvätta ur; spola ur
es deslavar; deslavazar
it scavare (dall'acqua)

2064
LEACHING
fr lessivage *m*
ne uitspoeling
de Abschwemmung *f*;
    Auswaschung *f*
da udspulning
sv urlakning
es lavado *m*; lixiviación *f*
it scavare (dall' acqua)

2065
LEAD ARSENATE
fr arséniate *m* de plomb
ne loodarsenaat *n*
de Bleiarsenat *n*
da blyarsenat
sv blyarsenat
es arseniato *m* de plomo
it arseniato *m* di piombo

LEADER, 2066

2066
LEADING SHOOT;
LEADER
fr pousse *f* principale;
    flèche *f*
ne hoofdscheut
de Haupttrieb *m*
da ledeskud *n*
sv ledskott *n*; toppskott *n*
es tallo *m* principal
it rampollo *m* principale

2067
LEADWORT
fr dentelaire *f*
ne loodkruid *n*
de Bleiwurz *f*
da blyrod
sv blyblomma
es dentelaria *f*; teleza *f*
it Plumbago *m*
    Plumbago L.

2068
LEAF
fr feuille *f*
ne blad *n*
de Blatt *n*
da blad *n*
sv blad *n*
es hoja *f*
it foglia *f*

LEAF, electronic, 1208
—, fern, 3875
—, seed, 866

2069
LEAF ANALYSIS
fr diagnostic *m* foliaire
ne bladonderzoek *n*;
    bladanalyse

de Blattanalyse *f*
da bladanalyse
sv bladanalys
es diagnóstico *m* foliar;
    análisis *m* foliar
it analisi *f* fogliale

2070
LEAF AND POD SPOT
(PEA)
fr anthracnose *f*
ne vlekkenziekte
de Brennfleckenkrankheit *f*
da ærtesyge
sv ärtfläcksjuka
es antracnosis *f*; 'rabia' *f*
it antracnosi *f*; seccume *f*;
    nebbia *f*
    Ascochyta pisi Lib.;
    Didymella pinodes;
    Ascochyta pinodella
    L.K. Jones

2071
LEAF AXIL
fr aisselle *f* des feuilles
ne bladoksel
de Blattacksel *f*
da bladakse
sv bladaxel
es axila *f* de la hoja
it ascella *f*

2072
LEAF BASE;
PHYLLOPODIUM
fr base *f* du limbe
ne bladvoet
de Blattgrund *m*
de bladfod
sv bladfot
es base *f* del limbo
it piede *m* fogliale

2073
LEAF BEETLE;
LADY-BIRD
fr coccinelle *f*
ne lieveheersbeestje *n*
de Marienkäfer *m*
da bladbille
sv bladbagge
es crisomelido *m*; mariquita *f*
it crisomelide *m*
    Chrysomelidae

LEAF BEETLE, alder, 66
—, cranberry tree, 879
—, red poplar, 3069

2074
LEAF BLIGHT
(CYDONIA)
fr entomosporiose *f*
ne bladvlekkenziekte
de Blattfleckenkrankheit *f*
da vildingsvamp
sv vildstamsvamp
es punteado *m* de las hojas
it imbrunimento *m* delle
   foglie
   Diplocarpon maculatum

2075
LEAF BLIGHT
(POPULUS)
fr taches *f pl* noires sur les
   feuilles
ne bladvlekkenziekte
de Blattfleckenkrankheit *f*
da bladpletsyge
sv bladfläcksjuka
es antracnosis *f*
it macchie *f pl* fogliari
   Marssonina spp.

2076
LEAF BLIGHT (THUJA)
fr taches *f pl* sur tiges
ne taksterfte
de Laubbräune *f*
da thuja-bladpletsvamp
sv Didymascella thujina
es moho *m* de las ramas
it macchie *f pl* fogliari
   Didymascella thujina
   (Durand) Maire

LEAF BLISTER, yellow,
4225

2077
LEAF BLISTER MITE
fr phytopte *m*
ne pokziekte
de Pockenmilbe *f*; Blatt-
   pocken-Gallmilbe *f*
da galmide
sv gallkvalster *n*

es sarna *f* de las hojas
it eriofide *m*
   Phytoptus spp.

2078
LEAF BLOTCH
(AESCULUS)
fr taches *f pl* foliaires
ne bladvlekkenziekte
de Blattfallkrankheit *f*
da bladpletsyge
sv bladfläcksjuka
es punteado *m* de las hojas
it seccume *m* fogliare
   Guignardia aesculi (Peek)
   Stew.

2079
LEAF BUD
fr bourgeon *m* foliaire
ne bladknop
de Blattknospe *f*
da bladknop
sv bladknopp
es yema *f* foliar; yema *f* de
   madera
it gemma *f*

LEAF CACTUS, 1234

LEAF CURL, 4188

2080
LEAF CUTTING
fr bouture *f* de feuille
ne bladstek
de Blattsteckling *m*
da bladstikling
sv bladstickling
es esqueje *m* de hoja
it talea *f* fogliale

2081
LEAF DAMAGE
fr dommage *m* du feuillage
ne bladbeschadiging
de Blattbeschädigung *f*
da bladskade
sv bladskada
es deterioro *m* de las hojas
it danno *m* fogliale

2082
LEAF DISEASE
(PLATANUS)
fr anthracnose *f*
ne bladvlekkenziekte
de Anthracnose *f*; Blattfleckenkrankheit *f*
da sortnæb
sv bladbränna
es antracnosis *f*; hojas *f pl*
   secas
it seccume *m* delle foglie
   Gnomonia veneta (Sacc.
   et Speg.) Kleb.

LEAF DROP, 1019

LEAF EATING WEEVIL,
447
LEAF GALL, azalea, 209

2083
LEAF GALL AND
FASCIATION;
CAULIFLOWER DISEASE
OF STRAWBERRY
fr fasciation *f* des tiges
ne rozetgal
de Blattgallen *f*; Fasziation *f*
da Corynebacterium
sv knippebakterios
es agalla *f* de roseta; fasciación
it fasciazione *f*
   Corynebacterium fascians
   (Tilford) Dowson

LEAFHOPPER, 682
—, rose, 3188

2084
LEAFLESS
fr sans feuilles
ne bladerloos
de blattlos
da bladløs
sv bladlös
es nudicaulo; deshojado
it senza foglia

2085
LEAF LETTUCE
fr laitue *f* à cueillir
ne pluksla

de Pflücksalat *m*
da pluksalat
sv bladsallad
es lechuga *f*
it lattuga *f* da taglio
    Lactuca sativa L.

2086
LEAF MARGIN
fr contour *m* du limbe
ne bladrand
de Blattrand *m*
da bladrand
sv bladrand
es borde *m* de las hojas
it limbo *m*

2087
LEAF MINER
fr mineuse *f*
ne mineervlieg
de Minierfliege *f*
da minerflue
sv minerarfluga
es mosca *f* minadora
it agromizido *m*
    Phytomyza spp.

LEAF MINER, azalea, 210
—, beach, 286
—, beet, 295
—, chrysanthemum, 681
—, holly, 1805
—, laburnum, 2019
—, lilac, 2165

LEAF-MINING-MIDGE,
1520

LEAF MOLD, 1116

2088
LEAF-MOULD
fr terreau *m* de feuille
ne bladaarde
de Lauberde *f*
da bladjord
sv lövjord
es mantillo *m* de hojas
it terra *f* vegetale

2089
LEAF MOULD
(TOMATO)
fr cladosporiose *f*;
    moisissure *f* olivâtre des
    feuilles et des fruits
ne bladvlekkenziekte
de Braunfleckigkeit *f*;
    Samtfleckenkrankheit *f*
da fløjlsplet
sv sammetsfläcksjuka
es abigarrado *m*
it cladosporiosi *f*;
    ticchiolatura *f*
    Cladosporium fulvum
    Cooke

2090
LEAF NARROWING
fr rétrécissement *m* du limbe
ne smalbladigheid
de Schmalblättrigkeit *f*
da smalbladet
sv smalbladighet
es lanceolado *m* de las hojas
it restringimento *m* della
    foglia; stenofillia *f*

LEAF ROLL, potato, 2870

2091
LEAF ROLLER
fr tordeuse *f*
ne bladroller
de Blattwickler *m*
da (blad)vikler
sv (blad)vecklare
es tortrix *m*
it torcitore *m* fogliale
    Tortricidae

LEAF ROLLER MOTH,
3891

2092
LEAF ROLLING
fr enroulement *m* foliaire
ne bladrolling
de Blattrollen *n*
da bladrulning
sv bladrullning
es enrollamiento *m* de hojas
it accartocciamento *m*;
    arrotolamento *m* della
    foglia

2093
LEAF-ROLLING ROSE
SAWFLY;
ROSE LEAF SAWFLY;
fr tenthrède *f* des feuilles du
    rosier
ne kleinste rozebladwesp
de kleinste Rosenblatt-
    rollwespe *f*
da lille rosenbladhveps
sv liten rosenbladrullstekel
es blenocampa *f* chiquita;
    oruga *f* menor del rosal
it tentredine *f* arrotolatrice
    delle foglie dei rosai
    Blennocampa pusilla
    Klug

2094
LEAF SCAR
fr cicatrice *f* foliaire
ne bladmerk *n*
de Blattnarbe *f*
da bladar *n*
sv bladärr *n*
es cicatriz *f* de la hoja
it cicatrice *f* fogliale

2095
LEAF SCORCH (ROSA);
ROSE LEAF SPOT;
ROSE ANTHRACNOSE
fr taches *f pl* foliaires;
    anthracnose *f* du rosier
ne bladvlekkenziekte
de Anthracnose *f*; Blatt-
    fleckenkrankheit *f*
da bladpletsyge
sv bladfläcksjuka
es punteado *m* de las hojas;
    antracnosis *f* del rosal
it septoriosi *f pl*; maculatura
    delle foglie
    Sphaerulina rehmiana
    Jaap; *Phyllosticta rosa-*
    rum

2096
LEAF SCORCH
(STRAWBERRY)
fr maladie *f* des taches
    pourpres
ne rode-vlekkenziekte
de Rotfleckenkrankheit *f*
da bladpletsyge

sv bladbränna
es punteado *m* rojo
it macchie *f pl* fogliari
   Diplocarpon earlianum
   (Ell. et Everh.) Wolf

2097
LEAF SHAPE
fr forme *f* du limbe
ne bladvorm
de Blattform *f*
da bladform
sv bladform
es forma *f* de las hojas
it forma *f* della foglia

2098
LEAF-SPINE
fr épine *f* foliaire
ne bladdoorn
de Blattdorn *m*
da bladtorn
sv bladtorn
es espina *f* foliar
it spino *m* fogliale

2099
LEAF SPOT
fr taches *f pl* foliaires
ne bladvlekkenziekte
de Blattfleckenkrankheit *f*
da bladpletsyge
sv bladfläcksjuka
es punteado *m* de las hojas;
   abigarrado *m* de las hojas
it macchie *f pl* fogliari;
   ticchiolatura *f*
   Alternaria spp.; *Ramularia* spp.; *Septoria* spp.;
   Phyllosticta spp.

LEAF SPOT, 1726
—, angular, 104
—, bacterial, 222, 223
—, dark, 986
—, lettuce, 410
—, rose, 2095

2100
LEAF SPOT (APIUM)
fr septoriose *f*
ne bladvlekkenziekte
de Blattfleckenkrankheit *f*

da bladpletsyge
sv bladfläcksjuka
es abigarrado *m* de las hojas
it septoriosi *f pl*
   Septoria apii-graveolentis

2101
LEAF SPOT (BETA)
fr cercosporiose *f*
ne bladvlekkenziekte
de Blattfleckenkrankheit *f*
da bladpletsyge
sv bladfläcksjuka
es cercospora *f*; enroya *f*;
   chamuscado *m*
it vaiolatura *f* delle foglie
   Cercospora beticola Sacc.;
   Ramularia beticola Fautr.
   et Lamb.

2102
LEAF SPOT (DAHLIA)
fr charbon *m* foliaire
ne bladvlekkenziekte
de Blattfleckenkrankheit *f*
da bladbrand
sv fläcksot
es punteado *m* de las hojas
it carbone *m* fogliare
   Entyloma dahliae Syd.

2103
LEAF SPOT (DIANTHUS CARYOPHYLLUS)
fr septoriose *f*
ne bladvlekkenziekte
de Blattfleckenkrankheit *f*
da bladpletsyge
sv bladfläcksjuka
es manchas *f* amarillas
it macchie *f pl* gialle
   Septoria dianthi Desm.;
   Heteropatella valtellinensis
   (Trav.) Wr.

2104
LEAF SPOT (HELLEBORUS)
fr taches *f* noires
ne zwarte-bladvlekkenziekte
de Schwarzfleckenkrankheit *f*
da bladpletsyge
sv svartfläcksjuka

es salpicón *m* negro de las
   hojas
it seccume *m* fogliare
   Coniothyrium hellibori
   Cke. et Massee

2105
LEAF SPOT (PRUNUS)
fr taches *f pl* foliaires
ne bladvlekkenziekte
de Sprühfleckenkrankheit *f*
da kirsebær-bladpletsyge
sv körsbär-bladfläcksjuka
es socarrina *f*
it seccume *m* fogliare
   Phloeosporella padi

2106
LEAF SPOT (RIBES)
fr anthracnose *f*
ne bladvalziekte
de Blattfallkrankheit *f*
da skivesvamp
sv bladfallsjuka
es caída *f* de las hojas
it seccume *m* delle foglie
   Drepano peziza ribis

2107
LEAF SPOT (SPINACIA)
fr cladosporiose *f*
ne bladvlekkenziekte
de Blattfleckenkrankheit *f*
da bladpletsyge
sv bladfläcksjuka
es punteado *m* de las hojas
it cladosporiosi *f*
   Cladosporium variabile;
   Cladosporium macrocarpum Preuss.

2108
LEAF SPOT (STRAWBERRY)
fr taches *f pl* blanches
ne witte-vlekkenziekte
de Weissfleckenkrankheit *f*
da bladpletsyge
sv ögonfläcksjuka
es salpicón *m* blanco
it vaiolotura *f*
   Mycosphaerella fragariae
   (Tul.) Lindau

2109
LEAF STALK; PETIOLE
fr pétiole *m*
ne bladsteel
de Blattstiel *m*
da bladstilk
sv bladskaft *n*
es peciolo *m*
it stelo *m*

2110
LEAF SURFACE
fr surface *f* foliaire
ne bladoppervlak *n*
de Blattoberfläche *f*
da bladoverflade
sv bladyta
es superficie *f* de la hoja
it superficie *f* fogliale

2111
LEAF TIP;
APEX OF A LEAF
fr sommet *m* du limbe
ne bladtop
de Blattspitze *f*
da bladspids
sv bladspets
es ápice *m* de la hoja
it apice *m* fogliale

2112
LEAF TISSUE;
MESSPHYLL
fr mésophylle *m*
ne bladmoes *n*
de Blattgewebe *n*
da bladkød *n*
sv bladkött
es mesófilo *m*
it parenchima *f* fogliale

2113
LEAF VEGETABLES
fr légumes *m pl* verts;
    légumes *mpl* feuillus
ne bladgroente
de Blättgemüse *n*
da bladgrønsager
sv (blad)grönsaker
es hortaliza *f* verde;
    verdura *f*
it erbaggi *m pl*

2114
LEAF WEEVIL
fr charançon *m* vert;
    curculionidé *m* vert
ne groene bladsnuitkever
de Grünrüssler *m*
da løvsnudebille
sv lövvivel
es gorgojo *m* verde del fresal
it ceutorrinco *m* verde
    Phyllobius urticae de G.

LEAF WEEVIL, brown,
447
—, pea, 2621

2115
LEAFY; FOLIATE
fr feuillé; feuillu
ne bebladerd
de beblättert
da med blade
sv med blad; lövad
es foliado
it fogliato

2116
LEAFY GALL
fr galles *f pl* foliaires;
    fasciations *f pl* des tiges
ne bloemkoolziekte
de blättrige Galle *f*;
    Blumenkohlkrankheit *f*
da knippebakteriose
sv knippebakterios
es fasciación *f*; florecido *m*
it fasciazione *f* batterica
    Corynebacterium fascians
    (Tilf) Dows.

2117
LEAFY GALL
(VARIOUS CROPS)
fr fasciation *f*
ne wratten
de blättrige Galle *f*
da knippebakteriose
sv knippebakterios
es verrugosidad *f*
it fasciazione *f* batterica
    Corynebacetrium fascians

2118
LEAN-TO GLASSHOUSE
fr serre *f* adossée
ne muurkas

de Pultdeckgewächshaus *n*
da halvtagshus *n*
sv kast; 'ensidigt' växthus *n*
es invernadero *m* adosado
it serra *f* murale

2119
LEASE
fr bail *m*
ne pacht
de Pacht *f*
da forpagtning
sv arrende *n*
es arriendo *m*
it affitto *m*

2120
LEASE CONTRACT
fr bail *m* (à ferme)
ne pachtovereenkomst
de Pachtvertrag *m*
da forpagtningskontrakt
sv arrendekontrakt *n*
es contrato *m* de arrenda-
    miento
it contratto *m* d'affitto

2121
LEASEHOLD ACT
fr loi *f* sur les fermages
ne pachtwet
de Pachtgesetz *n*
da lejelov
sv arrendelag
es ley *f* de arriendos
it legge *f* d'affito

LEATHER JACKET, 880

2122
LEATHERY (bot.)
fr coriace
ne leerachtig
de lederartig
da læderagtig
sv läderaktig
es coriáceo
it coriaceo

2123
LEAVE OFF
FLOWERING, TO
fr passer; défleurir
ne uitbloeien

de verblühen
da afblomstre
sv vissna; blomma ut
es dejar de florecer;
   marchitarse
it sfiorire

2124
LEEK
fr poireau *m*
ne prei
de Breitlauch *m*; Porree *m*
da porre
sv purjolök; purjo
es puerro *m*
it porro *m*

LEEK, house, 1841

2125
LEEK MOTH
fr teigne *f* du poireau
ne preimot
de Lauchmotte *f*
da porrmøl *n*
sv lökmal *n*
es polilla *f* del puerro
it tignola *f* della cipolla
   Acrolepia assectella Zell.

2126
LEGUME; POD
fr gousse *f*
ne peul
de Hülse *f*
da bælg
sv balja
es vaina *f*
it siliqua *f*

2127
LEGUMES
fr légumineuses *fpl*
ne peulvruchten
de Hülsenfrüchte *fpl*
da bælgfrugt *mfpl*
sv baljväxter
es leguminosas *f pl*
it legumi *mpl*

LEGUMINOUS PLANTS,
2600

2128
LEMON
fr citron *m*
ne citroen
de Zitrone *f*
da citron
sv citron
es limón *m*
it limone *m*

2129
LEMON TREE
fr citronnier *m*
ne citroenboom
de Zitronenbaum *m*
da citrontræ *n*
sv citronträd *n*
es limonero *m*
it limone *m*
   Citrus limonia Osbeck

2130
LENDER
fr prêteur *m*
ne kredietgever
de Kreditgeber *m*; Gläubiger *m*
da kreditor
sv långivare
es prestamista *m*
it prestatore *m*

2131
LENGTHENING OF
DAYLIGHT
fr allongement *m* du jour
ne dagverlenging
de Tagverlängerung *f*
da dagforlængelse
sv dagförlängning
es alargamiento *m* del día
it protrazione *f* della giornata

2132
LENGTH OF DAY
fr longueur *m* du jour
ne daglengte
de Taglänge *f*
da daglængde
sv dagslängd
es día *m* natural
it durata *f* della giornata

2133
LENTICEL
fr lenticelle *f*
ne lenticel
de Lentizelle *f*
da lenticel; barkpore
sv lenticell
es lenticula *f*; lentejilla *f*
it lenticellula *f*

2134
LENTICEL SPOT
fr taches *f pl* lenticellaires
ne lenticelvlekken
de Lentizellenfleckenkrank-
   heit *f*
da lenticelpletter
sv lenticelfläck
es manchas *f* lenticulares
it macchia *f* delle lenticellule

2135
LEOPARD MOTH
fr zeuzère *f* du poirier;
   coquette *f*
ne gele houtrups
de Blausieb *n*
da plettet træborer
sv blåfläckig träfjäril
es taladro *m* amarillo de los
   troncos
it zeuzera *f*; rodilegno *m*
   giallo; perdilegno *m*
   bianco
   Zeuzera pyrina L.

2136
LEOPARDS BANE
fr doronic *m*
ne voorjaarszonnebloem
de Gemswurz *f*
da gemserod
sv vårkrage
es matalobos *m*
it doronica *f*
   Doronicum L.

2137
LESION
fr lésion *f*
ne lesies
de Verletzung *f*
da sår
sv läsioner

es lesión *f*
it lesione *f*

2138
LESION NEMATODE
fr anguillule *f* libre
ne vrijlevend wortelaaltje *n*
de wandernde Wurzelnema-
   tode *f*
da Pratylenchus
sv ängsnematod
es nematado *m* de los prados
it nematode *m* dei prati
   Pratylenchus spp.

LESSER BULBFLY, 3444

2139
LETTUCE
fr laitue *f*
ne sla
de Salat *m*
da salat
sv sallad
es lechuga *f*
it lattuga *f*

LETTUCE, cabbage, 513
—, cos, 1866
—, cutting, 1190
—, early, 1190
—, heading, 513
—, Iceberg, 1866
—, leaf, 2085
—, young, 1190

2140
LETTUCE BIG VEIN
fr maladie *f* des grosses
   nervures
ne bobbelblad
de Breitadrigkeit *f*
da store nerver
sv stora nerver
it imbianchimento *m* nervale

LETTUCE LEAF SPOT,
410

2141
LETTUCE MOSAIC
fr mosaïque *f* de la laitue
ne slamozaïek *n*
de Salatmosaik *n*

da salat-mosaik
sv salladsmosaik
es mosaico *m* de la lechuga
it mosaico *m* della lattuga

2142
LETTUCE ROOT APHID;
ARTICHOKE TUBER
APHID
fr puceron *m* de la laitue
ne wortelluis
de Salatwurzellaus *f*;
   Salaterdlaus *f*
da Pemphygus
sv rotlus
es pulgón *m* de la raíz de la
   lechuga
it afide *m* ceroso
   Pemphigus lactucarius L.

2143
LEVEL
fr niveau *m*
ne niveau *n*
de Niveau *n*; Höhe *f*
da niveau *n*; højde
sv nivå
es nivel *m*
it livello *m*

2144
LEVEL, TO
fr égaliser; niveler
ne nivelleren; egaliseren;
   waterpassen
de ebnen; planieren;
   nivellieren
da udjævne; nivellere
sv planera; nivellera;
   avväga
es igualar; nivelar
it uguagliare; livellare

2145
LEVEL OF UNDER-
GROUND WATER
fr niveau *m* de la nappe
   phréatique
ne grondwaterstand
de Grundwasserstand *m*
da grundvandstand
sv grundvattenstånd *n*
es superficie *f* del agua
   subterránea

it livello *m* dell'acqua
   sotterrana

2146
LIABILITIES
fr passif *m*
ne passiva
de Passiva *n pl*
da passiver
sv passiva; skulder
es pasivo *m*
it passività *f*

2147
LIATRIS;
BUTTON SNAKE ROOT
fr Liatris *m*
ne Liatris
de Prachtscharte *f*
da pragtskær
sv rosenstav
es serratula *f*
it Liatris *m*
   Liatris Schreb

2148
LIBERALIZATION
fr libéralisation *f*
ne liberalisatie
de Liberalisierung *f*
da liberalisering
sv liberalisering
es liberalización *f*
it liberalizzazione *f*

2149
LIE FALLOW, TO;
TO FALLOW
fr être en friche
ne braak liggen
de brach liegen
da ligge brak
sv ligga i träde
es estar en barbecho
it essere incolto

LIFE, 1169

2150
LIFE CYCLE
fr cycle *m* vital
ne levenscyclus
de Lebenszyklus
da livscyclus

sv levnadscykel
es ciclo *m* vital
it ciclo *m* della vita

LIFT, TO, 1067

2151
LIGHT ABSORPTION
fr absorption *f* de la lumière
ne lichtabsorptie
de Lichtabsorption *f*
da lysabsorbtion
sv ljusabsorbtion
es absorción *f* de la luz
it assorbimento *m* della luce

2152
LIGHT DEFICIENCY
fr manque *m* de lumière
ne lichtgebrek *n*
de Lichtmangel *m*
da lysmangel
sv ljusbrist; ljusunderskott
es falta *f* de luz
it dificienza *f* di luce

2153
LIGHT DEFICIENT
fr faible en luminosité
ne lichtarm
de lichtarm
da lysfattig
sv ljusfattig
es pobre de luz
it privo di luce

2154
LIGHT DISTRIBUTION
fr répartition *f* de la lumière
ne lichtverdeling
de Lichtverteilung *f*
da lysfordeling
sv ljusfördelning
es distribución *f* de luz
it distribuzione *f* dello
    spettro luminoso

2155
LIGHT ENTRY
fr angle *m* d'incidence
ne lichtinval
de Lichteinfall *m*
da lysindstråling
sv ljusinfall *n*

es caída *f* de luz
it incidenza *f* della luce

2156
LIGHTING;
ILLUMINATION
fr éclairage *m*
ne belichting
de Beleuchtung *f*
da belysning
sv belysning
es iluminación *f*
it delucidazione *f*

LIGHTING, additional, 26
—, intensity of, 1934

2157
LIGHTING FITTING
fr appareil *m* d'éclairage
ne armatuur (van lichtbron)
de Leuchte *f*
da lysarmatur
sv belysningsarmatur
es aparato *m* de ilumina-
    ción
it apparecchio *m*
    d'illuminazione

2158
LIGHT INTENSITY
fr intensité *f* lumineuse
ne lichtintensiteit; licht-
    sterkte
de Lichtintensität *f*; Licht-
    stärke *f*
da lysintensitet
sv ljusintensitet
es intensidad *f* luminosa
it intensità *f* luminosa

2159
LIGHT INTENSITY
DISTRIBUTION CURVE
fr courbe *f* de distribution de
    l'intensité d'une source
ne lichtsterktekromme
de Lichtverteilungskurve *f*
da lysfordelingskurve
sv ljusfördelningskurva
es curva *f* de distribución de
    intensidad de una fuente
    luminosa
it curva *f* di distribuzione
    dell'intensità luminosa

2160
LIGHT SPIKED CHAIN
HARROW
fr herse *f* souple
ne onkruideg
de Unkrautstriegel *m*
da ukrudtsharve
sv ogräsharv
es grada *f* flexible
it erpice *m* a catena

2161
LIGHT SUPPLY
fr disposition *f* de lumière
ne lichtvoorziening
de Lichtvorsorge *f*
da lystilførsel
sv ljustillförsel
es surtido *m* de luz
it provvedimento *m*
    luminoso

2162
LIGHT TRANSMISSION
fr transmission *f* lumineuse
ne lichtdoorlatendheid
de Lichtdurchlässigkeit *f*
da lysgennemgang
sv ljusgenomsläpplighet
es transmisión *f* luminosa
it trasmissione *f* della luce

2163
LIGULATE RAY
FLOWER
fr ligule *f*; fleur *f* en
    languette
ne lintbloem
de Zungenblüte *f*
da tungeblomst
sv tungformig blomma
es flor *f* ligulada
it fiore *m* ligulato

2164
LILAC
fr lilas *m*
ne sering
de Flieder *m*
da syrén
sv syren
es lila *f*
it lilla *f*
    Syringa L.

LILAC BLIGHT, 215, 219

**2165
LILAC LEAF MINER**
fr teigne *f* des lilas
ne seringemot
de Fliedermotte *f*
da syrénmøl *n*
sv syrenmal *n*
es polilla *f* minadora de las
 lilas; minadora *f* de las
 hojas de la lila
it minatrice *f* delle foglie di
 lilla
 *Xanthospilapteryx syrin-
 guella* F.; *Caloptilia s.*
 (UK); *Gracillaria s.*
 (US)

**2166
LILY**
fr lis *m*
ne lelie
de Lilie *f*
da lilje
sv lilja
es lirio *m*; azucena *f*
it giglio *m*
 Lilium L.

LILY, arum, 162
—, day, 991
—, gold band, 1606
—, golden rayed, 1606
—, Peruvian, 2689
—, plaintain, 1503
—, regale, 3087
—, tiger, 3846

**2167
LILY BEETLE**
fr criocère *f* du lis
ne leliehaan
de Lilienhähnchen *n*
da liljebille
sv liljebagge
es criócero *m* de la azucena
it criocera *f* del giglio
 Crioceris lilii Scop.;
 Lilioceris l. (UK)

**2168
LILY DISEASE (LILIUM)**
fr Botrytis *m*
ne vuur *n*

de Grauschimmelkrankheit *f*
da lilje-gråskimmel
sv gråmögel
es encendido *m*
it Botrytis *m*
 Botrytis elliptica (Berk.)
 Cke.

**2169
LILY-OF-THE-VALLEY**
fr muguet *m*
ne lelietje-van-dalen *n*
de Maiblume *f*;
 Maiglöckchen *n*
da liljekonval
sv liljekonvalje
es lirio *m* de los valles;
 muguete *m*
it mughetto *m*
 Convallaria majalis L.

**2170
LIME**
fr chaux *f*
ne kalk
de Kalk *m*
da kalk
sv kalk
es cal *f*
it calce *f*

LIME, deficient in, 1016

**2171
LIME, TO**
fr chauler
ne bekalken
de kalken
da kalke
sv kalka
es encalar
it arricciare; incalcinare

**2172
LIME CONTENT**
fr teneur *f* en chaux
ne kalkgehalte *n*
de Kalkgehalt *m*
da kalkindhold *n*
sv kalkhalt
es contenido *m* en cal
it tenore *m* di calce

**2173
LIME FACTOR**
fr factor-chaux *m*
ne kalkfactor
de Kalkfaktor *m*
da kalkfaktor
sv kalkfaktor
es factor *m* cal
it fattore *m* calcareo

**2174
LIME MARL**
fr marne *f* calcaire
ne kalkmergel
de Kalkmergel *m*
da kalkmergel
sv märgel
es marga *f* calcárea
it marna *f* calcarea

**2175
LIME STATUS**
fr teneur *f* en chaux
ne kalktoestand
de Kalkzustand *m*
da kalktilstand
sv kalktillstånd *n*
es contenido *m* en calcio
it tenore *m* di calce

**2176
LIMESTONE**
fr pierre *f* calcaire;
 pierre *f* à chaux
ne kalksteen
de Kalkstein *m*
da kalksten
sv kalksten
es piedra *f* calcárea; piedra *f*
 de cal
it pietra *f* calcarea

LIMESTONE, ground,
1671
—, ground magnesian,
1672

**2177
LIME SULPHUR**
fr bouillie *f* sulfocalcique
ne Californische pap
de Schwefelkalkbrühe *f*
da svovlkalk
sv svavelkalkvätska

es caldo *m* sulfocálcico
it mistura *f* calce sulfureo

2178
LIME TREE
fr tilleul *m*
ne linde
de Linde *f*
da lind
sv lind
es tilo *m*
it tiglio *m*
   Tilia L.

2179
LIMING MATERIAL
fr amendement *m* calcaire
ne kalkmeststof
de Kalkdünger *m*
da kalkningsmiddel
sv kalkningsmedel *n*
es enmienda *f* calcárea
it concime *m* calcareo

2180
LINE; PLANTING LINE
fr cordeau *m*; ligne *f* de
   plantation
ne pootlijn
de Pflanzleine *f*
da line; snor; plantelinie
sv lina; rep
es cuerda *f*
it corda *f* a piantare

2181
LINEAR
fr linéaire
ne lijnvormig
de linear
da linieformet
sv linjeformig
es lineal
it lineare

2182
LINE PATTERN
(PEACH; PRUNUS)
fr arabesques *f pl*
ne figuurbont *n*
de Linien- und Kurvenmuster *n*
da båndmosaik
sv bandmönster *n*;
   bandkloros

es arabescos *m pl*
it lineolatura *f*; maculatura
   f lineare

2183
LING; HEATHER
fr bruyère *f*
ne struikheide
de Besenheide *f*
da lyng
sv ljung
es brezo *m* común
it Calluna *f*; brentoli *m*
   Calluna Salisb.

2184
LINING
(PACKING CASE)
fr rembourrage *m*
ne bekleding
de Isolierschutz *m*
da foret
sv lådbeklädning
es forro *m*
it rivestimento *m*

2185
LINTEL
fr imposte *f*
ne kalf (bouwk.) *n*
de Kämpferriegel *m*
da rigel
sv överstycke över dörr
   eller fönster
es dintel *m*; puente *m*;
   umbral *m*
it architrave *m*

2186
LIP
fr lèvre *f*
ne lip
de Lippe *f*
da læbe
sv läpp
es labio *m*
it labbro *m*

2187
LIPSTICK MOULD;
RED GEOTRICHEM
fr moisissure *f* rouge
   à lèvres; 'lipstick' *m*
ne lippenstiftschimmel

de Lippenstiftschimmel *m*
da læbestift (skimmel)
sv läppstiftsmögel;
   'lipstick'
es moho *m* rojo
it muffa *f* rosata
   Sporendonema sp.

2188
LIQUID
fr liquide *m*
ne vloeistof
de Flüssigkeit *f*
da vædske
sv vätska
es líquido *m*
it liquido *m*

LIQUID, covering, 870

2189
LIQUID ASSETS
fr avoirs *m* liquides
ne liquide middelen *n pl*
de flüssige Mittel *n pl*
da likvide midler *n pl*
sv likvida medel *n pl*
es bienes *m pl* líquidos
it capitali *m pl* liquidi

2190
LIQUID DISINFECTANT
fr liquide *m* désinfectant
ne natontsmetter
de Nassbeizmittel *n*
da vådafsvampning
sv våtbetningsmedel *n*
es líquido *m* desinfectante
it liquido *m* disinfettante

2191
LIQUID FUEL
fr combustible *m* liquide
ne vloeibare brandstof
de flüssiger Brennstoff *m*
da flydende brændstof *n*
sv flytande bränsle *n*
es combustible *m* líquido
it combustibile *m* liquido

2192
LIQUIDITY
fr liquidité *f*
ne liquiditeit

de Liquidität *f*
da likviditet
sv likviditet
es liquidez *f*
it liquidità *f*

2193
LIQUID MANURE
fr engrais *m* liquide;
   purin *m*
ne gier
de Jauche *f*
da ajle
sv gödselvatten *n*
es purín *m*
it letame *f* liquido

2194
LIST OF VARIETIES
fr catalogue *n* des variétés
ne rassenlijst
de Sortenliste *f*
da sortliste
sv sortlista
es rallo *m*; rallador *m*
it elenco *m* di razze

2195
LITMUS
fr tournesol *m*
ne lakmoes *n*
de Lackmus *n*
da lakmus
sv lackmus
es girasol *m*; tornasol *m*
it tornasole *m*

2196
LITTER
fr litière *f*; paillis *m*
ne strooisel *n*
de Streu *f*
da strøelse
sv strö *n*
es cama *f* de paja; litera *f*
it lettiera *f*

LITTER, coniferous, 806
—, forest, 1431
—, peat, 2656
—, pine needle, 2734
—, straw, 3688

2197
LIVERWORT
fr hépatique *f*
ne leverbloempje *n*
de Leberblümchen *n*
da blå anemone
sv blåsippa
es hierba *f* del hígado;
   hepática *f*
it Hepatica *f*
   Hepatica Dill. ex L.

2198
LIXIVIATE, TO
fr lixivier
ne uitlogen
de auslaugen
da udlude
sv laka ur; luta
es lixiviar
it lisciviare

2199
LOADING BAY;
LOADING. RAMP
fr pont *m* de chargement
ne laadbrug
de Laderampe *f*
da læsserampe
sv lastbrygga
es puente *m* de carga
it ponte *m* di caricamento

2200
LOAM
fr argile; terre *f* glaise;
   limon *m*
ne leem
de Lehm *m*
da lerjord
sv sandblandad lera
es barro *m*; arcilla *f*
it terra *f* grassa; argilla *f*

2201
LOAM LAYER
fr couche *f* de limon;
   couche *f* de glaise
ne leemlaag
de Lehmschicht *f*
da lerlag *n*
sv lerskikt *n*
es capa *f* de limo
it strato *m* d'argilla

2202
LOAN
fr prêt *m*
ne lening
de Darlehen *n*
da lån *n*
sv lån *n*
es préstamo *m*
it prestito *m*

2203
LOBE (bot.)
fr lobe *m*
ne slip
de Lappen *m*
da flig; lap
sv flik
es lóbulo *m*
it lobo *m*

LOBED, pinnately, 2742

2204
LOBELIA
fr lobélie *f*
ne Lobelia
de Lobelie *f*
da Lobelia
sv Lobelia
es Lobelia *f*
it Lobelia *f*
   Lobelia L.

2205
LOCKER PLANT
fr installation *f* à casiers
ne vrieskluis
de Gemeinschaftsgefrieranlage *f*
da fællesfryseri
sv frysfack *n*
es instalación *f* de congelación con compartimien-
   tos individuales
it locale *m* refrigerante

2206
LOESS; AEOLIAN SOIL
fr loess
ne löss
de Löss *m*
da løss
sv löss
es loess *m*
it loess *m*

2207
LONG-DAY PLANT
fr plante f de jour long;
   plante f héméropériodique
ne lange-dagplant
de Langtagpflanze f
da langdagsplante
sv långdagsväxt
es planta f de día largo;
   planta f macrohémera
it pianta f di giornata
   diuturna

2208
LONG-DAY
TREATMENT
fr traitement m de longs
   jours
ne langedagbehandeling
de Langtagsbehandlung f
da langdagsbehandling
sv långdagsbehandling
es tratamiento m de días
   largos
it trattamento m longidiurno

LONGHORN, large poplar,
2035

2209
LONGHORN BEETLE;
ROUNDHEADED WOOD
BORER (US)
fr longicorne m
ne boktor
de Bockkäfer m
da træbuk
sv långhornig (trä)bock
es taladro m; capricornio m
it cerambice m; capricorno m
   Cerambycidae

2210
LONGITUDINAL
GROWTH
fr croissance f en longueur
ne lengtegroei
de Längenwachstum n
da længdevækst
sv längdtillväxt
es desarollo m longitudinal
it crescita f longitudinale

LONG OUTFALL PIPE,
1227

2211
LONG SHOOT
fr rameau m à bois
ne langlot
de Langtrieb m
da langskud n
sv långskott
es brote m largo
it lamburda f

2212
LOOPER CATERPILLAR
fr chenille f arpenteuse
ne spanrups
de Spanner m
da målerlarve
sv mätarlarv
es cruga f geómetra
it geometra f; misuratore m
   Geometridae

2213
LOOSE
fr vrac
ne onverpakt; los
de in loser Schüttung;
   unverpackt
da løs
sv opackat (lös vikt)
es granel
it alla rinfusa

2214
LOOSE BUD (HYACINTH)
fr 'cracheur' m;
   avortement m
ne spouwer
de 'Spucker'
da kastere
sv spottning
es 'gargola' f
it 'boccio m sputato'

2215
LOOSEN (THE TOPSOIL),
TO
fr biner; ameublir
ne losmaken
de lockern (der Krume)
da jordløsning

sv ytluckring
es binar
it staccare

2216
LOOSESTRIFE
fr salicaire f commune
ne kattestaart
de Weiderich m
da kattehale
sv fackelblomster n
es lisimaquia f roja;
   salicaria f
it salcerella f
   Lythrum L.

LOP, TO, 2963

2217
LOSS OF LIGHT
fr perte f en luminosité;
   perte f en intensité
   lumineuse
ne lichtverlies n
de Lichtverlust m
da lystab n
sv ljusförlust
es pérdida f de luz
it perdita f luminosa

2218
LOT; PARCEL OF LAND
fr lot m; parcelle f
ne kavel; perceel
de Los n; Parzelle f; Flurstück n
da parcel; jordstykke
sv parcell; jordlott
es lote m; parcela f
it appezzamento m

2219
LOVAGE
fr livèche f
ne lavas
de Liebstöckel n; Maggi-
   kraut n
da løvstikke
sv libsticka
es apio m de montaña
it levistico m
   Levisticum officinale Koch

2220
LOVE-IN-A-MIST
fr nigelle *f* de Damas
ne juffertje in 't groen
de Jungfer im Grünen *f*
da jomfru i det grønne
sv jungfrun i det gröna
es neguilla *f*
it nigella *f*
Nigella damascena L.

LOW-GRADE PRODUCT,
1903

2221
LOW IN CHLORINE
fr pauvre en chlore
ne chloorarm
de chlorarm
da klorfattig
sv klorfattig
es pobre en cloro
it privo di cloro

2222
LOW PRESSURE AREA
fr aire *f* de basse pression
ne lagedrukgebied
de Tiefdruckgebiet *n*; Tief *n*
da lavtryksområde *n*
sv lågtrycksområde *n*

es región *f* ciclonal
it settore *m* di bassa
    pressione atmosferica

2223
LOW TEMPERATURE
BREAKDOWN
fr brunissement *m* interne
    par des basses tempéra-
    tures
ne lage-temperatuurbederf *n*
de Fruchtfleischbräune *f*;
    Kaltefleischbräune *f*
da kuldeskade *mf*
sv kylskada
es descomposición *f* interna
    de las frutas por efecto
    del frío
it corruzione *f* da tempera-
    tura bassa

2224
LUCERNE
fr luzerne *f*
ne lucerne
de Luzerne *f*
da lucerne
sv blåluzern
es alfalfa *f*
it alfalfa *f*

LUMP, 714

2225
LUNGWORT
fr pulmonaire *f*
ne longkruid *n*
de Lungenkraut *n*
da lungeurt
sv lungört
es Pulmonaria *f*
it polmonaria *f*
Pulmonaria L.

2226
LUPIN
fr lupin *m*
ne lupine
de Lupine *f*
da lupin
sv lupin
es altramuz *m*; lupino *m*
it lupino *m*
Lupinus L.

2227
LUX-METER
fr luxmètre *m*
ne luxmeter
de Beleuchtungsmesser *m*
da lysmåler
sv luxmeter; belysnings-
    mätare
es luxámetro *m*
it luxmetro *m*

M

2228
MACEDONIAN PINE
fr Pinus *f* peuce
ne balkanden
de rumelische Weymouths-
kiefer *f*
da silkefyr
sv makedonisk tall
es pino *m* mugho
it Pinus *f* peuce
Pinus peuce Gris

MACHINERY
CONTRACTOR, 820

2229
MAGGOT; GRUB
fr ver *m*; larve *f*
ne made; larve
de Made *f*; Larve *f*
da maddike
sv larv; mask
es oruga *f*
it verme *m*; bruco *m*

MAGGOT, cabbage, 515
—, onion, 2524
—, seed-corn, 275

MAGGOTTY, 4214

2230
MAGNESIA
fr magnésie *f*
ne magnesia
de Magnesia *f*
da magnesia
sv magnesia
es magnesia *f*
it magnesia *f*

2231
MAGNESIUM
fr magnésium *m*
ne magnesium *n*
de Magnesium *m*
da magnesium *n*
sv magnesium *n*
es magnesio *m*
it magnesio *m*

2232
MAGNESIUM
DEFICIENCY
fr carence *f* en magnesium
ne magnesiumgebrek *n*
de Magnesiummangel *m*
da magnesiummangel
sv magnesiumbrist
es carencia *f* en magnesio
it deficienza *f* di magnesio

2233
MAGNESIUM
SULPHATE
fr sulfate *m* de magnésie
ne bitterzout *n*
de Bittersalz *n*
da bittersalt *n*; magnesiumsulfat *n*
sv magnesiumsulfat *n*;
bittersalt *n*
es sal *f* de higuera; sulfato
m de magnesia
it sal *m* amaro

2234
MAGNIFYING GLASS
fr loupe *f*
ne vergrootglas *n*
de Vergrösserungsglas *n*
da forstørrelsesglas *n*
sv förstoringsglas *n*
es lente *m* de aumento
it lente *f* d'ingrandimento

2235
MAGNOLIA
fr magnolier *m*
ne Magnolia
de Magnolie *f*
da magnolie *mf*
sv Magnolia
es Magnolia *f*
it Magnolia *f*
Magnolia L.

2236
MAGPIE
fr pie *f*
ne ekster

de Elster *f*
da skade
sv skata
es urraca *f*
it gazza *f*
Pica pica L.

2237
MAGPIE MOTH;
CURRANT MOTH
(WORM)
fr phalène *f* du groseillier
ne bonte bessevlinder
de Stachelbeerspanner *m*
da stikkelsbærmåler; Harle-
kin
sv krusbärsmätare
es falena *f* del grosellero
it geometra *f* del ribes;
falena *f* dell'uva spina
Abraxas grossulariata L.

2238
MAHONIA
fr Mahonia *m*
ne Mahonia
de Mahonie *f*
da mahonie
sv Mahonia
es Mahonia *f*
it Mahonia *f*
Mahonia Nutt.

2239
MAIDENHAIR FERN
fr capillaire *m*; cheveux *m*
pl de Vénus *n*
ne venushaar *n*
de Frauenhaarfarn *m*
da venushår *n*
sv venushår *n*
es culantrillo *m* de poza
it capelvenere *m*
Adiantum L.

2240
MAIDENHAIR TREE
fr arbre *m* aux 40 écus
ne Japanse noteboom
de Ginkgobaum *m*

da tempeltræ *n*
sv Ginkgo
es Ginkgo *m*
it Ginkgo *m* biloba
   Ginkgo biloba L.

2241
MAIN AXIS (bot.)
fr rachis *m*
ne hoofdas
de Spindel *f*
da hovedakse
sv huvudaxel
es eje *m* principal
it perno *m* principale

2242
MAIN BRANCH
fr branche *f* principale;
   branche *f* charpentière
ne hoofdtak; gesteltak
de Hauptast *m*;
   Hauptzweig *m*
da høvedgren
sv huvudgren
es rama *m* principal lateral
it ramo *m* principale;
   branca *f* primaria

2243
MAIN BUD
fr bourgeon *m* principal
ne hoofdknop
de Mittelknospe *f*
da hovedknop
sv huvudskott
es yema *f* principal;
   botón *m* principal
it bottone *m* principale

2244
MAIN DRAIN
fr collecteur *m*
ne hoofddrain
de Hauptstrang *m*
da hovedledning *mf*
sv huvudledning;
   stamdike *n*
es colector *m*
it drenaggio *m* principale

2245
MAIN LEAD
fr branche *f* principale
ne hoofdtak

de Hauptast *m*
da hovedgren
sv huvudgren
es rama *f* principal
it ramo *m* maestro

2246
MAIN PIPE
fr conduite *f* principale
ne hoofdbuis
de Hauptrohr *n*
da hovedledning
sv stamledning;
   huvudledning
es conducto *m* principal
it tubo *m* principale

2247
MAIN ROOT
fr racine *f* principale
ne hoofdwortel
de Hauptwurzel *f*
da hovedrod
sv huvudrot
es raíz *f* principal
it radice *f* principale

2248
MAIN STEAM PIPE
fr tuyau *m* à vapeur principale
ne hoofdstoomleiding
de Hauptdampfrohr *n*
da hoveddamprør *n*
sv huvudångrör *n*
es tubo *m* de vapor principal
it tubo *m* a vapore principale

2249
MAIN STEM
fr tige *f* principale
ne hoofdstengel
de Hauptstengel *m*
da hovedstængel
sv huvudstängel
es tallo *m* principal
it fusto *m* principale

2250
MAIN STOP VALVE
fr soupape *f* d'arrêt principale
ne hoofdafsluiter
de Hauptabsperrventil *n*

da hovedventil
sv huvudventil
es válvula *f* de cierre
it valvola *f* d'arresto principale

2251
MAINTENANCE
fr taille *f* d'entretien
ne onderhoudssnoei
de Erhaltungsschnitt *m*
da fornyelsesbeskæring
sv underhåll-beskärning
es poda *f* de conservación
it potatura *f* di produzione;
   potatura *f* di manuten-
   zione

MAIZE, 3763

MALE, 3603

2252
MALFORMATION;
DEFORMATION
fr malformation *f*
ne misvorming
de Missbildung *f*; Bildungsabweichung *f*; Deformation *f*
da misformering
sv missbildning; deforma-
   tion
es malformación *f*
it malformazione *f*; defor-
   mazione *f*

2253
MALLOW
fr mauve *f*
ne malve; kaasjeskruid *n*
de Malve *f*
da katost
sv Malva
es Malva *f*
it Malva *f*
   Malva L.

MALLOW, curled, 941
—, Jew's, 1994
—, marsh, 2296
—, rose, 3189

2254
MALTOSE
fr maltose *f*
ne moutsuiker
de Malzzucker *m*
da maltsukker *n*
sv maltsocker *n*
es maltosa *f*
it maltosio *m*

2255
MALT SPROUTS
fr germes *m pl* de malt;
   touraillons *mpl*
ne moutkiemen
de Malzkeime *m pl*
da maltspirer
sv maltgroddar
es germen *m pl* de malta
it germogli *m pl* di malto

MAMMOTH TREE, 4127

2256
MANGANESE
fr manganèse *m*
ne mangaan *n*
de Mangan *n*
da mangan
sv mangan
es manganeso *m*
it manganese *m*

MANGANESE, excess of,
1254

2257
MANGANESE
DEFICIENCY
fr carence *f* en manganèse
ne mangaangebrek *n*
de Manganmangel *m*
da manganmangel
sv manganbrist
es carencia *f* en manganeso
it deficienza *f* di manganese

2258
MAN HOLE
fr trou *m* d'homme
ne mangat *n*
de Mannloch *n*
da mandhul *n*
sv manlucka

es agujero *m* de hombre
it passo *m* d'uomo;
   passaggio *m* a misora
   d'uomo

2259
MAN HOUR
fr heure *f* de travail
   d'homme
ne manuur *n*
de Arbeitsstunde *f*
da mandstime
sv arbetstimma
es peonada *f* de una hora
it ora *f* lavorativa per un
   operaio

MANNER OF GROWTH,
1703

2260
MANUAL LABOUR;
HAND WORK
fr main d'oeuvre *f*
ne handarbeid
de Handarbeit *f*
da håndarbejde
sv manuellt arbete
es trabajo *m* manual;
   mano *f* de obra
it mano *f* d'opera

2261
MANURE; DUNG
fr fumier *m*; engrais *m*
ne mest
de Dung *m*; Dünger *m*;
   Mist *m*
da gødning; møg
sv gödsel
es estiércol *m*; abono *m*
it concime *m*

MANURE, artificial, 161
—, basic, 258
—, chicken, 656
—, cow, 877
—, farmyard, 1297
—, heating, 1759
—, horse, 1821
—, liquid, 2193
—, organic, 2544
—, pig, 2725
—, poultry, 2883

—, sheep, 3369
—, well rotted, 4125

2262
MANURE, TO; TO FEED;
TO DUNG; TO FERTILIZE
fr fertiliser; fumer
ne bemesten
de düngen
da gøde
sv gödsla
es abonar; fertilizar
it concimare

2263
MANURE FORK
fr fourche *f* à fumier
ne mestvork
de Mistgabel *f*
da møggreb
sv gödselgrep
es horquilla *f* para estiércol
it forca *f* da letame;
   forcone *m*

2264
MANURE PILE;
MANURE HEAP;
DUNGHILL (ROUGH)
fr tas *m* de fumier
ne mesthoop
de Düngerhaufen *m*
da mødding; gødningsbunke
sv gödselhög
es pila *f*; montón *m*
it stiva *f* di letame

2265
MANURE SHREDDER
fr broyeur *m* à engrais
ne mestmachine
de Düngeraufbereitungs-
   maschine *f*
da gødningsblander
sv gödselsönderdelare
es desmenuzador *m* de
   abonos
it trincialetame *m*

2266
MANURE SPREADER
fr épandeur *m* de fumier
ne mestverspreider
de Miststreuer *m*

da staldgødningsspreder
sv gödselspridare
es esparcidor *m* de
  estiércol
it spargitore *m* di concime

2267
MANURE TURNER;
COMPOSTING
MACHINE
fr machine *f* à retourner
ne mestmachine
de Düngerwender *m*
da gødningsonderdeler;
  komposteringsmaskine
sv komposteringsmaskin
es máquina *f* de dar vueltas
  al estiércol
it macchina *f* per girare
  letame

2268
MANURIAL
EXPERIMENT
fr essai *m* de fumure
ne bemestingsproef
de Düngungsversuch *m*
da gødningsforsøg *n*
sv gödslingsförsök *n*
es ensayo *m* de abonos
it esperimento *m* da
  concimazione

2269
MANURING; FEEDING
fr fumure *f*; fertilisation *f*
ne bemesting
de Düngung *f*
da gødning
sv gödsling
es abonado *m*
it concimazione *f*

MANURING, basic, 257
—, green, 1653

2270
MAP, TO; TO SURVEY
fr cartographier
ne karteren
de kartieren
da kortlægge
sv kartlägga

es cartografiar
it rilevare la carta;
  cartografia *m*

2271
MAPLE
fr érable *m*
ne esdoorn
de Ahorn *m*
da ahorn; løn; nav
sv lönn
es arce *m*
it acero *m*
  Acer L.

MAPLE, purple Japanese,
2989

MARCH FLY, 308

2272
MARGINAL COSTING
fr calcul *m* du coût marginal
ne differentiële kostprijs
de Grenzkosten *pl*
da differensomkostninger
sv marginellt självkostnadspris *n*
es costa *f* de producción
  parcial
it costo *m* di produzione
  proporzionale (differen-
  ziale)

2273
MARGINAL COSTS
fr coûts *m pl* marginaux
ne grenskosten
de Marginalkosten *pl*;
  Grenzkosten *pl*
da grænseomkostninger
sv marginalkostnad;
  gränskostnad
es costes *m pl* marginales
it costi *m pl* marginali

2274
MARGINAL HOLDING
fr exploitation *f* marginale
ne grensbedrijf *n*
de Grenzbetrieb *m*
da grænse-virksomhed
sv marginalföretag
es explotación *f* marginal
it azienda *f* marginale

2275
MARGINAL UTILITY
fr utilité *f* marginale
ne grensnut *n*
de Grenznutzen *m*
da grænsenytteværdi
sv marginalnytta;
  gränsnytta
es utilidad *f* marginal
it utilitá *f* marginale

2276
MARGUERITE
fr anthémus *f*;
  chrysanthème *m* frutescent
ne margriet
de Wiesenmargerite *f*
da margarit
sv prästkrage
es margarita *f* grande
it margherita *f*
  *Chrysanthemum
  leucanthemum* L.

2277
MARIGOLD
fr souci *m* des jardins
ne goudsbloem
de Ringelblume *f*
da morgenfrue
sv ringblomma
es maravilla *f*; caléndula *f*
it belfiore *m*
  Calendula L.

MARIGOLD, African, 44
—, Cape, 563
—, French, 44
—, marsh, 2297

2278
MARINE BOILER
fr chaudière *f* marine
ne scheepsketel
de Marinekessel *m*
da skibskedel
sv sjöfartspanna;
  marinpanna
es caldera *f* marina
it caldaia *f* marina

2279
MARITIME PINE
fr pin *m* maritime
ne zeeden
de Strandkiefer *f*
da strandfyr
sv strandtall
es pino *m* maritimo
it pino *m* selvatico;
    pinastro *m*
    Pinus pinaster Ait.

2280
MAJORAM
fr marjolaine *f*
ne majoraan
de Majoran *m*
da merian
sv mejram
es mejorana *f*
it maggiorana *f*
    Majorana hortensis
    Mnch.

2281
MARKER
fr rayonneur *m*
ne markeur
de Markeur *m*
da markør
sv markör
es marcador *m*
it marcatore *m*

2282
MARKET; OUTLET
fr débouché *m*; marché *m*
ne afzetgebied *n*
de Absatzgebiet *n*; Markt *m*
da markedsområde *n*
sv avsättningsområde *n*;
    marknad
es lugar *m* de ventas
it sbocco *m*

2283
MARKET DISTURBANCE
fr altération *f* du marché
ne marktverstoring
de Marktstörung *f*
da markedsforstyrrelse
sv marknadsstörning
es alteración *f* del mercado
it perturbazione *f* del mer-
    cato

2284
MARKET GARDEN
fr exploitation *f* horticole
ne tuinderij
de Gärtnerei *f*; Garten-
    baubetrieb *m*
da gartneri *n*
sv handelsträdgårdsmästeri;
    trädgårdsföretag *n*
es explotación *f* hortícola
it azienda *f* orticola

2285
MARKETING FUND
fr fonds *m* de commer-
    cialisation
ne afzetfonds *n*
de Absatzfonds *m*
da afsætningsfonds
sv avsättningsfond
es fondos *m pl* de comer-
    cialización
it fondi *m pl* di commer-
    cializzazione

2286
MARKET PLANNING
fr plan *m* de commercialisa-
    tion
ne marktordening
de Marktordnung *f*
da markedsordning
sv marknadsreglering
es planificación *f* del merca-
    do
it pianificazione *f* del mer-
    cato (della commer-
    cializzazione)

2287
MARKET PRICE
fr prix *m* courant du marché
ne marktprijs
de Marktpreis *m*
da markedspris
sv marknadspris
es precio *m* de mercado
it prezzo *m* di mercato

2288
MARKET REPORT
fr bulletin *m* de marché
ne marktbericht *n*
de Marktbericht *m*

da markedsberetning
sv marknadsrapport
es información *f* del merca-
    do
it bollettino *m* di mercato

2289
MARKET
STABILIZATION
fr stabilisation *f* des marchés
ne marktstabilisatie
de Marktstabilisierung *f*
da markedsstabilisering
sv marknadsstabilisering
es estabilización *f* del mer-
    cado
it stabilizzazione *f* del mer-
    cato

2290
MARKET STRUCTURE
fr structure *f* du marché
ne marktstructuur
de Marktstruktur *f*
da markedsstruktur
sv marknadsstruktur
es estructura *f* del mercado
it struttura *f* del mercato

2291
MARL
fr marne *f*
ne mergel
de Mergel *m*
da mergel
sv märgel
es marga *f*
it marna *f*

2292
MARLY SOIL
fr sol *m* marneux
ne mergelgrond
de Mergelboden *m*
da mergeljord
sv märgeljord
es suelo *m* margoso
it terreno *m* marnoso

2293
MARMALADE
fr confiture *f*
ne marmelade
de Marmelade *f*

da marmelade
sv marmelad
es mermelada *f*
it marmellatta *f*

MARROW, 1612

2294
MARROWFAT PEA
fr pois *m* à cosses violettes
ne schokker
de blauhülsige Auskern-
    erbse *f*
da blå marvært
sv blå märgspirtärt
es guisante *m* tierno
it pisello *m* morbido

MARSH, 3755

2295
MARSH HORSETAIL
fr prêle *f* des marais
ne lidrus
de Sumpfschachtelhalm *m*;
    Duwock *m*
da kær-padderocke
sv kärrfräken
es equiseto *m*
it coreggiola *f* minore
    Equisetum palustre L.

2296
MARSH MALLOW
fr guimauve *f* officinale
ne heemst
de Eibisch *m*
da læge-stokrose
sv läkemalva
es malvavisco *m*
it bismalva *f*
    Althaea officinalis L.

2297
MARSH MARIGOLD
fr calthe *f* des marais;
    populage *m*
ne dotterbloem
de Sumpfdotterblume *f*
da engkabbeleje
sv kabbeleka
es calta *f*; hierba *f* centella
it calta *f*
    Caltha palustris L.

2298
MARSH PLANT
fr plante *f* de marécage
ne moerasplant
de Sumpfpflanze *f*
da sumpplante
sv kärrväxt; sumpväxt
es planta *f* de pantano
it pianta *f* palustre

2299
MARSH WOUNDWORT
fr épiaire *f* des marais
ne moerasandoorn
de Sumpfziest *m*
da kærgaltetand
sv knölsyska
es salvia *f* de pantano
it erba *f* strega
    Stachys palustris L.

2300
MARSH YELLOW CRESS
fr raifort *m* d'eau
ne moeraskers
de Sumpfkresse *f*
da kær-guldkarse
sv sumpsenap
es berro *m* falso
it Rorippa *f*
    Rorippa islandica
    (Oeder) Barbás

2301
MARSHY SOIL
fr sol *m* marécageux
ne moerassige grond
de Sumpfboden *m*
da sumpjord
sv kärrmark
es suelo *m* pantanoso
it terreno *m* paludoso

MASH, 2985

2302
MASS SELECTION
fr sélection *f* massale
ne massaselectie
de Massenauslese *f*
da masseudvalg *n*
sv massurval *n*
es selección *f* en masa
it selezione *f* in massa

2303
MASTER OF SCIENCE
fr ingénieur *m*
ne ingenieur
de Ingenieur *m*; Diplom-
    ingenieur *m*
da ingeniør
sv hortonom; ingenjör
es ingeniero *m*
it ingegnere *m*

MATRIX, 2321

MATURE, 3144

2304
MATURE, TO; TO RIPEN
fr mûrir
ne rijpen
de reifen
da modne
sv mogna
es madurar
it maturare

MATURITY, 3146

2305
MAXIMUM PRICE
fr prix *m* maximum
ne maximumprijs
de Höchstpreis *m*
da højeste pris
sv maximipris *n*
es precio *m* máximo
it prezzo *m* massimo

2306
MAXIMUM
TEMPERATURE
fr température *f* maximum
ne maximum-temperatuur
de Maximumtemperatur *f*;
    Höchsttemperatur *f*
da maximumtemperatur
sv maximitemperatur
es temperatura *f* máxima
it temperatura *f* massima

2307
MAYWEED
fr camomille *f*
ne kamille
de Kamille *f*

da kamille
sv kamomill
es manzanilla f
it camomilla f
   Matricaria

MAYWEED, rayless,   3048

2308
MEADOW FESCUE
fr fétuque f des prés
ne beemdlangbloem
de Wiesenschwingel m
da engsvingel
sv ängssvingel
es sañuela f de prados
it Festuca f
   Festuca pratensis Huds.

2309
MEADOW FOX TAIL
fr vulpin m des prés
ne vossestaart
de Wiesenfuchsschwanz m
da eng-rævehale
sv ängskavle
es cola f de zorra
it coda f di volpe;
   alopecuro m
   Alopecurus pratensis L.

MEADOW GRASS,
annual,   109
—, rough-stalked,   3208
—, smooth-stalked,   3449
—, wood,   4197

2310
MEADOW NEMATODE
fr anguillule f des racines
ne wortelaaltje n
de Wurzelälchen n
da rodål
sv rotål
es anguilula f de raíz
it nematode m delle lesioni
   radicali
   Pratylenchus spp.

2311
MEADOW RUE
fr pigamon m
ne ruit
de Wiesenraute f

da frøstjerne
sv ruta
es ruda f
it ruta f
   Thalictrum L.

2312
MEADOW SAFFRON;
AUTUMN CROCUS
fr colchique m d'automne
ne droogbloeier; herfsttijloos
de Herbstzeitlose f
da tidløs (høst)
sv tidlöse
es colquico m de otoño;
   quitameriendas m
it colchico m
   Colchicum autumnale L.

MEADOW SPITTLE BUG,
769

2313
MEALY
fr farineux
ne melig
de mehlig
da melet
sv mjölig
es harinoso
it farinoso

2314
MEALY BUG
fr cochenille f farineuse
ne wolluis
de Schmierlaus f; Wollaus f
da uldlus
sv ullus
es pulgón m lanigero
it pseudococco m;
   cocciniglia f farinosa
   Phenacoccus spp.

2315
MEAN PRICE
fr prix m moyen
ne middenprijs
de Mittelpreis m
da gennemsnitspris
sv mitpris n; medelpris n
es precio m medio
it prezzo m medio

2316
MEASURING ROD
fr règle f graduée
ne meetlat
de Messlatte f
da målestok
sv måttstock
es reglón m de medir
it pertica f da misurare

2317
MEASURING TAPE;
TAPE MEASURE
fr mètre m à ruban
ne meetlint n
de Messschnurf; Bandmass n
da målebånd n
sv måttband n
es cinta f de medir
it nastro m da misurare

2318
MECHANIZATION
fr mécanisation f
ne mechanisatie
de Mechanisierung f
da mekanisering
sv mekanisering
es mecanización f
it meccanizzazione f

2319
MEDICINAL HERBS
fr plantes f pl médicinales;
   simples f pl
ne geneeskruiden n pl
de Heilkräuter n pl
da lægeplanter
sv medicinalväxter
es plantas f pl medicinales
it erbe f pl medicinali

2320
MEDITERRANEAN
FRUITFLY
fr mouche f méditerranéen-
   ne des fruits
ne middellandsezeevlieg
de Mittelmeerfruchtfliege f
da middelhavsfrugflue
sv Medelhavsfruktfluga
es mosca f mediterránea de
   las frutas
it mosca f della frutta
   Ceratitis capitata Wied.

2321
MEDIUM SUBSTRATE;
MATRIX
fr milieu *m* nutritif
ne voedingsbodem
de Nährboden *n*
da næringsjord; substrat *n*
sv odlingssubstrat *n*
es substrato *m*; base *f*;
   suelo *m* nutritivo
it sottosuolo *m* nutritivo

2322
MEDIUM VIGOROUS
ROOTSTOCK
fr porte-greffe *m* semi-fort
ne matig zwakke onderstam
de mässige Unterlage *f*
da middelkraftig grund-
   stamme
sv mycket svagt växande
   underlag *n*
es patrón *m* de vigor
   mediano
it soggetto *m* moderato
   vigoroso

2323
MEDLAR
fr nèfle *f*; chaperon *m* de
   moine
ne mispel
de Mispel *f*
da mispel
sv mispel
es nispero *m*
it nespolo *m*; napello *m*

2324
MEDLAR TREE
fr néflier *m*
ne mispel
de Mispelbaum *m*
da mispeltræ *n*
sv mispelträd *n*
es nispero *m*
it nespolo *m*
   Mespilus L.

MEIOSE, 3075

2325
MELON
fr melon *m*
ne meloen

de Melone *f*
da melon
sv melon
es melón *m*
it popone *m*; melone *m*

MELON, water, 4087

2326
MEMBRANOUS
fr membraneux
ne vliezig
de häutig
da hindeagtig
sv hinnartad
es membranoso
it membranoso

2327
MERCHANTS' CREDIT
fr crédit *m* à l'achât
ne leverancierskrediet *n*
de Lieferantenkredit *m*
da leverandørkredit
de Lieferantenkredit *m*
da leverandørkredit
sv leveranskredit
es crédito *m* de proveedores
it credito *m* mercantile

2328
MERCURY;
QUICKSILVER
fr mercure *m*; vif-argent *m*
ne kwik *n*
de Quecksilber *m*
da kviksølv *n*
sv kvicksilver *n*
es mercurio *m*
it mercurio *m*

2329
MERCURY
COMPOUNDS
fr composés *pl* mercuriques
ne kwikhoudende verbin-
   dingen
de Quecksilberverbindun-
   gen *f pl*
da kviksølvforbindelser
sv kvicksilverföreningar
es compuestos *m pl* mercu-
   rios
it composizioni *f pl*
   chimique di mercurio

2330
MERISTEM
fr méristème *m*
ne meristeem *n*
de Meristem *n*
da meristem *n*
sv meristem *n*
es meristemo *m*
it meristemo *m*

2331
MESEMBRIANTHEMUM
fr ficoïde *f*
ne middagbloem
de Mittagsblume *f*
da middagsblomst
sv mesembrianthemum;
   middagsblomma
es escarchada *f*; escarchosa
   f; hierba *f* de la plata
it fiore *m* di mezzogiorno
   Mesembrianthemum L.

2332
MESH; TISSUE
fr tissu *m*
ne weefsel (kunststof) *n*
de Gewebe *n*
da netvæv *n*
sv väv *n*; vävnad
es tejido *m* sintético
it tessuto *m* artificiale

MESSPHYLL, 2112

2333
METAMORPHOSIS
fr métamorphose *f*
ne gedaanteverwisseling
de Metamorphose *f*
da forvandling; metamor-
   fose
sv metamorfos; förvandling
es metamórfosis *f*
it metamorfosi *f*

2334
MICHAELMAS DAISY
fr aster *m* d'automne
ne herfstaster
de Herbstaster *f*
da høstaster
sv höstaster

es aster *m*
it astero *m* d'autunno
   Aster spec.

2335
MICROBES
fr microbes *m pl*
ne microben
de Mikroben *f pl*
da mikrober
sv mikrober
es microbios *m pl*
it microbi *m pl*

2336
MICROCLIMATE
fr microclimat *m*
ne microklimaat *n*
de Mikroklima *n*
da mikroklima *n*
sv mikroklimat *n*
es microclima *m*
it microclima *m*

2337
MICROSCOPIC
fr microscopique
ne microscopisch
de mikroskopisch
da mikroskopisk
sv mikroskopisk
es microscópico
it microscopico

MIDGE, 1520
—, black currant leaf, 334
—, brassica pod, 424
—, gall, 1520
—, leaf-mining, 1520
—, pea, 2635
—, pear, 2644
—, pear leaf, 2642
—, raspberry cane, 3037
—, swede, 3761

2338
MIDRIB
fr nervure *f* médiane
ne middennerf; hoofdnerf
de Mittelrippe *f*; Hauptrippe *f*
da midtribbe; hovednerve
sv mittnerv; huvudnerv
es nervio *m* central
it nervo *m* mediano

2339
MIGNONETTE
fr réséda *m*
ne Reseda
de Reseda *f*
da Reseda
sv Reseda
es réseda *f*
it Reseda *f*; amorino *m*
   Reseda L.

2340
MILDEW
fr mildiou *m*
ne meeldauw
de Mehltau *m*
da meldug
sv mjöldagg
es mildio *m*
it peronospora *f*

MILDEW, American goose-
berry, 84
—, downy, 1115, 1116
—, powdery, 2885

2341
MILD HUMUS
fr humus *m* doux
ne milde humus
de milder Humus *m*
da mild humus
sv mild humus
es humus *m* dulce
it umus *m* dolce

2342
MILFOIL; YARROW
fr millefeuille *f*
ne duizendblad *n*
de Schafgarbe *f*
da alm; røllike
sv rölleka
es milenrama *f*
it Achillea *f*; millefoglie *m*
   Achillea millefolium L.

2343
MILLEGRAMME
EQUIVALENT
fr équivalent-milligramme *m*
ne milligram-equivalent *n*
de Milligramm-Äquivalent *n*
da milligram-ækvivalent

sv milli-ekvivalent; mval;
   me
es miligramo-equivalente *m*
it equivalente-milligrammo
   m

2344
MIMOSA
fr mimosa *m*; mimeuse *f*
ne mimosa
de Mimosa *f*
da sølvacacia; mimose
sv akacia
es mimosa *f*
it mimosa *f*
   Acacia spec.

2345
MINE; CAVE
fr carrière *f*; cave *f*
ne grot
de Stollen *m*; Felsenkeller *m*;
   Höhle *f*
da hule; grotte
sv grotta; håla
es cantera *f*; mina *f*; cueva *f*;
   bodega *f*; gruta *f*
it grotta *f*

2346
MINE, TO
fr miner
ne mineren
de minieren
da minere
sv minera
es minar
it minare

2347
MINERAL (adj.)
fr minéral
ne mineraal
de mineralisch
da mineralsk
sv mineralisk
es mineral
it minerale

2348
MINERALIZATION
fr minéralisation *f*
ne mineralisatie
de Mineralisierung *f*

da mineralisering
sv mineralisering
es mineralización f
it mineralizzazione f

2349
MINERALS
fr minéraux m pl
ne mineralen n pl
de Minerale n pl
da mineraler n pl
sv mineraler n pl
es minerales m pl
it minerali m pl

2350
MINIMUM PRICE
fr prix m minimum
ne minimumprijs
de Mindestpreis m
da mindstepris
sv minimipris n
es precio m mínimo
it prezzo m minimo

2351
MINIMUM
TEMPERATURE
fr température f minimum
ne minimum-temperatuur
de Minimumtemperatur f;
    Mindesttemperatur f
da minimumtemperatur
sv minimitemperatur
es temperatura f mínima
it temperatura f minima

2352
MISCELLANEOUS BULBS
fr produits m pl secondaires;
    bulbes m pl divers
ne bijgoed n
de Nebenware f
da bikultur
sv bikultur
es productos m pl anejos;
    productos m pl accessorios
it prodotti m pl secondari

2353
MISTBLOWER;
COMPRESSION
SPRAYER
fr atomiseur m
ne nevelspuit

de Sprühgerät n
da tågesprøjte
sv dimspruta;
    koncentratspruta
es atomizador m
it atomizzatore m

MISTBLOWER, knapsack,
1999
—, power-driven knapsack,
2889

2354
MISTLETOE
fr gui m
ne maretak; vogellijm
de Mistel f
da mistelten
sv mistel
es muérdago m
it vischio m; visco m
    Viscum album L.

2355
MIST PROPAGATION
fr multiplication f avec
    brouillard artificiel
ne stekken onder waternevel
de Sprühnebelvermehrung f
da tågeformering
sv dimförökning; sticklings-
    förökning i vattendimma
es multiplicación f con nie-
    bla artificial

2356
MITE; TICK
fr acarien m
ne mijt
de Milbe f
da mide
sv kvalster n
es ácaro m
it acaro m
    Acarina spp.

MITE, broad,   438
—, bulb,   485
—, bulb scale,   489
—, conifer spinning,   807
—, cyclamen,   3682
—, dryberry,   3038
—, European red,   1487
—, leaf blister,   2077

—, nut gall,   2499
—, plum leaf,   2804
—, plum rust,   2804
—, raspberry leaf and bud,
3038
—, red-legged earth,   3066
—, red spider,   3071
—, spruce spider,   807
—, strawberry,   3682
—, two-spotted spider,
3071
—, wheat curl,   4133
—, winter grain,   3066

2357
MIX, TO; TO BLEND
fr mêler; mélanger
ne mengen
de mischen
da blande
sv blanda
es mezclar
it mescolare

2358
MIXED BUD
fr bourgeon m mixte
ne gemengde knop
de gemischte Knospe f
da blandet knop
sv blandad knopp
es lamburda f; yema f mixta
it gemma f promiscua

2359
MIXER
fr agitateur m
ne mengwoeler
de Grubber m; Kultivator m
da jordblander
sv djupkultivator
es cavadora-mezcladora f
it (moto) coltivatore m;
    grufolatrice f a mescolare

2360
MIXING VALVE
fr vanne f de mélange
ne mengklep
de Mischventil n
da blandeventil
sv blandningsventil
es válvula f de mezcla
it valvola f miscelatrice

2361
MOBILE GLASSHOUSE
fr serre *f* roulante
ne rolkas
de fahrbares Gewächshaus *n*;
   Rollhaus *n*
da rullehus *n*
sv flyttbart växthus *n*
es invernadero *m* móvil sobre vías; estufa *f* rodante
it serra *f* mobile

2362
MOBILE STEAMING
GRID
fr herse *f* pour désinfection
   à la vapeur
ne traprek *n*
de Dampfegge *f*
da dampharve
sv ångharv
es grada *f* para desinfección
   con vapor
it erpice *m* per disinfezione
   a vapore

2363
MOCK ORANGE
fr seringa
ne boerenjasmijn
de Pfeifenstrauch *m*
da pibeved; uægte jasmin
sv schersmin
es jeringuilla *f*; filadelfo *m*;
   jazmín *m*
it gelsomino *m*
   Philadelphus L.

2364
MODIFICATION
fr modification *f*
ne modificatie
de Abänderung *f*
da modifikation
sv modifikation
es modificación
it modificazione *f*

2365
MOIST
fr humide
ne vochtig
de feucht
da fugtig

sv fuktig
es húmedo
it umido

2366
MOISTEN, TO
fr humecter; mouiller
ne bevochtigen
de anfeuchten
da fugte; befugte
sv fukta
es humectar
it inumidire

2367
MOISTURE BARRIER;
VAPOUR BARRIER
fr écran *m* d'étanchéité à la
   vapeur (d'eau)
ne dampdichte laag
de Dampfsperre *f*
da dampspærre
sv ångspärr
es barrera *f* antivapor;
   pantalla *f* hidrófuga
it barriera *f* al vapore

2368
MOISTURE CONTENT
fr teneur *f* en eau
ne vochtgehalte *n*
de Wassergehalt *m*;
   Feuchtigkeitsgehalt *m*
da fugtighedsindhold *n*
sv fuktighetshalt
es contenido *m* en agua
it contenuto *m* d'umidità

2369
MOISTURE
DEFICIENCY;
MOISTURE SHORTAGE
fr pénurie *f* d'eau;
   manque *m* d'eau
ne vochttekort *n*
de Feuchtigkeitsdefizit *n*
da mangel på fugtighed
sv brist på fuktighet
es carencia *f* de humedad
it deficienza *f* d'umidità

2370
MOISTURE RETAINING
fr retenant l'humidité
ne vochthoudend

de feuchthaltig
da vandholdende
sv fuktighetshållande
es hidrófilo
it conservando l'umidità

MOISTURE SHORTAGE,
2369

MOLD, leaf,   1116

2371
MOLE
fr taupe *f*
ne mol
de Maulwurf *m*
da muldvarp
sv mullvad
es topo *m*
it talpa *f*

2372
MOLE CRICKET
fr courtilière *f*; taupe-
   grillon *m*
ne veenmol
de Maulwurfsgrille *f*;
   Werre *f*
da jordkrebs
sv mullvadssyrsa
es grillotalpa *m*; alacrán *m*
   cebollero; grillo *m* real;
   calluezo *m*
it grillotalpa *m*
   Gryllotalpa gryllotalpa L.

2373
MOLE DRAINAGE
fr drainage-taupe *m*
ne moldrainage
de Maulwurfsdränung *f*
da muldvarpedræning
sv mullvadsdränering
es drenaje *m* sistema topo
it drenaggio *m* da talpa

2374
MOLE PLOUGH
fr charrue *f* taupe
ne molploeg
de Maulwurfdränpflug *m*
da muldvarpedrænplov
sv tubulator, mullvadsplog
es arado *m*
it aratro *m* talpa

2375
MOLYBDENUM
fr molybdène *m*
ne molybdeen *n*
de Molybdän *n*
da molybden *n*
sv molybden *n*
es molibdeno *m*
it molybdenia *f*

2376
MONKEY FLOWER
fr mimule *m*
ne maskerbloem
de Gauklerblume *f*
da abeblomst
sv gyckelblomster *n*
es mimulo *m*
it mimula *f*
    Mimulus L.

2377
MONKSHOOD
fr aconit *m*; chaperon *m* de
    moine
ne monnikskap
de Eisenhut *m*
da stormhat
sv stormhatt
es acónito *m*
it aconito *m*; napello *m*
    Aconitum L.

2378
MONOBASIC
fr monobasique
ne eenbasisch
de einbasisch
da monobasisk
sv enbasisk
es monobásico
it monobasico

2379
MONOCULTURE
fr monoculture *f*
ne monocultuur
de Monokultur *f*
da monokultur
sv monokultur
es monocultivo *m*
it monocoltura *f*

2380
MONOECIOUS
fr monoïque
ne eenhuizig
de einhäusig
da enbo; sambo
sv sambyggare
es monóico
it monoico

2381
MONOPETALOUS
fr monopétale; gamopétale
ne eenbladig
de einblättrig
da enbladet
sv enbladig
es monopétalo
it monopetalo

2382
MOOR; BOG
fr tourbe *f*
ne veen *n*
de Moor *n*
da mose
sv mosse
es turbera *f*
it torba *f*

MOOR, peat,   2657

2383
MORELLO
fr griotte *f*
ne morel
de Morelle *f*; Sauerkirsche *f*
da morel; surkirsebær *n*
sv surkörsbär *n*; morell
es guinda *f*
it visciolo *m*; amarasco *m*
    Prunus cerasus L.

2384
MORE THAN ONE YEAR
OLD (STORED SEED)
fr de plus d'un an
ne overjarig
de überjährig
da mere end ét år gammelt
sv överårig (t)
es de más de un año
it di piú d'un anno

2385
MORNING GLORY
fr ipomée *f*
ne dagbloem
de Prunkwinde *f*
da tragtonerle
sv blomman för dagen
es enredadera *f* de
    campanilla; ipomea *f*
it fiore *m* diurno
    Ipomoea L.

2386
MORTAR
fr mortier *m*
ne metselspecie
de Mörtel *m*
da mørtel
sv murbruk *n*
es mortero *m*; argamasa *f*
it malta *f*

2387
MORTGAGE
fr hypothèque *f*
ne hypotheek
de Hypothek *f*
da prioritet
sv hypotek *n*
es hipoteca *f*
it ipoteca *f*

2388
MOSAIC
fr mosaïque *f*
ne mozaïek *n*
de Mosaik *n*
da mosaik
sv mosaik
es mosaico *m*
it mosaico *m*

MOSAIC, bean yellow,
278
—, beet,   296
—, cauliflower,   612
—, common bean,   766
—, cucumber,   921
—, interveinal,   1946
—, lettuce,   2141
—, narcissus,   2446
—, pea,   2636
—, pea enation,   2630
—, poplar,   2843

—, rhubarb, 3126
—, rose, 3190
—, tulip, 3955
—, vein, 4016
see also: aucuba mosaic

2389
MOSAIC (HYACINTH)
fr mosaique *f*
ne grijs *n*
de Mosaikkrankheit *f*;
  Gelbfleckigkeit *f*
da mosaiksyge
sv mosaik
es mosaico *m*
it malattia *f* mosaica

2390
MOSAIC DISEASE
fr mosaïque *f*
ne mozaïekziekte
de Mosaikkrankheit *f*
da mosaiksyge
sv mosaiksjuka
es enfermedad *f* del mosaico
it mosaico *m*

2391
MOSS PEAT
fr tourbe *f* à sphaigne
ne mosveen *n*
de Moostorf *m*
da sphagnummose
sv mosstorv
es esfagno *m*
it torba *f* muscosa

MOSSY-ROSE-GALL
WASP, 282

MOTH, angle shades, 102
—, apple fruit, 130
—, brown tail, 451
—, cabbage, 514
—, clothes, 716
—, codling, 736
—, currant, 2237
—, currant clearwing, 945
—, dart, 987
—, diamond-back, 1050
—, dusky clearwing, 1173
—, garden swift, 1538
—, gipsy, 1572
—, goat, 1604

—, gold-tailed, 1654, 4230
—, juniper webber, 1987
—, lackey, 2021
—, leaf roller, 3891
—, leek, 2125
—, leopard, 2135
—, magpie, 2237
—, mottled umber, 2400
—, parsnip, 2613
—, pea, 2637
—, pear leaf blister, 2641
—, plume, 2802
—, plum fruit, 2803
—, privet hawk, 2935
—, satin, 3250
—, silver, 3417
—, small ermine, 3442
—, sphinx, 1740
—, tomato, 3874
—, winter, 4178
—, yellow-tailed, 4230
see also: bud moth;
  hawk moth; pine
  shoot moth;
  tortrix moth

2392
MOTHER BULB
fr bulbe *m* mère
ne moederbol
de Mutterzwiebel *f*
da moderløg *n*
sv moderlök
es bulbo *m* madre
it bulbo *m* madre

2393
MOTHER CELL
fr cellule-mère *f*
ne moedercel
de Mutterzelle *f*
da modercelle
sv modercell
es célula *f* madre
it cellula *f* madre

2394
MOTHER OF
THOUSANDS;
SAXIFRAGE
fr saxifrage *f* sarmenteuse
ne moederplantje *n*
de Judenbart *m*
da jødeskæg *n*

sv aronsskägg *n*
es Saxifraga *f* sarmentosa
it sassifraga *f* sarmentosa
  Saxifraga sarmentosa L.

2395
MOTHER PLANT;
STOCK PLANT
fr pied-mère *m*
ne moederplant; moerplant
de Mutterpflanze *f*
da moderplante
sv moderplanta
es planta *f* madre
it pianta *f* madre

2396
MOTION STUDY
fr étude *f* des mouvements
ne bewegingsstudie
de Bewegungsstudie *f*
da bevægelsestudie
sv rörelsestudie
es estudio *m* de los
  movimientos
it studio *m* dai movimenti

2397
MOTOR HOE
fr motohoue *f*
ne motorhak *f*
de Motorhacke *f*
da motorhakke
sv motorhacka
es motocultor *m*
it motozappa *f*

2398
MOTOR MOWER
fr motofaucheuse *f*
ne motormaaier
de Motormäher *m*
da motorgræsslåmaskine;
  motorgræsklippare
sv motorslåttermaskin
es motosegadora *f*
it motofalciatrice *f*

2399
MOTOR SPRAYER
fr pulvérisateur *m* à moteur
ne motorspuit
de Motorspritze *f*
da motorsprøjte

sv motorspruta
es pulverizador *m* de motor
it polverizzatore *m* a motore

MOTTLE, 2401
—, potato stem, 2875

2400
MOTTLED UMBER
MOTH
fr phalène *f* défeuillante
ne grote wintervlinder
de grosser Frostspanner *m*
da store frostmåler
sv lindmätare
es falena *f* mayor de los
　frutales
it defogliatrice *f* degli alberi
　fruttiferi; falena *f* defog-
　liatrice
　Erannis defoliaria Clerck

2401
MOTTLING, MOTTLE
fr marbrure *f*
ne vlekkerigheid
de diffuse Fleckung *f*
da spætning
sv mottle; diffusa
es moreado *m*
it maculatura *f*; mosaicatura
　f; mosaico *m*

MOULD, black, 341
—, brown plaster, 448
—, garden, 1535
—, grey, 1657, 1658, 1659,
1660, 1661, 1662
—, leaf, 2089
—, lipstick, 2187
—, olive green, 2521
—, sooty, 341, 3511
—, white, 4143
—, white plaster, 4144
—, yellow, 4226

2402
MOULDBOARD
fr versoir *m*
ne rister *n*
de Streichblech *n*
da muldfjæl
sv vändskiva
es vertedera *f*
it versoio *m*

2403
MOULD DISEASE
fr maladie *f* cryptogamique
ne schimmelziekte
de Pilzkrankheit *f*
da skimmelsygdom
sv mögelangrepp
es enfermedad *f* por hongo
it malattia *f* da muffa

2404
MOUNTAIN ASH
fr sorbier *m*
ne lijsterbes
de Eberesche *f*
da røn
sv rönn
es serbal *m*
it sorbo *m*
　Sorbus aucuparia L.

MOUNTAIN
CRANBERRY, 876

2405
MOUNTAIN CURRANT
fr Ribes *m* alpinum
ne alpenbes
de Alpen-Johannisbeere *f*
da alperibs
sv måbär
es calderilla *f*
it Ribes *m* alpinum
　Ribes alpinum L.

MOUNTAIN LAUREL,
533

2406
MOUNTAIN PINE
fr pin *m* des montagnes
ne bergden
de Bergkiefer *f*; Bergföhre *f*
da bjergfyr
sv bergtall
es pino *m* negro
it pino *m* alpestre; pino *m*
　nano
　Pinua mugo Turra

2407
MOUNTED QUARTER
TURN PLOUGH
fr charrue *f* portée de tour
ne aanbouwkantelploeg
de Anbauwinkelpflug *m*

da ophængt plov
sv påhängväxelplog för ¼
　varvs vridning
es arado *m* de ¼ vuelta
it aratro *m* portato a ¼ di
　giro

2408
MOUSE
fr souris *f*
ne muis
de Maus *f*
da mus
sv mus
es ratón *m*
it topo *m*
　Mus spp.

2409
MOUSE-EAR
fr céraiste *m*
ne hoornbloem
de Hornkraut *n*
da hønsetarm
sv arv
es cerastino *m*; morguelina *f*
it cerasta *f*
　Cerastium (Dill.) L.

2410
MOUSE-EAR
CHICKWEED ;
fr céraiste *m*
ne hoornbloem
de Hornkraut *n*
da hønsetarm
sv hönsarv
es oreja *f* de ratón
it cerastio *m*
　Cerastium holosteiodes Fr.

2411
MOW, TO; TO CUT
fr faucher; tondre
ne maaien; afmaaien
de mähen; abmähen
da slå; meje
sv meja; slå
es segar; cortar
it mietere

2412
MOWER
fr tondeuse *f*; faucheuse *f*
ne maaimachine

de Mähmaschine *f*
da slåmaskine
sv slåttermaskin
es segadora *f*; tundidora *f*
it mietitrice *f*; falciatrice *f*

MOWER, bank, 246
—, motor, 2398
—, rotary, 3199

2413
MUCK
ne meermolm
de Mischung *f* organischer
   und mineralischer Boden-
   substanz
sv dyjord
es cieno *m*
it fango *m* organico

2414
MUD
fr boue *f*; vase *f*
ne slijk; modder
de Schlick *m*
da slik
sv dy
es lodo *m*; vaso *m*
it melma *f*; fango *m*

2415
MUD BARGE
fr bateau-drague *m*;
   bateau-dragueur *m*
ne baggerschuit
de Baggerboot *n*;
   Schwimmbagger *m*
da mudderpram
sv mudderpråm
es lancha *f* dragadora
it cavafango *m*

2416
MUDDY; SLUDGY
fr boueux; limoneux; vaseux
ne slempig
de schlammig; verschlammt
da dyndet; mudret
sv slammig; gyttjig
es limoso; cenagoso
it melmoso; limaccioso

2417
MUD HOLE
fr trou *m* d'ébouage
ne slijkgat *n*
de Schlammloch *n*
da slamhul
sv slamlucka; renslucka
es agujero *m* de limo;
   agujero *m* de lodo;
   agujero *m* de barre
it foro *m* di spurgo

2418
MULBERRY
fr mûre *f*
ne moerbei
de Maulbeere *f*
da morbær
sv mullbär *n*
es mora *f*
it mora *f*
   Morus L.

2419
MULBERRY TREE
fr mûrier *m*
ne moerbeiboom
de Maulbeerbaum *m*
da morbærtræ *n*
sv mullbärsträd *n*
es moral *m*
it gelso *m*

2420
MULCH
fr couverture *f* du sol
ne bodembedekking
de Bodenbedeckung *f*
da jorddækning
sv marktäckning
es cobertura *f* del suelo
it copertura *f* del suolo

2421
MULCHER-
TRANSPLANTER
fr dérouleuse-planteuse *f* de
   film plastique
ne folielegger-plantmachine
de Folienlege-Pflanz-
   maschine *f*
da folielægge-plantemaskine
sv maskin för plantering och
   utläggning av folie

es desenrolladora *f* situadora de película plástica
it macchina *f* per situare e
   stendere il foglio di plastica

2422
MULLEIN
fr molène *f*
ne toorts
de Königskerze *f*
da kongelys *n*
sv kungsljus *n*
es verbasco *m*; gordolobo *m*
it verbasco *m*
   Verbascum L.

2423
MULTI FURROW
PLOUGH
fr charrue *f* poly-soc
ne meerscharige ploeg
de Mehrscharpflug *m*
da flerfuret plov
sv flerskärig plog
es arado *m* multireja
it aratro *m* multivomere

2424
MULTILOCULAR
fr pluriloculaire
ne meerhokkig
de mehrfächerig
da flerrummet
sv flerrummig
es plurilocular
it pluriloculare

2425
MULTIPLICATION;
PROPAGATION
fr multiplication *f*
ne vermenigvuldiging
de Vermehrung *f*;
   Vervielfältigung *f*
da formering
sv förökning
es multiplicación *f*
it moltiplicazione *f*

MULTIPLICATION,
asexual, 165
—, sexual, 3357

2426
MULTIPOT
fr multipot *m*
ne multipot
de Multitopf *m*
da multipotte
sv multipot
es multitiesto *m*
it multivaso *m*

2427
MULTISPAN
GLASSHOUSE
fr serre *f* multichapelles
ne complexkas
de Gewächshausblock *n*
da væksthusblok
sv blockväxthus *n*
es invernadero *m* multicu-
erpe
it serra *f* a piú campate

2428
MUMMY DISEASE
fr maladie *f* de la momification
ne mummieziekte
de Mumienkrankheit *f*
da 'mummy'-sygdom
sv mummy-sjuka
es enfermedad *f* de la
momificación
it mummificazione *f*

2429
MURIATE OF POTASH
fr chlorure *f* de potassium
ne chloorkali
de Chlorkali *n*
da klorsur kali *n*
sv kalisalt *n*
es cloruro *m* potásico
it cloruro *m* di calce

2430
MUSHROOM COMPOST
fr engrais *m* de champignon
ne champignonmest
de Champignonmist *m*
da champignongødning
sv champinjongödsel
es estiércol *m* usado para el
cultivo de la champiñón
it concime *m* da fungo

2431
MUSHROOM DIE BACK
fr dépérissement *m* du
champignon
ne afstervingsziekte van de
champignon
de Mumienkrankheit *f* des
Champignons
da d-b-syge (champignon)
es languidez *f* de la seta
it moria *f* dei prataioli

2432
MUSHROOM SPAWN
fr blanc *m* de champignon
ne champignon-broed *n*
de Champignonbrut *f*
da mycelium *n*
sv mycelium *n*
es micelio *f*; semilla *m*
it bianco *m*

2433
MUSK RAT
fr rat *m* musqué
ne bisamrat
de Bisamratte *f*
da moskusrotte *mf*
sv bisamråtta
es rata *f* almizclada
it topo *m* muschiato
Ondatra zibethicus L.

2434
MUSSEL SCALE;
OYSTERSHELL SCALE
(US)
fr cochenille *f* virgule
ne kommaschildluis
de gemeiner Kommaschild-
laus *f*
da kommaskjoldlus
sv kommasköldlus
es serpeta *f* del manzano
it cocciniglia *f* a virgola
degli alberi da frutto
Lepidosaphes ulmi L.

2435
MUST
fr moût *m*
ne most
de Most *m*
da most

sv must
es mosto *m*
it mosto *m*

2436
MUTAGENIC
fr mutagène
ne mutageen
de mutagen
da mutagen
sv mutagen
es mutageno
it mutageno

2437
MUTANT
fr mutant *m*
ne mutant
de Mutant *m*
da mutant
sv mutant
es mutante *m*
it mutazione *f*

2438
MUTATION
fr mutation *f*
ne mutatie
de Mutation *f*
da mutation
sv mutation
es mutación *f*
it mutazione *f*

MUTATION, bud, 468

2439
MUTATION BREEDING
fr sélection *f* par mutation;
production *f* de mutation
ne mutatieveredeling
de Mutationszüchtung *f*
da mutations-forædling
sv mutations-förädling
es mejoramiento *m* por
mutación
it miglioramento *m* da
mutazione

2440
M-VIRUS DISEASE
(POTATO)
fr maladie *f* à virus M
(pomme de terre)
ne M-virusziekte (aard-
appel)

de M-Viruskrankheit *f*
(Kartoffel)
da M-virose (kartoffel)
sv M-virus sjuka (potatis)
es virus *m* M (patata)
it virus *m* M (patata)

2441
MYCELIAL STRAND
(MUSHROOM
GROWING);
STRINGLY MYCELIUM
fr cordon *m* mycélien;
filements *m pl* épais
ne dikke streng
de Myzelstrang *m*; Myzel-
strähne *f*
da grov hyfe
sv tjock mycelietråd
es cordones *m pl*; filamentos
m pl espesos
it filo *m* grosso del micelio

2442
MYCELIUM
fr mycélium *m*
ne mycelium *n*
de Myzel *n*
da mycelium *n*
sv mycelium *n*
es mycelio *m*
it micelio *m*

MYCELIUM, stringly,
2441

2443
MYCELIUM-GROWTH
(MUSHROOM
GROWING); SPAWN
RUN; IMPREGNATION;
COLONISATION
fr croissance *f* du mycélium;
croissance du blanc; prise
f du blanc; incubation *f*

ne myceliumgroei
de Myzelwachstum *n*; Brut-
durchspinnung *f*; Vorkultur *f*
da løbning
sv myceliespinning
es desarollo *m* del micelio;
agarre *m*
it presa *f* del micelio

2444
MYRTLE
fr myrte *m* commun
ne mirt
de Myrte *f*
da myrte
sv myrten
es mirto *m*
it mirto *m*
Myrtus communis L.

N

NARCISSUS, 973
—, trumpet, 3936

2445
NARCISSUS BULB FLY;
LARGE NARCISSUS FLY
fr mouche *f* des narcisses
ne grote narcisvlieg
de grosse Narzissenfliege *f*
da store narcisflue
sv stora narcissflugan
es mosca *f* del narciso
it mosca *f* del narciso
    Lampetia equestris L.

2446
NARCISSUS MOSAIC
fr mosaïque *f* du narcisse
ne grijs (narcis)
de Hyazinthenmosaik *n*;
    Weisstreifigkeit *f* der
    Narzisse
da narcis-mosaik
sv narcissmosaik
es mosaico *m* del narciso
it mosaico *m* del narciso

2447
NARROW LEAVE
(TOMATO)
fr feuille *f* aciculaire
ne naaldblad *n*
de Nadelblatt *n*
da trådblad *n*; bregneblad *n*
sv trådlika blad *n*
es aciculifolie *m*
it foglia *f* d'ago

2448
NARROW LEAVED
fr à feuilles étroites
ne smalbladig
de schmalblättrig
da smalbladet
sv smalbladig
es acicular
it con foglie strette

2449
NASTURTIUM
fr capucine *f*
ne Oostindische kers
de Kapuzinerkresse *f*
da blomsterkarse; tropaeo-
    lum
sv krasse
es capuchina *f*
it cappuccina *f*
    Tropaeolum L.

NATIVE, 1893

2450
NATURAL ENEMY
fr ennemi *m* naturel
ne natuurlijke vijand
de natürlicher Feind *m*
da naturlige fjende
sv naturliga fiender
es enemigo *m* natural
it nemico *m* naturale

2451
NATURAL GAS
fr gaz *m* naturel
ne aardgas *n*
de Erdgas *n*
da jordgas
sv jordgas
es gas *m* natural
it metano *m*

2452
NATURALIZE, TO
fr devenir agreste; planter
    en demeure
ne verwilderen
de verwildern
da 'wilgarden'
sv förvildas
es degenerarse
it diventare selvatico

2453
NATURE OF THE SOIL
fr nature *f* du sol
ne bodemgesteldheid
de Bodenbeschaffenheit *f*

da jordensbeskaffenhed
sv jordbeskaffenhet; mark-
    beskaffenhet
es naturaleza *f* del suelo
it natura *f* del suolo

2454
NECK ROT (ALLIUM)
fr pourriture *f* du collet
ne bodemrot *n*
de Grauschimmelfäule *f*
da gråskimmel
sv gråmögel
es mal *m* del esclerocio
it marciume *f*
    Botrytis allii Munn.

NECK ROT (ONION),
1659

2455
NECROSIS
fr nécrose *f*
ne necrose
de Nekrose *f*
da nekrose
sv nekros
es necrosis *f*
it necrosi *f*

NECROSIS, bark,   551
—, bud,   469
—, cucumber, 922
—, tobacco,   3869
—, top,   3887

NECROSIS VIRUS,
tobacco,   3870

2456
NECTAR GLAND
fr glande *f* nectarifère;
    nectaire *m*
ne nektarklier
de Nektarrüse *f*
da nektarie
sv nektarie
es glándula *f* nectarífera
it nettario *m*

2457
NEEDLE (bot.)
fr aiguille *f*
ne naald
de Nadel *f*
da nål
sv barr
es aguja *f*
it foglia *f* aghiforme

2458
NEEDLE CAST
fr maladie *f* rouge du pin
ne denneschotziekte
de Kiefernschütte *f*;
   Kiefernritzenschorf *m*
da fyrrens sprækkesvamp
sv skyttesvamp
es mal *m* rojo de las agujas
   del pino
it arrossamento *m* e caduta
   f delle foglie di pino
   Lophodermium pinastri
   (Schrad.) Chev.

2459
NEEDLE CAST
(PSEUDOTSUGA)
fr chute *f* des aiguilles
ne Rhabdoclineziekte
de Rhabdocline-Schütte *f*;
   rostige Douglasien-
   schütte *f*
da nålefald
sv barrfall; skyttesvamp
es defoliación *f* del abeto
   Douglas
it defogliazione *f* delle
   duglasie
   Rhabdocline pseudotsugae
   Syd.

2460
NEEDLE-SHAPED
fr aciculaire
ne naaldvormig
de nadelförmig
da nåleformet
sv nålformig
es acicular
it aghiforme

2461
NEEDLE VALVE
fr pointeau *m* de fermeture
ne naaldafsluiter
de Absperrspindel *m*; Nadel-
   ventil *n*
da nåleventil
sv nålventil
es punta *f* de cierre
it valvola *f* a spilio

2462
NEGATIVE WELL
fr source *f* négative
ne negatieve bron
de Senkbrunnen *m*; Ver-
   sickerungsschacht *m*;
   Sickerschacht *m*
da 'forsvindingsbrønd'
sv 'negativ kalla';
   dagvattenbrunn
es pozo *m* absorbente
it sorgente *f* negativa

2463
NEMATICIDE
fr nématicide *f*
ne nematicide *n*
de Nematizid *n*
da nematicid
sv nematicid
es nematicida *f*
it nematocida *n*

NEMATODE, 1200
—, lesion, 2138
—, meadow, 2310
—, potato tuber-rot, 2877
—, root knob, 3175
—, spiral, 3557
—, spring crimp, 3573
—, stem and bulb, 3635
See also: cyst nematode

NEMATODES, meadow,
3176
—, root lesion, 3176

2464
NERVATION; VENATION
fr nervation *f*
ne nervatuur
de Nervatur *f*
da nervatur

sv ådring
es nervadura *f*
it nervatura *f*

2465
NERVE; VEIN
fr côte *f*; nervure *f*
ne nerf
de Rippe *f*; Nerv *m*
da nerve; ribbe
sv nerv
es nervio *m*
it nervo *m*

2466
NET
fr filet *m*
ne net *n*
de Netz *n*
da net *n*
sv nät *n*
es red *f*
it rete *f*

2467
NETTLE
fr ortie *f*
ne brandnetel
de Brennessel *f*
da brændenælde
sv brännässla; etternässla
es ortiga *f*
it ortica *f*
   Urtica

NETTLE, day 1779
—, hamp 1779

2468
NET WEIGHT
fr poids *m* net
ne nettogewicht *n*
de Netto-gewicht *n*
da nettovægt
sv nettovikt
es peso *m* neto
it peso *m* netto

2469
NEUTRAL
fr neutre
ne neutraal
de neutral
da neutral

sv neutral
es neutro
it neutrale

2470
NEW VALUE
fr valeur *f* à neuf
ne nieuwwaarde
de Neuwert *m*
da anskaffelsesværdi
sv nyvärde *n*
es valor *m* en nuevo
it valore *m* a nuove

2471
NEW ZEALAND
SPINACH
fr tétragone *f*
ne Nieuwzeelandse spinazie
de neuseeländischer Spinat *m*
da Nyzeelandsk spinat
sv nyzeeländsk spenat
es espinaca *f* de Nueva Zelandia
it spinaci *m pl* di Nuova Zelanda
Tetragonia tetragonoides (Pall.) Ktze.

2472
NICOTIANA; TOBACCO
fr tabac *m*
ne (sier) tabak
de Tabak *m*
da tobak
sv tobak
es tabaco *m*
it tabacco *m* d'ornamento *Nicotiana* Trn.

2473
NICOTINE
fr nicotine *f*
ne nicotine
de Nikotin *n*
da nicotin
sv nikotin
es nicotina *f*
it nicotina *f*

2474
NIGHT FROST
fr gelée *f* nocturne
ne nachtvorst

de Nachtfrost *m*
da nattefrost
sv nattfrost
es hielo *m* nocturno
it gelo *m* notturno

2475
NIGHT INTERRUPTION
fr interruption *f* de la nuit
ne nachtonderbreking
de Nachtunterbrechung *f*
da natafbrydelse
sv nattavbrott *n*
es interrupción *f* de la noche
it interruzione *f* notturna

2476
NICHT SOIL
fr engrais *m* fécal
ne beer (mest)
de Abtrittsdünger *m*; Fäkaldünger *m*
da latrin
sv latrin
es excrementos *m pl* humanos; abono *m* de deyecciones
it sterco *m*; escrementi *pl*

2477
NITRATE
fr nitrate *m*
ne nitraat *n*
de Nitrat *n*
da nitrat *n*
sv nitrat *n*
es nitrato *m*
it nitrato *m*

2478
NITRATE OF LIME
fr nitrate *m* de chaux
ne kalksalpeter
de Kalksalpeter *m*
da kalksalpeter
sv kalksalpeter
es nitrato *m* de cal
it nitrato *m* di calce

2479
NITRATE OF POTASH
fr nitrate *m* de potasse
ne kalisalpeter
de Kalisalpeter *m*

da kalisalpeter
sv kalisalpeter
es nitrato *m* de potasa
it nitrato *m* di potassa

2480
NITRIFICATION
fr nitrification *f*
ne nitrificatie
de Nitrifikation *f*
da nitrifikation
sv nitrifikation
es nitrificación *f*
it nitrificazione *f*

2481
NITRIFY, TO
fr nitrifier
ne nitrificeren
de nitrifizieren
da nitrificere
sv nitrificera
es nitrificar
it nitrificare

2482
NITROGEN
fr azote *m*
ne stikstof
de Stickstoff *m*
da kvælstof *n*
sv kväve *n*
es nitrógeno *m*
it azoto *m*

NITROGEN, excess off;
1255

2483
NITROGEN DEFICIENCY
fr manque *m* d'azote; carence *f* en azote
ne stikstofgebrek *n*
de Stickstoffmangel *m*
da kvælstofmangel
sv kvävebrist
es falta *f* de nitrógeno
it deficienza *f* d'azoto

2484
NITROGEN FIXATION
fr fixation *f* de l'azote
ne stikstofbinding
de Stickstoffbindung *f*

da kvælstofbindelse
sv kvävebindning
es fijación *f* del nitrógeno
it fissazione *f* d'azoto

2485
NITROGENOUS
FERTILIZER
fr engrais *m* azoté
ne stikstofmeststof
de Stickstoffdünger *m*
da kvælstofgødning
sv kvävegödselmedel *n*
es abono *m* nitrogenado
it concime *m* azotato

2486
NOCTUID
(larva = cutworm)
fr noctuelle *f*; phalène *f*
ne uil (vlinder)
de Eule *f*; Eulenschmetter-
    ling *m*; Eulenfalter *m*
da ugle
sv nattfly; nattfjäril
es noctuela *f*
it agrotide *f*; nottua *f*
    Agrotidae

NODE, 1978

2487
NODULE BACTERIA
fr bactéries *f pl* nodulaires
ne knolletjesbacteriën
de Knöllchenbakterien *f pl*
da knoldbakterier
sv knölbakterie
es bacterias *f pl* nodulares
it batteri *m pl* nodulari

2488
NON RETURNABLES
fr emballage *m* perdu
ne eenmalige verpakking
de verlorene Verpackung *f*
da engangsemballage
sv engångsemballage *n*
es embalaje *m* perdido
it imballagio *m* vuoto a per-
    dere

2489
NON RETURN VALVE
fr clapet *m* de retenue
ne keerklep
de Rückschlagklappe *f*
da tilbageslagsventil
sv backventil; vändventil
es válvula *f* de retención
it valvola *f* di ritenuta

NORWAY SPRUCE, 776

NOSEGAY, 414

2490
NOXIOUS; INJURIOUS
fr nuisible
ne schadelijk
de schädlich
da skadelig
sv skadlig
es perjudicial
it dannoso

2491
NOXIOUS ANIMALS;
INJURIOUS ANIMALS
fr animaux *mpl* nuisibles;
    parasites *mpl*
ne schadelijke dieren
de Schädlinge *mpl*
da skadedyr *npl*
sv skadedjur *npl*
es animales *mpl* daniños
it animali *mpl* dannosi

2492
NOZZLE
fr buse *f*; jet *m*
ne spuitdop
de Düse *f*
da dysse
sv spridare; sprutmunstycke
es boquilla *f*
it ugello *m*

2493
NUCLEUS
fr nucléus *m*; noyau *m* de la
    cellule
ne celkern
de Zellkern *m*
da cellekerne

sv cellkärna
es núcleo *m*
it nucleo *m*

2494
NURSERY
fr pépinière *f*
ne kwekerij; boomkwekerij
    plantenkwekerij
de Gärtnerei *f*; Baumschule *f*
    Pflanzschule *f*
da planteskole
sv plantskola
es vivero *m*
it vivaio *m*

2495
NURSERY BED
fr bâche *f* à multiplication
ne kweekbed *n*
de Anzuchtbeet *n*
da formeringsbed *n*
sv förökningsbädd *n*
es semillero *m*
it aiuola *f*

2496
NURSERY MAN
fr pépiniériste *m*
ne boomkweker
de Baumschuler *m*; Baum-
    züchter *m*
da planteskolemand
sv plantskoleman
es arboricultor *m*
it coltivatore *m* da alberi

2497
NURSERY STOCK
fr produits *m pl* de pépinière
ne boomkwekerijproduk-
    ten *n pl*
de Baumschulerzeugnisse
    npl
da planteskoleprodukter *npl*
sv plantskolealster *npl*
es productos *m pl* del vivero
it prodotti *mpl* vivaioli

2498
NUT
fr noix *f*
ne nootvrucht

de Nuss *f*
da nød
sv nötfrukt
es nuez *f*
it noce *f*

NUT, wing,   4174

2499
NUT GALL MITE;
FILBERT BUD MITE (US)
fr phytopte *f* du noisetier
ne rondknop
de Haselnussknospengall-
   milbe *f*
da hasselmide
sv hasselgallkvalster
es botón *m* de avellano
it eriofide *m* galligeno del
   nocciolo
   Phytoptus avellanae Nal.;
   Eriophyes

2500
NUTRIENT MEDIUM
fr milieu *m* nutritif
ne voedingsbodem
de Nährboden *m*
da næringsjord
sv näringssubstrat *n*
es medio *m* nutritivo
it ambiente *m* nutritivo

2501
NUTRIENT SOLUTION
fr solution *f* nutritive
ne voedingsoplossing
de Nährlösung *f*
da næringsopløsning
sv näringslösning
es solución *f* nutritiva
it soluzione *f* nutritiva

2502
NUTRITION; FEEDING
fr nutrition *f*; alimentation *f*
ne voeding

de Ernährung *f*
da føde; ernæring
sv näring
es nutrición *f*
it nutrizione *f*

2503
NUTRITIVE
fr nutritif
ne voedzaam
de nahrhaft
da næringsrig
sv närande
es nutritivo
it nutritivo; nutriente

2504
NUTRITIVE VALUE
fr valeur *f* nutritive
ne voedingswaarde
de Nährwert *m*
da næringsværdi
sv näringsvärde *n*
es valor *m* nutritivo
it valore *m* nutritivo

O

2505
OAK
fr chêne *m*
ne eik
de Eiche *f*
da eg
sv ek
es encina *f*
it querce *f*
  Quercus L.

OAK, common, 772
—, durmast, 1172
—, red, 3067

2506
OAK COPPICE
fr taillis *m* de chêne
ne eikehakhout *n*
de Eichenschälwald *m*
da egekrat *n*; egepur *n*;
  egeungskov *n*
sv ungskog av ek; eksnår *n*
es chaparral *m*
it bosco *m* ceduo di quercia

2507
OAK-LEAF PATTERN
fr taches *f pl* digitées;
  dessin *m* en feuilles de
  chêne
ne eikebladpatroon *n*
de Eichenblattmuster *n*
da egebladmønster *n*
sv ekbladsmönster *n*
es manchas *f pl* digitadas
it maculatura *f* a foglia di
  quercia

2508
OBEDIENT PLANT
fr Physostegia *m*
ne scharnierplant
de Gelenkblume *f*
da drejeblomst
sv drakmynta
es planta *f* de bisagra
it Physostegia *f*
  Physostegia Benth.

2509
OBLIQUE PALMET
fr palmette *f* à branches
  obliques
ne schuine palmet
de schräge Palmette *f*
da skrå palmette
sv snedvinklig palmett
es palmeta *f* oblicua
it palmetta *f* a branche
  oblique

2510
OBLONG
fr oblong
ne langwerpig
de länglich
da aflang
sv avlång
es oblongo
it oblungo

2511
OBSERVATION STUDY;
PRELIMINARY CHECK
fr essai *m* d'uniformité
ne blancoproef
de Blankoversuch *m*
da blindforsøg *n*
sv blindförsök *n*
es ensayo *m* de uniformidad
it esperimento *m* senza
  testimonio; esperimento
  in bianca

2512
OBSOLESCENDE (ec.)
fr obsolescence *f*; usure *f*
  économique
ne economische slijtage
de wirtschaftliche Entwertung *f*
da værditab *n*
sv ekonomisk värdeminsk-
  ning
es obsolescencia *f*; envejecimiento *m* económico
it obsolescenza *f*

2513
OFF-FLAVOUR
fr arrière-goût *m*
ne smaakafwijking
de Geschmacksfehler *m*
da afsmag
sv smakavvikning
es gusto *m* diferente
it gusto *m* anomalo;
  differenza *f* di sapore

2514
OFFSET
fr déporté *m*
ne verstek *n*
de 'Offset'-form *f*
da forgrening udhøber
sv offset-typ; sidförskjuten
es deportado *m*
it disassamento *m*

OFFSET, large, 2034

2515
OFFSET BULB
fr caïeu *m*
ne klister
de Brutzwiebel *f*; Zeh *m*
da sideløg *n*
sv sidolök
es bulbillo *m*
it bulbicino *m*

OFFSHOOT, 3217

OFFSPRING, 2948

2516
OIL
fr mazout *m*; huile *f*
ne olie
de Öl *n*; Oel *n*
da olie; fyringsolie
sv olja; eldningsolja
es aceite *m* pesado; gas-oil *m*
it olio *m*

2517
OIL BURNER
fr brûleur *m* d'huile
ne oliebrander

de Oelbrenner *m*
da oliefyr *n*
sv oljebrännare
es quemador *m* de aceite
it bruciatore *m* ad olio

2518
OIL FIRED BOILER
fr poêle *m* à l'huile
ne oliekachel
de Ölofen *m*
da oliekamin
sv oljekamin
es estufa *f* de aceite
it stufa *f* d'olio

OLD MAN'S BEARD,
703

2519
OLEANDER; ROSE BAY
fr laurier-rose *m*
ne oleander
de Oleander *m*
da nerie
sv oleander
es adelfa *f*
it oleandro *m*
   Nerium oleander L.

2520
OLIVE GREEN
fr vert olive
ne olijfgroen
de olivgrün
da olivengrøn
sv olivgrön
es verde oliva
it verde olivo

2521
OLIVE GREEN MOULD
fr moisissure *f* verte d'olive
ne olijfgroene schimmel
de olivgrüner Schimmel *m*
da Chætomium *m*
sv Chaetomium
es moho *m* verde;
   chaetomio *m*
it muffa *f* olivastra;
   Chaetomium *m*
   Chaetomium olivaceum

ONE-CELLED, 3976

2522
ONE MAN FARM;
ONE MAN HOLDING
fr exploitation *f* à UT
ne eenmansbedrijf *f*
de Einzelwirtschaft *f*
da enkeltmandsbedrift
sv enmansföretag *n*
es explotación *f* uni-
   personal
it mestiere *m* lavorata per
   una persona

2523
ONION
fr ognon *m*; oignon *m*
ne ui
de Zwiebel *f*
da løg *n*
sv lök
es cebolla *f*
it cipolla *f*
   Allium cepa L.

ONION, silverskin, 3414
—, spring, 3575

2524
ONION FLY;
ONION MAGGOT (US)
fr mouche *f* de l'oignon
ne uievlieg
de Zwiebelfliege *f*
da løgflue
sv lökfluga
es mosca *f* de las cebollas
it mosca *f* delle cipolle
   Hylemya antiqua Meig.;
   Delia a. (UK)

2525
ONION HARVESTER
fr arracheuse d'oignons
ne uienrooier
de Zwiebelroder *m*
da løgoptager
sv upptagare för lök
   (matlök)
es cosechadora *f* de cebollas
it raccoglitrice *f* per cipolle

ONION MAGGOT, 2524

2526
ONION SET
fr oignon *m* à planter
ne plantui
de Steckzwiebel *f*; Pflanz-
   zwiebel *f*
da stikløg *n*
sv sättlök; sticklök
es cebollita *f* para plantar
it cipollina *f* da piantare

2527
ONION TOPPER
fr équeteuse *f* d'oignon
ne uienafstaarter
de Zwiebelputzmaschine *f*
da løg-toppemaskine
sv lökblastare
es cosechadora *f* de cebollas
it scollettatore *m* per cipolle

2528
ONION YELLOW
DWARF
fr bigarrure *f* de l'oignon
ne uiemozaïekvirus *n*
de Gelbstreifigkeit *f* der
   Zwiebel
da løg-mosaik
sv lökmosaik; Alliummosaik
es mosaico *m* de la cebolla
it mosaico *m* della cipolla

2529
OÖSPHERE; EGG
fr ovule *m*
ne eicel
de Eizelle *f*
da ægcelle
sv äggcell
es célula *f* huevo; óvulo *m*
it ovulo *m*

2530
OPEN DRAINAGE
fr rabattement *m* à ciel
   ouvert
ne open drainage
de Einzeldränung *f*
da grøftedræning
sv gravavvattning; öppet
   dräneringssystem *n*
es drenaje *m* sin fin
it drenaggio *m* aperto

2531
OPEN HOTBED
fr couche *f* chaude
ne broeiveur
de Mistbeet *n*
da gødningsfure
sv varmbänk
es zanja *f* para cama-caliente
it solco *m* riscaldante

'OPEN' MUSHROOM,
1369

2532
OPEN STACKING
(MUSHROOM
GROWING)
fr caisses *f pl* en étagères
ne kisten op stellages
de offenes Stapeln *n*;
stellagenartiges Stapeln *n*
da kasser på hylder
sv lådor på 'stellage'
es cajas *f pl* en estantes
it cassa *f* su palchi

OPEN VEILS, 1724

2533
OPERATING COSTS;
RUNNING COSTS;
INPUT
fr coût *m* de fonctionne-
ment; coût *m* d'utilisation
ne bewerkingskosten;
exploitatiekosten
de Arbeitserledigungskosten
f pl; Kosten *f pl* der Ar-
beitserfertigung; Be-
triebsaufwand *m*
da arbejdsudgifter
sv driftskostnader
es costes *m pl* de funciona-
miento
it costi *m pl* aziendali

2534
OPERATING
STATEMENT
fr compte *m* d'exploitation
ne exploitatierekening
de Kostenrechnung *f*;
Betriebsabrechnung *f*
da driftsregnskab *n*

sv företagets resultatkonto *n*
es cuenta *f* de la explotación
it contabilità *f* aziendale
(d'esercizio)

2535
OPPOSITE (bot.)
fr opposé
ne tegenoverstaand
de gegenständig
da modsat
sv motsatt
es opuesto
it opposto

2536
OPUNTIA
fr ragnette *f*
ne schijfcactus
de Opuntia *f*; Feigenkak-
tus *m*
da figenkaktus
sv Opuntia
es higuera *f* chumba
it Opuntia *f*
Opuntia (Tourn.) Mill.

2537
ORACHE
fr arroche *f*
ne melde
de Melde *f*
da mælde
sv målla
es armuelle *m* de huerta
it atriplice *f*
Atriplex L.

ORACHE, common,   773

2538
ORANGE
fr orange *f*
ne sinaasappel
de Apfelsine *f*
da appelsin
sv apelsin
es naranja *f*
it arancia *f*

ORANGE, mock,   2363

2539
ORANGERY
fr orangerie *f*
ne orangerie

de Orangerie *f*
da orangeri *n*
sv orangeri *n*
es invernadero *m*
it aranciera *f*

2540
ORANGE TREE
fr oranger *m*
ne sinaasappelboom
de Apfelsinenbaum *m*
da appelsintræ *n*
sv apelsinträd *n*
es naranjo *m*
it arancio *m*
Citrus sinensis Osbeck

2541
ORCHARD
fr verger *m*; jardin *m* fruitier
ne boomgaard
de Obstgarten *m*; Obstan-
lage *f*
da frugthave
sv fruktträdgård
es huerto *m* frutal
it frutteto *m*

2542
ORCHID
fr orchidée *f*
ne orchidee
de Orchidee
da orkidé
sv orkidé
es orquídea *f*
it orchidea *f*
Orchidaceae

2543
ORDER OF WORK;
WORK SEQUENCE
fr déroulement *m* du travail
ne werkvolgorde
de Arbeitsablauf *m*;Arbeids-
folge *f*; Arbeitsordnung *f*
da arbejdsrækkefølje
sv operationsföljd; arbets-
ordning
es sucesión *f* de los trabajos
it successione *f* dei lavori

2544
ORGANIC MANURE
fr fumier *m*
ne organische mest
de organischer Dünger *m*
da organisk gødning
sv organisk gödsel
es estiércol *m*
it concime *m* organico

2545
ORGANIC MATTER
fr matière *f* organique
ne organische stof
de organischer Stoff *m*
da organisk stof *n*
sv organiskt ämne *n*
es materia *f* orgánica
it materia *f* organica

2546
ORGANISM
fr organisme *m*
ne organisme *n*
de Organismus *m*
da organisme
sv organism
es organismo *m*
it organismo *m*

2547
ORGANIZATION (FARM)
fr organisation *f*
ne bedrijfsinrichting
de Betriebsorganisation *f*
da driftsorganisation
sv organisation; företags-
    inriktning
es organización *f* de la explo-
    tación
it organizzazione *f* (azien-
    dale)

2548
ORIENTAL POPPY
fr pavot *m* oriental
ne Oosterse papaver
de Feuermohn *m*
da kæmpevalmue
sv jättevallmo
es adormidera *f* oriental
it papavero *m* orientale
    Papaver orientale L.

2549
ORIENTATION PRICE
fr prix *m* d'orientation
ne oriëntatieprijs
de Richtpreis *m*
da vejledende pris
sv orienteringspris *n*
es precio *m* de orientación
it prezzo *m* d'orientamento

2550
ORIGIN; PROVENANCE
fr origine *f*; provenance *f*
ne herkomst; afkomst
de Herkunft *f*
da herkomst; oprindelse
sv härkomst; ursprung
es origen *m*
it provenienza *f*; origine *f*

2551
ORIGINAL COSTS (ec.)
fr coûts *m pl* initiaux
ne stichtingskosten
de Anlagekosten *f pl*
da etableringsomkostninger
sv initialkostnad
es costes *m pl* de instalación
it costi *m pl* iniziali

2552
ORIGINAL SEED
fr semence *f* d'origine
ne origineel zaad *n*
de Originalsaat *f*
da originalfrø *n*
sv orginalutsäde *n*
es semilla *f* original
it seme *m* orginale

2553
ORIGINAL VALUE
fr valeur *f* initiale
ne aanschaffingswaarde
de Anschaffungswert *m*
da anskaffelsesværdi
sv anskaffningsvärde
es valor *m* de adquisición
it valore *m* iniziale

ORNAMENTAL
FOLIAGE, 956

2554
ORNAMENTAL
GARDEN;
DECORATIVE GARDEN
fr jardin *m* d'agrément
ne siertuin
de Ziergarten *m*
da prydhave
sv prydnadsträdgård
es jardín *m* de adorno
it giardino *m* d'ornamento

2555
ORNAMENTAL PLANT
fr plante *f* d'ornement
ne siergewas *n*
de Zierpflanze *f*
da prydplante
sv prydnadsväxt
es planta *f* de adorno
it pianta *f* d'ornamento

2556
ORNAMENTAL SHRUB;
DECORATIVE SHRUB
fr arbuste *f* d'ornement
ne sierheester
de Zierstrauch *m*
da prydbusk
sv prydnadsbuske
es arbusto *m* de adorno
it arbusto *m* d'ornamento

2557
ORSAT APPARATUS
fr appareil *m* Orsat
ne Orsat-toestel *n*
de Orsat-apparat *m*
da orsatapparat *n*
sv gasavskiljare;
    orsatapparat *n*
es aparato *m* de Orsat
it apparato *m* d'Orsat

OSIER, 263

2558
OSMOSIS
fr osmose *f*
ne osmose
de Osmose *f*
da osmose
sv osmos
es ósmosis *f*
it osmosi *f*

OSMUNDA, 3213

2559
OUTDOOR CROP;
FIELD CROP
fr culture *f* en plein air
ne opengrondsteelt
de Freilandkultur *f*
da frilandsdyrkning
sv frilandskultur
es cultivos *m pl* al aire libre
it coltura *f* all'aperto

2560
OUTFALL PIPE
fr drain *m*; tuyau *m* de drainage
ne afvoerpijp
de Abflussrohr *m*
da afløbsrør
sv avloppsrör
es tubo *m* de drenaje
it tubo *m* di scolo

OUTLET, 2282

2561
OUTSIDE LABOUR
fr main d'oeuvre *f* salariée
ne vreemde arbeidskrachten
de Fremdarbeitskraft *f*; Lohnarbeitskraft *f*
da fremmed arbejdskraft
sv anställd arbetskraft; lejd arbetskraft
es mano *f* de obra salariada
it mano *f* d'opera salariata

2562
OVAL
fr ovale
ne ovaal
de oval
da oval
sv oval; äggrund
es oval; óvalo
it ovale

2563
OVARY
fr ovaire *m*
ne vruchtbeginsel *n*
de Fruchtknoten *m*
da frugtknude

sv fruktämne
es ovario *m*
it ovaia *f*; ovario *m*

2564
OVATE
fr oviforme
ne eirond
de eiförmig
da ægformet
sv äggrund
es oviforme; oval
it ovale; oviforme

2565
OVERALL WIDTH SPAN
fr portée *f*
ne overspanning
de Stützweite *f*; Spannweite *f*
da spænvidde
sv spännvidd
es anchura *f* de nave
it larghezza *f* totale della campata

2566
OVERCROPPING
fr culture *f* épuisante
ne roofbouw
de Raubbau *m*
da rovdrift
sv rovdrift
es cultivo *m* esquilmante
it coltura *f* in eccesso; sovrapproduzione *f*

2567
OVERLAY
(ON MUSHROOM BEDS)
fr mycelium *m* en surface
ne wol (op champignonbed)
de oberflächig wachsendes Myzel *n*; Wolle *f*
da strimelignende vaekst
sv overlay
es micelio *m* en superficie
it tela *f* del micelio

2568
OVER-RIPE
fr blet; trop mûr
ne beurs; buikziek
de überreif
da overmoden

sv övermogen
es pasado; sobre-maduro
it stramaturo

2569
OVERSALT, TO
fr se saliniser
ne verzilten
de versalzen
da gøre salt
sv försalta
es salobrarse
it salare

2570
OVERWINTER, TO
fr hiverner
ne overwinteren
de überwintern
da overvintre
sv övervintra
es invernar
it svernare; passar l'inverno

2571
OVERWINTERED
CABBAGE SEEDLINGS
fr plants *m pl* de choux hivernés
ne weeuwen
de überwinterte Kohlpflanzen *fpl*
da overvintrede kålplanter
sv övervintrande kålplantor
es plantitas *fpl* de col invernantes
it semenzali *mpl* svernati

2572
OVICIDAL
fr ovicide
ne eidodend
de ovizid
da ægdræbende
sv äggdödande
es ovicido
it ovicido

2573
OVIPOSITOR
fr oviscapte *m*
ne legboor
de Legestachel *m*; Legeröhre *m*

da læggebrod
sv äggläggningsrör *n*
es oviscapto *m*
it ovodepositore *m*

2574
OVULE
fr ovule *m*
ne eitje *n*
de Eichen *n*
da æg *n*
sv fröämne *n*
es óvulo *m*
it ovulo *m*

2575
OWN CAPITAL
fr capital *m* propre
ne eigen vermogen *n*
de Eigenkapital *n*
da egenkapital *n*
sv eget kapital *n*

es capital *m* propio
it capitale *m* proprio

2576
OXALIS; WOOD SORREL
fr oseille-de-bois *f*; surelle *f*
ne klaverzuring
de Sauerklee *m*
da surkløver
sv Oxalis
es acerderilla *f*
it acetosella *f*
   Oxalis L.

2577
OXHEART CABBAGE
fr chou *m* cabus de
   printemps
ne spitskool
de Spitzkohl *m*
da spidskål
sv spetskål

es col *f* tipo corazón de
   buey; col *f* picuda
it cavolo *m* cappuccio
   Brassica oleracea L. cv.
   Conica (DC) Boom

2578
OYSTER SHELL SCALE;
EUROPEAN FRUIT
SCALE (US)
fr cochenille *f* ostréiforme
ne oestervormige schildluis
de gemeine Austernschild-
   laus *f*; austernförmige
   Schildlaus *f*
sv ostronsköldlus
es cochinilla *f* ostriforme
it cocciniglia *f* ostreiforme
   Quadraspidiotus
   ostreaeformis (Curt.);
   Quadraspidiotus pyri
   (Lichtenstein)

P

PACK, TO, 4217

2579
PACKING
fr emballage m
ne verpakking
de Verpackung f
da pakning; pakke
sv förpackning
es embalaje m
it imballaggio m;
   confezionamento m

2580
PACKING CLOTH
fr toile f d'emballage
ne paklinnen n
de Packleinen n
da sækkelærred n
sv säckväv
es tela f de embalar
it tela f d'imballagio

2581
PACKING SHED
fr hangar m d'emballage
ne pakloods
de Packschuppen m
da pakrum n
sv packbod; packlada
es almacén m de embalaje
it tettoia f d'imballagio

2582
PACKING STATION
fr lieu m de conditionne-
   ment
ne pakstation n
de Packstation f
da pakkerum n
sv packhus n; packlokal
es sala f de embalaje
it sala f di confezione

2583
PALATABILITY
fr saveur f
ne smakelijkheid
de Geschmack m

da smag
sv smaklighet
es sabor m; sapidez f
it saporosità f

2584
PALISSADE TISSUE
fr tissu m palissadique
ne palissadenweefsel n
de Palisadengewebe n
da pallisadevæv n
sv palissadvävnad
es tejido m de empalizada
it tessuto m palizzato

2585
PALLET
fr palette f
ne stapelbord n; pallet
de Palette f
da pallet
sv pall
es paleta f
it paletta f

2586
PALLET BOX; BULK BIN
fr caisse f à palette;
   palette-caisse f
ne stapelkist
de Grosskiste f; Paletten-
   kiste f
da kæmpekasse
sv stapellåda
es paleta f caja
it cassa f paletta

2587
PALLET TRUCK;
HAND FORK TRUCK
fr chariot m à main
ne handheftruck
de Gabelhubwagen m;
   Handhubwagen
da løftevogn
sv gaffeltruck
es transportador m de
   paletas
it sollevatore m a mano

2588
PALMATE
fr palmatinervé
ne handnervig
de handnervig
da håndnervet
sv handnervigt
es palminervado
it nervato palmatiforme

2589
PALMATIFID
fr palmatifide
ne handspletig
de handförmig gespalten
da håndsnitdelt
sv handkluven
es palmatifido
it scisso palmatiforme

2590
PALMATILOBATE
fr palmatilobé
ne handlobbig
de handförmig gelappt
da håndlappet
sv handflikigt
es palmatilobulado
it logato m palmatiforme

2591
PALMATIPARTITE
fr palmatipartite
ne handdelig
de handförmig geteilt
da hånddelt
sv handdelat
es palmatipartido
it diviso m palmatiforme

2592
PALMET
fr palmette f
ne palmet
de Palmette f
da palmette
sv palmett
es palmeta f
it palmetta f

PALMET, oblique, 2509

2593
PAMPAS GRASS
fr herbe f des Pampas
ne pampasgras n
de Pampasgras n
da pampasgræs n
sv pampasgräs n
es ginerio m; hierba f de las
   pampas
it erba f delle pampe
   Cortaderia selloana
   A. et G.

2594
PANE
fr verre m
ne glas (ruit)
de Glasscheibe f
da rude
sv glas (rute) n
es vidrio m
it lastra f di vetro

2595
PANE WIDTH
fr largeur f de verre
ne glasbreedte
de Glasbreite f
da rudebredde
sv glasbredd
es anchura f de vidrio
it larghezza f della lastra di
   vetro

2596
PANICLE
fr panicule f
ne pluim
de Rispe f
da top
sv vippa
es panícula f
it pannocchia f

2597
PANSY
fr pensée f
ne driekleurig viooltje n
de Stiefmütterchen n
da stedmoderblomst
sv pensé

es pensamiento m
it Viola f tricolore
   Viola tricolor L.

PANSY, field, 1332

2598
PAPHIOPEDILUM
fr sabot m de Vénus
ne venusschoentje n;
   snij-orchidee
de Venusschuh m
da Venussko; snitorkidé
sv venussko
es zapatillo m de Venus
it cipripedio m
   Paphiopedilum Pfitz.

2599
PAPILIONACEOUS
fr papilionacé
ne vlindervormig
de schmetterlingsförmig
da ærteblomstret
sv fjärillik
es papilonáceo; cigomorfo
it papilioniforme

2600
PAPILIONACEOUS
PLANTS;
LEGUMINOUS PLANTS
fr papilionacées f pl
ne vlinderbloemigen
de Schmetterlingsblütler m pl
da ærtblomstrede
sv ärtväxter
es papilonáceos f pl
it leguminose f pl

2601
PAPPUS
fr pappe m
ne haarkelk; zaadpluis n
de Haarkrone f; Federkelch m
da fnug n
sv pappus; fruktfjun
es papo m; vilano m
it pappo m

2602
PARALLELIZATION
fr diversification f
ne parallellisatie

de Erhöhung f der Vielseitig-
   keit
da parallellisering
sv ta upp nya produkter
es diversificación f
it diversificazione f

2603
PARALLEL VEINED
fr rectinervé; à nervures
   parallelles
ne rechtnervig
de parallelnervig
da ligenervet
sv parallellnervig
es de nervadura paralela;
   de nervios rectos
it di nervatura parallela

2604
PARASITE
fr parasite m
ne parasiet
de Schmarotzer m; Parasit m
da snylter
sv parasit
es parásito m
it parassito m

2605
PARASITIC PLANT
fr plante f parasite
ne woekerplant
de Schmarotzer m
da snylteplante
sv parasit; snyltväxt
es planta f parásita
it pianta f parassita

2606
PARCEL OF LAND
fr parcelle f de terre
ne perceel n
de Flurstück n; Parzelle f
da arael n (parcel)
sv skifte n; parcell
es parcela f
it parcella f; appezzamento m

2607
PARCEL OUT, TO
fr lotir
ne verkavelen

de parzellieren
da udstykke
sv stycka; utstycka
es parcelar
it dividere in lotti

2608
PARE OFF TURF, TO;
TO CUT TURF
fr découper un gazon en
   plaques
ne zoden wegnemen
de Soden abheben
da skære græstorv
sv upptagning av grästorvor
es cortar el césped
it tagliare piote

2609
PARK
fr parc *m*
ne park *n*
de Park *m*
da park
sv park
es parque *m*
it parco *m*

2610
PARSLEY
fr persil *m*
ne peterselie
de Petersilie *f*
da persille
sv persilja
es perejil *m*
it prezzemolo *m*;
   petrosello *m*

2611
PARSLEY PIERT
fr alchémille *f* perce-pierre
ne akkerleeuweklauw
de Acker-Frauenmantel *m*
da almindelig dværgløvefod
sv jungfrukam
es alquimila *f*; pie *m* de léon
it alchemilla *f* dei campi
   Alchemilla arvensis L.

2612
PARSNIP
fr panais *m*
ne pastinaak

de Pastinake *f*
da pastinak
sv palsternacka
es pastinaca *f*
it pastinaca *f*
   Pastinaca sativa L.
   cv. Hortensis (Mill.)
   Ehrh.

2613
PARSNIP MOTH;
PARSNIP WEBWORM
(US)
fr teigne *f* du panais
ne pastinaakmot
de Bärenklaumotte *f*
da skærmplantemøl
sv palsternaksmal
es polilla *f* de la chirivía
it tignola *f* della pastinaca
   Depressaria heracliana L.

2614
PARTED (bot.)
fr divisé
ne gedeeld
de geteilt
da delt
sv delad
es partido
it diviso

2615
PARTNERSHIP FIRM
fr société *f* en nom collectif
ne firma
de Firma *f*
da firma
sv handelsbolag *n*
es sociedad *f* colectiva
it società *f* in nome collettivo

2616
PASSION FLOWER
fr passiflore *f*
ne passiebloem
de Passionsblume *f*
da passionsblomst
sv passionsblomma;
   kristikorsblomma
es pasionaria *f*
it passiflora *f*
   Passiflora Juss.

PASTE, 2985

PASTEURIZING ROOM,
2633

2617
PATCHES, IN
fr localisé; par endroits
ne pleksgewijs
de stellenweise
da pletvis
sv fläckvis
es alternativamente
it ordinato al luogo

2618
PATHOGENIC
fr pathogène
ne ziekteverwekkend
de krankheiterregend
da sygdomsfremkaldende
sv sjukdomsalstrande
es patógeno
it patogenico

2619
PATTERN OF DAMAGE
fr tableau *m* des méfaits
ne schadebeeld *n*
de Schadbild *n*
da skadebillede *n*; sygdoms-
   billede *m*
sv skadebild
es imagen *f* del daño
it sintomo *m* del danno

2620
PEA
fr pois *m*
ne erwt
de Erbse *f*
da haveært; ært
sv ärt
es guisante *m*; arveja *f*
it cece *m*
   Pisum sativum L.

PEA, field, 1333
—, garden, 1536
—, green, 1536
—, marrowfat, 2294
—, round seeded, 3209
—, split, 3559
—, sugar, 3720
—, sweat, 3767

2621
PEA AND BEAN WEEVIL;
PEA LEAF WEEVIL (US)
fr sitone *m* des pois
ne bladrandkever
de gestreifter Blattrand-
käfer *m*
da stribet bladrandbille
sv randig ärtvivel
es sitona *f* del guisante;
brugo *m* de las arvejas
it sitona *f* lineato del fagiolo; sitona *f* del legumi
Sitona lineatus L.

2622
PEA APHID
fr puceron *m* du pois
ne erwtebladluis
de grüne Erbsenblattlaus *f*
da ærtelus
sv ärtbladlus
es pulgón *m* verde del guisante
it afidone *m* verdastro del
pisello; afide *m* del pisello
Acyrthosiphon pisum
Harris

2623
PEA BEETLE
fr charançon *m* des pois
ne erwtekever
de Erbsenkäfer *m*
da ærtebille
sv ärtvivel
es gorgojo *m* del guisantes
it tonchio *m* dei piselli
Bruchus pisorum L.

2624
PEACH
fr pêche *f*
ne perzik
de Pfirsich *m*
da fersken
sv persika
es melocotón *m*
it pesca *f*

2625
PEACH-POTATO APHID;
GREEN PEACH APHID
(US)
fr puceron *m* vert du pêcher
ne groene perzikluis

de grüne Pfirsichlaus *f*
da ferskenlus
sv persikobladlus
es pulgón *m* verde del melocotonero
it afide *m* verde del pesco
Myzus persicae Sulz.

2626
PEACH STUNT
fr rabougrissement *m* du
pêcher
ne dwergziekte
de Pfirsichstauche *f*
da fersken dværgsyge
sv dvärgsjuka (persika)
es nanismo *m* del melocotonero
it nanismo *m* del pesco

2627
PEACH TREE
fr pêcher *m*
ne perzikboom
de Pfirsichbaum *m*
da perskentræ *n*
sv persikoträd *n*
es melocotonero *m*
it pesco *m*
Prunus persica L. Batsch

2628
PEACH TWIG BORER
fr teigne *f* du pêcher
ne perzikscheutboorder
de Pfirsichmotte *f*
da ferskenmål *n*
sv persikomal *n*
es barreno *m* de los retoños
del melocotonero
it anarsia *f*
Anarsia lineatella Zoll.

2629
PEA CYST NEMATODE
fr anguillule *f* du pois
ne erwtecystenaaltje *n*
de Erbsenzystenälchen
da ærtål; ærtcystenematod
sv ärtcystnematod
es heterodera *f* del guisante
it nematode *m* del pisello
Heterodera goettingiana
Liebsch.

2630
PEA ENATION MOSAIC
fr mosaïque *f* énation du
pois
ne enatiemozaïek *n* van erwt
de scharfes Adernmosaik *n*
der Erbse
da ært-enationsmosaik
sv enationsmosaik (ärt)
es mosaico *m* enación
it mosaico *m* con enazioni
del pisello

2631
PEA HARVESTER
fr moissonneuse *f* à pois
ne erwtenoogstmachine
de Erbsenerntemaschine *f*
da ærtehøstmaskine
sv ärtskördemaskin
es máquina *f* de recollección
de grisantes
it macchina *f* per il raccolto
per piselli

2632
PEAK-HEAT
(MUSHROOM
GROWING);
INDOOR-COMPOSTING;
TO PASTEURIZE; TO
SWEAT; TO COOK OUT
fr fermentation *f* dirigée
et contrôlée (fdc);
pasteuriser
ne uitzweten *n*
de Ausschwitzen *n*; Pasteuri-
sierung *f*; kontrollierte
Erhitzung *f*
da pasteurisering; peak-
heate
sv pastörisering;
peak-heating
es fermentación *f* dirigida y
controlada; pasteuriza-
ción *f*
it pastorizzazione *f*

2633
PEAK-HEATING ROOM
(MUSHROOM
GROWING);
PASTEURIZING ROOM
fr chambre *f* de pasteurisa-
tion
ne zweetcel

de Schwitzraum *m*; Pasteuri-
    sierung *m*
da pasteuriseringsrum *n*
sv pastöriseringsrum *n*
es cámara *f* de pasteuriza-
    ción
it camera *f* di pastorizza-
    zione

2634
PEAK LOAD
fr pointe *f* au travail
ne arbeidspiek
de Arbeitsspitze *f*
da arbejdstop
sv arbetstopp
es punta *f* de trabajo
it punta *f* di lavoro

PEA LEAF WEEVIL,
2621

2635
PEA MIDGE
fr cécidomyie *f* du pois
ne erwtegalmug;
    erwteknopmade
de Erbsengallmücke *f*
da ærtegalmyg
sv ärtgallmygga
es mosca *f* de los guisantes
it cecidomia *f* del pisello
    Contarinia pisi Winn.

2636
PEA MOSAIC
fr mosaïque *f* commune du
    pois
ne erwtenmozaïek *n*
de gewöhnliches Erbsen-
    mosaik
da ært-mosaik
sv ärtmosaik
es mosaico *m* del guisante
t mosaico *m* del pisello

2637
PEA MOTH
fr tordeuse *f* du pois
ne erwtepeulboorder
de Erbsengallwickler *m*
da ærtevikler
sv ärtvecklare
es polilla *f* del guisante

it tortrice *f* del pisello
    Enarmonia nigricana F.;
    Laspeyresia n. (UK)

2638
PEAR
fr poire *f*
ne peer
de Birne *f*
da pære
sv päron
es pera *f*
it pera *f*

2639
PEAR-BEDSTRAW
APHID
fr puceron *m* cendré du
    poirier
ne rose pereluis
de mehlige Birnenblattlaus *f*
da pærebladlus
sv päronbladlus
es pulgón *m* rosa del peral
it afide *m* sanguigno del
    pero
    Sappaphis pyri Fonsc.;
    Dysaphis p. (UK)

2640
PEAR-GRASS APHID
fr puceron *m* noir du poirier
ne zwarte pereluis
de braune Birnenblattlaus *f*
da pærebladlus
sv päronbladlus
es pulgón *m* negro del peral
it afide *m* nero del pero
    Longiunguis pyrarius
    Pass.

2641
PEAR LEAF BLISTER
MOTH
fr 'oeil *m* de perdrix'
ne appeldamschijfmot
de Fleckenminiermotte *f*
sv fruktträdsfläckminerar-
    mal
es minadora *f* de las hojas
    del manzano
it minatrice *f* concentrica
    delle foglie del melo;
    cemiostoma *m*
    Leucoptera scitella Zell
    (Cemiostoma)

2642
PEAR LEAF MIDGE;
PEAR LEAF-CURLING
MIDGE
fr cécidomyie *f* des feuilles
    du poirier
ne perebladgalmug
de Birnenblattgallmücke *f*
da pæreblad-galmyg
sv päronbladgallmygga
es mosca *f* de agalla del
    peral
it cecidomia *f* delle foglie
    del pero
    Dasyneura pyri Bouché

2643
PEARLWORT
fr sagine *f* courbée;
    lance *f* d'eau
ne liggende vetmuur
de niederliegendes Mast-
    kraut *n*
da almindelig firling
sv krypnarv
es sagina *f* tendida
it Sagina *f* procumbens
    Sagina procumbens L.

2644
PEAR MIDGE
fr cécidomyie *f* des poirettes
ne peregalmug
de Birnengallmücke *f*
da pæregalmyg
sv pärongallmygga
es cecidomia *m* de las peras
it cecidomia *f* delle perine
    Contarinia pyrivora Riley

PEAR PSYLLA, 2648

2645
PEAR SAWFLY
fr hoplocampe *m* du poirier
ne perezaagwesp
de Birnensägewespe *f*
da pærehveps
sv päronstekel
es hoplocampa *f* del peral
it tentredine *f* della perine
    Hoplocampa brevis Kl.

PEAR-SLUG, 2646

2646
PEAR SLUG SAWFLY;
PEAR-SLUG (US);
(larva = pear and cherry
slugworm)
fr tenthrède *f* limace
ne vruchtboombladwesp;
(larve: slakvormige
bastaardrups)
de schwarze Kirschblatt-
wespe *f*
da frugttræ-bladhveps
sv fruktbladsstekel
es babosilla *f* del peral
it limacina *f* del pero
Caliroa cerasi L.; *C.
limacina*

2647
PEAR STONY PIT
fr poire *f* pierreuse; gravelle
f du poirier
ne stenigheid
de Steinfrüchtigkeit *f*
da sten i pærer
sv stensjuka
es piedra *f*
it litiasi *f* contagiosa

2648
PEAR SUCKER;
PEAR PSYLLA (US)
fr psylle *m* du poirier
ne perebladvlo
de gelber Birnenblattsäuger
m
da pære-bladloppe
sv päronbladloppa
es tingido *m* del peral; psila
m del peral
it psilla *f* piricola
Psylla pyricola Först;
P. simulans (UK)

2649
PEAR TREE
fr poirier *m*
ne pereboom
de Birnbaum *m*
da pæretræ *n*
sv päronträd *n*
es peral *m*
it pero *m*
Pyrus communis L.

2650
PEA STICKS
fr rames *f pl* à pois
ne erwtenrijs *n*
de Erbsenreisig *n*
da ærteris *n*
sv ärtris *n*
es tutor *m* para arveja
it ramoscelli *mpl* da cece

2651
PEAT
fr tourbe *f*
ne turf
de Torf *n*
da tørv
sv torv
es turba *f*
it torba *f*

PEAT, black,   336
—, frozen,   1468
—, granulated,   1634
—, moss,   2391
—, reed,   3079
—, rough,   3206
—, sedge,   3307
—, transition,   3908

PEAT-BLOCK,   2659

2652
PEAT BOG
fr tourbière *f* basse
ne laagveen *n*
de Niedermoor *n*;
Niederungsmoor *n*
da lavmose
sv lågmosse
es turbera *f* baja
it aggallato *m*

2653
PEAT COMPOST
fr compost *m* tourbeux
ne veencompost
de Torfkompost *m*
da tørvekompost
sv torvkompost
es mezcla *f* di turba y compost
it concime *m* di torba

2654
PEAT DUST
fr poussière *f* de tourbe
ne turfmolm
de Torfmull *m*
da tørvemuld
sv torvmull
es polvo *m* de turba
it polvere *f* di torba

2655
PEA THRIPS
fr thrips *m* des pois
ne erwtetrips
de Erbsenblasenfuss *m*
da ærtethrips
sv ärttrips
es thrips *m* del guisante
it tripide *m* del pisello
Kakothrips robustus
(Uzel)

2656
PEAT LITTER
fr tourbe *f* mottière
ne turfstrooisel
de Torfstreu *f*
da grov tørvestrøelse;
hundekød *n*
sv torvströ *n*
es turba *f* molida; compost
m de turba
it polvere *f* di torba

2657
PEAT-MOOR
fr tourbière *f* haute
ne hoogveen *n*
de Hochmoor *n*
da højmose
sv högmosse
es turbera *f* alta
it torbiera *f*

PEAT MOSS,   3548

2658
PEA TOP YELLOWS
fr jaunisse *f* du pois
ne topvergeling (erwt)
de Blattrollkrankheit *f*
da ært-bladrullesyge
sv bladrullsjuka (ärt);
gultoppighet (ärt)

es enrollado *m* de la hoja del
    guisante
it accartocciamento *m* del
    pisello; (giallume *m* api-
    cale del pisello)

2659
PEAT-POT; PEAT-BLOCK
fr godet *m* de sphaigne
ne turfpot
de Torftopf *m*
da tørvepotte
sv torvkruka
es maceta *f* de turba prensa-
    da
it vaso *m* di torba

2660
PEA TREE
fr acacia *m* jaune
ne erwteboompje *n*
de Erbsenstrauch *m*
da ærtetræ *n*
sv caragan; ärtbuske
es acacia *f* de Rusia;
    Caragana *f*
it Caragana *f*
    Caragana Lam.

2661
PEAT SOIL; BOG SOIL
fr sol *m* marécageux; sol *m*
    tourbeux
ne veengrond
de Moorboden *m*
da mosejord *mf*
sv mossjord; torvjord
es suelo *m* turboso
it terra *m* torbosa

2662
PEBBLE
fr caillou *m*; galet *m*
ne kiezelsteen
de Kieselstein *m*
da perlegrus *n*
sv kiselsten
es guija *f*; guijo *m*
it ciottolo *m*; sasso *m*

PEDICLE,   1405

2663
PEDOGENESIS
fr développement *m* du sol
ne bodemvorming
de Bodenbildung *f*
da jordbundsdannelse
sv jordmånsbildning
es formación *f* del suelo
it formazione *f* del suolo

PEDOLOGY,   3495

2664
PEDUNCLE
fr tige *f* florale
ne bloemstengel
de Blütenstengel *m*
da blomsterstængel
sv blomstängel
es tallo *m* de la flor
it stelo *m* del fiore

PEEL,   3432

PELARGONIUM, ivy
leaved,   1966
—, regal,   3088
—, zonal,   4240

2665
PENDULOUS
fr pendant; retombant
ne afhangend
de herabhängend
da hængende
sv hängande
es pendiente; colgante
it pendente

2666
PEONY
fr pivoine *f*
ne pioen
de Päonie *f*
da pæon
sv pion
es peonía *f*
it peonia *f*
    Paeonia L.

PEONY, TREE,   3919

2667
PEPPERMINT
fr menthe *f* poivrée
ne pepermunt
de Pfefferminze *f*
da pebermynte
sv pepparmynta
es menta *f*
it menta *f* piperita
    Mentha piperita L.

2668
PERCOLATE, TO
fr s'infiltrer
ne doorsijpelen
de durchsickern
da gennemsive
sv sila igenom
es infiltrar
it filtrare

2669
PERCOLATING WATER;
RISING GROUNDWATER
IN A POLDER
fr eau *f* de suintement;
    eau *f* de pression
ne drangwater *n*;
    kwelwater *n*
de Qualmwasser *n*; Druck-
    wasser *n*
da trykvand *n*
sv stigande grundvatten i en
    polder
es agua *f* viva; agua *f* sub-
    terránea ascendente en
    un polder
it acqua *f* di pressa; acqua *f*
    d'infiltrazione

2670
PERENNIAL
fr vivace; pérenne
ne overblijvend
de ausdauernd
da overvintrende
sv perenn
es perenne; vivaz
it perenne

2671
PERENNIAL ASTER
fr Aster *m*
ne Aster

de Staudenaster *f*; Herbstaster *f*
da høstaster; Aster
sv Aster; höstaster
es estraña *f*; reina *f* Margarita
it astero *m*
  Aster L.

2672
PERENNIAL RYEGRASS
fr ray-grass *m* anglais
ne Engels raaigras *n*
de Raigras *n*; deutsches
  Weidegras *n*
da almindelig rajgræs *n*;
  Engelsk rajgræs *n*
sv engelskt rajgräs *n*;
  rajgräs *n*
es ballico *m* perenne;
  raygrass *m*
it loglio *m*
  Lolium perenne L.

2673
PERENNIALS
fr plantes *f pl* vivaces
ne vaste planten
de Stauden *f pl*
da stauder
sv perenna växter
es plantas *f pl* vivaces
it piante f *pl* perenni

PERENNIALS, herbaceous,
1784

2674
PERENNIAL
SOWTHISTLE
fr laiteron *m* des champs
ne akkermelkdistel
de Ackergänsedistel *f*
da ager-svinemælk
sv fettistel
es hierba *f* del sacre;
  lechuguilla *f*
it grespino *m* dei campi
  Sonchus arvensis L.

2675
PERENNIAL WEED
fr mauvaise herbe *f* (vivace)
ne overblijvend onkruid *n*;
  wortelonkruid *n*

de Unkraut *n*
da flerårigt ukrudt *n*
sv flerårigt agräs *n*
es mala hierba *f*
it erba *f* infestante

2676
PERFOLIATE
fr perfolié
ne doorgroeid
de durchwachsen
da gennemvokset
sv genomvuxen
es retoñado
it penetrato; perforato

2677
PERFORATED
SPRINKLER TUBE
fr tube *m* souple perforé
ne foliegietdarm
de Folien-Sprühschlauch *m*
da perforeret vardingsslange
sv perforerad bevattnings-
  slang av plast
es tubo *m* flexible perforado
it tubo *m* forato per aspersione

2678
PERGOLA
fr pergola *f*
ne pergola
de Laubengang *m*; Pergola *f*
da pergola
sv pergola
es pérgola *f*
it pergola *f*

2679
PERIANTH
fr périanthe *m*
ne bloemdek; bloembe-
  kleedsels
de Blütenhülle *f*
da blomsterdække; bloster
sv hylle; blomkalk
es periantio *m*; perigonio *n*
it perianto *m*; perianzio *m*

2680
PERIODICITY
fr périodicité *f*
ne periodiciteit

de Periodizität *f*
da periodicitet; rytme
sv periodicitet
es periodicidad *f*
it periodicità *f*

PERISH, TO, 1056

2681
PERISHABLE GOODS
fr produits *m pl* périssables
ne bederfelijk produkt *n*
de leicht verderbliche Ware *f*
da letfordærvelige varer
sv (lätt-)förstörbar vara
es productos *m pl* perecederos
it beni *m pl* deperibili

2682
PERIWINKLE
fr pervenche *f*
ne maagdepalm
de Immergrün *n*
da singrøn *n*
sv vintergröna
es hierba *f* doncella; Vinca *f*
it pervinca *f*
  Vinca L.

2683
PERMANENT TREE
fr arbre *m* permanent
ne blijver
de Bleiber *m*
da blivende træ *n pl*
sv permanenta träd *n*
es árbol *m* frutal permanente
it albero *m* permanente;
  pianta *f* della piantagione
  principale

2684
PERMEABILITY;
POROSITY
fr perméabilité *f*
ne doorlatendheid
de Durchlässigkeit *f*
da permaebilitet;
  gennemtrængelighed
sv genomtränglighet
es permeabilidad *f*
it permeabilità *f*

2685
PERMEABLE (SOIL);
PERVIOUS (SOIL);
POROUS (SOIL)
fr perméable; poreux
ne doorlatend
de durchlässig
da gennemtrængelig; porøs
sv genomsläpplig
es permeable
it permeabile

2686
PERPETUAL
FLOWERING
fr remontant
ne doorbloeiend; remonte-
rend
de remontierend
da remonterende
sv remonterande
es remontante
it fioritura f perpetua

2687
PERPETUAL FRUITING
STRAWBERRY
fr fraisier m des quatre saisons
ne maandbloeier
de Monatserdbeere f
de månedsjordbær n
sv måndssmultron n
es fresal m de las cuatro
estaciones
it fragola f di tutte le stagi-
oni

2688
PERSISTENT (bot.)
fr persistant
ne blijvend
de bleibend
da blivende
sv bestående
es permanente
it permanente

2689
PERUVIAN LILY
fr alstroemère
ne Incalelie
de Inkalilie f
da inkalilje

sv alströmeria
es peregrina f de Lima
it Alstroemeria f
Alstroemeria L.

PERVIOUS, 2685

PEST CONTROL, 1082

2690
PESTICIDES; CROP
PROTECTION
CHEMICALS
fr produits m pl antiparasi-
taires
ne bestrijdingsmiddelen n pl
de Bekämpfungsmittel n pl
da bekæmpelsesmidler n pl
sv bekämpningsmedel n pl
es medios m pl de lucha
it mezzi m pl antiparasitari

2691
PETAL
fr pétale m
ne kroonblad n
de Kronblatt n
da kronblad n
sv kronblad n
es pétalo m
it petalo m

2692
PETIOLATE; STALKED
fr pétiolé
ne gesteeld
de gestielt
da stilket
sv skaftad
es peciolado
it gambuto

PETIOLE, 2109

2693
PETROLEUM
CARBURETTOR
fr carburateur m à pétrole
ne petroleumvergasser
de Petroleumvergaser m
da petroleumsbrænder;
petroleumsmotor
sv förgasare
es carburador m de petroles
it carburatore m a petrolio

2694
PETROVA RESINELLA
fr tordeuse f des galles
résineuses
ne harsbuilrups
de Kiefernharzgallenwickler
m
da harpiksgalvikler
sv hartsgallvecklare
es piral f de la resina
it evetria f del pino
Evetria resinella L.;
Petrova r. (UK)

2695
PETTY SPURGE
fr euphorbe f des jardins
ne tuinwolfsmelk
de Gartenwolfsmilch f
da gaffel-vortemælk
sv rävmjölkstörel
es euforbia f
it euforbia f minore;
calensola f piccola
Euphorbia peplus L.

2696
PETUNIA
fr pétunia m
ne Petunia
de Petunia f
da petunie
sv Petunia
es Petunia f
it Petunia f
Petunia Juss.

2697
PHACELIA
fr phacélia m
ne Phacelia
de Phacelie f
da honningurt
sv facelia
es facelia f
it facelia f
Phacelia Juss.

2698
PHENOMENON DUE TO
EXCESS
fr accident m dû à un excès
ne overmaatverschijnsel n
de Schaden m durch Über-
mass

da overmålssymptom *n*
sv överskottsföreteelse
es síntoma *m* de exceso
it fenomeno *m* da eccesso

2699
PHENOTYPE
fr phénotype *m*
ne fenotype
de Phænotypus *m*
da fænotype
sv fenotyp; fransningstyp
es fenotipo *m*
it fenotipo *m*

2700
PHLOX
fr Phlox *m*
ne flox (vlambloem)
de Flammenblume *f*
da floks
sv flox
es flox *m*
it Phlox *m*
    Phlox L.

2701
PHORID FLY;
HUMPBACKED FLY (US)
fr phoride *f*
ne champignonvlieg
de Buckelfliege *f*
da champignonflue
sv champinjonfluga
es mosca *f* del champiñon
it foride *m*
    Phoridae

2702
PHOSPHATE
fr phosphate *m*
ne fosfaat *n*
de Phosphat *n*
da fosfat *n*
sv fosfat *n*
es fosfato *m*
it fosfato *m*

2703
PHOSPHATE ROCK;
ROCK PHOSPHATE
fr phophate *m* naturel
ne natuurlijk fosfaat *n*
de Rohphosphat *n*

da råfosfat *n*
sv råfosfat *n*
es fosfato *m* natural
it fosfato *m* naturale

2704
PHOSPHATIC
FERTILIZERS
fr engrais *m pl* phosphatés
ne fosforzuurhoudende
    meststoffen
de Phosphorsäuredünger *m*
    pl
da fosforsyregødninger
sv fosforgödselmedel *n pl*
es abonos *m pl* fosfatados
it materie *f pl* da ingrasso
    fosfate

2705
PHOSPHORUS
fr phosphore *m*
ne fosfor *n*
de Phosphor *m*
da fosfor *n*
sv fosfor
es fósforo *m*
it fosforo *m*

2706
PHOTOCELL
fr cellule *f* photo-électrique
ne fotocel
de Photozelle *f*
da fotocelle
sv fotocell
es fotocélula *f*
it cellula *f* fotoelettrica

2707
PHOTOPERIODICITY
fr photopériodicité *f*
ne fotoperiodiciteit
de Photoperiodizität *f*
da fotoperiodicitet
sv fotoperiodicitet
es fotoperiodicidad *f*; foto-
    periodismo *m*
it fotoperiodicità *f*

2708
PHOTOSYNTHESIS
fr photosynthèse *f*
ne fotosynthese

de Photosynthese *f*
da fotosyntese
sv fotosyntes
es fotosíntesis *f*
it fotosintesi *f*

PHYLLOPODIUM, 2072

PHYLLOTAXIS, 154

2709
PHYSICAL PLANNER
fr planificateur *m*
ne planoloog
de Planer *m*
da planlægger
sv planläggare
es planeador *m*
it pianologo *m*

2710
PHYSIOGENICAL
fr physiogène
ne fysiogeen
de physiogen
da fysiogen
sv fysiogen
es fisiogeno
it fisiogeno

2711
PHYSIOLOGY
fr physiologie *f*
ne fysiologie
de Physiologie *f*
da fysiologi
sv fysiologi
es fisiología *f*
it fisiologia *f*

2712
PHYTOPHTHORA
FRUITROT (APPLE)
fr mildiou *m* des pommes
ne Phytophthora-rot *f*
de Phytophtora-Fäule *f*
da stammebasisråd
sv phytophthora-röta
es gangrena *f* del pie
it Phytophthora *f*
    Phytophthora cactorum
    (Leb. et Cohn) Schroet.

2713
PHYTOPHTHORA ROOT
ROT (CHAMAECYPARIS)
fr pourriture f des racines
ne wortelrot n
de Wurzelfäule f
da rodråddenskab
sv rotröta
es podredumbre f de la raíz
it marciume m radicale
Phytophthora cinnamomi
Rands.

2714
PHYTOTRON m
fr phytotron m
ne klimaatkas
de Klimagewächshaus n
da klimahus n
sv klimatkammare; klimat-
hus n
es invernadero m climático
it serra f ambientale

2715
PICK, TO
fr cueillir
ne plukken
de pflücken
da plukke
sv plocka
es cosechar; recolectar
it cogliere

PICKING, 1548

2716
PICKING BASKET
fr cueilloir m
ne plukmand
de Pflückkorb m
da plukkekurv
sv plockkorg
es cesta f de recolección;
recogedor m
it cesta f da cogliere;
raccoglitore m

2717
PICKING BUCKET
fr sac m de cueillette
ne plukemmer
de Pflückeimer m
da plukkespand

sv plockningslåda
es saco m de recolección
it sacco m di raccolta

2718
PICKING PLATFORM
fr plate-forme f
ne plukstelling
de Pflückgerüst n
da plukkestilling;
plukkeplatform
sv plockningsställning;
plockningsplattform
es plataforma f recogedora
it piattaforma f

2719
PICKING SEASON
fr époque f de la cueillette
ne pluktijd
de Pflückzeit f
da plukketidspunkt
sv plockningstid
es época f de recolección
it stagione f della raccolta

2720
PICKING SLEDGE
fr traîneau m
ne plukslee
de Pflückschlitten m
da plukkeslæde
sv plockningssläde
es trineo m; rastra f
it traino m

2721
PICKING STAND
fr plate-forme f
ne plukstandaard
de Pflückgestell f
da plukkestandard
sv plockningsställe n
es escalera f para cosechar
frutas
it erpice m livellatore

2722
PICKLES
fr conserves f pl au vinaigre
ne tafelzuren n pl
de Essiggemüse n pl
da surt (asie, rødbede)
sv pickles

es conservas f pl en vinagre
it salamoia f

2723
PICKLING VINEGAR
fr vinaigre m à confire
ne inmaakzaijn
de Einmachessig m
da sylteeddike
sv konserveringsättika
es vinagre m para conservas
it aceto m a conservare

2724
PIECE WORK RATE
fr salaire m à la tâche
ne akkoordloon n
de Akkordlohn m
da akkordløn
sv ackordslön
es salario m a destajo
it salario m a cottimo

2725
PIG MANURE
fr fumier m de porc
ne varkensmest
de Schweinemist m
da svinegødning
sv svingödsel
es estiércol m de puerco
it sterco m di porco

2726
PIGMY MANGOLD
BEETLE; BEET BEETLE
fr atomaire m de la betterave
ne bietekever
de Moosknopfkäfer m
da runkelroebille
sv liten betbagge
es escarabajo m de la remolacha
it atomaria f della barbabietola
Atomaria linearis Steph.

2727
PILE AND PLANK, TO;
TO CAMPSHED (US)
fr endiguer à l'aide de
planches
ne beschoeien
de bekleiden

da beklæde
sv bekläda; brädfodra
es revestir con (madera etc.)
it arginare mediante assi

2728
PILE FOUNDATION
fr fondation *f* sur pieux
ne palenfundering
de Pfahlgründung *f*;
    Pfahlfundament *n*
da pælefundament *n*
sv grundpålning
es pilastra *f* de cimiento
it fondazione *f* su palo

PILEUS, 561

2729
PILING AND PLANKING;
CAMPSHEDDING (US)
fr revêtement *m* de planches
    (sur digue)
ne beschoeiing
de Uferbekleidung *f*;
    Uferbefestigung *f*
da beklædning; faskiner
sv beklädning
es revestimento *m*
it arginatura *f*

PILOSE, 1706

2730
PINCH, TO; TO TOP
fr pincer
ne innijpen; toppen
de entspitzen
da knibe
sv pincera
es despuntar
it spuntare

2731
PINCH OUT SHOOTS, TO
fr supprimer les drageons
ne dieven
de entgeizen; ausgeizen
da fjerne sideskud; udtynde
sv tjuvning; knipning
es desyemar; despampanar;
    destallar
it staccare i germogli
    laterali

PINE, Arolla, 151
—, Austrian, 195
—, Bhutan, 307
—, Chile, 660
—, Macedonian, 2228
—, maritime, 2279
—, mountain, 2406
—, Scots, 3283

2732
PINEAPPLE
fr ananas *m*
ne ananas
de Ananas *f*
da ananas
sv ananas
es piña *f*; ananá *m*
it ananasso *m*
    Ananas comosus L. Merr.

PINEAPPLE WEED, 3048

2733
PINE BARK ADELGES;
DOUGLAS FIR ADELGES;
COOLEY SPRUCE GALL
APHID (US)
fr chermès *m* du pin Wey-
    mouth; chermès *m* des
    conifères
ne wolluis
de Douglasienwollaus *f*;
    Strobenrindenlaus *f*
da uldlus
sv douglasgranbarrlus
es pulgón *m* lanigero de la
    corteza del pino
it cherme *m* lanigero del
    pino
    Gilletteella cooleyi Gilette;
    Adelges c.; Pineus strobi
    (UK); *Chermes* L. (US)

2734
PINE NEEDLE LITTER;
PINE NEEDLE CARPET
fr litière *f* forestière de
    conifères
ne naaldengrond
de Nadelstreu *f*; Waldstreu
    f; Nadelerde *f*
da grannålejord
sv barrträdsförna
es lecho *m* forestal de coní-
    feras; tierra *f* de agujas
it terra *f* di conifere

2735
PINE SAWFLY
fr lophyre *m* du pin
ne dennebladwesp
de Kiefernbuschhornblatt-
    wespe *f*; gemeine Busch-
    hornblattwespe *f*
da nåletræ bladhveps
sv tallstekel
es lofiro *m*; falsa oruga *f* del
    pino; avispa *f* del abeto
it lofiro *m*; tentredine *f* del
    pino
    Diprion pini L.

2736
PINE SHOOT MOTH;
EUROPEAN PINE SHOOT
MOTH (US)
fr tordeuse *f* des pousses du
    pin
ne dennelotrups
de Kiefern(knospen)trieb-
    wickler *m*
da fyrreskudviklere
sv tallkottsvecklare
es brugo *m* de los pinos
it tortrice *f* delle gemme
    apicali dei pini
    Evetria buoliana Schiff.;
    Rhyacionia b. (UK, US)

2737
PINE WEEVIL
fr grand charançon du pin;
    hylobe *f* du pin
ne grote dennesnuitkever
de grosser schwarzer Rüssel-
    käfer *m*
da store brune snudebille
sv snytbagge
es hilobio *m* del abeto;
    gorgojo *m* del abeto
it ilobio *m* dell'abette
    Hylobius abietis L.

2738
PINHEADING (MUSH-
ROOM GROWING);
FRUCTIFICATION (US);
PINNING (US)
fr fructification *f*; formation
    f des grains
ne knopvorming

de Pilzknotung f
da frugtifikation
sv fruktifikation
es fructificación f; formación f de los granos
it fruttificazione f

2739
**PINHEADS (MUSH-
ROOM GROWING); PINS
(US)**
fr petits boutons m pl
ne knopjes
de Stecknadelköpfe m pl;
'Pinheads' m pl
da knappenålshoveder n pl;
'pinheads'
sv 'pinheads'; svampämne n
es granos m pl
it puntini m pl

2740
PINK
fr oeillet m mignardise
ne grasanjer
de Federnelke f
da fjernellike
sv fjädernejlika
es clavel m coronado;
clavellina f de pluma
it garofano m piumato
Dianthus plumarius L.

2741
PINNATE
fr penné
ne gevind
de gefiedert
da finnet
sv parbladig
es pinado
it pennato

2742
PINNATELY LOBED
fr pennatilobé
ne veerlobbig
de fiederlappig
da fjerlappet
sv parflikig
es pinadolobulado
it lobato penniforme

2743
PINNATIFID
fr pennatifide
ne veerspletig
de fiederspaltig
da fjersnitdelt
sv parkluven
es pinatifido
it scisso penniforme

2744
PINNATIPARTITE
fr pennatipartite
ne veerdelig
de fiederteilig
da fjerdelt
sv pardelad
es pinadopartido
it diviso penniforme

2745
PINNIFORM
fr en forme de plume;
penniforme
ne veervormig
de fiederförmig
da fjerformet
sv fjäderformig
es peniforme
it penniforme

PINNING, 2738

2746
**PINNIVEINED;
FEATHER VEINED**
fr penninervé
ne veernervig
de fiedernervig
da fjernervet
sv fjädernervig
es peninerviado
it nervato penniforme

PINS, 2739

2747
PIP; KERNEL
fr noyau m
ne pit
de Kern m
da kærne; sten
sv kärna
es pepita f
it nocciolo m

PIPE COLLAR, 747

2748
**PIPE HEATING; TUBE
HEATING**
fr chauffage m par tuyaux;
chauffage m par radia-
teurs
ne buisverwarming
de Rohrheizung f
da røropvarmning
sv röruppvärmning
es calefacción f por tubos;
calefacción f por radia-
dores
it riscaldamento m con tubi

2749
PISTIL
fr pistil m
ne stamper
de Stempel m
da støvvej
sv pistill
es pistillo m
it pistillo m

2750
PISTON PUMP
fr pompe f à piston
ne zuigerpomp
de Kolbenpumpe f
da sugepumpe
sv kolvpump
es bomba f de pistón
it pompa f a pistoni

2751
PIT
fr fosse f
ne kuil
de Grube f
da kule
sv grop
es fosa f
it fossa f; buco m

2752
PITCH; ROOF SLOPE
fr pente f
ne dakhelling
de Dachneigung f
da taghældning
sv taklutning

es pendiente *f*
it pendenza *f* del tetto

2753
PITH (bot.)
fr moelle *f*
ne merg *n*
de Mark *n*
da marv
sv märg
es médula *f*
it midollo *m*

2754
PITH RAY
fr rayon *m* médullaire
ne mergstraal
de Markstrahl *m*
da marvstråle
sv märgstråle
es rayo *m* médular
it raggio *m* midolloso

2755
PLAGUE
fr fléau *m*
ne plaag
de Plage *f*; Angriff *m*
da plage; angreb
sv plåga; angrepp
es plaga *f*
it piaga *f*

2756
PLAINTAIN LILY;
FUNKIA
fr Hosta *f*; funkia *f*
ne Funkia
de Funkie *f*; Trichterlilie *f*
da funkia
sv funkia; blålilja
es finquia *f*; hermosa *f* de
día
it Funkia *f*
Hosta Tratt.

2757
PLANE TREE
fr platane *m*
ne plataan
de Platane *f*
da platan
sv platan

es plátano *m*
it platano *m*
Platanus L.

PLANK, 389

2758
PLANNING
fr planning *f*; programme *m*
ne planning
de Planung *f*
da planlægning
sv planering; planläggning
es planificación *f*; programa
m
it pianificazione *f*

2759
PLANT; CROP
fr plante *f*; végétal *m*; culture *f*
ne gewas *n*
de Gewächs *n*; Pflanze *f*;
Kultur *f*
da vækst; plante; kultur
sv planta; växt
es planta *f*
it pianta *f*

2760
PLANT, TO; TO SET
fr planter
ne poten; aanplanten; planten
de anpflanzen; setzen; pflanzen
da beplante; plante
sv plantera
es plantar
it piantare

2761
PLANTAIN
fr plantain *m*
ne weegbree
de Wegerich *m*
da vejbred
sv groblad; svartkämpar
es llantén *m*; zaragatona *f*
it piantaggine *f*;
cinquenervi *f*
Plantago L.

PLANTAIN, common,
774

2762
PLANT ASSOCIATION
fr association *f* végétale
ne plantengemeenschap
de Pflanzengemeinschaft *f*
da plantesamfund *n*
sv växtsamhälle *n*
es comunidad *f* de plantas
it associazione *f* vegetale

2763
PLANTATION
fr plantation *f*
ne aanplant; plantopstand
de Pflanzung *f*; Anpflanzung
f; Pflanzenbestand *m*
da plantning; plantage
sv plantering
es plantación *f*
it piantata *f*

2764
PLANT BORDER
fr bordure *f*
ne randbed *n*
de Randbeet *n*
da rabat
sv kantrabatt; bård
es arriate *m* marginal;
canastillo *m*; platabanda *f*
it aiuola *f* d'orlo

2765
PLANT BREEDER
fr producteur *m*
ne kweker
de Züchter *m*
da tiltrækker; dyrker; avler
sv uppdragare
es cultivador *m*
it coltivatore *m*

2766
PLANT BREEDING
fr amélioration *f* de plantes;
sélection *f*
ne plantenteelt
de Pflanzenzucht *f*
da plantedyrkning
sv växtodling
es cultivo *m* de plantas
it miglioramento *m* delle
piante

2767
PLANT DENSITY
fr densité *f* de plantation
ne beplantingsdichtheid
de Bepflanzungsdichte *f*
da plantetæthed
sv planteringstätthet
es intensidad *f* de planta-
    ción
it densità *f* della piantatura;
    fitezza *f* d'impianto

2768
PLANT DISEASE
fr maladie *f* des plantes
ne planteziekte
de Pflanzenkrankheit *f*
da plantesygdom
sv växtsjukdom
es enfermedad *f* de las plan-
    tas
it malattia *f* delle piante

2769
PLANT FOOD;
PLANT NUTRITION
fr éléments *m pl* nutritifs
    (des plantes)
ne plantenvoedsel *n*
de Pflanzennahrung *f*
da planteernæring
sv växtnäring
es alimento *m* para plantas;
    nutrientes *m pl* para plan-
    tas
it elementi *m pl* nutritivi per
    le piante

2770
PLANT HOLE
fr trou *m* de plantation
ne plantgat *n*
de Pflanzloch *n*
da plantehul *n*
sv planteringsgrop
es agujero *m* para plantar
it buco *m* da piantare

2771
PLANTING DISTANCE;
PLANTING SPACE
fr espacement *m*; distance *f*
    de plantation
ne plantafstand

de Pflanzweite *f*; Pflanzab-
    stand *m*
da planteafstand
sv plantavstånd
es distancia *f* entre plantas
it distanza *f* di piantatura

PLANTING LINE,   1534

2772
PLANTING SCHEME
fr plan *m* de culture
ne plantschema *n*
de Pflanzystem *n*; Bepflan-
    zungsplan *m*
da kulturplan; beplantnings-
    plan
sv odlingsplan; planterings-
    plan
es plan *m* de cultivo
    esquena *m* de plantación
it schema *m* di coltivazione
    modello *m* di piantatura

2773
PLANTING SEASON;
TIME OF PLANTING
fr époque *f* de plantation
ne planttijd
de Pflanzzeit
da plantetid
sv planteringstid
es época *f* de plantación
it stagione *f* per piantare

PLANTING SPACE,   2771

2774
PLANTING STOCK
(BULBS)
fr plants *mpl*
ne plantgoed *n*
de Pflanzgut *n*
da læggeløg
sv planteringslökar
es plantas *f pl*
it materiale *m pl* a piantare

2775
PLANT KINGDOM;
VEGETABLE KINGDOM
fr règne *m* végétal
ne plantenrijk
de Pflanzenreich *n*

da planterige *n*
sv växtrike *n*
es reino *m* vegetal
it regno *m* vegetale

PLANT LOUSE,   125

2776
PLANT MATERIAL
fr plants *m pl*
ne plantgoed *n*
de Jungpflanzen *f pl*
da plantemateriale *n*
sv plantor
es plantulas *f pl*
it sementa *f*

PLANT NUTRITION,
2769

2777
PLANT OUT, TO
fr repiquer
ne uitplanten
de auspflanzen
da udplante
sv utplantera
es plantación *f* de asiento
it piantare

2778
PLANT PROTECTION
PRODUCT
fr produit *m* pesticide;
    produit *m* phytosanitaire
ne bestrijdingsmiddel *n*
de Bekämpfungsmittel *n*
da bekæmpelsesmiddel *n*
sv bekämpningsmedel *n*
es medio *m* de control
it prodotto *m* per la prote-
    zione della piante; mezzo
    m di lotta

2779
PLANT SAMPLE
fr échantillon *m* végétal
ne gewasmonster *n*
de Probe *f* pflanzlichen
    Materials
da planteprøve
sv blad (vävnads-) ana-
    lysprov *n*

es muestra *f* de material
   vegetal
it campione *f* vegetale

2780
PLANT SYSTEM
fr système *m* de plantation
ne plantsysteem *n*
de Pflanzsystem *n*
da plantningssystem *n*
sv planteringssystem *n*
es sistema *m* de plantación
it disposizione *f* d'impianto

2781
PLANT TUB
fr bac *m*; caisse *f* à plantes
ne plantekuip
de Pflanzenkübel *m*
da plantebalje
sv planteringsbalja
es cubo *m* para plantas
it tino *m* di piante

2782
PLASTIC
fr matière *f* plastique
ne kunststof
de Kunststoff *m*
da plastmateriale *n*; kunststof *n*
sv plast
es plástico *m*
it plastica *f*

2783
PLASTIC CLADDING
fr revêtement *m* de plastique
ne bekleden met folie
de Folienauskleidung *f*
da plastfoliebeklædning
sv foliebeläggning;
   plastbeläggning
es revestimiento *m* de
   plástico
it rivestimento *m* in plastica

PLASTIC FILM, 1929

2784
PLASTIC FOAM
fr mousse *f* plastique
ne kunststofschuim *n*;
   schuimplastic *n*

de Kunststoffschaum *m*
da kunststofskum *n*
sv skumplast
es espuma *f* de plástico
it plastica *f* schiumosa

2785
PLASTIC FOIL
fr feuille *f* de plastique
ne plasticfolie
de Plastikfolie *f*
da plastikfolie
sv plastfolie
es tela *f*
it foglio *m* di plastico

2786
PLASTIC LINING
fr doublage *f*
ne foliewand
de Innenbespannung *f* mit
   Folie
da folievæg
sv plastfodrad
es forro *m*; revestimiento *m*
it rivestimento *m*

2787
PLASTIC MULCH LAYER
fr dérouleur *m* de film plastique
ne folielegger
de Folienlegevorrichtung *f*
da folielæggemaskine
sv maskin för utläggning av
   folie
es desenrollador *m* de película
it stenditrice *f* di foglio in
   plastica

2788
PLASTIC POT
fr godet *m* de plastic
ne kunststofpot
de Kunststofftopf *m*
da plastpotte; fiberpotte
sv plastkruka
es maceta *f* de plástico *m*
it vaso *m* di plastica

2789
PLASTIC SHEET FOR
SOIL STEAMING
fr bâche *f* pour la stérilisation du sol par vapeur
ne stoomzeil *n*
de Dampfplane *f*
da dampsejl *n*
sv presenning för ångning
es cubierta *f* para esterilización del suelo con vapor
it foglio *m* per la sterilizzazione del suolo

2790
PLASTIC TIE
fr bande *f* de plastic
ne plasticband *n*
de Plastikband *n*
da plastikbånd *n*
sv plastsnöre *n*
es banda *f* de plástico;
   cinta *f* de plástico
it legaccio *m* di plastica

2791
PLASTIC TREE-TIE
fr plastilien *m*
ne boomband (plastic)
de Kunststoffbaumband *m*
da plast-træbånd *n*
sv träduppbindningsband
   (plast)
es palo *m* sujetador (plástico)
it plastilina *f*

2792
PLEDGE
fr gage *m*; garantie *f*
ne onderpand *n*
de Pfand *n*
da pant *n*
sv pant
es prenda *f*
it pegno *m*; garanzia *f*

—, multi furrow, 2423
—, reversible, 3121
—, single furrow, 3420
—, steaming, 3626
—, trailed, 3902
—, trenching, 1091

2793
PLOUGH, TO
fr labourer
ne ploegen
de pflügen
da pløje
sv plöja
es arar
it arare

2794
PLOUGH BEAM
fr âge m
ne ploegboom
de Grindel m
da plovbom
sv plogås
es árbol m; eje m de arado
it bure m

2795
PLOUGH IN, TO
fr enfouir
ne onderploegen
de unterpflügen
da nedploje
sv plöja ned
es enterrar con arado
it ricoprire di terra arando

2796
PLOUGHING WITH SOIL
MIXING
fr soussoler
ne mengploegen
de mischpflügen
da blandepløje
sv blandplöja
es subsolar
it arare da mescolanza

2797
PLOUGH LAYER
fr couche f arable; sillon f
ne bouwvoor
de Pflugfurche f
da agerfure

sv översta skiktet
es capa f de tierra arable
it solco m arabile

2798
PLOUGH OUT, TO
fr arracher à la charrue
ne uitploegen
de herauspflügen
da oppløjning
sv upptaga med plog
es arrancar con arado
it sarchiare per mezza di
un aratro

2799
PLOUGH SHARE
fr soc m
ne ploegschaar
de Pflugschar f
da plovskær n
sv plogbill
es reja f
it vomere f

2800
PLOUGH UP, TO
(GRASSED SURFACE)
fr retourner; défricher
ne scheuren
de umbrechen; aufreissen;
umpflügen
da pløje græsland; omlægge
sv plöja upp
es roturar
it dissodare

2801
PLUM
fr prune f
ne pruim
de Pflaume f
da blomme
sv plommon
es ciruela f
it susina f

PLUM, cherry, 653

2802
PLUME MOTH
fr ptérophore m
ne vedermot
de Federmotte f

da fjermøl
sv fjädermott
es polilla f plumosa
Pterophoridae

2803
PLUM FRUIT MOTH
fr carpocapse m des prunes
ne pruimemot
de Pflaumenwickler m
da blommevikler
sv plommonvecklare
es gusano m del ciruelo
it verme m delle susine;
tortrice f delle susine
Enarmonia funebrana Tr.;
Grapholita f. (UK)

2804
PLUM LEAF MITE;
PLUM RUST MITE (US)
fr phytopte f
ne pruimegalmijt; galmijt
de Gallmilbe f
ad galmide
sv gallkvalster n
es agalla f del ciruelo
it eriofide m delle drupacee
Vasates fockeui Nal. et
Trt.

2805
PLUM SAWFLY
fr hoplocampe m commun
du prunier
ne gele pruimezaagwesp
de gelbe Pflaumensäge-
wespe f
da blommehveps
sv plommonstekel
es hoplocampa f del ciruelo
it tentredine f delle susine
Hoplocampa flava L.

PLUM SAWFLY. black,
343

2806
PLUM TREE
fr prunier m
ne pruimeboom
de Pflaumenbaum m
da blommetræ n
sv plommonträd n

es cirolero *m*; ciruelo *m*
it susino *m*
Prunus domestica L.

2807
PLUMULE
fr plumule *f*
ne pluimpje *n*
de Federchen *n*
da midterknop
sv stamknopp
es plúmula *f*
it piumetta*f*; plumula *f*

2808
PLUNGER PUMP
fr pompe *f* à immersion
ne dompelpomp
de Tauchpumpe *f*; Unter-
wasserpumpe *f*
da undervandspumpe
sv dränkbar pump; under-
vattenspump
es bomba *f* de sumersión
it pompa *f* ad attuffare

2809
PLUNGING BED;
PIT; ROOTING BED
fr silo *m*
ne kuil
de Einschlag *m*
da kule
sv lökgrav
es silo *m*; fosa *f*
it fossa *f*

2810
POD; LEGUME
fr gousse *f*
ne peul
de Hülse *f*
da bælg
sv balja
es vaina *f*
it guscio *m*

2811
POINSETTIA;
CHRISTMAS STAR
fr poinsettie *m*
ne kerstster
de Weihnachtsstern *m*;
Poinsettie *f*

da julestjerne
sv julstjärna
es poinsetia *f*; flor *f* de navi-
dad
it euforbia *f*
Euphorbia pulcherrima
Willd.

2812
POINTER; INDICATOR;
NEEDLE
fr aiguille *f*
ne wijzer
de Zeiger *m*
da viser
sv visare
es ajuga *f*
it lancetta *f*

2813
POISON
fr poison *m*
ne vergif *n*
de Gift *n*
da gift
sv gift *n*
es veneno *m*
it veleno *m*

POISON, contact, 814

2814
POISON, TO
fr empoisonner
ne vergiftigen
de vergiften
da forgifte
sv förgifta
es envenenar
it avvelenare

2815
POISONING
fr empoisonnement *m*
ne vergiftiging
de Vergiftung *f*
da forgiftning
sv förgiftning
es envenenamiento *m*
it avvelenamento *m*

2816
POISONOUS
fr vénéneux; toxique
ne vergiftig

de giftig
da giftig
sv giftig
es venenoso
it velenoso

2817
POLDER;
EMPOLDERING
fr polder *m*
ne polder; droogmakerij
de Polder *m*; Binnensee-
polder *m*
da polder; inddæmning;
tørlægning
sv 'polder'; torrlagd sjö;
indämning
es pólder *m*
it polder *m*

POLDER DISTRICT,
1119

2818
POLDER DRAINAGE
fr épuisement *m* par pompage du polder
ne polderbemaling
de Polderentwässerung *f*
da polderafvanding
sv 'polder'-törrlaggning;
'polder'-dränering
es mantenimiento del nivel
del agua en un polder
it prosciugamento *m* del
polder

2819
POLE; STAKE; POST
fr pieu *m*; piquet *m*
ne paal
de Pfahl *m*
da pæl
sv påle
es palo *m*; estaca *f*
it palo *m*

2820
POLE SLICING BEAN
fr haricot *m* sabre à rames
ne stoksnijboon
de Stangenschwertbohne *f*
da stangsnittebønne
sv stångskärböna

es judia *f* verde para cortar trepadora
it fagiolino *m* a bastone

2821
POLE SNAP BEAN
fr haricot *m* à rames mangetout
ne stokslaboon
de Stangenbrechbohne *f*
da stangbrækbønne
sv stångbrytböna
es judia *f* verde trepadora
it fagiolo *m* comune a bastone
Phaseolus sp.

2822
POLLEN
fr pollen *m*
ne stuifmeel *n*
de Blütenstaub *m*
da blomsterstøv *n*; pollen
sv frömjöl; pollen
es polen *m*
it polline *m*

2823
POLLEN GERMINATION
fr germination *f* de pollen
ne stuifmeelkieming
de Pollenkeimung *f*
da pollenspiring
sv pollengroning
es germinación *f* del polen
it germinazione *f* del polline

2824
POLLEN GRAIN
fr grain *m* de pollen
ne stuifmeelkorrel
de Pollenkorn *n*
da pollenkorn *n*
sv pollenkorn *n*
es grano *m* de polen
it granello *m* di polline

2825
POLLINATE, TO
fr polliniser
ne bestuiven
de bestäuben
da bestøve

sv pollinera
es polinizar
it impollinare

2826
POLLINATION
fr pollinisation *f*
ne bestuiving
de Bestäubung *f*
da bestøvning
sv pollination; befruktning
es polinización *f*
it impollinazione *f*

POLLINATION, cross,
909

2827
POLLINATOR
fr transporteur *m* de pollen
ne bestuiver
de Bestäuber *m*
da bestøver
sv pollinator
es polinizador *m*
it impollinatore *m*

2828
POLLUTION
fr salissure *f*; salissement *m*
ne verontreiniging; vervuiling
de Verschmutzung *f*
da tilsmudse; forurene
sv förorening
es ensuciamiento *m*
it insudiciamento *m*

POLLUTION, air, 59
—, glass, 1591

2829
POLYAMIDE
fr polyamide *f*
ne polyamide
de Polyamid *n*
da polyamid
sv polyamid; amidplast; nylon
es poliamida *f*
it poliamide *f*

2830
POLYESTER
fr polyester *m*
ne polyester

de Polyester *m*
da polyester
sv polyester; esterplast
es poliester *m*
it poliestere *m*

2831
POLYETHYLENE
fr polyéthylène *m*
ne polyethyleen
de Polyäthylen *n*
da polyætylen
sv polyeten; etenplast
es polietileno *m*
it polietilene *m*

2832
POLYMER
fr polymère *m*
ne polymeer
de Polymer *m*
da polymer
sv polymer
es polimere *m*
it polimero *m*

2833
POLYPETALOUS
(COROLLA)
fr dialypétale
ne veelbladig
de vielblättrig
da mangebladet
sv polypetalus; mångbladig
es polipétalo
it polipetalo

2834
POLYPODIUM
fr polypode *m*
ne eikvaren
de Tüpfelfarn *m*
da Polypodium; engelsød
sv stensöta
es polipodio *m*
it polipodio *m*; felce *f*
Polypodium L.

2835
POLYPROPYLENE
fr polypropylène *m*
ne polypropyleen
de Polypropylen *n*
da polypropylen

sv polypropylen
es polipropileno *m*
it polipropilene *m*

2836
POLYSEPALOUS
(CALYX)
fr dialysépale
ne veelbladig
de vielblättrig
da mangebladet
sv polysepalus; mångbladig
es polisépalo
it polisepalo

2837
POLYSTYRENE
fr polystyrène *m*
ne polystyreen
de Polystyrol *n*
da polystyren
sv polystyren; styrenplast
es polistireno *m*
it polistirene *m*

2838
POLYTETRAFLUOR-
ETHYLENE (PTFC)
fr polytétrafluoréthylène *m*
ne polytetrafluoretheen
de Polytetrafluoräthylen *n*
da polytetrafluorætylen
sv polyfluoreten;
fluoretenplast
es politetrafluoretileno *m*
it politetrafluoroetilene *m*

2839
POLYVINYL CHLORIDE
fr chlorure *m* de polyvinyle
ne polyvinylchloride; pvc
de Polyvinylchlorid *n*
da polyvinylchlorid
sv polyvinylklorid;
vinylkloridplast; pvc
es cloruro *m* de polivinilo
it clorisso *m* di polivinile

2840
POME (FRUIT)
fr fruit *m* à pépins
ne pitvrucht
de Kernfrucht *f*
da kernefrugt

sv kärnfrukt
es fruta *f* de pepita
it pomo *m*

2841
POMEGRANATE TREE
fr grenadier *m*
ne granaatappelboom
de Granatapfelbaum *m*
da granatæbletræ *n*
sv granatäppelträd *n*
es granado *m*
it melagrano *m*
    Punica granatum L.

2842
POND
fr étang *m*
ne vijver
de Teich *m*
da dam
sv damm
es estanque *m*
it stagno *m*

POPLAR, grey,   1663
—, white,   4145

POPLAR AND WILLOW
BORER,   4162

2843
POPLAR MOSAIC
fr mosaïque *f* du peuplier
ne populieremozaïek *n*
de Pappelmosaik *n*
da poppel-mosaik
sv poppelmosaik
es mosaico *m* del álamo
it mosaico *m* del pioppo

2844
POPPY
fr pavot *m*
ne papaver
de Mohn *m*
da valmue
sv vallmo
es adormidera *f*
it papavero *m*
    Papaver L.

POPPY, Californian,   534
—, Iceland,   1867
—, oriental,   2548

2845
POPULATION
fr population *f*
ne populatie
de Sippe *f*; Zuchtstamm *m*;
    Population *f*
da population
sv population
es población *f*
it popolazione *f*

2846
PORE
fr pore *m*
ne porie
de Pore *f*
da pore
sv por
es poro *m*
it poro *m*

2847
PORE SPACE
fr volume *m* des pores
ne poriënvolume *n*
de Porenvolumen *n*
da porevolumen *n*
sv porvolym
es volumen *m* de poros
it volume *m* di pori

2848
POROSITY
fr porosité *f*
ne poreusheid; porositeit
de Porosität *f*
da porøsitet
sv porositet
es porosidad *f*
it porosità *f*

2849
POROUS
fr poreux
ne poreus
de porös
da porøs
sv porös
es poroso
it poroso

2850
PORPHYRY KNOTHORN
fr tordeuse *f*
ne bladroller

de Blattwickler *m*
da vikler
sv vecklare
es tortrix *m*
it tortrice *f*
   Eurhodope suavella
   Zinck.

PORTER'S TROLLEY,
3227

2851
POSITION; SITUATION
(OF A CROP)
fr condition *f*
ne stand
de Stand *m*
da tilstand
sv tillstånd *n*
es posición *f*; nivel *m*;
   situación *f*
it stato *m*

POST, 2819, 3923

2852
POST-EMERGENCE
APPLICATION
fr traitement *m* en post-
   émergence
ne na-opkomst-toepassing
de Nach-Auflaufbehand-
   lung
da behandling efter opspiring
sv behandling efter upp-
   komst
es aplicación *f* después de la
   emergencia *f*
it applicazione *f* in post-
   emergenza

2853
POST-SOWING
APPLICATION
fr traitement *m* en post-
   semis
ne na-zaaien-toepassing
de Vor-Auflaufbehandlung *f*
da behandling efter såning
sv behandling efter sådd
es aplicación *f* después de la
   siembra
it applicazione *f* in post-
   semina

2854
POSTURE
fr position *f* du corps
ne werkhouding
de Körperstellung *f*; Arbeitsfähigkeit *f*
da arbejdsstilling
sv arbetsställning
es posición *f* del cuerpo
it posizione *f* del corpo

2855
POT, TO
fr empoter
ne oppotten
de eintopfen
da potte
sv inplatera i kruka;
   'kruka'
es plantar en macetas
it piantare in un vaso

2856
POTASH
fr potasse *f*
ne kali; kaliumcarbonaat;
   potas
de Pottasche *f*; Kaliumkar-
   bonaat *n*; Kali *n*
da kali *n*
sv kali *n*
es potasa *f*
it potassa *f*

2857
POTASH ALUM
fr alun *m* de potasse
ne kalialuin
de Kalialaun *m*
da kalialun
sv alun
es alumbre *m* potásico
it allume *m* di potassa

2858
POTASH FERTILIZER
fr engrais *m* potassique
ne kalimeststof
de Kalidünger *m*
da kaligødning
sv kaliumgödselmedel
es abono *m* potásico
it concime *m* di potassa

2859
POTASH SALT
fr sel *m* de potasse
ne kalizout *n*
de Kalisalz *n*
da kalisalt *n*
sv kalisalt *n*
es sal *f* de potasa
it sale *m* di potassa

2860
POTASSIUM
fr potassium *m*
ne kalium *n*
de Kalium
da kalium *n*
sv kalum *n*
es potasio *m*
it potassio *m*

2861
POTASSIUM CONTENT
fr teneur *f* en potasse
ne kaligehalte *n*
de Kaliumgehalt *m*
da kaliumindhold *n*
sv kalihalt
es contenido *m* en potasio
it tenore *m* di potassa

2862
POTASSIUM
DEFICIENCY
fr carence *f* potassique
ne kaligebrek *n*
de Kalimangel *m*
da kalimangel
sv kalibrist
es carencia *f* de potasa
it deficienza *f* di potassa

2863
POTASSIUM FIXATION
fr fixation *f* de potasse
ne kalifixatie
de Kaliumfestlegung *f*
da kalibinding
sv kaliumfixering
es fijación *f* de potasio
it fissazione *f* di potassa

POTATO, seed, 3324
—, ware, 4061

2864
POTATO APHID
fr puceron *m* de la pomme
 de terre
ne aardappeltopluis
de Kartoffelblattlaus *f*
da Macrosiphum solanifolii
sv potatisbladlus
es pulgón *m* de la patata
it afide *m* della patata
 Macrosiphum euphorbiae
 Thos.

POTATO APHID,
glasshouse, 1586

2865
POTATO AUCUBA
MOSAIC
fr mosaïque *f* aucuba
 (pomme de terre)
ne aucubabont (aardappel)
de Aucuba-Mosaik *n*
 (Kartoffel)
da aucubamosaik
sv aucubamosaik
es 'mosaico *m* aucuba'
 (patata)
it mosaico *m* aucuba
 (patata)

2866
POTATO CYST
NEMATODE
fr anguillule *f* de la pomme
 de terre
ne aardappelcysteaaltje *n*
de Kartoffelnematode *m*
da kartoffelcysten nematod
sv potatisål
es anguilula *f* dorada
it nematode *m* dorato della
 patata
 Heterodera rostochiensis

2867
POTATO DIGGER
fr arracheuse *f* de pommes
 de terre
ne aardappelrooier
de Kartoffelroder *m*
da kartoffeloptager
sv potatisupptagare
es cosechadora *f* de patatas
it cavapatata *f*

2868
POTATO DIGGER WITH
WINDROWER
fr arracheuse *f* aligneuse de
 pommes de terre
ne voorraadrooier
de Vorratsroder *m*
da optager med magasin
sv potatisupptagare med
 strängläggare
es cosechadora *f* en línea de
 patatas
it cavatuberi *f* con andana-
 tore

2869
POTATO HARVESTER
fr arracheuse-chargeuse *f*
ne verzamelrooier
de Sammelroder *m*
da kartoffelopsamler
sv potatisskördemaskin
es cargadora *f* de patatas
it raccoglitrice *f* per patata

2870
POTATO LEAF ROLL
fr enroulement *m*
 (pomme de terre)
be aardappelbladrol *n*;
 bladrolziekte
de Blattrollkrankheit *f*
 (Kartoffel)
da kartoffel-bladrullesyge
sv bladrullsjuka (potatis)
es enrollamiento *m* de las
 hojas (patata)
it accartocciamento *m*
 (patata)

2871
POTATO LIFTER
fr souleveuse *f* de pommes
 de terre
ne aardappellichter
de Kartoffelheber *m*
da kartoffeloplager
sv potatisupptagare
es cosechador *m* de patatas
it elevatore *m* per patate

2872
POTATO PLANTER
fr planteuse *f* de pommes de
 terre
ne aardappelpootmachine

de Kartoffellegemaschine *f*
da kartoffellæggemaskine
sv potatissättare
es sembradora *f* de patatas
it piantatrice *f* per patate:
 piantapatate *f*

2873
POTATO SCAB
fr gale *f* de la pomme de
 terre
ne aardappelschurft
de Kartoffelschorf *m*
da kartoffelskurv *mf*
sv potatisskorv
es sarna *f* de la patata
it scabbia *f* della patata
 Streptomyces scabies
 (Thaxter)

2874
POTATO SORTER
fr trieuse *f* de pommes de
 terre
ne aardappelsorteermachine
de Kartoffelsortiermaschine *f*
da kartoffelsorteremaskine
sv potatissorterare
es máquina *f* clasificadora
 de patatas
it macchina *f* per selezione
 di patates

2875
POTATO STEM MOTTLE
fr tacheture *f* de la tige
 (pomme de terre)
ne stengelbont *n* (aardappel)
de Stengelbuntkrankheit *f*
 (Kartoffel)
da rattle (kartoffel)
sv stengelbont
es abigarramiento *m* del
 tallo (patata)
it virus *m* del rattle (patata)

2876
POTATO STIPPLE
STREAK
fr bigarrure *f* de la pomme
 de terre
ne stippelstreep (aardappel)
de Strichelkrankheit *f* der
 Kartoffel

da Y-virose (kartoffel)
sv strecksjuka (potatis)
es Y-virus *m*
it necrosi *f* lineare della
    patata
    Y-virus

2877
POTATO TUBER-ROT
NEMATODE
fr Ditylenchus *m* destructor
ne destructoraaltje (Iris)
de Krätzeälchen *n*
da stængelnematod (i løg)
sv rötnematod
es anguilula *f* de la raíz de
    la patata
it nematode *m* del mar-
    ciume della patata
    Ditylenchus destructor
    Thorne

2878
POT BALL
fr plante *f* en motte
ne potkluit
de Topfballen *m*
da potteklump
sv krukklump
es terrón *m* de tiesto; terrón
    m de maceta
it zolla *f* di vaso

POTENTILLA,   686

2879
POT OUT, TO; TO REPOT
fr rempoter
ne verpotten
de umtopfen; verpflanzen
da ompotte
sv omplantera i kruka;
    omkruka
es mudar de tieste
it rinvasare

2880
POT PLANT
fr plante *f* en pot
ne potplant
de Topfpflanze *f*
da potteplante
sv krukväxt
es planta *f* de tiesto
it pianta *f* in vaso

2881
POTTING SOIL;
POTTING COMPOST
fr terre *f* de rempotage
ne potgrond
de Topferde *f*
da pottejord
sv krukjord
es tierra *f* para maceta
it terra *f* da piantare in
    vaso

2882
POT TRAY
fr plateau *m* à pots; pot *m*
    multiple
ne potblad
de Multitopf *m*
da multipot
sv multipot
es apoyo *m* para macetas;
    bote *m* multiple
it vaso *m* multiplo;
    sostegno *m* da vasi

2883
POULTRY MANURE
fr galline *f*
ne vogelmest
de Geflügelmist *m*
da guano; fuglegødning
sv guano; fågelgödsel
es gallinaza *f*
it letame *m* di gallina

2884
POUR INTO, TO
fr verser
ne ingieten
de eingiessen
da vande til
sv vattna ut; indränka
es regar
it versare (inqualcosa)

2885
POWDERY MILDEW
fr Oïdium *m*
ne meeldauw
de Mehltau *m*
da meldug
sv mjöldagg
es mildio *m*; oidio *m*
it oidio *m*
    Sphaerotheca macularis
    spp.; *Oidium* spp.

2886
POWDERY MILDEW
(QUERCUS)
fr Oïdium *m*
ne eikemeeldauw
de Eichenmehltau *m*
da ege-meldug
sv (ek)mjöldagg
es mal *m* blanco
it oidio *m*
    Microsphaera spp.

2887
POWDERY SCAB
fr gale *f* poudreuse
ne poederschurft
de Pulverschorf *m*
da pulverskurv
sv pulverskorv
es podredumbre *f* seca
it scabbia *f* secca
    Spongospora subterranea
    (Wallr.) Lagerh.

2888
POWER-DRIVEN
KNAPSACK DUSTER
fr poudreuse *f* à dos à
    moteur
ne motorrugverstuiver
de Motorrückenstäuber *m*
da motorrygpudderblæser
sv motorryggpudra;
    motorryggpuderspridare
es espolvoreadora *f*
    de dos a motor
it impolveratrice a spalla a
    motore

2889
POWER-DRIVEN
KNAPSACK
MISTBLOWER
fr atomiseur *m* à dos
ne motorrugnevelspuit
de Motor-Rücken-
    Sprühgerät *n*
da motorrygtågesprøjte
sv motorryggsspruta
es atomizador *m* de espalda
it atomizzatore *m* a spalla a
    motore

2890
POWER INPUT
fr puissance *f*
ne elektrisch vermogen *n*
de Leistung *f*
da energi
sv effekt
es potencia *f*
it potenza *f*

POWER LIFT, 1857

2891
POWER TAKE OFF; PTO
fr prise *f* de force
ne aftakas
de Zapfwellenschluss *m*
da kraftudtag *n*
sv kraftuttag *n*
es toma *f* de fuerza
it presa *f* di forza

2892
POWER TAKE-OFF
SHAFT GUARD
fr protection *f* de la prise de force
ne aftakasbescherming
de Gelenkwellenschutz *m*
da kraftudtagsskærm
sv skydd kring kraftöverföringsaxel
es protección *f* de la toma de fuerza
it protezione *f* della presa di forza

2893
PRACTICAL TRIAL;
PRACTICAL
EXPERIMENT
fr essai *m* de pratique; examen *m* de pratique
ne praktijkproef
de Praktikversuch *m*
da lokalforsøg *n*
sv provedyrkning
es ensayo *m* práctico
it esperimento *m* pratico

2894
PRECEDING CROP
fr pré-culture *f*
ne voorvrucht
de Vorfrucht *f*
da forfrugt; forkultur
sv förkultur; förfrukt
es cultivo *m* precedente
it preraccolta *f*

2895
PRECIPITATE (chem.)
fr précipité *m*; dépôt *m*
ne neerslag
de Präzipitat *n*
da bundfald
sv bottensats; fällning
es precipitado *m*
it sedimento *m*

PRECOCITY, 1187

2896
PRECOOLING
fr pré-réfrigération *f*
ne voorkoeling
de Vorkühlung *f*
da forkøling
sv förkylning
es prerefrigeración *f*
it prerafreddamento *m*

2897
PREDATOR
fr prédateur *m*; mite *f* suceur
ne roofinsekt *n*; roofmijt
de Prädator *m*; Raubmilbe *f*
da rovinsekt *n*; rovmide
sv rovinsekt *n*; rovkvalster *n*
es predator *m*; ácaro *m* predator
it predatore *m*; acaro *m* predatore
praedator

2898
PRE-EMERGENCE
APPLICATION
fr traitement *m* en préémergence
ne voor-opkomst-toepassing
de Vor-Auflaufbehandlung *f*
da behandling før opspiring
sv behandling före uppkomst
es applicación *f* antes de la emergencia
it applicazione *f* in pre-emergenza

2899
PRE-GERMINATE, TO
fr pré-germer
ne voorkiemen
de vorkeimen
da forspire
sv förgro
es apitonar (melón)
it pregerminare

2900
PREHEAT, TO
fr préchauffer
ne voorstoken
de erstes Trockenheizen nach dem Roden
da forbehandling
sv värmebehandling
es precalentar
it prescaldare

2901
PRE-HEATER
fr préchauffeur *m*
ne voorverwarmer
de Vorwärmer *m*
da forvarmer
sv förvärmare
es pre-calentador *m*
it preriscaldamento *m*

PRELIMINARY CHECK, 2511

2902
PREMATURELY RIPE
fr mûr prématurement
ne noodrijp
de notreif
da nødmoden
sv brådmogen
es maduro prematuro
it prematuro

2903
PRE-PACKING
fr pré-emballage *m*
ne voorverpakking
de Vorverpackung *f*
da prepakning
sv föremballering
es preembalaje *m*
it pre-imballaggio *m*

2904
PREPARATION
fr préparation f
ne preparaat n
de Präparat n
da præparat n
sv preparat n
es preparado m
it preparato m

2905
PREPARATION FOR
SCATTERING
fr produit m d'épandage
ne strooimiddel n
de Streumittel n
da strømiddel n
sv spridningsmedel n;
   vidhäftningsmedel n
es medio m distribuidor
it prodotto m da spargere

2906
PREPARE, TO (BULBS)
fr préparer
ne prepareren
de präparieren
da præparere
sv preparera
es preparar
it preparare

2907
PRESENT VALUE (ec.)
fr valeur f actuelle
ne huidige waarde
de Gegenwartswert m;
   Zeitwert m
da dagsværdi
sv nuvärde n; diskonterat
   nuvärde n
es valor m actual
it valore m attuale

2908
PRESERVATION
fr conservation f
ne conservering
de Konservierung f
da konservering
sv konservering
es conservación f
it preservazione f;
   conservazione f

2909
PRESERVATIVE
fr produit m de protection
ne conserveringsmiddel
   n (hout)
de Holzschutzmittel n
da træbeskyttelsesmiddel n
sv konserveringsmedel
   n (för trä)
es producto m de protección
it prodotto m di protezione

2910
PRESERVATIVES
fr agents mpl de conservation
ne conserveermiddelen n pl
de Konservierungsmittel
   mpl
da konserveringsmiddel n pl
sv konserveringsmedel n pl
es medios mpl de conservación
it preservativi mpl;
   mezzi mpl de conserva-
   mento

2911
PRESERVE, TO
fr conserver; mettre en conserve
ne inmaken
de einmachen; einkochen
da konservere; henkoge
sv konservera; inlägga
es conservar
it preservare; salvare; conservare

2912
PRESERVED FRUIT
fr conserves f pl de fruits
ne vruchtenconserven
de Obstkonserven f pl
da frugtkonserves
sv fruktkonserver
es conservas f pl de frutas
it conserve fpl di frutti

2913
PRESERVER
fr produit m de conservation
ne conserveringsmiddel n
de Schutzmittel n

da conserveringsmiddel n
sv konserveringsmedel n;
   impregneringsmedel n
es preservando m
it preservativo m

2914
PRESERVES
fr conserves fpl
ne conserven
de Konserven f pl
da konserves
sv konserver
es conservas fpl
it conserve f pl

2915
PRESERVING
INDUSTRY
fr industrie f de la conserve
ne conservenindustrie
de Konservenindustrie f
da konserveindustri
sv konservindustri
es industria f conservera
it industria f di conserva

2916
PRE-SOAK, TO
fr tremper au préalable
ne voorweken
de vorweichen
da sætte i blød
sv blötlägga
es remojar
it mettere in molle;
   ammollare

2917
PRE-SOWING
APPLICATION
fr traitement m en présemis
ne voor-zaaien-toepassing
de Vor-Saatbehandlung
da behandling før såning
sv behandling före sådd
es applicación f antes de la
   siembra
it applicazione f in pre-
   semina

2918
PRESS JUICE
fr jus m pressé
ne perssap n

de Pressaft *m*
da pressesaft
sv pressaft
es jugo *m*
it succhio *m* spremitura

2919
PRESSURE ATOMIZING
BURNER
fr brûleur *m* à pulvérisation
   par pression d'huile
ne drukverstuivingsbrander
de Druckölbrenner *m*
da trykforstøvningsfyr *n*
sv tryckoljebrännare
es quemador *m* de pulverización por presión de aceite
it bruciatore *m* atomizzatore a pressione

2920
PRESSURE GAUGE
fr manomètre *m*
ne manometer
de Manometer *n*
da manometer
sv manometer
es manómetro *m*
it manometro *m*

2921
PRESTRESSED
CONCRETE
fr béton *m* précontraint
ne voorgespannen beton *n*
de Spannbeton *m*
da spændbeton
sv förspänd betong
es hormigón *m*
it calcestruzzo *m* precompresso

2922
PRICE CONTROL
fr régulation *f* des prix
ne prijsregeling
de Preisregelung *f*
da prisregulering
sv prisreglering; priskontroll
es regulación *f* de precios
it controllo *m* dei prezzi

2923
PRICE INDEX
fr indice *m* de prix
ne prijsindex
de Preisindex *m*
da prisindeks
sv prisindex
es índice *m* de precios
it indice *m* dei prezzi

2924
PRICE LIST
fr catalogue *m* de prix
ne prijscourant
de Preisliste *f*; Preisverzeich-
   nis *n*
da prisliste
sv prislista; priskurant
es lista *f* de precios
it catalogo *m* dei prezzi

2925
PRICKING OFF TOOL
fr petit plantoir *m* à repiquer
ne verspeenhout *n*
de Pikierholz *n*
da priklepind
sv skolningspinne
es transplantador *m*; espá-
   tula *f*

2926
PRICKLE
fr épine *f*; aiguillon *m*
ne stekel
de Stachel *m*
da barktorn
sv tagg
es copina *f*
it spina *f*

2927
PRICKLY; SPINY;
THORNY
fr épineux
ne stekelig
de stachelig
da stikkende
sv taggig; stickande
es espinoso
it spinoso

2928
PRICK OUT, TO;
TO PRICK OFF
fr repiquer
ne pikeren; verspenen
de pikieren; vereinzeln;
   auspflanzen
da prikle
sv skola
es repicar; picar; entresacar
it trapiantare

2929
'PRIMARY';
BOTRYTIS (TULIP)
fr tulipe *f* infectée par
   Botrytis
ne steker
de 'Stecker' *m*
da stikkere
sv gråmögelsangripna skott
es bulbo *m* infecto de
   Botritis
it Botritis *m*

2930
PRIME, TO
fr abreuver
ne voorstrijken
de grundieren
da grunde
sv grundmåla
es abrevar (prelisado del
   hormigón)
it dare la mano di fondo

2931
PRIMROSE
fr primevère *f*
ne Primula; sleutelbloem
de Primel *f*; Schlüsselblume *f*
da Primula; kodriver
sv Primula; viva
es Primula *f*; primavera *f*
it Primula *f*
   Primula L.

PRIMROSE, evening, 1250

2932
PRINCIPLES OF FARM
MANAGEMENT
fr théorie *f* de l'exploitation
   agricole
ne bedrijfsleer (landbouw-)

de landwirtschaftliche
  Betriebslehre *f*
da landøkonomisk drifts-
  lære
sv lantbrukets företagseko-
  nomi
es teoría *f* de la explotación
  agrícola
it teoria *f* economica aziendale

2933
PRIVATE LAW, ON
fr de droit *m* privé
ne privaatrechtelijk
de privatrechtlich
da privatretligt
sv privaträttslig
es de derecho privado
it di diritto privato

2934
PRIVET
fr troène *m*
ne liguster
de Liguster *m*; Rainweide *f*
da liguster
sv liguster
es ligustro *m*; aligustre *m*
it ligustro *m*
  Ligustrum L.

2935
PRIVET HAWK MOTH
fr sphinx *m* du troène
ne ligusterpijlstaart
de Ligusterschwärmer *m*
da ligustersværmer
sv ligustersvärmare
es esfinge *f* del aligustre
it sfinge *f* del ligustro
  Sphinx ligustri L.

2936
PROCEEDS
fr produit *m*
ne opbrengst (geldelijk)
de Erlös *m*
da udbytte *n*
sv intäkt; avkastning
es producto *m*
it rendita *f*

2937
PROCESSING
fr transformation *f*
ne verwerking
de Verarbeitung *f*
da forarbejdning
sv förädling
es transformación *f*
it attività *f* di trasformazione

2938
PROCESSING INDUSTRY
fr industrie *f* transformatrice
ne verwerkingsindustrie
de Verwertungsindustrie *f*
da forarbejdningsindustri
sv förädlingsindustri
es industria *f* de transformación
it industria *f* trasformatrice

2939
PROCUMBENT
fr couché
ne liggend
de niederliegend
da nedliggende
sv nedliggande
es echado
it coricato

2940
PRODUCE, TO
fr produire
ne produceren
de produzieren
da producere
sv producera
es producir
it produrre

2941
PRODUCTION
DIRECTION
fr direction *f* de production;
  direction *f* technique
ne produktierichting
de Produktionsrichtung *f*
da produktionsretning
sv produktionsriktning
es dirección *f* de producción
it direzione *f* della produzione

2942
PRODUCTION MEANS
fr moyen *m* de production
ne produktiemiddel
de Produktionsmittel *n*
da produktionsmiddel *n*
sv produktionsmedel *n*
es medio *m* de producción
it mezzo *m* di produzione

PRODUCTION MEANS,
durable, 1167

2943
PRODUCTIVITY
fr productivité *f*
ne produktiviteit
de Produktivität *f*
da produktivitet
sv produktivitet
es productividad *f*
it produttività *f*

PROFESSIONAL
SCHOOL, 3273

2944
PROFILE
fr profil *m*
ne profiel *n*
de Profil *n*
da profil
sv profil
es perfil *m*
it profilo *m*

2945
PROFILE-EXAMINATION
fr analyse *f* du profil
ne profielonderzoek *n*
de Profiluntersuch *m*
da profilundersøgelse
sv profilundersökning
es investigación *f* del perfil
  del suelo
it ricerca *f* del profilo

2946
PROFILE-HOLE
fr fosse *f* de profil
ne profielkuil
de Profilgrube *f*
da profilkule
sv profilgrop

es hoyo *m* para perfil
it fossa *f* da profilo

2947
PROFILE PIT
fr puits *m* de profilage
ne profielkuil
de Profilgrube *f*
da profilhul *n*
sv profilgrop
es calicata *f*; zanja *f* perfilada
it buca *f* da profilo

2948
PROGENY; OFFSPRING
fr progéniture *f*; descendance *f*
ne nakomelingschap
de Nachkommenschaft *f*
da afkom *n*
sv avkomma
es descendencia *f*; progenie *f*
it discendenza *f*

2949
PROPAGATE, TO
fr multiplier
ne vermenigvuldigen; vermeerderen
de vermehren
da formere
sv föröka
es multiplicar
it moltiplicare; propagare

2950
PROPAGATING CASE
fr coffre *m*
ne kweekbak
de Vermehrungskasten *m*
da formeringsbænk
sv förökningsbänk
es campinera *f* de multiplicación
it letto *m*

2951
PROPAGATING HOUSE
fr serre *f* à multiplication
ne kweekkas
de Vermehrungshaus *n*; Anzuchtshaus *n*
da formeringshus *n*

sv förökningshus *n*
es invernadero *m* de multiplicación; estufa *f* de multiplicación
it serra *f* da moltiplicazione; serra *f* da propagazione

PROPAGATION, 2425

PROPAGATION,
generative, 1554

2952
PROPAGATION ROOM (HYACINTHS)
fr chambre *f* de multiplication
ne holkamer
de Brutansatzraum *m*
da hulkammer
sv förökningsrum
es sala *f* de preparación
it luogo *m* a scavare

2953
PROPANE
fr propane *m*
ne propaan *n*
de Propan *n*
da propan *n*
sv propan *n*
es propano *m*
it propano *m*

2954
PROPANE SPRAYER
fr pulvérisateur *m* à propane
ne propaanspuit
de Propanspritze *f*
da propansprøjte
sv propanspruta
es pulverizador *m* de propano
it irroratrice *f* a propano

2955
PROP ROOT
fr racine-crampon *f*; racine *f* adventive
ne steunwortel
de Stutzwurzel *f*
da støtterod
sv stödrot
es raíz *f* aéra de sostén
it radice *f* di sostegno

2956
PROTANDRY
fr protandrie *f*
ne protandrie
de Protandrie *f*
da førsthannethed; protandri
sv protandri; försthandlighet
es protandria *f*
it protandrie *f*

2957
PROTECT, TO
TO SHELTER;
TO SCREEN
fr abriter; protéger; couvrir
ne beschutten; beschermen
de schützen
da beskytte
sv beskydda
es proteger
it proteggere

2958
PROTECTANTS;
PESTICIDES
fr mesures *f pl* de protection
ne afweermiddelen *n pl*
de Abwehrmittel *m pl*
da beskyttelsesmiddel *n pl*
sv skyddsmedel *n pl*
es medios *m pl* de protección
it mezzi *m pl* protettivi

2959
PROTECTED FROM
FROST; FROST-PROOF
fr à l'abri de la gelée
ne vorstvrij
de frostfrei
da frostfri
sv frostfri
es protegido de la helada
it protetto contro la gelata; antigelo

2960
PROTHALLIUM
fr prothallé *m*
ne voorkiem
de Vorkeim *m*
da forkim
sv prothallium; förgrodd
es prótalo *m*
it protallo *m*

2961
PROTOGYNY
fr protogynie *f*
ne protogynie
de Protogynie *f*
da førsthunnethed
sv protogyni; försthonlighet
es protoginia *f*
it protogynie *f*

PROVENANCE,   2550

2962
PRUINOSE; HOARY
fr pruineux
ne berijpt
de bereift
da dugget; melet
sv belagd; med dagg
es cubierto de helada blanca
it pruinoso

2963
PRUNE, TO; TO LOP;
TO CUT; TO TRIM
fr tailler; élaguer
ne snoeien
de beschneiden; stutzen;
   schneiden
da beskære
sv beskära
es podar
it tagliare; potare

2964
PRUNING CUT
fr coupe *f*; plaie *f* de taille
ne snoeiwond
de Schnittwunde *f*
da beskæresår *n*
sv snitt
es herida *f*
it lesione *f* di potatura

2965
PRUNING KNIFE
fr serpette *f*
ne snoeimes *n*
de Gartenhippe *f*; Baum-
   messer *n*
da beskærekniv
sv beskärningskniv;
   plantskolekniv
es podadera *f*
it potatoio *m*

2966
PRUNINGS
fr bois *m* de taille
ne snoeihout *n*
de Schnittholz *n*
da afklipning
sv ris *n*
es leña *f* proveniente de la
   poda
it legno *m* di potatura;
   sarmenti *mpl* di potatura

2967
PRUNING SAW
fr scie *f* de jardinier;
   égoïne *f*
ne snoeizaag
de Gartensäge *f*; Stutzsäge *f*
da grensav; beskæresav
sv grensåg; trädgårdsåg
es sierra *f* podadera
it sega *f* da potare

2968
PRUNING SHEARS
fr sécateur *m*; ébranchoir *m*
ne snoeischaar; takkeschaar
de Rosenschere *f*; Gartenschere *f*; Aufastungsschere *f*
da beskæresaks
sv sekatör; grensax
es tijeras *f pl* de podar;
   podadera *f*
it forbici *f pl* da potare

2969
PSEUDO FRUIT
fr faux fruit *m*
ne schijnvrucht
de Scheinfrucht *f*
da falsk frugt
sv skenfrukt
es fruto *m* aparente
it pseudo frutto *m*

PSEUDO WHORL,   1287

PSYLLID,   3712
—, boxwood,   417

PTO,   2891

2970
PTO DRIVEN SPRAYER
fr pulvérisateur *m* sur prise
   de force
ne aftakas-spuit
de Zapfwellenspritze *f*
da traktordrevet sprøjte
sv kraftuttagsdriven spruta
es pulverizador *m* sobre
   toma de fuerza
it polverizzatore *m* su
   presa di forza

2971
PTO PUMP
fr pompe *f* sur prise de force
ne aftakaspomp
de Zapfwellenpumpe
da kraftudtagspumpe
sv kraftuttagspump
es bomba *f* sobre tomado
   fuerzo
it pompa *f* a presa di potenza

PUBESCENT,   1114

2972
PUBLIC LAW, ON
fr de droit public
ne publiekrechtelijk
de öffentlich-rechtlich
da offentligretlig
sv offentlig-rättslig
es de derecho público
it di diritto pubblico

2973
PUCKERED (bot.)
fr bosselé; raboteux
ne bobbelig
de höckerig
da rynket
sv knölig
es abollamiento
it bernoccoluto

PUFFER,   302

2974
PULP
fr pulpe *f*
ne pulp
de Pulpe *f*
da pulp

sv pulpa
es pulpa f
it polpa f

PULP, FRUIT, 1480

2975
PULSE BEETLE;
SEED BEETLE (US)
fr bruche f
ne zaadkever
de Samenkäfer m
da frøbille
sv fröbagge; 'smyg'
es gorgojo m
it tonchio m; brucio
Bruchus spp.
(Bruchidae)

2976
PULVERIZE, TO
fr pulvériser; poudrer
ne verstuiven
de zerstäuben
da forstøve
sv pudra
es pulverizar; espolvorear
it polverizzare

2977
PUMP DRAINAGE;
PUMPING FROM WELLS
fr rabattement m par puits
ne putbemaling
de Brunnenentwässerung f;
offene Wasserhaltung f
da afvanding
sv brunnirrigation
es bomba f de pozo
it prosciugamento m da
pozzo

2978
PUMPING
(MAKING DRY)
fr assèchement d'un polder
ne bemaling
de Entwässerung f
da oppumpning
(ved afvanding)
sv torrläggning
es manteniemento m del nivel del agua en un polder
it prosciugamento m

2979
PUMPING STATION
fr station f de pompage
ne gemaal n
de Pumpwerk n
da pumpestation
sv pumpstation; vattenverk
n; kvarnverk n
es estación f de bombeo
it macchina f prosciugatrice

2980
PUMPKIN
fr potiron m; giraumon m
ne pompoen
de Zentnerkürbis m
da centnergræskar n
sv jättepumpa
es calabaza f grande
it cocomero m
Cucurbita maxima Des.

2981
PUMP UP, TO;
TO RAISE BY PUMPING
fr pomper
ne oppompen
de aufpumpen
da pumpe op
sv pumpa upp
es bombear
it tirare su colla pompa

PUNNET, 666

2982
PUPA; CHRYSALIS
fr nymphe f; chrysalide f;
pupe f
ne pop
de Puppe f
da puppe
sv puppa
es ninfa f
it pupa f; crisalide f

2983
PUPATE, TO
(INSECT)
fr se chrysalider
ne zich verpoppen
de sich verpuppen
da forpuppe sig
sv förpuppa sig

es crisalidarse
it incrisalidarsi

2984
PURE CULTURE
fr culture f pure; bouillon m
de culture
ne reincultuur
de Reinkultur f
da renkultur
sv renkultur
es cultivo m puro
it coltura f pura

2985
PUREE; PASTE; MASH
fr purée f
ne puree
de Mark n; Brei m; Puree n
da puré
sv puré
es puré m
it puré m; passato m

PURIFY, TO, 698

2986
PURITY
fr pureté f
ne zuiverheid
de Reinheit f
da renhed
sv renhet
es pureza f
it purezza f

2987
PURLIN
fr panne f
ne gording
de Pfette f
da ås
sv ås
es jabalcón m
it arcareccio m

2988
PURPLE BLOTCH
(BLACKBERRY)
fr taches f pl sur tiges
ne bruine-stengelvlekken-
ziekte
de Brombeerrankenkrank-
heit f

da stængelplet
sv stamfläcksjuka
es salpicón *m* pardo del tallo
it Septocyta *f* ramealis
Septocyta ramealis Sacc.

2989
PURPLE JAPANESE
MAPLE
fr érable *m* japonais rouge
ne zwarte acer
de rotblättriger Fächerahorn
m
da purpurløn
sv japansk blodlönn
es arce *m* palmato atropur-
púrea
it acero palmatum *m*
Acer palmatum cv.
Atropurpureum

2990
PURSLANE
fr pourpier *m*
ne postelein
de Portulak *m*
da portulak
sv portlak
es verdolaga *f*
it porcellana *f*; portulaca *f*

PURSLANE, winter, 4179

2991
PUT IN PITS, TO;
ENSILAGE
fr enfouir; ensiler
ne inkuilen; kuilen
de einmieten; einsenken

da kule
sv förvara i stuka
es enterrar
it infossare

2992
PUT INTO A CLAMP, TO
fr 'mettre en tombe'
ne inkuilen
de einschlagen
da nedkule
sv nedgräva
es enterrar
it sotterrare

PUTREFACTION, 999

2993
PUTREFY, TO;
TO DECOMPOSE
fr se décomposer
ne verteren
de verzehren; vermodern
da dekomponere; rådne
sv ruttna
es putreficarse; descomponerse
it putrefarsi

2994
PUTRID; ROTTEN
fr pourri; décomposé
ne bedorven
de faul
da råddent
sv rutten
es descompuesto
it putrido

2995
PUTTY
fr mastic *m*
ne kit
de Kitt *m*
da kit *n*
sv kitt *n*
es mástic *m*
it mastice *m*

PYRACANTHA, 1358

2996
PYRAMID HOLLY
fr houx *m* baccifère
ne beshulst
de Stechpalme *f*
da kristtorn; pyramidefor-
met järnek
sv järnek
es agrifolio *m*; acebo *m* con
bayas
it agrifoglio *m*
Ilex aquifolium L. cv.
'Pyramidalis'

2997
PYRETHRUM
fr pyrèthre *m*
ne pyrethrum
de Pyrethrum *m*; kaukasische
Kamille *f*
da pyrethrum
sv mattram; rosenkrage
es piretro *m*; pelitre *m*
it piretro *m*
Chrysanthemum coccine-
um Willd.; *Chrys. parthe-*
nisum Bernh.

Q

QUACKGRASS, 867

2998
QUALITY CONTROL
fr contrôle *m* de qualité
ne kwaliteitscontrole
de Qualitätskontrolle *f*
da kvalitetskontrol
sv kvalitetskontroll
es control *m* de calidad
it controllo *m* di qualità

QUALITY GRADING,
3002

2999
QUALITY INSPECTION
fr contrôle *m* de qualité
ne kwaliteitscontrole
de Qualitätskontrolle *f*
da kvalitetskontrol
sv kvalitetskontroll
es control *m* de calidad
it controllo *m* qualitativo

3000
QUALITY LOSS
fr perte *f* de qualité
ne kwaliteitsverlies *n*
ed Qualitätsverlust *m*
da kvalitetsforringelse
sv kvalitetsförlust
es pérdida *f* de calidad
it perdita *f* di qualità

3001
QUALITY NORM
fr norme *f* de qualité
ne kwaliteitsnorm
de Qualitätsnorm *f*
da kvalitetsnorm
sv kvalitetsnorm

es norma *f* de calidad
it norma *f* di qualità

3002
QUALITY SORTING;
QUALITY GRADING
fr triage *m*
ne kwaliteitssortering
de Qualitätssortierung *f*
da kvalitetssortering
sv kvalitetssortering
es clasificatión *f* por calidad
it graduamento *m*

3003
QUARANTINE
fr quarantaine *f*
ne quarantaine
de Quarantäne *f*
da karantæne
sv karantän
es cuarentena *f*
it quarantena *f*

3004
QUEEN BEE
fr reine *f*
ne bijenkoningin
de Bienenkönigin *f*;
Weisel *m*
da bidronning
sv bidrottning
es abeja reina *f*
it ape regina *f*

3005
QUICK COUPLING PIPE
fr tuyau *m* à accouplement
ne snelkoppelbuis
de Schnellkupplungsrohr *n*
da lynkoblingsrør *n*
sv snabbkopplingsrör *n*

es tubo *m* de acoplamiento
it manicotto *m*

3006
QUICK FREEZING
fr surgélation *f* rapide
ne snelvriezen
de schnellgefrieren
da lynfrysning
sv snabbfrysa
es congelación *f* rápida
it congelazione *f* rapide

3007
QUICKLIME
fr chaux *f* clacinée;
chaux *f* vive
ne gebrande kalk
de gebrannter Kalk *m*;
Aetzkalk *m*
da brændt kalk
sv bränd kalk
osläckt kalk
es cal *f* calcinada;
cal *f* viva
it calce *f* viva

QUICKSILVER, 2328

3008
QUINCE
fr cognassier *m*
ne kwee
de Quitte *f*
da kvæde
sv kvitten
es membrillero *m*
it cotogno *m*
Cydonia Mill.

QUINCE, Japanese, 1973

R

3009
RABBIT
fr lapin *m* de garenne
ne konijn *n*
de Kaninchen *n*
da kanin
sv kanin
es conejo *m*
it coniglio *m* selvatico
Cucumus sativus L.

3010
RACEME (BULBS)
fr grappe *f*; racème *m*
ne tros
de Traube *f*
da klase
sv klase
es racimo *m*
it grappolo *m*

RACK, 3844

3011
RADIATING SURFACE
fr surface *f* de rayonnement
ne stralend oppervlak *n*
de Strahlungsoberfläche *f*
da udstrålingsflade
sv utstrålningsyta
es superficie *f* radiante
it superficie *f* raggiante

3012
RADIATION
fr radiation *f*; rayonnement
m
ne straling
de Strahlung *f*
da stråle
sv strålning
es radiación *f*
it irraggiamento *m*;
radiazione *f*

RADIATION, black body,
330
—, intensity of, 1935
—, ionizing, 1950

3013
RADIATOR
fr radiateur *m*
ne radiator
de Heizkörper *m*
da radiator
sv radiator
es radiador *m*
it radiatore *m*

3014
RADIOACTIVE
FALL-OUT
fr retombée *f* radioactive
ne radioactief neerslag
de radioaktiver Nieder-
schlag *m*
da radioaktivt nefald
sv radioaktivt nedfall
es precipitación *f* radioactiva
it precipitazione *f* radioactiva

3015
RADISH
fr radis *m*
ne radijs
de Radies *m*
da radis
sv rädisa
es rábano *m*
it ravanello *m*; rafano *m*
Raphanus sativus L.

RADISH, black, 344
—, horse, 1822
—, wild, 4159

3016
RAFFIA
fr raphia *m*
ne raffia
de Raphiabast *m*
da raffiabast
sv (raffia)-bast
es rafia *f*
it rafia *f*
Raphia spp.

3017
RAFTER
fr chevron *m*
ne spant (balk)
de Sparren *m*
da spær *n*
sv takröste *n*
es cabrio *m*
it puntone *m*

3018
RAINFALL
fr pluie *f*; précipitation *f*
atmosphérique
ne neerslag
de Niederschlag *m*
da nedbør
sv nederbörd
es lluvia *f*; precipitación *f*
atmosférica
it precipitazione *f pl* .

3019
RAIN GAUGE
fr pluviomètre *m*
ne regenmeter
de Regenmesser *m*
da regnmåler
sv regnmätare
es pluviometro *m*
it pluviometro *m*

3020
RAISE, TO (GLASS)
fr enlever les châssis
ne lichten
de abheben
da løfte; fjerne
sv lyfta; upptaga
es destapar; quitar bastido-
res
it levare il vetro

3021
RAISE, TO (ROOT CROP)
fr soulever
ne lichten
de ausheben; herausnehmen
da løfte (optage)

sv lyfta; rothugga
es levantar; cavar; desenterrar
it sarchiare; sradicare

RAISE BY PUMPING, TO,
2981

3022
RAKE
fr râteau *m*
ne hark
de Rechen *m*; Harke *f*
da rive
sv kratta
es rastrillo *m*
it rastrello *m*

3023
RAKE, TO
fr ratisser
ne harken
de rechen; harken
da rive
sv kratta
es rastrillar
it rastrellare

3024
RAKE IN, TO
fr faire pénétrer au rateau
ne inharken
de einrecken
da nedrive
sv kratta ned; hacka ned
es rastrillar
it rastrellare

3025
RAKE OFF, TO
fr râteler
ne afharken
de abrechen; abharken
da rive sammen
sv kratta samman
es rastrillar
it rastrellare

3026
RAMIFICATION
fr ramification *f*
ne vergaffeling
de Verästelung *f*
da forgrening

sv förgrening
es ramificación *f*
it ramificazione *f*

3027
RAMIFIED; BRANCHED
fr ramifié
ne vertakt
de verästelt; verzweigt
da grenet
sv förgrenad
es ramificado
it ramificato

3028
RAMIFY, TO;
TO BRANCH OUT
fr se ramifier
ne vergaffelen; zich vertakken
de sich verästeln; sich verzweigen
da forgren sig
sv förgrena sig
es ramificarse
it ramificarsi

3029
RAMPANT; CREEPING
fr rampant
ne kruipend
de kriechend
da krybende
sv krypande
es rastrero
it rampicante

3030
RANDOM SAMPLE
fr échantillon *m*
ne steekproef
de Stichprobe *f*
da stikprøve
sv stickprov
es muestra *f*
it prova *f* a caso

3031
RANDOM SOWING;
SCATTER, TO
fr semer à la volée
ne uitstrooien
de aussäen; ausstreuen
da udstrøning

sv strö ut
es diseminar
it spargere

3032
RANUNCULUS
fr renoncule *f*
ne ranonkel
de Ranunkel *f*
da ranunkel
sv ranunkel
es ranúnculo *m*
it ranuncolo *m*
   Ranunculus L.

RASP, 1645

3033
RASPBERRY
fr framboise *f*
ne framboos
de Himbeere *f*
da hindbær *n*
sv hallon
es frambuesa *f*
it lampone *m*

3034
RASPBERRY APHID
fr puceron *m* vert du framboisier
ne kleine frambozeluis
de kleine Himbeerblattlaus *f*
da lille hindbærbladlus
sv hallonbladlus
es pulgón *m* de la frambuesa
it afide *m* verde minore del lampone
   Aphis idaei v.d. Goot.

3035
RASPBERRY BEETLE
fr ver *m* des framboises
ne frambozekever
de Himbeerkäfer *m*
da hindbærbille
sv hallonänger; hallonmask
es gorgojo *m* de la frambuesa
it verme *m* del lampone
   Byturus tomentosus de G.

3036
RASPBERRY BUSH
fr framboisier *m*
ne frambozestruik

de Himbeerstrauch *m*
da hindbærbusk
sv hallonbuske
es frambueso *m*
it lampone *m*
    Rubus idaeus L.

3037
RASPBERRY CANE
MIDGE
fr cécidomie *f* de l'écorce
ne frambozeschorsgalmug
de Himbeerrutengallmücke*f*
da hindbærstængelgalmyg
sv hallongallmygga
es mosca *f* de agalla de la
    corteza del frambuesa
it cecidomia *f* dei rami del
    lampone
    Thomasiniana theobaldi
    Barnes

3038
RASPBERRY LEAF AND
BUD MITE; DRYBERRY
MITE (US)
fr Phyllocoptes *m* gracilis
ne frambozegalmyt
de Himbeerblattgallmilbe *f*
da hindbærgalmide
sv hallongallkvalster *n*
es ácaro *m* de la frambuesa
it eriofide *m* del lampone
    Phyllocoptes gracilis Nal.;
    Aceria g. (UK)

3039
RAT
fr rat *m*
ne rat
de Ratte *f*
da rotte
sv råtta
es rata *f*
it ratto *m*
    Mus rattus L.

RAT, MUSK, 2433

3040
RATE; DOSAGE
fr dosage *m*
ne dosering
de Dosierung *f*

da dosering
sv dosering
es dosificación *f*
it dosatura *f*

3041
RATE OF GERMINATION
fr vitesse *f* germinative
ne kiemsnelheid
de Keimschnelligkeit *f*;
    Keimernergie *f*
da spirehastighed
sv groningshastighet
es vigor *m* germinativo
it rapidità *f* germinativa

3042
RATSBANE
fr mort *f* aux rats
ne muizetarwe
de Mäusekorn *n*
da musekorn *n*
sv råttgift *n*
es veneno *m* para ratones
it topicida *f*; rodenticida *f*

3043
RATTLE (DISEASE)
fr maladie *f* du rattle
ne ratel
de Mauke *f*
da rattle
sv rattle
es rayado *m* y encrespa-
    miento *m*
it (malattia *f* del) rattle

3044
RATTLE VIRUS
fr virus *m* rattle
ne ratelvirus *n*
de Ratelvirus *n*
da rattlevirus *n*
sv rattlevirus
es virus *m* de sonajero

3045
RAW MATERIALS
fr matières *f* premières
ne grondstoffen
de Grundstoffe *m pl*;
    Rohstoffe *m pl*
da råstoffer *n pl*
sv råmaterial *n pl*

es materias *f pl* primas
it materie *f pl* prime

3046
RAY BLIGHT
(CHRYSANTHEMUM)
fr Ascochyta *f*
ne ascochyta-ziekte
de Ascochyta-Krankheit *f*
da chrysanthemum-sortråd
sv Ascochyta
es Ascochyta *f*
it Ascochyta *f*
    Didymella ligulicola

3047
RAY FLORET
fr demi-fleuron *m*; fleur *f*
    ligulée
ne straalbloem
de Strahlenblüte *f*
da stråleblomst
sv strålblomma
es flor *f* radial
it fiore *m* raggiato

3048
RAYLESS MAYWEED;
PINEAPPLE WEED
fr matricaire *f*
ne schijfkamille
de strahllose Kamille *f*
da skive-kamille
sv gatkamomill
es falsa manzanilla *f*
it camomilla *f* selvatica
    Matricaria matricarioides
    (Less.) Porter

3049
READY MONEY
fr encaisse *f*
ne kasgeld *n*
de Bargeld *n*
da kontant beholdning
sv kontanter
es dinero *m* en metálico
it denaro *m* in cassa

RE-ALLOCATION, 2027

3050
REAL VALUE (ec.)
fr valeur *f* d'usage
ne gebruikswaarde
de Gebrauchswert *m*

da brugsværdi
sv bruksvärde *n*
es valor *m* de uso
it valore *m* reale

REAP, TO, 1734

**3051
RECEPTACLE; TORUS**
fr réceptacle *m*
ne bloembodem
de Blütenboden *m*
da blomsterbund
sv blombotten
es receptaculo *m*
it ricettacolo *m*

**3052
RECESSIVE**
fr récessif
ne recessief
de rezessiv
da recessiv
sv recessiv
es recesivo
it recessivo

**3053
RECIPROCATING
SPOUT FERTILIZER
BROADCASTER**
fr distributeur *m* d'engrais à
    tube projecteur
ne pendelstrooier
de Pendelstreuer *m*
da pendul-kunstgødnings-
    spreder
sv gödselspridare med
    pendelrör
es distribuidor *m* de abonos
it spandiconcime *m* a brac-
    cio oscillante

**3054
RECIPROCATING
THINNER**
fr pré-démarieuse*f* à couteau
    oscillant
ne slingerdunner
de Pendellichter *m*;
    Ausdünner *m*
da penduludtynder
sv uttunnare av pendeltyp

es separadora *f* de cuchillo
    oscillante; achoradora *f*
    de semillero
it diradatrile *f* a utensili
    oscillanti

**3055
RECIRCULATION
SYSTEM OF AIR
(MUSHROOM GROWING)**
fr système *m* de recirculation
    d'air; système *m* de bras-
    sage d'air
ne luchtcirculatiesysteem *n*
de Umluftanlage *f*;
    Zirkulationssystem *n*
da recirkulationsanlæg *n*
sv luftcirkulationssystem *n*
es sistema *f* de recirculación
    del aire
it sistema *f* di circolazione
    d'aria

**3056
RECLAIM, TO**
fr défricher; assécher
ne ontginnen; droogleggen
de urbar machen; gewinnen
da kultivere; tørlægge
sv uppodla; utnyttja;
    törlägga
es roturar; explotar; des-
    montar
it dissodare

RED BEET, 292

**3057
RED BUD BORER**
fr cécidomyie *f* des greffes
ne oculatiegalmug
de Okuliergallmücke *f*
da okuleringsgalmyg
sv okulagegallmygga
es agalla *f* de los injertos del
    rosal
it cecidomia *f* degli innesti
    Thomasiniana oculiperda
    Rübs.

**3058
RED CABBAGE**
fr chou-rouge *m*
ne rode kool

de Rotkraut *n*; Rotkohl *m*
da rødkål
sv rödkål
es lombarda *f*; repollo *m*
    morado
it cavolo *m* rosso
    Brassica oleracea L.
    cv. Rubra DC.

**3059
RED CLOVER**
fr trèfle *m* violet
ne rode klaver
de Rotklee *m*
da rødkløver
sv rödklöver
es trébol *m* encarnado
it trifoglio *m* rosso
    Trifolium pratense L.

**3060
RED CURRANT BLISTER
APHID; CURRANT
APHID (US)**
fr puceron *m* jaune du
    groseiller
ne bloedblaarluis
de Johannisbeerblasenlaus
da ribsbladlus
sv vinbärsbladlus
es pulgón *m* amarillo del
    grosellero
it afide *m* giallo del ribes
    Cryptomyzus ribis L.

**3061
REDDISH PURPLE**
fr pourpre
ne purperrood
de purpurrot
da purpurrød
sv purpurröd
es púrpura
it rosso porporino

**3062
RED FESCUE**
fr fétuque *f* rouge
ne rood zwenkgras *n*
de roter Schwingel *m*
da rød svingel
sv rödsvingel
es festuca *f* roja
it Festuca *f* rubra
    Festuca rubra L.

3063
RED FIRE DISEASE
(AMARYLLIS)
fr maladie f des taches rou-
    ges
ne vuur n
de Stengelbrenner m
da bladpletsyge
sv rödfläcksjuka
es 'fuego' m rojo
it macchie m pl rosse

RED GEOTRICHEM,
2187

3064
RED HOT POKER
fr tritoma m; faux-aloès m
ne vuurpijl
de Fackellilie f
da raketblomst
sv fackellilja
es tritoma f
it tritoma f
    Kniphofia Moench.

3065
RED LEAD
fr minium m
ne menie
de Mennig m
da mønje
sv mönja
es minio m; azascón m
it minio m

3066
RED-LEGGED EARTH
MITE; WINTER GRAIN
MITE (US)
fr Penthaleus m major
ne andijviemijt
de Wintergetreidemilbe f
da mide
sv endiviakvalster
es ácaro m de la escarola
it acaro m dell'insalata
    Penthaleus major Dug.

3067
RED OAK
fr chêne m d'Amérique
ne Amerikaanse eik
de amerikanische Roteiche f

da Amerikansk rødeg
sv rödek
es roble m americano
it quercia f americana
    Quercus rubra L.

3068
RED PEPPER
fr piment m
ne Spaanse peper
de spanischer Pfeffer m;
    Paprika m
da spansk peber
sv spansk peppar
es guindilla f; ají m
it pimento m
    Capsicum annuum L.

3069
RED POPLAR LEAF
BEETLE
fr chrysomèle f du peuplier
ne populierehaan
de Pappelblattkäfer m
da pilebladbille
sv aspglansbagge
es crisomela f del chopo
it crisomela f del pioppo
    Melasoma populi L.

3070
REDSHANK; WILLOW
WEED
fr renouée f persicaire
ne perzikkruid n
de Floh-Knöterich m
da ferskenpileurt
sv akerpilört
es pimentilla f; durazmillo f
it persicaria f
    Polygonum persicaria L.

RED SPIDER, fruit tree,
1487

3071
RED SPIDER MITE;
TWO-SPOTTED SPIDER
MITE (US)
fr araignée f rouge;
    tétranyche m tisserand
ne rode spin; bonespintmijt;
    spint

de rote Spinne f; gemeine
    Spinnmilbe f; Bohnen-
    spinnmilbe f
da rødt spind; spindemide
sv rött spinn; växthusspinn-
    kvalster n
es ácaro m blanco de las
    habas; araña f roja;
    arañita f de las legumbres
it ragno m rosso; ragnetto
    m giallo; tetranichide m
    Tetranychus urticae Auct.

3072
REDUCE, TO
fr réduire
ne reduceren
de reduzieren
da reducere
sv reducera; minska
es reducir
it ridurre

3073
REDUCING VALVE
fr vanne f réductrice
ne reduceerklep
de Reduzierventil n
da reduktionsventil
sv reduceringsventil
es válvula f reductora
it valvola f di riduzione

3074
REDUCTION
fr réduction f
ne reductie
de Reduktion f
da reduktion
sv reduktion
es reducción f
it riduzione f

3075
REDUCTION DIVISION;
MEIOSE
fr meiose f
ne reductiedeling
de Reduktionsteilung f
da reduktionsdeling
sv reduktionsdelning
es división f reductora
it meiose f

3076
REED
fr roseau m
ne riet n
de Rohr n; Schilf n
da rør n; siv n
sv vass; bladvass
es junco m
it canna f
Phragmites communis
Trin.

REED, giant, 1570
—, great 1570

3077
REED MACE
fr massette f
ne lisdodde
de Rohrkolben m
da dunhammer
sv kaveldun n
es anea f espadaña
it canna f pannocchiuta
Typha L.

3078
REED MAT
fr paillasson m en roseau
ne rietmat
de Rohrdeckef; Schilfmattef
da rørmåtte
sv vassmatta
es estera f de junco
it stuoia f

3079
REED PEAT
fr tourbe f à roseau
ne rietveen n
de Schilftorf m
da tagrørsmose
sv vasstorv
es turba f tifacea
it torbiera f di canne

3080
REEL
fr dévidoir m; bobine f
ne haspel
de Haspel m
da rulle; vinde
sv haspel

es devanadera f
it argano m

3081
REFERENCE PRICE
fr prix m de référence
ne referentieprijs
de Referenzpreis m
da referencepris
sv referenspris n
es precio m de referencia
it prezzo m di referenza

3082
REFLECTION
fr réflexion f
ne reflectie; terugkaatsing
de Reflexion f
da reflection
sv reflexion, återkastning
es reflexión f
it reflessione f

3083
REFLECTOR
fr réflecteur
ne reflector
de Reflektor m
da reflektor
sv reflektor
es reflector m
it riflettore m

REFRACTORY, 1354

3084
REFRIGERATED
DISPLAY CASE;
REFRIGERATED
SHOW-CASE
fr presentoir m frigorifique
ne koelvitrine
de gekühlte Schauvitrine f
da køledisk
sv kyldisk
es vitrina f frigorífica
it banco (frigorifero) m

3085
REFRIGERATED HOLD
fr cale f frigorifique
ne koelruim n
de Kühlraum m
da kølerum n

sv kylrum n
es bodega f refrigerada
it cella f frigorifica;
stiva f frigorifica

REFRIGERATED ROOM,
745

REFRIGERATED
SHOW-CASE, 3084

3086
REFRIGERATING
MACHINE
fr machine f frigorifique
ne koelmachine
de Kühlmaschine f
da kølemaskine
sv kylmaskin; kylaggregat n
es máquina f frigorífica
it macchina f frigorifera

3087
REGAL LILY
fr lis m royal
ne koningslelie
de Königslilie f
da kongelilje
sv kungslilja
es azucena f real
it giglio m regale
Lilium regale Wils.

3088
REGAL PELARGONIUM
fr 'Pelargonium' m
ne Franse geranium
de Edelpelargonie f
da Engelsk pelargonie
sv vanlig pelargon
es pelargonio m
it geranio m francese
Pelargonium grandiflorum
Willd.

3089
REGIONAL DEVELOP-
MENT CAMPAIGN
fr campagne f d'améliora-
tion régionale
de Gegendverbesserung f;
Gegendumgestaltung f
da regional forbedrings-
aktie

sv regional förbättrings-
kampanj
es mejora f regional
it miglioramento m regio-
nale

3090
REGULAR WORKER
fr ouvrier m permanent
ne vaste werknemer
de Arbeiter m in Dauer-
stellung
da faste medarbejdere
sv fast arbetskraft
es obrero m permanente
it salariato m fisso

3091
REGULAR WORKMAN
fr ouvrier m fixe; ouvrier m
à la journée
ne vaste arbeider
de ständige Arbeitskraft f
da fast (ansat) arbejder
sv fast arbetare
es obrero m fijo
it salariato m fisso

3092
REINFORCED PLASTIC
FILM
fr plastique m armé
ne gewapend folie
de verstärkte Kunststof-
folie f
da armeret folie
sv amerad folie
es plástico m armado
it foglio m in plastica rin-
forzata

3093
REJECT (PROCESSING)
fr écart m
ne afwijkend
de Ausschuss m
da afviger
sv avvikande
es destrío m; pacotilla f
it scarto m

3094
RELATIVE HUMIDITY
fr humidité f relative
ne relatieve vochtigheid

de relative Feuchte f;
relative Feuchtigkeit f
da relative fugtighed
sv relativ fuktighet
es humedad f relativa
it umidità f relativa

3095
RELAY
fr relais m
ne relais n
de Relais n
da relae n
sv relä n
es ralé m
it relè m

3096
REMEDY
(DISEASE CONTROL)
fr moyen m de lutte; moyen
m de prophylaxie; produit
m antiparasitaire; princi-
pe m de lutte; remède m
ne middel m; bestrijdingsmiddel m
de Mittel n
da middel n; bekæmpelses-
middel n
sv medel n (bekämpnings-
medel)
es medio m; remedio m
it mezzo m

3097
REMOVABLE
fr amovible; démontable
ne afneembaar
de abnehmbar
da aftagelig
sv avtagbar
es desmontable
it amovibile

3098
REMOVE TURF, TO;
TO CUT TURF
fr découper des plaques de
gazon
ne afplaggen
de abplaggen; Plaggen ab-
schälen
da fjerne græstørv
sv flåhacka

es cortado m de céspedes
it scavare zolle

3099
RENEWED GROWTH
fr recrû m
ne hergroei
de Wiederwuchs m
da genvækst
sv nyttoväxt
es retoño m
it ricrescita f

3100
RENIFORM;
KIDNEY-SHAPED
fr réniforme
ne niervormig
de nierenförmig
da nyreformet
sv njurformig
es reniforme
it reniforme

3101
REPLACEMENT VALUE
fr valeur f de remplacement
ne vervangingswaarde
de Wiederbeschaffungswert
m
da genanskaffelsesværdi
sv återanskaffningsvärde n
es valor m de sustitución
it valore m di sostituzione

3102
REPLACE PLANTS, TO
fr combler; compléter
ne inboeten
de nachpflanzen
da efterplante
sv komplettera
es rellenar
it rimettere

REPOT, TO, 2879

3103
REQUIREMENTS FOR
ELIGIBILITY
fr conditions f pl d'admission
ne erkenningseisen
de Anerkennungs-
bedingungen f pl

da adgangsbetingelser
sv inträdesfordringar
es condiciones *fpl* de
   admisión
it esigenze *fpl* di riconos-
   cenza

3104
RESEARCH
fr recherche *f*
ne onderzoek *n*
de Untersuchung *f*
da undersøgelse
sv forskning; undersökning
es investigación *f*
it ricerca *f*

RESEARCHER, 3106

3105
RESEARCH STATION;
EXPERIMENTAL
STATION
fr station *f* de recherches
ne proefstation *n*
de Forschungsanstalt *f*
da forsøgsstation
sv försöksstation
es estación *f* experimental
it stazione *f* sperimentale

3106
RESEARCH WORKER;
RESEARCHER
fr chercheur *m* scientifique
ne onderzoeker
de Untersucher *m*
da forsker
sv forskare
es investigador *m*
it ricercatore *m*

3107
RESIDUAL VALUE
fr valeur *f* résiduelle
ne residu-waarde
de Restwert *m*
da restværdi
sv restvärde *n*
es valor *m* residual
it valore *m* residuo

3108
RESIDUE
fr résidu *m*
ne residu *n*
de Rückstand *m*
da rest; restkoncentration
sv bottensats; rest
es residuo *m*
it residuo *m*

3109
RESIDUE OF IGNITION
fr cendres *f pl*
ne gloeirest
de Glührückstand *m*
da gløderest
sv glödrest
es cenizas *f pl*
it resto *m* dell'incandescen-
   za

3110
RESISTANCE
fr résistance *f*
ne resistentie
de Resistenz *f*; Widerstands-
   fähigkeit *f*
da resistens
sv resistens
es resistencia *f*
it resistenza *f*

3111
RESISTANT
fr résistant
ne resistent
de resistent
da resistent
sv resistent
es resistente
it resistente

3112
RESPIRATION
fr respiration *f*
ne ademhaling
de Atmung *f*
da ånding
sv andning; respiration
es respiración *f*
it respirazione *f*

3113
RESTORE, TO;
(ORCHARD)
fr restaurer
ne verjongen
de neubepflanzen
da forynge
sv föryngra
es replantar
it rinnovare un frutteto

3114
RESULTS OF
EXPERIMENTS; TRIAL
RESULTS
fr résultats *m pl* expérimentaux
ne proefresultaten *n pl*
de Versuchsergebnisse *n pl*
da forsøgsresultater *n pl*
sv försöksresultater *n pl*
es resultados *m pl* de
   ensayos
it esiti *m pl* sperimentali

3115
RETAILER
fr détaillant *m*
ne kleinhandelaar
de Einzelhändler *m*
da detailhandler
sv detaljist; detaljhandlare
es detallista *m*
it dettagliante *m*

3116
RETARD (IN GROWTH)
fr retard *m*
ne achterstand
de Rückstand *m*
da væksthæmning
sv växthämning
es retraso *m*
it vegetazione *f* arretrata

3117
RETARD, TO
fr retarder
ne verlaten
de verspäten
da forsinke
sv fördröja; retardera
es retrasar
it ritardare

3118
RETARDED BULBS
fr bulbes *m pl* retardés
ne geremde bollen
de zurúckgehaltene Zwiebeln
    f pl
da hæmmede løg *n*
sv retarderade lökar
es bulbos *m pl* retardados
it bulbi *m pl* ritardati

RETURN, 4232

3119
RETURNABLES
fr emballage *m* consigné;
    emballage *m* réutisable
ne meermalig fust *n*
de Dauerverpackung *f*;
    Leihpackung *f*
da returemballage;
    flergangsemballage
sv flergångsförpackning;
    returemballage
es embalaje *m* recuperable
it imballagio *m* vuoto a
    rendere

3120
RETURN VALVE
fr soupape *f* de décharge
    tardée
ne terugslagklep
de Rückschlagventil *n*
da tilbageløbsventil
sv backventil
es válvula *f* de retorno
it valvola *f* di scarico

3121
REVERSIBLE PLOUGH
fr brabant *m*
ne wentelploeg
de Wendepflug *m*
da vendeplov
sv vändplog
es arado *m* reversible
it aratro *m* voltolante

3122
RHIZOME
fr rhizome *m*
ne wortelstok
de Wurzelstock *m*

da rodstok
sv rotstock
es rizoma *m*
it rizoma *m*

3123
RHODODENDRON BUG;
RHODODENDRON LACE
BUG (US)
fr tigre *m* du rhododendron
ne Japanse vlieg
de amerikanische Rhodo-
    dendronwanze *f*
da Stephanitis
sv rhododendron-nätstink-
    fly
es chinche *f* del bujo
it tigre *m* del rododendro;
    stephanitis *m*
    Stephanitis rhododendri
    Horv.

3124
RHOMBIC
fr rhomboïdal
ne ruitvormig
de rautenförmig
da rudeformet
sv rugformig
es romboidal
it rombico; romboidale

3125
RHUBARB
fr rhubarbe *f*
ne rabarber
de Rhabarber *m*
da rabarber
sv rabarber
es ruibarbo *m*
it rabarbero *m*

RHUBARB, Chinese, 664

3126
RHUBARB MOSAIC
fr mosaïque *f* de la rhubarbe
ne virusvlekkenziekte
    (rabarber)
de Rhabarbermosaik *n*
da rabarber-mosaik
sv rabarbermosaik
es mosaico *m* del ruibarbo
it mosaico *m* del rabarbaro

3127
RIBBED
fr nervé
ne geribd
de gerippt
da ribbet
sv ådrig; nervig
es estriado; con nervadura
it costoluto

3128
RIDDLE; SIZE
fr calibre *m*
ne ziftmaat
de Siebgrösse *f*
da løgstørrelse
sv storlek
es calibre
it misura *f* d'affinamento

3129
RIDDLING SCREEN
fr crible *m*
ne ziftplaat
de Siebplatte *f*
da sorteringssold *n*
sv sorteringssåll *n*
es criba *f*
it tavolone *m* d'affinamento

3130
RIDGE
fr faîtage *m*
ne nok
de First *m*
da tagrygning
sv nock; takås
es caballete *m*
it colmo *m*

3131
RIDGE BED
fr meule *f*
ne heuvelbed *n*
de Hügelbeet *n*
da fransk bed
sv fransk bädd
es caballón *m*
it cumulo *m*

3132
RIDGE CAPPING
fr faîtière *f*
ne nokafdekking

de Firstpfette *f*
da rygplanke
sv nocktäckning
es cimera *f*; cobertura *f* de caballete
it copertura *f* di colmo

3133
RIDGE CUCUMBER; GHERKIN
fr cornichon *m*
ne augurk
de Einlegegurke *f*; Essiggurke *f*
da agurk
sv frilandsgurka; druvgurka
es pepino *m*
it cetriolo *m*
Cucumis sativus L.

3134
RIDGE CULTURE
fr culture *f* sur sillons
ne ruggenteelt
de Dammanbau *m*
da dyrke på rygge
sv odling på drill
es cultivo *m* en lomos
it coltivazione *f* in elevati strati di terra

3135
RIDGER
fr butteur *m*
ne aanaarder
de Häufler *m*
da hyppeplov
sv kupbill; kupningsaggregat
es arado *m* para aporcar
it rincalzatore *m*

RIDGE UP, TO, 1193

3136
RIGID TINE CULTIVATOR
fr cultivateur *m* à dents rigides
ne cultivator met stijve tanden
de Grubber *m* mit starren Zinken
da stivtandetkultivator

sv styvpinnkultivator; kultivator med stela pinnar
es cultivador *m* de dientes rígidos
it coltivatore *m* a denti rigidi

3137
RING FORMATION (SPRAING OR CORKY RINGSPOT IN POTATO TUBERS)
fr formation *f* de cernes d'anneaux (tubercule de pomme de terre)
ne kringerigheid
de Kringelbildung *f*; Ringbildung *f*
da kartoffel-rustring
sv rostringar
es asalchichonado *m*
it maculatura *f* anulare tuberosa

3138
RINGING
fr marcottage *m* aérien
ne marcotteren
de abschnüren mit Draht
da aflægning
sv avläggning
es acodo *m* al aire
it margottare

3139
RING SPOT (BRASSICA)
fr taches *f pl* foliaires
ne bladvlekkenziekte
de Blattfleckenkrankheit *f*
da kålens bladpletsyge
sv bladfläcksjuka
es punteado *m* de las hojas
it macchie *f pl* fogliari
Mycosphaerella brassicicola Ces. et De Not.

3140
RING SPOT (DIANTHUS)
fr Didymellina *f*
ne spat
de Nelkenschwärze *f*
da ringplet

sv nejliksvärta
es punteado *m*
it nero *m*
Didymellina dianthi C. C. Burt

3141
RING SPOT (ENDIVE)
fr taches *f pl* brunes
ne vuur *n*
de Blattfleckenkrankheit *f* und Fäulnis *f*
da bladpletsyge
sv bladfläcksjuka
es encendido *m*
it marciume *m*
Marssonina panattoniana (Berl.) Magn.

3142
RINGSPOT (VARIOUS PLANTS)
fr maladie *f* de taches *f pl* en anneau (de taches annulaires)
ne kringvlekkenziekte
de Ringfleckenkrankheit *f*
da ringplet
sv ringfläcksjuka
es manchas *f pl* anulares
it maculatura *f* anulare

RINSE, TO, 4064

3143
RINSE WELL, TO
fr rincer
ne doorspoelen
de durchspülen
da gennemspule
sv genomspola
es enjuagar
it pulire

3144
RIPE; MATURE
fr mûr
ne rijp
de reif
da moden
sv mogen
es maduro
it maturo

3145
RIPEN, TO
fr mûrir; aoûter
ne afrijpen
de ausreifen
da afmodne; modne
sv avmogna
es madurar
it maturare

3146
RIPENESS; MATURITY
fr maturité f
ne rijpheid
de Reife f
da kvist
sv mognad
es madurez f; sazón f
it maturità f

3147
RIPENING (SOIL)
fr maturation f
ne rijping
de Reifung f
da modning
sv 'mogning'
es maduración f
it maturazione f

RIPENING, blotchy,
379, 380

3148
RIPENING ROOM
(PROCESSING)
fr mûrisserie f
ne stokerij; rijperij
de Reifekammer m
da modningsrum m
sv mogningsrum n
es cámara f de maduración
it camera f di maturazione

3149
RIVER CLAY
fr argile f alluvionnaire
ne rivierklei
de Flusston m
da flodler n
sv flodlera
es arcilla f aluvial
it argilla f fluviale

3150
ROCK; STONE
fr roches f pl
ne gesteenten n pl
de Gesteine n pl
da stenart; bjergart
sv bergarter
es rocas f pl
it pietrame m

3151
ROCKERY
fr rocaille f
ne rotspartij
de Felsenpartie f
da stenparti n
sv stenparti n
es jardín m rocoso
it scalata f

3152
ROCK GARDEN
fr jardin m de rocaille
ne rotstuin
de Steingarten m
da stenhave
sv klippträdgård
es jardín m rocoso
it giardino m di rocca

3153
ROCK PHOSPHATE
fr phosphate m naturel
ne ruw fosfaat n
de Rohphosphat n
da råfosfat n
sv råfosfat n
es fosfato m natural
it fosfato m naturale

ROCKPLANT, 73

ROD, 247

3154
RODENT
fr rongeur m
ne knaagdier n
de Nagetier n
da gnaver mf
sv gnagare
es roedor m
it roditore m
Rodentia

3155
ROGUE
(BULB GROWING)
fr mélange
ne dwaling
de Fehltyp m
(nicht sortenecht)
da iblandet
(anden sort)
sv inblandning
es mezcla f
it semenzale f inferiore

3156
ROGUE, TO
(BULB GROWING)
fr épurer
ne zuiveren
de Fehlfarben entfernen
da udvælge
sv avlägsna
es purificar
it eliminare le falsità

3157
ROGUING TUBE FOR
DISEASED BULBS
fr tarière f
ne snotkoker
de Zwiebelstecher m
da løgstikker
sv jordborr för upptagning
av sjuka lökar
es instrumento m para
arrancar bulbos
it strumento m per dissot-
terrare bulbi

3158
ROLL, TO
fr rouler
ne rollen
de walzen
da tromle
sv välta
es rodar; pasar el sodillo
it rotolare

3159
ROLLER
fr rouleau m
ne wals; rol
de Rolle f; Walze f
da tromle
sv vält

es rollo *m*; rodillo *m*
it rotolo *m*

3160
ROLLER CONVEYOR
fr convoyeur *m* à rouleaux
ne rollenbaan
de Rollenbahn *f*
da rullebane
sv rullbana
es transportador *m* de
   rodillos
it trasportatore *m* a rulli

3161
ROLLER PUMP
fr pompe *f* à rouleaux
ne rollenpomp
de Rollenpumpe *f*
da rotationspumpe
sv kapselpump; pump med
   rullar
es bomba *f* rotativa
it pompa *f* rotativa

3162
ROLLER-TYPE HAND
HOE
fr binette *f* roulante
ne rolschoffel
de Rollhacke *f*
da hjulhakke
sv handhacka med vält
   (rulle); skjuthacka
es binadora *m* de mano rodado; almocafre *m* de
   jardinero
it rullo *m* frangizolle

3163
ROOF CLADDING
fr voligeage *m*
ne dakbeschot *n*
de Verschalung *f*
da forskalling
sv takbeläggning
es entarimado *m*; artesena-
   do *m*
it lastra *f*

3164
ROOFING FELT
fr feutre *m* toiture
ne dakvilt *n*
de Dachfilz *m*

da tagfilt *n*
sv takpapp *n*
es cartón *m* embreado
it feltro *m* bituminato per
   copertura; cartone *m*
   cartramato

ROOF SLOPE, 2752

ROOM OUT, TO, 3523

3165
ROOT
fr racine *f*
ne wortel (bot.)
de Wurzel *f*
da rod
sv rot
es raíz *f*
it radice *f*

ROOT, aerial, 37
—, bean black, 273
—, breathing, 428
—, button snake, 2147
—, club, 719
—, contractile, 819
—, corky, 845
—, hair, 1705
—, lateral, 2044
—, main, 2247
—, prop, 2955
—, sucking, 3714
—, tap, 3791

3166
ROOT, TO
fr prendre racine
ne aanslaan (van planten);
   wortelschieten
de einwurzeln; Wurzel fassen
fa slå rod
sv rota sig
es arraigar
it radicarsi

3167
ROOT ACTIVITY
fr activité *f* radiculaire
ne wortelactiviteit
de Wurzelaktivität *f*
da rodaktivitet
sv rotaktivitet
es actividad *f* radicular
it attivitá *f* radicale

3168
ROOT BALL
fr motte *f*
ne wortelkluit
de Erdballen *m*
da rodklump
sv rotklump
es cepellón *f* champa
it motta *f* della pianta

ROOT BUD, 3172

3169   ●
ROOT CAP
fr coiffe *f* de la racine
ne wortelmutsje *n*
de Wurzelhaube *f*
da rodhætte
sv rotmössa
es cofia *f*; pilorriza *f*
it berrettina *f* radicale

3170
ROOT COLLAR;
ROOT NECK
fr collet *m* de la racine
ne wortelhals
de Wurzelhals *m*
da rodhals
sv rothals
es cuello *m* de la raíz
it colletto *m*

3171
ROOT CUTTING
fr bouture *f* de racine
ne wortelstek *n*
de Wurzelsteckling *m*
da rodstikling
sv rotstickling
es estaca *f* de raíz;
   esqueje *m* de raíz
it talea *f* radicale

3172
ROOT EYE;
ROOT BUD
fr drageon *m*
ne wortelknop
de Wurzelknospe *f*
da rodknop
sv rotknopp
es yema *f* radical
it occhio *m* radicale

3173
ROOT HAIRS
fr poils *m pl* radiculaires;
poils *m pl* absorbants
ne wortelharen *n pl*
de Wurzelhaare *n pl*
da rodhår *n pl*
sv rothår *n pl*
es pelos *m pl* radicales
it peli *m pl* succiatori

3174
ROOTING
fr enracinement *m*
ne beworteling
de Bewurzelung *f*
da roddannelse
sv rotning
es arraigar
it radicazione *f*

ROOTING (MUSHROOM
GROWING), 3913
—, deep, 1010

ROOTING BED, 2809

ROOTING THROUGH,
3178

3175
ROOT KNOB
NEMATODE
fr anguillule *f* des racines
noveuses
ne wortelknobbelaaltje *n*
de Krätzeälchen *n*
da rodgallenematod
sv rotgallnematod
es anguilula *f* de las agallas
radiculares
it nematode *m* galligeno
delle radici
Meloidogyne spp.

3176
ROOT LESION NEMATO-
DES; FREE LIVING EEL-
WORMS; MEADOW NE-
MATODES
fr nématodes *m pl* libres;
anguillules *f pl* des
prairies
ne vrijlevende aaltjes *n pl*
de freilebende Aelchen *n pl*;
Wiesennematodes *f pl*

da fritlevende nematoder
sv fritt levanda nematoder
es nematodos *m pl* de los
prados
it anguillulas *f pl* dei prati;
nematodi *m pl* liberi
Pratylenchus spp.

ROOT NECK, 3170

3177
ROOT PATTERN
fr système *m* radiculaire
ne wortelbeeld *n*
de Wurzelbild *n*
da rodmønster *n*
sv rotmönster *n*
es imagen *f* radicular
it disegno *m* radicale

3178
ROOT PENETRATION;
ROOTING THROUGH
fr enracinement *m*;
pénétration *f* des racines
ne doorworteling
de Durchwurzelung *f*
da gennemtrængningsevne
sv genomrotning
es penetración *f*
(de las raíces)
it penetrazione *f* radicale

3179
ROOT ROT
fr pourriture *f* des racines;
piétin *m*; fonte *f*
ne wortelrot *n*; wortelbrand
de Wurzelfäule *f*
da rodråddenskab
sv rotröta
es podredumbre *f* radical;
quemazón *m* de las raíces
it marciume *f* radicale;
putrefazione *f* radicale
Cylindrocarpon radicicola
Wr.; *Pythium ultimum*
Trow.

3180
ROOT ROT
(ASPARAGUS)
fr fusariose *f*
ne voetziekte
de Fusskrankheit *f*

da slimskimmel
sv stenmögel
es mal *m* del pie
it fusariosi *f*
Fusarium spp.

ROOT ROT,
armillaria, 149
—, black, 345
—, phytophthora, 2713
—, violet, 4039

3181
ROOTSTOCK
fr sujet *m*; porte-greffe *m*
ne onderstam
de Unterlage *f*
da grundstamme; underlag *n*
sv grundstam; underlag *n*
es portainjerto *m*; patrón *m*
it portinnesto *m*; soggetto *m*

ROOTSTOCK, medium
vigorous, 2322
—, vigorous, 4033
—, weak, 4104

3182
ROOTSTOCK SHOOT
(MELON)
fr tirsève *f* du sujet
ne onderstamscheut
de Unterstammspross *m*
da vildskud *n*; grundstam-
meskud *n*
sv vildskott *n*
es hijuelo *m*
it germoglio *m* del portinnesto

3183
ROOT SUCKER
fr drageon *m*
ne wortelscheut
de Wurzeltrieb *m*
da rodskud *n*
sv rotskott *n*
es brote *m* radicular
it pollone *m*

3184
ROOT SYSTEM
fr sytème *m* de racine;
système *m* radiculaire
ne wortelbeeld *n*; wortelgestel *n*; beworteling

de Wurzelbild *n*; Wurzel-
  werk *n*; Bewurzelung *f*
da rodbillede
sv rotsystem *n*; rötter
es sistema *f* radicular;
  imagen *f* de la raíz
it sistema *f* radicale;
  figura *f* radicale

3185
ROSE
fr rose *f*
ne roos
de Rose *f*
da rose
sv ros
es Rosa *f*
it Rosa *f*
  Rosa L.

3186
**ROSE (OF WATERING
CAN)**
fr pomme *f* d'arrosoir
ne broes
de Brause *f*
da bruse
sv stril
es roseta *f* de la regardera
it cipolla *f* (d'annafiatoio)

ROSE, Christmas,   678
—, dog,   1097
—, sun,   3736

ROSE ANTHRACNOSE,
2095

ROSE BAY,   2519

3187
ROSE   BUSH
fr rosier *m*
ne rozestruik
de Rosenstrauch *m*
da rosenbusk
sv rosbuske
es rosal *m*
it rosaio *m*
  Rosa L.

3188
ROSE LEAFHOPPER
fr cicadelle *f* du rosier
ne rozecicade

de Rosenzikade *f*
da rosencikade
sv rosenstrit
es polilla *f* de la cascara del
  manzano
it cicadina *f* delle rose
  Typhlocyba rosae L.

ROSE LEAF SAWFLY,
2093

ROSE LEAF SPOT,   2095

3189
ROSE MALLOW;
CHINA ROSE
fr ketmie *f*; hybiscus *m*
ne Chinese roos; Hibiscus
de Eibisch *m*
da Hawaiiblomst
sv hibiskus
es rosa *f* de China;
  hibisco *m*
it rosa *f* sinensis
  Hibiscus L.

3190
ROSE MOSAIC
fr mosaïque *f* de la rose
ne rozemozaïek *n*
de Rosenmosaik *n*
da rose-mosaik
sv rosmosaik
es mosaico *m* de la rosa
it mosaico *m* della rosa

3191
ROSE TORTRIX MOTH
fr tordeuse *f* verte
ne heggebladroller
de Heckenwickler *m*
da cacoecia rosana
sv häckvecklare
es oruga *f* cigarrera de los
  frutales
it cacecia *f*
  Cacoecia rosana L.;
  Archips r. L. (UK)

3192
ROSETTING
fr formation *f* de rosettes
ne rozetvorming
de Rosettenbildung *f*
da rosetformet

sv rosettbildning
es formación *f* de rosetas
it sviluppo *m* a rosetta;
  formazione *f* di rosette

3193
ROSY APPLE APHID
fr puceron *m* rose du
  pommier
ne rose appelluis
de rosige Apfelblattlaus *f*
da røde æblebladlus
sv röd äppelbladlus
es pulgón *m* rosado del
  manzano
it afide *m* grigio-rosa del
  melo
  Sappaphis plantaginea
  Pass.; *Dysaphis p.* (UK)

3194
ROT
fr pourriture *f*
ne rot *n*
de Fäule *f*
da råd *n*
sv röta
es putrefacción *f*
it putrefazione *f*

ROT, bacterial,   215, 219
—, basal,   256
—, bitter,   324
—, black, 346, 347, 348, 3276
—, blossom-end,   380
—, Botrytis,   410
—, brown,   449
—, crown,   915
—, dry,   1153, 1154, 1155
—, foot,   1423
—, fomes,   1419
—, fusarium,   1511, 1512,
1513
—, fusarium corm,   1510
—, grey bulb,   1656
—, hard,   1726
—, heart,   1751, 1752
—, neck,   2454
—, phytophtora fruit,   2713
—, sclerotinia,   3276, 3277
—, slimy,   3438,
—, soft,   3438, 3468, 3469
—, storage,   3667

—, tuber, 3948
—, white, 3438, 4146
see also: collar rot;
root rot; stem rot

3195
ROT, TO; TO DECOMPO-
SE; TO PUTREFY
fr pourrir
ne bederven; rotten; verrot-
    ten
de faulen; verotten; verwe-
    sen
da rådne
sv ruttna
es descomponer; pudrir
it putrefarsi

3196
ROTARY ATOMIZING
BURNER
fr brûleur *m* à pulvérisation
    par coupelle rotative
ne roterende verstuivings-
    brander
de Drehbecherbrenner *m*;
    Rotationszerstäuber *m*
da rotationsfyr *n*
sv rotationsbrännare
es quemador *m* de atomiza-
    ción rotativo
it bruciatore *m* atomizzato-
    re rotativo

3197
ROTARY CULTIVATOR
fr houe *f* rotative
ne frees
de Rotakrümler *m*; Fräse *f*
da fræser
sv fräs; jordfräs
es cultivador *m* rotativo
it coltivatore *m* con utensili
    rotanti

3198
ROTARY DUSTER
fr poudreuse *f* à turbine
ne borstverstuiver
de Bauchstäuber *m*
da pudderblæser
sv pudra (puderspridare)
    som bäres på bröstet

es espolvoreador *m* de tur-
    bina
it impolveratrice *f* rotativa

3199
ROTARY MOWER;
ROTO SCYTHE
fr mototondeuse *f* à couteau
    horizontal
ne cirkelmaaier
de Sichelrasenmäher *m*
da rotationsplæneklipper;
    rotationsslåmaskine
sv rotorklippare; motor-
    gräsklippare
es segadora *f*; guadañadora
    f rotativa
it falciatrice *f* circolare

3200
ROTARY SCYTHE
fr faucheuse *f* rotative
ne cirkelmaaimachine
de Kreiselmäher *m*
da græsslåmaskine;
    roterende skær
sv gräsklippare rotor
es segadora *f* rotativa
it falciatrice *f* rotativa

3201
ROTARY SPRINKLER
fr arroseur *m* rotatif
ne draaiende sproeier
de Drehstrahlregner *m*
da sprinkler
sv roterande bevattnare
es aspersor *m* rotativo
it irrigatore *m* rotativo

3202
ROTARY TILLAGE
fr fraiser
ne frezen
de fräsen
da fræse
sv fräsa
es fresar
it fresare

ROTO SCYTHE,   3199

3203
ROTTEN; PUTRID
fr pourri
ne rot

de faul
da rådden
sv rutten
es podrido
it putrido

3204
ROUGH
fr rugueux; rude
ne ruw
de rauh
da rå
sv sträv
es áspero
it rozzo

3205
ROUGH (TOMATO)
fr fruit *m* cotelé; fruit *m* fascié
ne bonk
de Bunken *m*
da 'abe'; riflet tomatfrugt
sv räfflade frukter
es fruto *m* grueso
it qualità *f* membruta

3206
ROUGH PEAT
fr tourbe *f* brute
ne turfmolm
de Torfmull *m*
da tørvestrøelse
sv torvmull
es polvo *m* de turba
it polvere *f* di torba

3207
ROUGH SKIN (APPLE)
fr taches *f pl* liégeuses
    (pommier)
ne ruwschilligheid (appel)
de Rauhschaligkeit *f* (Apfel)
da skrubbethed; stukrevner
sv strävskallighet; rough skin
es aspereza *f* de la piel
    (manzana)
it rugginosità *f* ulcerosa;
    suberosità *f* della buccia
    (del melo)

3208
ROUGH-STALKED
MEADOWGRASS
fr paturin *m* commun
ne ruw beemdgras

de gemeines Rispengras *n*
da almindelig rapgræs *n*
sv betesgroë; kärrgroë
es espiguilla *f*
it Póa *f*
  Poa trivialis L.

ROUNDHEADED WOOD
BORER, 2209

ROUND SCOOP, 950

3209
ROUND SEEDED PEA
fr petit pois *m* lisse
ne rondzadige erwt
de Palerbse *f*
da rundfrø ært
sv trindfröig ärta
es guisante *m* verde liso
it cece *m* sferico

3210
ROW; LINE
fr rangée *f*; ligne *f*
ne rij
de Reihe *f*
da række
sv rad; räcka
es fila *f*; línea *f*; hilera *f*
it fila *f*

3211
ROW PLANTING
fr plantation *f* en rangées
ne rijenbeplanting
de Reihenpflanzung *f*
da rækkebeplantning
sv radplantering
es plantación *f* en hileras
it impianto *m* a filari

3212
ROW SPACING
fr écartement *m* des lignes
ne rijenafstand
de Reihenentfernung *f*;
  Reihenabstand *m*
da rækkeafstand
sv radavstånd
es distancia *f* entre líneas
it distanza *f* fra le files

3213
ROYAL FERN;
OSMUNDA
fr osmonde *f* royale
ne koningsvaren

de Königsfarn *m*
da kongebregne
sv kungsbräken
es helecho *m* real
it felce *f* florida
  Osmunda regalis L.

3214
RUBBERY WOOD
(APPLE)
fr bois *m* souple; bois *m*
  caoutchouc (pommier)
ne rubberhoutziekte (appel)
de Gummiholzkrankheit *f*
  (Apfel)
da gummived *n*
sv gummived
es enfermedad *f* de la madera gomosa
it mal *m* del caucciù (del
  melo)

3215
RUBUS APHID
fr Amphorophora *f*
ne grote frambozeluis
de grosse Brombeerblatt-
  laus *f*
da stor hindbærbladlus
sv stor hallonbladlus
es pulgón *m* de la frambuesa
it afide *m* maggiore del
  lampone
  Amphorophora rubi Kltb.

3216
RUGOSE; WRINKLED
fr ridé; rugueux
ne rimpelig
de runzlig
da rynket
sv fårad; rynkig
es arrugado
it grinzoso; rugoso

RUNCH, 4159

3217
RUNNER; OFFSHOOT;
SUCKER; STOLON
fr rejet *m*; stolon *m*
ne uitloper
de Ausläufer *m*
da sideskud *n*; udløber

sv skott;utlöpare
es estolón *m*
it stolone *m*; rampollo *m*;
  pollone *m*

3218
RUNNER BEAN
fr haricot *m* d'Espagne
ne pronkboon
de Feuerbohne *f*; Prunk-
  bohne *f*
da pralbønne
sv blomsterböna; rosenböna
es judia *f* verde trepadora;
  judia *f* escarlata
it fagiolo *m* di Spagna
  Phaseolus coccineus L.

RUNNING COSTS, 2533

3219
RUN TO SEED, TO
fr monter (en graine);
  feuiller
ne doorschieten; in zaad
  schieten
de durchschiessen
da vokse; skyde igennem;
  gå i frø
sv skjuta igenom; gå i frö
es espigar
it spigare

3220
RUSH
fr jonc *m*
ne rus
de Binse *f*
da siv
sv tåg
es junco *m*
it giunco *m*
  Juncus

3221
RUSSETING
fr rugosité *f*
ne vruchtverruwing
de Rauhschaligkeit *f*
da skrub
sv (frukt)korkrostbildning
es aspereza *f* de los frutos
it rugginosità *f* del frutto

3222
RUST
fr rouille *f*
ne roest
de Rost *m*
da rust
sv rost
es roya *f*
it ruggine *f*
    i.a.: Puccinia spp.; *Uro-
    myces* spp.; *Phragmidium*
    spp.

3223
RUST (PEAR)
fr rouille *f* grillagée
ne pereroest
de Birnengittersort *m*
da gitterrust
sv päronrost; gelérost
es roya *f* del peral

it ruggine *f* del pero
    Gymnosporangium sabinae
    (Dicks.) Wint.

3224
RUST (RIBES)
fr rouille *f* du groseillier
ne bekerroest
de Becherrost *m*
da stikkelsbær-skåltrust
sv krusbärsrost
es roya *f* del cáliz
it ruggine *f*
    Puccinia pringsheimiana
    Kleb.

RUST, black,   349
—, white,   4147

3225
RUSTY
fr rouillé
ne roestig

de rostig
da rusten
sv rostig
es con roya; roñoso;
    herrumbroso
it rugginoso

RUTABAGA,   3760

3226
RYEGRASS
fr ray-grass *m*
ne raaigras *n*
de Weidelgras *n*
da rajgræs *n*
sv rajgräs *n*
es ballico *m*; vallice *m*
it loglio *m*

RYEGRASS, Italian,   1963
—, perennial,   2672

S

3227
SACK BARROW;
PORTER'S TROLLEY
fr diable *m*
ne steekwagen
de Sackkarre *f*
da trækvogn
sv säckkärra
es carretilla *f*
it carello *m*

3228
SAFETY MARGIN
fr délai *m* de sécurité
ne veiligheidstermijn
de Karenzzeit *f*
da behandlingsfrist; karenstid
sv karenstid
es plazo *m* de seguridad
it margine *m* di sicurezza

3229
SAFETY VALVE
fr soupape *f* de sécurité
ne veiligheidsklep
de Sicherheitsklappe *f*
da sikkerhedsventil
sv säkerhetsventil
es válvula *f* de seguridad
it valvola *f* di sicurezza

3230
SAGE
fr sauge *f*
ne salie
de Salbei *m*
da salvie
sv Salvia; kryddsalvia
es Salvia *f*
it Salvia *f*
Salvia officinalis L.

3231
SAGITTATE
fr sagitté
ne pijlvormig
de pfeilförmig
da pileformet

sv pilformig
es sagital
it sagittato

SAINTPAULIA, 3518

3232
SALEABLE SIZE
(BULBS)
fr bulbes *m pl* vendables
ne leverbaar
de lieferbare Zwiebeln
da drivløg *n*
sv säljbara lökar *n pl*
es suministrable (bulbos)
it misura *f* vendible (bulbi)

3233
SALE VALUE
fr valeur *f* à la vente; valeur *f* vénale
ne verkoopwaarde
de Verkaufswert *m*
da salgsværdi
sv försäljningsvärde *n*
es valor *m* en venta
it valore *m* di vendita

3234
SALINITY
fr état *m* salifère
ne zouttoestand
de Salzverhältnis *n*
da salttilstand
sv salttillstånd *n*
es situación *f* alcalina; condición *f* alcalina
it salso *m*; salsedine *f*

3235
SALPETER
fr salpêtre *m*
ne salpeter
de Salpeter *m*
da salpeter
sv salpeter
es salitre *m*; nitro *m*
it nitrato *m* di potassa

3236
SALSIFY
fr salsifis *m*
ne haverwortel
de Haferwurz *f*
da havrerod
sv havrerot
es salsifi *m*
it sassifraga *f*
Tragopogon porrifolius L.

3237
SALT
fr sel *m*
ne zout *n*
de Salz *n*
da salt *n*
sv salt *n*
es sal *f*
it sale *m*

SALT, total soluble, 1160

3238
SALT CONTENT
fr teneur *f* en sel
ne zoutgehalte *n*
de Salzgehalt *n*
da saltindhold
sv salthalt
se contenido *m* en sal; salsedumbre *f*
it salso *m*; salsedine *f*

3239
SALVER SHAPED;
HYPOCRATERIFORM
fr orbiculaire
ne schaalvormig
de tellerförmig
da skålformet
sv skålformig
es en forma de bandeja; orbicular
it piattiforme; orbicolare

3240
SAMPLE
fr échantillon *m*
ne monster *n*; steekproef

de Probe *f*
da prøve
sv sampel; stickprov *n*
es muestra *f*
it campione *m*

3241
SAMPLE, TO;
TO TAKE A SAMPLE
fr prélever des échantillons;
échantillonner
ne monsters nemen;
bemonsteren
de Proben ziehen
da tage prøver
sv taga prov
es tomar muestras
it disporre campioni

3242
SAND
fr sable *m*
ne zand *n*
de Sand *m*
da sand *n*
sv sand
es arena *f*
it sabbia *f*

SAND, cover,   874
—, dune,   1163

3243
SAND WEEVIL
fr cneorrhinus *m* plagiatus
ne grijze bolsnuitkever
de grauer Kugelrüssler *m*;
Sandgraurüssler *m*
da roegnavere
sv betvivel
es escarabajo *m* gris de jeta
redonda
it Philopedon *m* plagiatus
Philopedon plagiatus
Schall.; *Cneorrhinus pl.*

3244
SANDY
fr sablonneux
ne zandig
de sandig
da sandet
sv sandig
es arenoso
it sabbioso

3245
SANDY CLAY
fr gravier *m*
ne zavel
de Kleisand *m*
da sandblandet ler *n*
sv sandblandad lerjord
es suelo *m* arcillo-arenoso
it terreno *m* argilloso-
sabbioso

3246
SANDY SOIL
fr terre *f* sablonneuse
ne zandgrond
de Sandboden *m*
da sandjord
sv sandjord
es tierra *f* arenosa
it suolo *m* sabbioso

SAP,   1983

SAP, flow of,   1407

3247
SAPROPHYTES
fr saprophytes *m pl*
ne saprofyten
de Saprophyten *m pl*
da saprofyter
sv saprofyt
es saprófitos *m pl*
it saprophyte *m pl*

SAP STREAM,   1407

3248
SAPWOOD
fr aubier *m*
ne spinthout *n*
de Splint *m*
da splintved *n*
sv splintved *n*
es albura *f*
it alburno *m*

3249
SASH
fr châssis *m*
ne raam *n*
de Fenster *m*
da vindue *n*; bænkevindue
sv fönster *n*
es ventana *f*
it finestra *f*

3250
SATIN MOTH
fr bombyx *m* du saule
ne satijnvlinder
de Weidenspinner *m*
da atlaskspinder
sv videspinnare; pilvitgump
es bombix *m* del sauce;
mariposa *f* plateada de
los álamos
it bombice *m* bianco del
salice; farfalla *f* bianca
del pioppo
Leucoma salicis L.; *Stil-
pnotia s.* (US)

3251
SATURATE, TO
fr saturer
ne verzadigen
de sättigen
da mætte
sv mätta
es saturar
it saturare

3252
SATURATION
fr saturation *f*
ne verzadiging
de Sättigung *f*
da mætning
sv mättning
es saturación *f*
it saturazione *f*

SATURATION, degree of,
1026

3253
SATURATION POINT
fr point *m* de saturation;
point *m* de satiété
ne verzadigingspunt *n*
de Sättigungspunkt *m*
da mætningspunkt *n*
sv mättnadspunkt *n*
es punto *m* de saturación
it punto *m* di saturazione

3254
SAUERKRAUT
fr choucroute *f*
ne zuurkool

de Sauerkraut *n*
da surkål
sv surkål
es choucroute *f*; chucrut *n*
it salcraut *m*

3255
SAVOY CABBAGE
fr chou *m* de Milan;
   chou *m* cabus frisé
ne savooiekool; groene kool
de Wirsingkohl; Herzkohl;
   Grünkohl
da Savoykål; grønkål
sv savoykål; grönkål
es col *f* de Milán; col *f* rizada
it cavolo *m* di Milano; cavolo *m* verzotto; verza *f*

3256
SAWFLY
fr tenthrède *m*; hoplo-
   campe *m*
ne zaagwesp
de Sägewespe *f*; Blattwespe *f*
da savhveps
sv växtstekel; bladstekel
es hoplocampa *f*; mosca *f* de
   sierra
it oplocampa *f*; tentredini-
   da *f*
   Tenthredinidae spp.;
   Hoplocampa spp.

SAWFLY, banded rose,
243
—, common gooseberry,
770
—, curled rose, 243
—, dock, 1096
—, hazel, 67
—, large rose, 2036
—, leaf-rolling rose, 2093
—, pear, 2645
—, pear slug, 2646
—, pine, 2735
—, rose leaf, 2093
—, turnip, 3961
—, willow, 4163
see also: apple sawfly;
       plum sawfly

3257
SAWFLY LARVA
fr larve *f* de tenthrède
ne bastaardrups

de Afterraupe *f*
da bladhvepselarve
sv bladstekellarv
es oruga *f* falsa
it tentredine *f*
   Thenthredinidae (larva)

3258
SAXIFRAGE
fr sacifrage *f*
ne steenbreek
de Steinbrech *m*
da stenbræk
sv bräcka
es Saxifraga *f*
it sassifraga *f*
   Saxifraga L.

3259
SCAB (BETA)
fr galle *f* commune
ne gordelschurft
de Gürtelschorf *m*
da bælteskurv
sv skorv
es sarna *f* de la corona
it scabbia *f*
   Streptomyces spp.

3260
SCAB (SALIX)
fr tavelure *f*
ne schurft
de Schorf *m*
da pileskurv
sv pilskorv
es sarna *f*
it ticchiolature *f*
   Pollaccia saliciperda
   (All. & Tub.) v. Arx.

3261
**SCAB (VARIOUS FRUIT
SPECIES)**
fr tavelure *f*
ne schurft
de Schorf *m*
da skurv
sv skorv
es roña *f*
it ticchiolatura *f*
   Venturia spp.

SCAB, black, 350
—, common, 775

—, potato, 2873
—, powdery, 2887
—, twig, 3964

3262
SCABIOSA
fr scabieuse *f*
ne Scabiosa
de Skabiose *f*
da skabiose
sv vädd
es escabiosa *f*; viuda *f*
it Scabiosa *f*
   Scabiosa L.

3263
SCABIOUS
fr scabieuse *f*
ne Scabiosa
de Skabiose *f*; Grindkraut
da skabiose
sv vädd
es escabiosa *f*
it Scabiosa *f*
   Scabiosa L.

SCALD, 3279
—, superficial, 3738

3264
SCALE (bot.)
fr écaille *f*
ne schub
de Schuppe *f*
da skæl *n*
sv fjäll *n*
es escama *f*
it scaglia *f*

SCALE, beech, 287
—, cottony maple, 865
—, European fruit, 2578
—, mussel, 2434
—, oystershell, 2434, 2578
—, soft, 3258

3265
SCALE, TO
fr détartrer
ne afbikken
de ausklopfen
da afbanke
sv knacka bort
es desincrustar
it disincrostare

3266
SCALE INSECT;
SOFT SCALE (US)
fr cochenille *f*
ne dopluis; schildluis
de Schildlaus *f*
da skjoldlus
sv sköldlus
es lecanino *m*; cochinilla *f*
it cocciniglia *f*
  Coccidae

3267
SCAR
fr cicatrice *f*
ne litteken *n*
de Narbe *f*
da ar *n*
sv ärr *n*
es cicatriz *f*
it cicatrice *f*

SCARAB, 640

3268
SCARLET
fr écarlate
ne scharlaken
de scharlachrot
da skarlagen
sv scharlakansröd
es escarlata
it scarlatto

3269
SCARLET PIMPERNEL
fr mouron *f* rouge
ne guichelheil
de Ackergauchheil *n*
da rød arve
sv rödarv
es mujares *m*
it anagallide *f*
  Anagallis arvensis L.

3270
SCATTER, TO
fr épandre
ne uitstrooien
de ausstreuen
da strø; sprede
sv sprida; strö ut
es distribuir; esparcir;
  polvorear
it spargere

SCENERY, 2028

SCENT (OF FLOWER),
1445

SCENTED, 1446

3271
SCENTLESS
fr inodore
ne reukloos
de geruchlos
da lugtløs
sv luktfri
es inódoro
it inodoro

3272
SCHIZONERA
LANUGINOSA
fr puceron *m* de l'orme et du
  poirier
ne perebloedluis
de Birnenwurzellaus *f*;
  Ulmenblattlaus *f*
da blodlus
sv almgallus
es agalla *m* del olmo
it afide *m* galligeno maggiore dell'ormo; afide *m*
  lanigero delle radici del
  pero
  Eriosoma lanuginosum
  Htg.; *Schizonera lanuginosa* (UK)

3273
SCHOOL OF PRACTICAL
HORTICULTURE;
PROFESSIONAL SCHOOL
fr école *f* pratique d'horticulture; école *f* professionnelle
ne tuinbouwvakschool
de Gartenbaufachschule *f*
da havebrugsskole
sv trädgårdsskola
es escuela *f* de especialistas
  en horticultura
it scuola *f* pratica d'orticoltura

3274
SCION
fr greffon *m*
ne ent

de Edelreis *n*; Pfropfreis *n*
da podekvist; ædelris *n*
sv ädelris *n*; ympkvist
es injerto *m*
it innesto *m*

3275
SCLEROTIA
fr sclérotes *f pl*
ne sclerotiën
de Sklerotien *f pl*
da sklerotier *n*; hvileknolde
sv skleroser
es esclerocios *m pl*
it sclerotie *f pl*

3276
SCLEROTINIA ROT;
BLACK ROT
fr Sclerotinia *m*
ne sclerotiënrot *n*
de Sclerotinia-Fäule *f*
da sclerotinia
sv Sclerotinia-röta
es moniliosis *f*
it marciume *m* da sclerotinia
  Sclerotinia spp.

3277
SCLEROTINIA ROT
(BEAN)
fr Sclerotium *m*
ne schuimziekte
de Sclerotinia-Fäule *f*
da sclerotinia; bomulds-
  vamp
sv bomullsmögel
rs gomosis *f*
it marciume *m* da sclerotinia
  Sclerotinia spp.

3278
SCOOP, TO
(HYACINTH)
fr creuser
ne hollen
de aushöhlen des Zwiebel-
  bodens
da hule
sv urholka
es ahuecar
it scavare

SCOOPING KNIFE,   950

3279
SCORCH; BURN; SCALD
fr brûlure f; échaudage m
ne brandplek
de Brandmal n; Brandstelle f
da brandplet
sv brännfläck
es quemadura f
it bruciatura f

3280
SCORCH (GRAPES)
fr brûlure f du raisin
ne besverbranding
de Traubenverbrennung f
da bærsforbrænding
sv vinbärsförbränning
es quemadura f de los granos
     de uvas
it abbruciamento m della
     bacca

SCORCH, leaf,   2095, 2096

3281
SCORPIOID CYME
fr cyme f scorpioïde
ne schicht
de Wickel m
da svikkel
sv ensidigt knippe; sick-
     sackknippe
es inflorescencia f escorpioide
it cime m scorpioido

3282
SCORZONERA
fr scorsonère f; salsifis m
     noir
ne schorseneer
de Schorzonere f; Schwarz-
     wurzel f
da skorsonerrod
sv svartrot; skorzonera
es salsifi m negro; escorzo-
     nera f
it scorzonera f

SCOTCH ELM,   4222

3283
SCOTS PINE
fr pin m
ne grove den

de gemeine Kiefer f; Föhre f
da skovfyr
sv gran
es pino m silvestre
it pino m silvestre
     Pinus sylvestris L.

SCREEN,   1306

3284
SCREEN, TO
fr abriter
ne afschermen
de abschirmen
da afskærme
sv avskärma
es proteger
it parare

3285
SCRUBBER
fr scrubber m
ne scrubber
de Scrubber m
da skrubber
sv skrubber; tvättare
es scrubber m
it scrubber m

3286
SCRUB CUTTER
fr émondori m
ne snoeihoutversnipperaar
de Buschhacker m
da kvashugger
sv grenmyllare
es trituradora f de brozas y
     leñas
it tagliuzzatrice f di legno

3287
SCUFFLE TO; TO HOE
fr biner; ratisser
ne schoffelen
de hacken; schuffeln
da skuffe
sv skyffla; luckra
es escardar
it estirpare; rastrellare

3288
SCUFFLER
fr bineuse f
ne schoffelmachine

de Hackmaschine f
da skuffemaskine
sv skyffelmaskin
es binadora f de acción
     selectiva
it erpice f; zappatrice f;
     sarchiatrice f

3289
SCUTELLUM; SHIELD
fr scutellum m
ne schildje n
de Schildchen n
da skjold n
sv sköld
es escutelo m
it scudo m

3290
SCUTIGERELLA
fr blanc mille-pieds m poilu
ne wortelduizendpoot
de Wurzeltausendfuss m
da tusindben n
sv tusenfoting
es milpies m de las raíces
it scolopendra f radicale

3291
SCYTHE
fr faux f
ne zeis
de Sense f
da le
sv lie
es guadaña f
it falce f

3292
SCYTHE STONE
fr affûtoir m
ne strekel
de Streichholz n
da strygespån; strygesten
sv strykspån
es piedra m de afilar;
     guadaña f
it affilatoio m

3293
SEA BUCKTHORN
fr argousier m
ne duindoorn
de Sanddorn m

da havetorn
sv havtorn
es espino *m* falso; espino *m*
   amarillo
it ramno *m*
   Hippophae L.

3294
SEA CLAY; SEA SILT
fr argile *f* d'alluvion marine
ne zeeklei
de Meereston *m*; Seeton *m*
da marint ler *n*
sv sjölera
es arcilla *f* marítima
it alluvione *f* marittima

3295
SEA GREEN
fr vert d'eau
ne zeegroen
de meergrün; seegrün
da søgrøn
sv sjögrön
es verdemar
it verdemare

3296
SEA HOLLY
fr panicaut *m*
ne kruisdistel
de Edeldistel *f*; Mannestreu *f*
da mandstro
sv martorn
es cardo *m* corredor
it eringio *m*
   Eryngium L.

3297
SEAKALE
fr chou *m* marin
ne zeekool
de Meerkohl *m*
da strandkål
sv strandkål
es col *f* de mar
it Crambe *f* maritima
   Crambe maritima L.

3298
SEAL, TO
fr plomber
ne plomberen
de plombieren

da plombere
sv plombera
es precintar
it sigillare

SEA SILT,   3294

3299
SEASON, TO
fr assaisonner
ne kruiden
de würzen
da krydre
sv krydda
es sazonar; condimentar
it condire

3300
SEASONAL LABOUR
fr travail *m* saisonnier
ne seizoenarbeid
de Saisonarbeit *f*
da sæsonarbejde *n*
sv säsongsarbete *n*
es trabajo *m* estacional
it lavoro *m* stagionale

3301
SEA-WEED
fr algue *f*
ne wier
de Algen *f pl*
da alge
sv alg; tång
es alga *f*; varec *m*
it alga *f*; frico *m*

3302
SECATEUR
fr sécateur *m*
ne snoeischaar
de Gartenschere *f*
da beskæresaks
sv sekatör
es tijeras *f pl* de podar
it forbici *f pl* da potare

3303
SECOND CROP
fr culture *f* secondaire
ne nateelt
de Nachkultur *f*
da efterkultur
sv efterkultur

es cultivo *m* consecutivo
it coltura *f* bis

3304
SECRETION
fr sécrétion *f*
ne afscheidingsprodukt *n*
de Ausscheidungsstoff *m*
da afsondring; sekretion
sv sekret *n*; avsöndring
es secreción *f*
it materia *f* secretoria

3305
SECTIONAL BOILER
fr chaudière *f* sectionnée
ne ledenketel
de Gliederkessel
da element kedel
sv sektionspanna
es caldera *f* seccionada
it caldaia *f* a sezione

3306
SECURITY (ec.)
fr garantie *f*; caution *m*
ne zekerheidsstelling; waar-
   borgsom
de Gewährleistung *f*;
   Haftsumme *f*
da garanti
sv garanti; borgenssumma
es garantía *f*; fianza *f*
it garanzia *f*; cauzione *f*

3307
SEDGE PEAT;
CAREX PEAT SOIL
fr tourbe *f* à carex
ne zeggeveen
de Seggentorf *m*
da startørv
sv starrtorv
es turbera *f* de carizzo
it torba *f* di carice

3308
SEDIMENTARY
DEPOSITS
fr dépôt *m* sédimentaire;
   sédiment *m*
ne sedimentaire afzetting
de sedimentaire
   Ablagerung *f*
da sedimentær aflejring

sv sedimentär avlagring
es depósito *m* sedimentario
it sedimento *m*

3309
SEDIMENTARY ROCK
fr roche *f* sédimentaire
ne sedimentgesteente *n*
de Sedimentgestein *n*
da sedimentsten
sv sedimentär bergart
es roca *f* sedimentaria
it roccia *f* sedimentosa

3310
SEDUM; STONECROP
fr orpin *m*
ne vetkruid *n*
de Fetthenne *f*
da stenurt
sv fetblad *n*; fetknopp
es hierba *f* callera
it fabacia *f*; sedo *m*
    Sedum L.

3311
SEED
fr semence *f*; graine *f*;
ne zaad *n*
de Same(*n*) *m*
da frø
sv frö *n*
es semilla *f*; simiente *f*
it seme *m*

SEED, original, 2552

3312
SEED BED
fr planche *f* de semis
ne zaaibed *n*
de Saatbeet *n*
da såbed *n*
sv såbädd
es semillero *m*
it semenzaio *m*

SEED BEETLE, 2975
—, strawberry, 3685

3313
SEED CLEANER
fr nettoyeur *m* de semences
ne zaadreiniger

de Saatreiniger *m*
da frørenser
sv rensningsmaskin
   (för frö m. m.)
es limpiadora *f* de simientes
it pulitrice *f* per sementi

SEED COAT, 3801

SEED-CORN MAGGOT, 275

3314
SEED-DEALER
fr fournisseur *m* de graines
ne zaadleverancier
de Samenlieferant *m*
da frøleverandør
sv fröleverantör
es proveedor *m* de semillas
it fornitore *m* di semi

3315
SEED DISINFECTANT;
SEED DRESSING AGENT
fr produit *m* de désinfection
   des semences
ne zaadontsmettingsmiddel *n*
de Saatbeizmittel *n*
da afsvampningsmiddel *n*
sv betningsmedel *n* för frö
es desinfectante *m* para
   semillas
it disinfettante *m* per il seme

3316
SEED DISINFECTION;
SEED DRESSING
fr désinfection *f* des semences
ne zaadontsmetting
de Samenbeizung *f*
da frøafsvampning
sv fröbetning
es desinfección *f* de las
   semillas
it disinfezione *f* del seme

SEED DRESSING AGENT, 3315

3317
SEED DRILL
fr semoir *m*
ne zaaimachine

de Drillmaschine *f*
da såmaskine
sv såmaskin; radsånings-
   maskin
es sembradora *f*; sembrador
   m
it seminatrice *f*

SEED DRILL, hand, 1719

3318
SEE GROWING
fr culture *f* grainière
ne zaadteelt
de Samenbau *m*
da frøavl
sv fröodling
es cultivo *m* de semillas
it coltivazione *f* di sementi

3319
SEE LEAF; COTYLEDON
fr cotylédon *m*; feuille *f*
   primordiale; feuille *f*
   séminale
ne kiemblad
de Keimblatt *n*
da kimblad *n*
sv hjärtblad *n*
es cotiledón *m*; lóbulo *m*
it cotiledone *m*

3320
SEEDLING
fr plante *f* de semis; plantule *f*
ne zaailing; kiemplant
de Keimpflanze *f*; Sämling *m*
da frøplante; kimplante
sv fröplanta; groddplanta
es planta *f* de semillero;
   plantula *f*
it germinello *m*; seminello
   m; semenzale *m*

SEEDLINGS, overwintered
cabbage, 2571

3321
SEED PAN
fr terrine *f* à semis
ne zaaipan
de Saatschale *f*
da såbakke

sv såfat
es lebrillo *m* germinador
it bacino *m* da sementi

3322
SEED PLANT
fr phanérogame *f*; porte-
graines *m*
ne zaadplant; zaaddrager
de Samenpflanze *f*; Samenträger *m*
da frøplante
sv frödragare
es planta *f* de semilla
it pianta *f* da seme

3323
SEED POD
fr péricarpe *m*; capsule *f*
ne zaaddoos
de Samenkapsel *f*
da frøkapsel
sv fröhus *n*
es cápsula *f* de la semilla
it capsula *f*

3324
SEED POTATO
fr plant *m* de pomme de terre; pomme *m* de terre de
semence
ne pootaardappel
de Pflanzkartoffel *f*
da læggekartoffel *mf*
sv sättpotatis
es patata *f* de siembra
it patata *f* di trapianto

3325
SEEDS ACT
fr loi *f* sur la commercialisation de semences et grai-
nes
ne zaadwet
de Saatgutgesetz *n*
da frølov
sv frölag
es leyes *f pl* protectoras del
seleccionador
it legge *f* sulla commercializzazione di sementi

3326
SEED SAMPLE
fr échantillon *m* de graines
ne zaadmonster *n*

de Samenprobe *f*; Samen-
muster *n*
da frøprøve
sv fröprov *n*
es muestra *f* de semillas
it campione *m* di semi

3327
SEED SPACING DRILL
fr semoir *m* de précision
ne precisiezaaimachine
de Einzelkorndrillgerät *m*
da enkelkornsåmaskine
sv precisionssåmaskin
es sembrador *m* de precisión
it semenatrice *f* di precisione

3328
SEED STALK
fr homozygote *m*
ne zaadstengel
de Samenstengel *m*
da frøstængel
sv fröstängel
es pedúnculo *m*; bohordo *m*
it stelo *m* da seme

3329
SEED TESTING
fr contrôle *m* des semences
ne zaadcontrole
de Samenprüfung *f*
da frøkontrol
sv frökontroll
es control *m* de semillas
it controllo *m* delle semenze

3330
SEED TRADE
fr commerce *m* de graines
ne zaadhandel
de Samenhandel *m*;
Sämereienhandel *m*
da frøhandel
sv fröhandel
es comercio *m* de semillas
it negozio *m* di semi

3331
SEED TRAY
fr caissette *f* à semis
ne zaaibakje *n*
de Saatkiste *f*

da såkasse
sv sålåda
es caja *f* de semillero
it cassetta *f* da seminare

3332
SEED TREATMENT
fr traitement *m* des semences
ne zaadbehandeling
de Saatgutbehandlung *f*
da frøbehandling
sv fröbehandling
es tratamiento *m* de las semillas
it trattamento *m* delle sementi

3333
SELECT, TO
fr sélectionner
ne selecteren
de auswählen; selektieren
da udvalge
sv selektera
es seleccionar
it selezionare

3334
SELECTION
fr sélection *f*
ne selectie
de Auslese *f*; Selektion *f*
da selektion; udvalg *n*
sv urval *n*; selektion
es selección *f*
it selezione *f*

3335
SELECTIVE
fr sélectif
ne selectief
de selektiv; auswählend
da selektiv
sv selektiv
es selectivo
it selettivo

3336
SELF CLEANING DRILL
fr sonde *f* à clapet
ne puls
de Ventilbohrer *m*
da borerør

sv ventilborr
es sonda f con válvula
it sonda f con valvola

3337
SELF-FERTILE
fr autogame; auto-fertil
ne zelffertiel
de selbstfruchtbar
da servfertil
sv självfertil
es autofértil
it autofertile

3338
SELF-FERTILISATION;
SELF-POLLINATION
fr autofécondation f
ne zelfbestuiving
de Selbstbestäubung f
da selvbestøvning
sv självbefruktning
es autofecundación f
it autofecondazione f

3339
SELF-SERVICE
fr libre service m
ne zelfbediening
de Selbstbedienung f
da selvbetjening
sv självbetjäning
es autoservicio m
it self-service

3340
SELF-STERILE
fr autostérile
ne zelfsteriel
de selbstunfruchtbar
da selvsteril
sv självsteril
es autosteril
it autosterile

3341
SELLING PRICE
fr prix m de vente
ne verkoopprijs
de Verkaufpreis m
da salgspris
sv försäljningspris n
es precio m de venta
it prezzo m di vendita

3342
SEMI-AUTOMATIC
fr semi-automatique
ne halfautomatisch
de halbautomatisch
da halvautomatisk
sv halvautomatisk
es semi automático
it semiautomatico

3343
SENSITIVE PLANT
fr sensitive f
ne kruidje-roer-me-niet n
de Sinnpflanze f
da mimose
sv sensitiva
es sensitiva f; Mimosa f
it sensitiva f; Mimosa f pudica
    Mimosa pudica L.

3344
SENSITIVE TO SLAKING
fr sensible au glaçage
ne slempgevoelig
de zur Verschlämmung neigend
da følsom til tilslammes
sv ömtålig för slamma igen
es sensible a enceganarse;
    sensible a enfangarse
it sensibile per ammelmarsi

3345
SENSITIVITY
fr sensibilité f
ne gevoeligheid
de Empfindlichkeit f
da følsomhed; sensitivitet
sv känslighet
es sensibilidad f
it sensibilità f; sensitività f

3346
SEPAL
fr sépale m
ne kelkblad n
de Kelchblatt n
da bægerblad n
sv foderblad
es sépalo m
it sepalo m

3347
SEPARATE SOIL
CONSTITUENTS, TO
fr séparer
ne ontmengen
de entmischen
da ublande
sv avblanda; vertikal erosion
cs scparar
it demescolare

3348
SEPTORIA SPOT
(BLACKBERRY)
fr septoriose f
ne bladvlekkenziekte
de Septoria rubi
da bladpletsyge
sv bladfläckjuka
es septoriosis f de las hojas
it septoriosi f del rovo
    Mycosphaerella rubi
    Roark

3349
SEPTORIA SPOT (RUBUS)
fr septoriose f
ne bladvlekkenziekte
de Blattfleckenkrankheit f
da hindbær-bladpletsyge
sv bladfläcksjuka
es punteado m de las hojas
it septoriosi f del Rubus
    Mycosphaerella rubi
    Roark

3350
SEROLOGY
fr sérologie f
ne serologie
de Serologie f
da serologi
sv serologi
es serología f
it serologia f

3351
SERRADELLA
fr serradelle f; pied d'oiseau m
ne serradella
de Serradella f
da serradelle
sv serradella

es serradella *f*
it serradella *f*
Ornithopus sativus (Brot.)

3352
SERRATE
fr denticulé
ne gezaagd
de gesagt
da savtakket
sv sågtandad
es aserrado
it seghettato

3353
SESSILE
fr sessile
ne zittend (bot.)
de setzend
da siddende
sv oskaftad
es sésil; sentado
it sessile

SET, TO,   2760

3354
SETTLE, TO (SOIL);
TO CONSOLIDATE
fr se tasser
ne beklinken; inklinken
de sich setzen; sich senken
da sætte sig
sv sätta sig
es asentar; endurecerse;
   depositarse
it raffermare

3355
SEWAGE SLUDGE
fr gadoues *f pl*
ne rioolslib
de Klärschlamm *m*
da kloakslam *n*
sv kloakslam *n*
es lodo *m* de ciudades
it melma *f* di fogna

3356
SEWAGE SLUDGE COM-
POST
fr vase *f* de filtrage
ne zuiveringsslib
de Klärschlamm *n*

da rensningsslam
sv slamkompost
es limo *m* de alcantarilla
   purificado
it melma *f* di chiarificazione

3357
SEXUAL MULTIPLICA-
TION
fr multiplication *f* sexuée
ne generatieve vermeerde-
   ring
de generative Vermehrung *f*
da kønnet formering
sv sexuell förökning
es multiplicación *f* sexual;
   cualitación *f*
it propagazione *f* sessuale

3358
SHADE, TO
fr ombrer; ombrager; abriter
ne afschermen; beschadu-
   wen
de beschatten
da skygge; skærme
sv beskugga
es dar sombra; abrigar
it ombreggiare; parare

3359
SHADING MAT
fr claie *f* à ombrer
ne schermmat
de Schattendecke *f*
da skyggemåtte; skygge-
   tremme
sv skuggmatta
es zarzo *m* de cañas
it stoia *f* da ombra

3360
SHAKE, TO (TOMATO
PLANTS)
fr secouer
ne tikken
de ticken; tippen
da ryste for bestøvning
sv rista för befruktning
es tremolar (tomate)
it toccare leggermente

3361
SHALLOT
fr échalote *f*
ne sjalot
de Schalotte *f*
da skalotteløg *n*
sv schallottenlök
es chalote *m*
it cipollina *f*
   Allium ascalonicum L.

3362
SHALLOT APHID
fr puceron de l'échalote
ne sjalotteluis
da Schalottenlaus *f*
da løgbladlus
sv bladlus
es ácaro *m* de la chalota
it afide *m* dello scalogno
   Myzus ascalonicus Donc.

3363
SHALLOTT SET;
CLOVE
fr échalote *f* grise
ne plantsjalot
de Pflanzschalotte *f*
da læggeløg *n*
sv schallottenlöksutsäde
es escalona *f* para plantar
it bulbo *m* di cipolla

3364
SHALLOW
fr bas; peu profond
ne ondiep
de seicht; flach
da overfladisk
sv grund
es poco profundo; super-
   ficial
it poco profondo

3365
SHALLOW ROOTED
fr à racines superficielles
ne vlakwortelend
de flachwurzelnd
da øverligt rodnet
sv ytlig rotning
es de raíces superficiales
it radicandosi superficial-
   mente

3366
SHARE TENANCY
fr métayage m
ne deelbouw
de Teilbau m
da biproduktion
sv hälftenbruk n
es aparcería f
it compartecipazione f;
   (mezzadria f)

3367
SHARTA DAISY
fr Chrysanthemum m
   maximum
ne grootbloemige margriet
de Gartenmargerite f
da kæmpemargerit
sv jätteprästkrage
es artemisa f
it margherita f a fiore
   grande
   Chrysanthemum maximum
   Ram.

3368
SHEATH (bot.)
fr gaine f
ne bladschede
de Blattscheide f
da bladskede
sv bladslida
es vaina f
it guaina f d'una foglia

SHED, 252

3369
SHEEP MANURE
fr fumier m de mouton
ne schapemest
de Schafmist m
da fåregødning mf
sv fårgödsel
es estiércol m de oveja
it sterco m pecorino

3370
SHEEP'S FESCUE
fr fétuque f ovine
ne schapegras
de Schafschwingel m
da fåresvingel
sv färsvingel

es cañuela f de ovejas
it Festuca f ovina
   Festuca ovina L.

3371
SHEEP'S SORREL
fr petite oseille f
ne schapezuring
de kleiner Sauerampfer m
da rødknæ
sv bergsyra
es acederilla f
it acetosa f minore; erba f
   salamoia
   Rumex acetosella L.

3372
SHEET STEAMING
fr désinfection f du sol par
   la vapeur
ne stomen onder zeil
de Haubendämpfung f
da dampning under
   plastsejl
sv ångning under presen-
   ning
es desinfección f del suelo
   por vapor
it sterilizzazione f del
   terreno per vapore

SHELL, 1844

3373
SHELL, TO
fr écosser
ne doppen
de auslösen; enthülsen;
   auspalen
da bælge; afskalle
sv sprita
es desgranar; desvainar
it sgusciare

3374
SHELL LIME
fr calcaire m de coquille
ne schelpkalk
de Muschelkalk m
da muslingekalk
sv musselkalk
es cal f de conchas
it calcina f conchifera

SHELTER, TO, 2957

3375
SHELTER BELT; WIND
SCREEN; WIND BREAK
fr brise-vent m
ne windscherm n
de Windschirm m; Wind-
   schutz m
da vindskærm
sv vindskydd; läbälte
es zarzo m contra el viento;
   cortaviento m
it paravento m

3376
SHEPHERD'S PURSE
fr bourse f à pasteur; tabouret m; capselle f
ne herderstasje n
de Hirtentäschel n
da hyrdetaske
sv lomme
es bolsa f de pastor; zurrón
   de pastor; panique sillo m
it borse f del pastore
   Capsella bursa-pastoris L.
   Med.

SHIELD, 3289

3377
SHIELDLIKE
fr en forme d'écusson
ne schildvormig
de schildförmig
da skjoldformet
sv sköldformig
es en forma de escudo
it scudiforme

SHIFT, TO, 1326

3378
SHIFTING SAND
fr sable m mouvant
ne stuifzand n
de Flugsand m
da flyvesand n
sv flygsand n
es arena f muy fina
it polverino m; sabbia f in
   polvere

3379
SHOOT; SPROUT
fr pousse f; rejet m; rejeton
    m
ne scheut; spruit; loot; rijs
de Trieb m; Schoss m;
    Schössling m
da skud n
sv skott n
es vástago m; brote m;
    retoño m
it rampollo m

3380
SHOOT (BULBS)
fr pousse f
ne pen
de 'Pinn' m
da spire
sv lökskott
es brote m
it germoglio m

SHOOT, lateral,   2045
—, leading,   2066
—, long,   2211
—, side,   3398

3381
SHOOT FORMATION
fr formation f de pousses
ne scheutvorming
de Triebbildung f
da skuddannelse
sv skottbildning
es formación f de vástagos
it formazione f di germoglio

3382
SHOOT WILT (CHERRY)
fr moniliose f
ne taksterfte
de Monilia-Spitzendürre f
da toptørhed; grå monilia
sv grå monilia; blom-och
    grentorka
es moho m
it Sclerotinia f
    Sclerotinia laxa Aderh. et
    Ruhl

3383
SHORT DAY PLANT
fr plante f de jour court;
    plante f nyctipériodique
ne korte-dagplant

de Kurztagpflanze f
da kortdagsplante
sv kortdagsväxt
es planta f de día corto;
    planta f microhémera
it pianta f da giornata breve

3384
SHORT DAY TREAT-
MENT
fr allongement m de la nuit
ne korte-dagbehandeling
de Kurztagsbehandlung f
da kortdagsbehandling
sv kortdagsbehandling
es tratamiento m de día
    corto
it trattamento m a giornata
    breve

SHORTEN, TO,   954

3385
SHORTENING OF DAY-
LIGHT
fr réduction f du jour
ne dagverkorting
de Tagverkürzung f
da dagforkortning
sv dagförkortning
es acortamiento m del día
it accorciamento m della
    giornata

3386
SHORT SHOOT
fr dard m; lambourde f
ne kortlot
de Kurztrieb m
da kortskud n
sv kortskott
es dardo m; ramita f
it dardo m

SHOT-HOLE BORER,
2032
—, larger,   2032
—, small,   79

3387
SHOT-HOLE DISEASE;
GUM SPOT DISEASE
fr criblure f des amygda-
    lées; maladie f criblée du
    cerisier
ne hagelschotziekte

de Schrotschusskrankheit f
da haglskudsyge
sv hagelskottsjuka
es perforación f, mal m de la
    goma; gomosis m
it mal m della gomma dei
    rameti del pesco; arrossa-
    mento m del pesco
    Clasterosporium carpo-
    philum Ad.; Stigmina car-
    pophila (Lev.) M. B. Ellis

3388
SHOVEL
fr pelle f
ne schop
de Schaufel f
da skovl
sv skyffel
es pala f
it zappetta f

3389
SHRINK (CARNATIONS)
fr enroulement m des feuilles
ne krimpen
de einschrumpfen
da indkrølning
sv krympa samman
es abarquillado m de las ho-
    jas
it restringimento m

3390
SHRINK, TO (SOIL)
fr se rétrécir
ne krimpen
de schrumpfen
da krympning
sv krympa
es contracción f
it restringimento m

3391
SHRINKAGE FILM
fr film m de retrait
ne krimpfolie
de Schrumpffolie f
da krympefolie
sv krympfolie
es película f de contracción
it foglio m di contrazione

3392
SHRIVEL, TO
fr se ratatiner;
ne verschrompelen
de zusammenschrumpfen
da indskrumpe
sv skrumpna
es encorgerse; arrugarse
it raggrinzarsi

SHRIVELLED, 4191

3393
SHRUB; BUSH
fr arbuste *m*; arbrisseau *m*
ne heester
de Strauch *m*
da busk
sv buske
es arbusto *m*
it arbusto *m*; cespuglio *m*

SHRUB, ornamental, 2556

3394
SHRUBBY
fr frutescent
ne struikachtig
de strauchartig
da buskagtig
sv buskartad; buskig
es arbustivo
it cespuglioso

3395
SHRUBBY ALTHEA
fr mauve *f* en arbre;
ketmie *f* de Syrie
ne altheastruik
de syrischer Eibisch *m*
da frilandshibiscus
sv frilandshibiskus
es rosa *f* de Siria; granado
m blanco
it altea *f*
Hibiscus syriacus L.

3396
SHRUB FOR FORCING
fr arbuste *m* à forcer
ne trekheester
de Treibstrauch *m*
da drivbuske; prydbuske til
drivning

sv drivbuske
es arbusto *m* para forzar
it arbusto *m* da forzatura

3397
SIDE GRAFTING
fr greffe *f* en incrustation
latérale
ne enten ter zijde
de Seitenpfropfen *n*; seit-
liches Entspitzen *n*
da sidepodning
sv sidoympning
es injerto *m* lateral
it innesto *m* a spacco laterale

3398
SIDE SHOOT
fr tige *f* latérale
ne zijstengel
de Seitenstengel *m*
da sidestængel
sv sidostängel
es tallo *m* lateral
it gambo *m* laterale

3399
SIEVE; STRAINER;
RIDDLE
fr crible *m*; tamis *m*
ne zeef
de Sieb *n*
da si; sigte
sv såll *n*
es tamiz *m*
it vaglio *m*; crivello *m*

3400
SIEVE TUBE
fr vaisseau *m* criblé
ne zeefvat *n*
de Siebgefäss *n*
da sivrør *n*
sv silrör *n*
es tubo *m* criboso
it vaso *m* cribiforme

3401
SILICA
fr silice *f*
ne kiezel *m*
de Kiesel *m*
da kisel

sv kisel
es silicio *m*; cascajo *m*
it silicio *m*

3402
SILICIC ACID
fr acide *m* silique
ne kiezelzuur *n*
de Kieselsäure *f*
da kiselsyre
sv kiselsyra
es ácido *m* silicio
it acido *m* di silicio

3403
SILICIOUS
fr silicieux
ne kiezelhoudend
de kieselhaltig
da kiselholdig
sv kiselhaltig
es siliceo; cascajoso
it contenante silicio

3404
SILICLE
fr silicule *f*
ne hauwtje *n*
de Schötchen *n*
da kortskulpe
sv skida (kort)
es silicula *f*
it siliqua *f*

3405
SILICONE
fr silicone
ne siliconhars
de Silikonharze *f*
da silikone
sv silikon; silikonharts
es silicona *f*
it silicone *m*

3406
SILIQUA
fr silique *f*
ne hauw
de Schote *f*
da langskulpe
sv skida (lång)
es silicua *f*
it siliqua *f*

3407
SILT
fr limon *m*
ne slib *n*
de Schlamm *m*
da slam *n*
sv slam *n*; gyttja
es limo *m*; barro *m*
it melma *f*

3408
SILT, TO
fr colmater; envaser
ne aanslibben
de anschlämmen; auf-
   schwemmen
da tilslemme
sv avsätta slam
es formar aluviones
it far un'alluvione

3409
SILT UP, TO
fr s'envaser
ne dichtslibben
de verschlammen
da tilmudre
sv slamma igen; uppslamma
es embarrar; cubrir de barro
it amelmarsi

3410
SILVER FIR
fr sapin *m* blanc
ne zilverspar
de Weisstanne *f*
da ædelgran
sv ädelgran
es abeto *m* plateado
it abete *m* bianco
   Abies Mill.

3411
SILVER FIR (MIGRA-
TORY) ADELGES
fr Dreyfusia *f* musslini
ne zilversparwolluis
ne Tannentrieblaus *f*
da Chermes nordmannianæ;
   (Chermes nuesslini)
sv ädelgranbarrlus
es pulgón *m* lanigero;
   plateado *m* del abeto

it cherme *f* dell' abete
   Dreyfusia nordmannianae
   Ratzb.; *Adelges nuesslini*
   (UK)

3412
SILVER LEAF
fr galène *f*; alquifoux *m*
ne loodglans
de Bleiglanz *m*
da blyglans
sv silverglans
es galena *f*
it malattia *f* di piombo
   Stereum purpureum Fr.

3413
SILVER SCURF
(POTATO)
fr gale *f* argentée
ne zilverschurft
de Silberschorf *m*; Silber-
   grind
da sølvskurf
sv silverskorv
es tiña *f* plateada; manchas
   f pl plateadas; castras *f pl*
   plateadas
it scabbia *f* argentea;
   tigna *f* argentata
   *Spondylocladium atrovi-
   rens* Harz.

3414
SILVERSKIN ONION
fr petit oignon *m* blanc
ne zilverui
de Perlzwiebel *f*
da perleløg *n*
sv pärllök
es cebollino *m*
it cipolla *f* argentea
   Allium ampeloprasum L.

3415
SILVER STREAK
fr argent *m*; virus *m* argen-
   té; virus *m* 'panachure
   argentée'
ne zilver (virus)
de Silberstreifigkeit *f*
da sølvblad *n*
sv silvervirus

es argirofilosis *f*; estriación *f*
   plateada
it linea *f* bianca (virus)

3416
SILVERWEED
fr potentille *f* ansérine;
   argentine *f*
ne zilverschoon
de Gänsefingerkraut *n*
da gåse-potentil
sv gåsört
es argentina *f* plateada
it pié *m* d'oca
   Potentilla anserina L.

3417
SILVER Y-MOTH
fr noctuelle *f* gamma
ne gamma-uil
de Gammaeule *f*
da gammaugle
sv gammafly
es nóctua *f* plateada;
   nóctua *f* gamma
it nottua *f* gamma
   Autographa gamma L.;
   Plusia g. (UK)

3418
SIMPLE (bot.)
fr simple; entier
ne enkelvoudig
de einfach
da enkeltnervet
sv enkel
es simplo
it semplice

3419
SINGLE, TO
fr démarier
ne op één zetten
de vereinzeln
da udlynde; udskille
sv gallra
es aclarar
it scindere

3420
SINGLE FURROW
PLOUGH
fr charrue *f* monosoc
ne eenscharige ploeg

de Einscharpflug *m*
da enkeltskær plov
sv enskärig plog
es arado *m* monoreja
it aratro *m* monovomere

3421
SINGLE GRAIN
STRUCTURE
fr structure *f* granulaire
ne korrelstructuur
de Einzelkornstruktur *f*
da enkeltkornstruktur
sv kornstruktur
es estructura *f* en granos
it struttura *f* del granello

3422
SINGLE PHASE
(CURRENT)
fr monophasé (courant)
ne eenfase (stroom)
de Einphasen (Strom) *m*
da énfaset strøm
sv enfas ström
es monofásico *m*
it monofase *f*

3423
SINGLE PLANT
SELECTION
fr sélection *f* généalogique
ne stamselectie
de Stammauslese *f*; Stamm-
selektion *f*
da familieudvalg *n*
sv stamurval *n*
es seleccíon *f* individual;
selección *f* genealógica
it selezione *f* clonale-gene-
alogica

3424
SINGLE WHEEL
WALKING TRACTOR
fr monoroue *f*
ne eenwielige trekker
de Einradschlepper *m*
da énhjulet traktor
sv enhjulig traktor
es monorueda *f*
it motocoltivatore *m*;
monoruota *f*

3425
SINUATE; WAVY
fr sinué
ne gegolfd
de wellig
da bølget
sv buktig; vågformig
es ondulado; sinuoso
it ondulato

3426
SITKA SPRUCE
fr épicéa *m* de Sitka
ne Sitkaspar
de Sitkafichte *f*
da Sitkagran
sv sitkagran
es abeto *m* de Menzies;
Sitka *f*
it Picea *f* sitchensis
Picea sitchensis (Bong.)
Carr.

SITUATION (OF A CROP),
2851

3427
SIZE; GRADE
fr calibre *m*
ne maat
de Grössenmass *n*
da størrelse
sv storlek
es calibre *m*
it misura *f*

3428
SIZE, TO
fr calibrer
ne sorteren (naar grootte)
de sortieren (nach Grösse);
kalibrieren
da størrelsessortere
sv storlekssortera
es calibrar
it calibrare

SIZER, 1614

3429
SIZING
fr calibrage *m*
ne maatsortering

de Grössensortierung *f*
da størrelsessortering
sv sortering efter storlek
es calibrado *m*
it calibrazione *f*

3430
SKELETONIZE, TO
(LEAVES)
fr ronger le mésophylle;
squelettiser
ne skeletteren
de skelettieren
da skelettere
sv skelettera
es hacer esquelético
it fare scheletrito

SKELETONIZER,
apple-and-thorn, 132
apple leaf, 132

3431
SKIM COULTER
fr rasette *f*
ne voorschaar
de Vorschäler *m*
da forskær *n*
sv skumrist
es escardadera *f* de un arado
it avanvomere *m*

3432
SKIN; PEEL
fr pelure *f*; peau *f*
ne schil
de Schale *f*
da skin *n*; skal
sv skal *n*
es piel *f*; cáscara *f*
it buccia *f*

3433
SKIN DISEASE
fr maladie *f* de la pelure
ne huidziekte
de Zwiebelhautkrankheit *f*
da hudsyge
sv hudsjukdom
es enfermedad *f* de la piel;
dermatosis *f*
it ammalato *m* nel involucro

3434
SKY BLUE
fr azur
ne hemelsblauw
de azurblau
da himmelblå
sv himmelsblå; azurblå
es azul celeste
it azzurro celeste

3435
SLAKED LIME
fr chaux f éteinte; cendrée
    f de chaux
ne gebluste kalk
de Löschkalk m
da læsket kalk
sv släckt kalk; jordbruks-
    kalk
es cal f muerta; cal f apaga-
    da
it calce f spenta

3436
SLICING BEAN
fr haricot m sabre
ne snijboon
de Schwertbohne f
da snittebønne
sv skärböna
es judía f verde para cortar
it fagiolino m da tagliare
    Phaseolus sp.

SLICING BEAN, dwarf,
1184
—, pole,   2820

3437
SLIDE VALVE; GATE
VALVE
fr vanne f
ne schuifafsluiter
da Schieber m
da skydeventil
sv skjutventil
es válvula f de entrada
it valvola f a cassetto

3438
SLIMY ROT; GUMMOSIS;
SOFT ROT (US); WHITE
ROT (US)
fr pourriture f blanche
ne witsnot n

de Zwiebelnassfäule f;
    weisser Rotz m
da hvidbakteriose
sv vitblötröta
es podredumbre f blanca
it putrefazione f bianca

SLOE,   354

SLUDGE,   1130

SLUDGY,   2416

3439
SLUG; SNAIL
fr limace f; limaçon m; es-
    cargot m
ne slak
de Schnecke f
da snegl
sv snigel
es caracol m
it lumaca f
    Gastrapoda

SLUG, field,   1335
—, grey garden,   1335

3440
SMALL CUPPED
(NARCISSUS)
fr à petite couronne
ne kleinkronig
de kurzkronig
da småkronede
sv med liten krona
es de corona pequeña
it a corona piccola

3441
SMALL ELM BARK
BEETLE
fr petit scolyte m de l'orme
ne kleine iepespintkever
de kleiner Ulmensplint-
    käfer m
da lille elmebarkbille
sv lilla almsplintborre
es barrenillo pequeño m del
    olmo
it piccolo m scolitide dell'-
    ormo
    Scolytus multistriatus
    Marsh.

3442
SMALL ERMINE MOTH
fr hyponomeute m
ne spinselmot
de Gespinstmotte f
da spindemøl n
sv spinnmal n
es polilla f; hiladora
it iponomeuta m; ragna f
    del melo
    Hyponomeuta spp.;
    Yponomeuta (UK)

3443
SMALL FRUIT
fr petits fruits m pl
ne klein fruit
de Beerenobst n
da bærfrugt
sv bärfrukt
es bayas f pl
it frutta minore f; bacche
    f pl

3444
SMALL NARCISSUS FLY;
LESSER BULB FLY
fr mouche f des narcisses
ne kleine narcisvlieg
de Zwiebelmondfliege f
da lille narcisflue
sv mindre narcissfluga;
    taggig lökfluga
es mosca f pequeña del nar-
    ciso
it mosca f dei bulbi
    Eumerus strigatus Fall.

SMALL OFFSET,   665

SMALL SHOT-HOLE
BORER,   79

3445
SMALL WHITE BUTTER-
FLY; IMPORTED CAB-
BAGE WORM (US)
fr piéride f de la rave
ne klein koolwitje n
de Kohlweissling m
    (kleiner)
da lille kålsommerfugl
sv rovfjäril

es pequeña mariposa *f* blanca de la col
it rapaiola *f*
   Pieris rapae L.

3446
SMELL
fr odeur *f*
ne geur
de Geruch *m*
da lugt; duft
sv doft; lukt
es olor *m*
it odore *m*

SMOKEBOMB, 1494

3447
SMOKE BOX
fr boîte *f* à fumer
ne rookkast
de Rauchkammer *f*
da røgkammer *n*
sv rökkammare
es cámara *f* de humo
it camera *f* a fumo

SMOKE FLUE TUBE,
1359

3448
SMOOTH (bot.)
fr lisse
ne glad
de glatt
da glat
sv glatt
es liso
it liscio

3449
SMOOTH-STALKED
MEADOWGRASS
fr paturin *m* des prés
ne veldbeemdgras
de Wiesenrispengras *n*
da eng-rapgræs *n*
sv ängsgroë
es Poa *f*
it Poa *f* pratensis
   Poa pratensis L.

3450
SMOULDER
fr fonte *f*; infection *f* par
   botrytis
ne smeul *n*

de Stengelgrundfäule *f*
da narcissygdom *f*
sv narcissgråmögel
es Botrytis *m* (narciso)
it sclerotium *m* (narciso)
   Sclerotium perniciosum
   v. Stogt et Thom.

3451
SMUT
fr charbon *m*
ne brand
de Brandkrankheit *f*
da brandsygdomme
sv sot
es tizón *m*; carbón *m*
it carboni *m*
   Ustilaginaceae

SMUT, anther, 117

SNAIL, 3439

3452
SNAP BEAN
fr haricot *m* mange-tout
ne slaboon
de Brechbohne *f*
da brydbønne; brædbønne
sv brytböna
es judía *f* verde
it fagiolino *m* comune
   Phaseolus sp.

SNAP BEAN, dwarf, 1185
—, pole, 2821

SNAPDRAGON, 123

3453
SNOWBALL
fr viorne *f*
ne sneeuwbal
de Schneeball *m*
da kvalkved; snebolle
sv olvon *n*; snöbollsbuske
es bola *f* de nieve
it palla *f* di neve
   Viburnum L.

3454
SNOWBERRY
fr symphorine *f*
ne sneeuwbes
de Schneebeere *f*
da snebær
sv snöbär *n*

es bolitas *f* de nieve; perlas
   f; sinforina *f*
it Symphoricarpus *m*
   Symphoricarpus L.

3455
SNOWDROP
fr perce-neige *m*
ne sneeuwklokje *n*
de Schneeglöckchen *n*
da vintergæk
sv snödroppe
es rompenieve *m*; galanto *m*
   de nieve
it bucaneve *m*
   Galanthus L.

3456
SNOWY MESPILUS;
JUNE BERRY
fr amélanchier *m*
ne krenteboompje *n*
de Felsenbirne *f*
da bærmispel
sv häggmispel
es cornijuelo *m*
it Amelanchier *m*
   Amelanchier Med.

3457
SOAK, TO
fr tremper
ne weken
de tauchen
da udbløde
sv blötlägga
es remojar
it inzuppare; immergere

3458
SOAPWORT
fr saponaire *f*; savonnière *f*
ne zeepkruid *n*
de Seifenkraut *n*
da sæbeurt
sv såpnejlika
es saponaria *f* jabonera
it saponaria *f*
   Saponaria L.

3459
SOCIAL SECURITY
CHARGES
fr charges *m pl* sociales;
   dépenses *f pl* sociales
ne sociale lasten

de Sozialaufwand *m*; Sozial-
   lasten *f pl*
da sociale forpligtelser
sv sociala avgifter; sociala
   kostnader
es cargas *f pl* sociales
it spese *f pl* per la previden-
   za sociale

3460
SOD; TURF
fr couche *f* gazonnée
ne zode
de Sode *f*
da tørv
sv grästorva
es césped *m*
it zolla *f*; piota *f*

3461
SODIUM
fr sodium *m*
ne natrium *n*
de Natrium *n*
da natrium
sv natrium
es sodio *m*
it sodio *m*

3462
SODIUM CHLORATE
fr chlorate *m* de soude
ne natriumchloraat *n*
de Natriumchlorat *n*
da natriumklorat *n*
sv natriumklorat *n*
es clorato *m* de sodio
it clorato *m* di sodio

3463
SODIUM NITRATE
fr nitrate *m* de soude
ne natronsalpeter
de Natronsalpeter *m*
da natronsalpeter *n*
sv natriumnitrat *n*
es nitrato *m* sódico
it nitrato *m* di sodio

3464
SOFT
fr mou; tendre
ne week
de weich

da blød
sv mjuk
es blando
it molle

3465
SOFT DRINKS
fr limonade *f*
ne frisdranken
de Erfrischungsgetränke *npl*
da læskedrik
sv förfriskningsdryck
es limonada *f*; bebida *f*
   refrescante
it limonate *f pl*

3466
SOFTEN (WATER), TO
fr adoucir
ne ontharden
de enthärten
da blødgøre
sv avhärda
es ablandar; templar
it addolcire

3467
SOFT FRUIT
fr baies *f pl*
ne zacht fruit *n*
de Weichobst *n*
da blød frugt
sv bär
es frutos *m pl* carnosos
it frutta *f* molle

SOFT ROT, 3438

3468
SOFT ROT (CABBAGE)
fr pourriture *f* bactérienne
ne bacteriehartrot *n*; natrot *n*
de Nassfäule *f*; bakterielle
   Herzfäule *f*
da hvidbakteriose
sv blötröta; rothalsröta
es podredumbre *f* blanda
   bacteriana; podredumbre
   f húmeda; gangrena *f* húmeda
it cancrena *f* umida; marciume *m* molle
   Erwinia carotovora f. sp.
   parthenii Starr.

3469
SOFT ROT (POTATO)
fr pourriture *f* molle
ne natrot *n*
de Nassfäule *f*
da sortbensyge
sv vitbakterios
es podredumbre *f* blanda
it marciume *m* molle
   Erwinia carotovora
   (Jones) Holland

SOFT SCALE, 3266

SOIL, 1192
—, aeolian, 2206
—, brook silt, 441
—, calcareous, 524
—, casing, 601
—, clay, 697
—, forest, 1432
—, geest, 1550
—, glasshouse, 1587
—, gravel, 1649
—, heath, 1757
—, marly, 2292
—, marshy, 2301
—, night, 2476
—, peat, 2661
—, potting, 2881
—, sandy, 3246
—, sticky clay, 3650
—, top, 3890

3470
SOIL AERATION
fr aération *f* du sol
ne bodemventilatie
de Bodenbelüftung *f*
da udluftning af jord
sv genomluftning av jorden;
   jordventilering
es aireación *f* del suelo
it aerazione *f* di suolo

3471
SOIL ANALYSIS
fr analyse *f* de sol
ne grondanalyse
de Grundanalyse *f*; Bodenanalyse *f*
da jordanalyse
sv jordanalys

es análisis *m* de tierras
it analisi *m* del suolo

3472
SOIL ANALYSIS REPORT
fr résultats *m pl* d'analyse
ne analyserapport *n*
de Bodenbefund *m*; Untersuchungsbericht *m*
da analyseresultat *n*
sv analysrapport
es boletín *m* de análisis
it rapporto *m* analisi

3473
SOIL AUGER
fr tarière *f*; foreuse *f* de trou
ne grondboor
de Erdbohrer *m*
da jordbor *n*
sv jordborr *n*
es barreno *m* de suelo
it trivella *f*

3474
SOIL BLOCK
fr pot *m* en terre comprimée
ne perspot
de Presstopf *m*; Erdtopf *m*
da gødningspotte; jordpotte
sv jordkruka
es tiesto *m* de tierra prensada
it vaso *m* di terra compressa

3475
SOIL BLOCK MACHINE
fr machine *f* à faire des mottes
ne pottepers (machine)
de Erdtopfpresse *f*
da pottemaskine
sv krukpress (för jordkrukor)
es máquina *f* para hacer bloques de suelo
it blocchiera *f* per terreno

3476
SOIL-BORNE VIRUS
fr virus *m* transmis par (conservé dans) le sol
ne grondvirus *n*
de bodenübertragbares Virus *n*

da jordbårent virus
sv jordburet virus
es virus *m* transmisible par el suelo
it virus *m* trasmesso attraverso il terreno

3477
SOIL CONDITION
fr état *m* du sol
ne toestand van de grond
de Bodenbeschaffenheit *f*
da jordbundstilstand
sv jordbeskaffenhet
es características *f pl* del suelo
it condizione *f* del suolo

3478
SOIL COVER
fr paillage *m*; couverture *f* du sol
ne grondbedekking
de Grund (Boden) bedeckung *f*
da jorddækning
sv jordtäckning
es cobertura *f* de tierra
it coperta *f* del suolo

3479
SOIL DISEASE
fr maladie *f* du sol
ne bodemziekte
de Bodenkrankheit *f*
da jordsygdom
sv sjukdom hos jorden
es enfermedad *f* del suelo
it malattia *f* del suolo

3480
SOIL DISINFECTION
fr désinfection *f* du sol
ne grondontsmetting
de Bodenentseuchung *f*
da jorddesinfektion
sv jorddesinfektion
es desinfección *f* del suelo
it disinfezione *f* del suolo

3481
SOIL EXHAUSTION
fr fatigue *f* du sol; épuisement *m* du sol
ne bodemmoeheid

de Bodenmüdigkeit *f*
da jordtræthed
sv jordtrötthet
es fatiga *f* del suelo; agotamiento *m* del suelo
it esaurimento *m* del suolo

3482
SOIL FERTILITY
fr fertilité *f* du sol
ne bodemvruchtbaarheid
de Bodenfruchtbarkeit *f*
da jordensfrugtbarhed
sv bördighet; jordfertilitet
es fertilidad *f* de los suelos
it fertilità *f* del suolo

3483
SOIL HEAT; BOTTOM HEAT
fr chaleur *f* de fond
ne bodemwarmte
de Bodenwärme *f*
da jordvarme
sv jordvärme
es calor *m* del suelo
it calore *m* terrestre

3484
SOIL HEATING
fr chauffage *m* du sol
ne grondverwarming
de Bodenheizung *f*
da opvarmning af jord
sv jorduppvärming
es calentamiento *m* del suelo
it riscaldamento *m* del suolo

SOIL HEATING,
electrical, 1207

3485
SOIL IMPROVEMENT
fr amélioration *f* du sol
ne bodemverbetering; grondverbetering
de Bodenverbesserung *f*
da grundforbedring
sv jordförbättring
es mejoramiento *m* del suelo
it miglioramento *m* del suolo

3486
SOILING
fr encrassement; pollution *f*
ne vervuiling

de Verunreinigung *f*; Verschmutzung *f*
da tilsmudse; snavse til
sv nedsmutsning
es ensuciamiento *m*
it insudiciamento *m*

3487
SOIL LAYER
fr couche *f* de sol; strate *f*
ne bodemlaag
de Bodenschicht *f*
da jordlag *n*
sv jordlager; jordskikt *n*
es capa *f* de tierra
it strato *m* del suolo

3488
SOILLESS CULTURE;
HYDROPONICS
fr culture *f* sans sol
ne teelt zonder aarde
de erdlose Kultur *f*; Hydrokultur *f*
da dyrkning uden jord
sv jordfri odling
es cultivo *m* sin suelo
it coltivazione *f* senza terra

3489
fr SOIL MAP
fr carte *f* des sols
ne bodemkaart
de Bodenkarte *f*
da jordbundskort *n*
sv bonitetskarta
es mapa *m* geológico
it mappa *f* del suolo

3490
SOIL MIXTURE;
COMPOST
fr mélange *m* terreux
ne grondmengsel *n*
de Erdmischung *f*
da jordblanding
sv jordblandning
es mezcla *f* de tierra
it mistura *f* di terra

3491
SOIL MOISTURE
CONTENT
fr tension *f* de l'eau du sol
ne vochtspanning

de Bodenwasserspannung *f*
da jordfugtighedspænding
sv jordens sugförmåga
es fuerza *f* de absorción del suelo
it tensione *f* d'umidità

SOIL-MOISTURE
TENSION, 3716

3492
SOIL PARTICLES < 16 μ
fr particules *f pl* lévigables
ne afslibbare bestanddelen *npl*
de abschlämmbare Bestandteile *m pl*
da slembare bestanddele
sv slambara beståndsdelar
es arcilla *f* y limo; partículas *f pl* levigables
it elimenti *m pl* non alluvionali

3493
SOIL PROFILE
fr profile *m* du sol
ne bodemprofiel; grondprofiel *n*
de Bodenprofil *n*; Bodenquerschnitt *m*
da jordbundsprofil
sv markprofil
es perfil *m* del suelo
it profilo *m* del suolo

3494
SOIL SAMPLE
fr échantillon *m* de terre
ne grondmonster *n*
de Bodenprobe *f*
da jordprøve
sv jordprov *n*
es muestra *f* de tierra
it mostra *f* di terra

3495
SOIL SCIENCE;
PEDOLOGY
fr pédologie *f*; science *f* du sol
ne bodemkunde
de Bodenkunde *f*
da jordbundsvidenskab

sv marklära; jordlära
es edafología *f*; pedología *f*
it scienza *f* del suolo

3496
SOIL STERILIZATION
fr désinfection *f* par la vapeur
ne grondstomen
de Bodendämpfung *f*
da jorddampning
sv jordsterilisering; ångning
es desinfección *f* por vapor
it sterilizzazione *f* del terreno

SOIL STRUCTURE,
improvement of the, 1879

3497
SOIL STRUCTURE
DETERIORATION
fr détérioration *f* de la structure
ne structuurbederf *n*
de Strukturverschlechterung *f*
da strukturforringelse
sv strukturförsämring
es deterioración *f* de la estructura
it corruzione *f* della struttura

3498
SOIL SURFACE
fr terrain *m* naturel
ne maaiveld *n*
de Bodenoberfläche *f*
da jordoverflade
sv jordyta; markyta
es superficie *f* del suelo
it piano *m* superiore

3499
SOIL SURVEY
fr cartographie *f* pédologique
ne bodemkartering
de Bodenkartierung *f*
da jordbundskortlægning
sv bonitetskartering; markkartering
es cartografía *f* del suelo
it cartografia *f* del suolo

3500
SOIL TESTING
fr analyse *f* de sol; analyse *f*
de terre
ne grondonderzoek *n*
de Bodenuntersuchung *f*
da jordbundsundersøgelse
sv jordanalys
es análisis *m* de suelo; análisis *m* de tierra
it esame *m* della terra

3501
SOIL TREATMENT
fr traitement *m* du sol
ne grondbehandeling;
bodembehandeling
de Bodenbehandlung *f*
da jordbehandling
sv markbehandling; mark-
vård
es tratamiento *m* del suelo
it trattamento *m* del suolo

3502
SOIL TYPE
fr type *m* de sol
ne grondsoort; bodemtype*m*
ne Bodenart *f*
da jordtype
sv jordart
es tipo *m* de suelo
it tipo *m* di suolo

3503
SOLID BODIES
fr corps *m* solide
ne vaste stof
de feste Masse *f*
da fastmasse
sv fasta beståndsdelar
es cuerpo *m* sólido
it materia *f* solida

3504
SOLID FUEL
fr combustible *m* solide
ne vaste brandstof
de fester Brennstoff *m*
da fast brændstof *f*
sv fast bränsle *n*
es combustible *m* sólido
it combustibile *m* solido

3505
SOLUBLE
fr soluble
ne oplosbaar
de löslich
da opløselig
sv löslig
es soluble
it solubile

3506
SOLUTION
fr solution *f*
ne oplossing
de Lösung *f*
da opløsning
sv lösning
es solución *f*
it soluzione *f*

3507
SOLVENCY
fr solvabilité *f*
ne solvabiliteit
de Solvenz *f*; Zahlungsfähig-
keit *f*
da solvens
sv solvens; betalningsför-
måga
es solvencia *f*
it solvibilità *f*

3508
SOOT
fr suie *f*
ne roet
de Russ *m*
da sod
sv sot
es hollín *m*; tizne *m*
it fuliggine *f*

3509
SOOT DOOR
fr trappe *f* à suie
ne roetluik *n*
de Russtür *f*
da renselåge
sv sotlucka
es tropilla *f* de hollín
it porta *f* a fuliggine

3510
SOOTY BLOTCH (APPLE)
fr blotch *m* fumeux
ne roetvlekkenziekte

de Russfleckenkrankheit *f*
da sodplet
sv sotfläck
es negrón *m*
it macchie *f pl* nere
Gloeodes pomigena
(Schw.) Collby

3511
SOOTY MOULD
fr fumagine *f*
ne roetdauw
de Russtau *m*
da sodskimmel
sv sotdagg
es fumagina *f*
it peronospora *f* fuligginosa

3512
SORREL
fr oseille *f*
ne zuring
de Sauerampfer *m*
da syre
sv ängssyra
es acedera *f*
it acetosa *f*
Rumex acetosa L.

SORREL, sheep's, 3371
—, wood, 2576

3513
SORT, TO; TO GRADE
fr assortir; trier
ne sorteren
de sortieren
da sortere
sv sortera
es clasificar
it assortire

SORTER, 1614

3514
SORTING MACHINE
fr trieuse *f*
ne sorteermachine
de Sortiermaschine *f*
da sortermaskine
sv sorteringsmaskin
es escogadora *f*
it selezionatrice *f*

3515
'SOUR' (BULB
GROWING)
fr fusariose *f*
ne zuur *n*
de Saurer *n*; saure Tulpen *fpl*
da zure løg *n*
sv fusariumröta
es fusariosis *f*
it putrefazione *f* radicale;
   'acidume' *m*
   Fusarium oxysporum
   Schl.

3516
SOURCE OF INFECTION
fr source *f* d'infection
ne infectiebron
de Infektionsquelle *f*
da smittekilde
sv infektionshärd; smitto-
   källa
es foco *m* de infección
it origine *f* dell infezione

3517
SOUR CHERRY
fr cerisier *m*
ne zure kers
de Sauerkirsche *f*
da surkirsebær *n*
sv surkörsbär *n*
es guinda *f*; cereza *f* ácida
it ciliegio *m* acido
   Prunus cerasus L.

3518
SOUTH AFRICAN
VIOLET; SAINTPAULIA
fr violette *f* de l'Usambara
ne Kaaps viooltje *n*
de Usambaraveilchen *n*
da Usambaraviol
sv Usambaraviol; prinsess-
   blomma
es Saintpaulia *f*; violeta *f*
   africana
it violetta *f* di Sudafrica
   Saintpaulia ionantha
   Wendl.

3519
SOW, TO
fr semer
ne zaaien

de säen
da så
sv så
es sembrar
it seminare

SOWBUG,   4196

3520
SOWING
fr ensemencement *m*
ne uitzaai; zaaisel *n*
de Aussaat *f*
da så
sv så
es siembra *f*
it sementa *f*

3521
SOWING-DATE
fr date *f* de semis; date *f* de
   semaille
ne zaaidatum
de Säedatum *n*
da sådatum
sv såningsdatum
es fecha *f* de siembra
it data *m* della semina

SOWING IN ROWS,   1137

3522
SOWING SEASON;
TIME OF SOWING
fr semailles *f pl*; époque *f* du
   semis
ne zaaitijd
de Saatzeit *f*
da såtid
sv såningstid
es época *f* de siembra
it stagione *f* della semina

SOWTHISTLE, annual,
111
—, perennial,   2674
—, spiny,   3556

3523
SPACE, TO (FLOWER-
POTS); TO ROOM OUT
(FLOWERPOTS); TO
STAND OVER (FLOWER-
POTS)
fr distancer; écarter
ne omzetten

de umräumen; umsetzen
da udflytte; give afstand
sv glesa
es distanciar; cambiar
it cambiare

3524
SPACE TREATMENT
fr traitement *m* à l'air libre
ne ruimtebehandeling
de Raumbehandlung *f*
da rumbehandling
sv rumbehandling
es tratamiento *m* espacial
it trattamento *m* di ambiente; trattamento *m* diaria

3525
SPACING OF DRAIN
fr écartement *m* des drains
ne drainafstand
de Dränabstand *m*
da drænrørafstand
sv dikesavstånd *n*
es distancia *f* de los tubos de
   drenaje
it distanza *f* di fognatura

3526
SPACING OF TRUSSES
fr écartement *m* des fermes
ne spantafstand
de Binderabstand *m*
da spærfagsafstand
sv takstolsavstånd
es separación *f*
it intervallo *m* dalla travatura

3527
SPACING WITHIN THE
ROW
fr distance *f* sur le rang
ne afstand in de rij
de Abstand *m* in der Reihe
da rækkeafstand
sv radavstånd *n*
es distancia *f* en el surco
it distanza *f* sulla fila

3528
SPADE
fr bêche *f*
ne spade

de Spaten *m*
da spade
sv spade
es azada *f*
it zappa *f*

3529
SPADING MACHINE
fr charrue *f* à bêches
ne spitmachine
de Spatenmaschine *f*
da gravemaskine
sv spadmaskin; jordbear-
betningsmaskin
es máquina *f* de cavar
it vangatrice *f* rolante

3530
SPADIX
fr spadice *m*
ne kolf
de Kolben *m*
da kolbe
sv kolv
es mazorca *f*
it pannocchia *f*

3531
SPAN
fr chapelle *f*
ne kap
de Schiff *n*
da hus *n* (i væksthusblok)
sv skepp *n*; hus *n*
es capilla *f* (cuerpo *m* de
invernadero)
it campata *f*

3532
SPAN ROOFED GLASS-
HOUSE
fr serre *f* à deux versants
ne tweezijdige kas
da zweiseitiges Haus *n*
da sadeltagshus *n*; heltags-
hus *n*
sv sadeltakshus *n*
es invernadero *m* de dos
aguas
it serra *f* bilaterale

3533
SPANROOF FRAME;
DOUBLESPAN FRAME
fr châssis *m* double
ne dubbele bak

de Doppelkasten *m*
da dobbelt bænkevindue *n*;
tosidet bænk
sv drivbänk i sadeltaksform
es cajonera *f* doble; cajonera
f de dos aguas
it letto *m* bilaterale

3534
SPARK CATCHER
fr pare-étincelles *m*
ne vonkenvanger
de Funkenfanger *m*
da gnistfanger
sv gnistsläckare
es prendedor *m* de chispa
it parascintille *m*

3535
SPARMANNIA; AFRICAN
HEMP
fr Sparmannia *m*
ne kamerlinde
de Zimmerlinde *f*
da stuelind
sv Sparrmannia
es esparmania *f*
it Sparmannia *f*
Sparmannia africana L.

3536
SPARROW
fr moineau *m*
ne mus
de Sperling *m*; Spatz *m*
da spurv
sv sparv
es gorrión *m*
it passero *m*
Passer spp.

3537
SPATHE
fr spathe *f*
ne schede
de Scheide *f*
da skede; hylster *n*
sv (blom) hölster *n*
es vaina *f*
it spata *f*

3538
SPATULATE
fr spatulé
ne spatelvormig

de spatelförmig; spatelig
da spatelformet
sv spatelformig
es espatulado; trasovado
it spatolo

3539
SPAWN, TO (MUSH-
ROOM GROWING)
fr larder; ensemencer
ne enten
de spicken; impfen
da plante
sv späcka
es sembrar
it seminare

SPAN RUN, 2443

3540
SPAN RUNNING ROOM
(MUSHROOM
GROWING)
fr chambre *f* d'incubation
ne groeicel
de Anwachsraum *m*
da løberum *n*
sv tillväxtrum *n*; utväxtrum
n
es sala *f* (local) de germinación
it stanza *f* d'incubazione

3541
SPEARMINT
fr menthe *f* crépue
ne kruizemunt
de Krauseminze *f*
da krusemynte
sv krusmynta
es menta *f* crespa; hierba *f*
buena
it menta *f* crespa
Mentha spicata cv. Chris-
pata (Schrad.) Beck.

3542
SPECIES
fr espèce *f*
ne soort
de Art *f*
da art
sv art; sort
es especie *f*
it specie *f*

SPECKLED, 3564

3543
SPECTRAL DISTRIBU-
TION
fr courbe f de répartition
spectrale de l'énergie
ne energie-golflengte-
kromme
de Kurve f der spektralen
Energieverteilung
da spektral energifordeling
sv kurva för spektral ener-
gifördelning
es distribución f espectral
it ripartizione f dello spettro

3544
SPEED BALANCE
fr balance f automatique
ne snelweger
de Schnellwaage f
da vægt
sv våg
es báscula f
it bilancia f

3545
SPEED SPRAYER
fr arroseur m rapide
ne snelspuit
de Schnellspritze f
da tryksprøjte
sv snabbspruta
es pulverizador m acelerado
potente
it polverizzatore m ad alta
pressione

3546
SPEEDWELL; VERONICA
fr véronique f
ne ereprijs
de Ehrenpreis m
da ærenpris
sv Veronika; ärenpris
es verónica f
it Veronica f
Veronica L.

SPEEDWELL, field, 1336
—, ivy leaved, 1967
—, wall, 4059

3547
SPENT COMPOST
fr compost m usé; corps m
de meule
ne afgedragen mest
de abgetragener Kompost
m; Aushub m
da afdrevet kompost
sv förbrukad kompost
es estiércol m agotado
it letame m usato

3548
SPHAGNUM; PEAT MOSS
fr Sphagnum m; Sphaigne f
ne sphagnum n; veenmos n
de Torfmoos n
da tørvemos; hvidmos
sv vitmossa
es esfagno m; arañuela f roja
it muschio m di torbiera

SPHINX MOTH, 1740

3549
SPIKE
fr épi m
ne aar
de Aehre f
da aks n
sv ax n
es espiga f
it spiga f

3550
SPIKELET
fr épillet m
ne aartje n
de Aehrchen n
da små aks n
sv små ax n
es espiguilla f
it spigetta f

3551
SPINACH
fr épinard m
ne spinazie
de Spinat m
da spinat
sv spenat
es espinaca f
it spinaci mpl
Spinacia oleracea L.

SPINACH, New Zealand,
2471

3552
SPINACH BEET; SWISS
CHARD (US)
fr poirée f; bette f
ne snijbiet
de Mangold m; Beisskohl m
da bladbede
sv mangold
es acelga f
it bietola f capuccio
Beta vulgaris cv. Cicla L.

3553
SPINDLE (BUSH)
fr fuseau m
ne spil
de Spindel f
da topform; spindel
sv spindelpyramid
es árbol m en forma de huso
it fusaggine f

3554
SPINDLE TREE;
EUONYMUS
fr fusain m
ne kardinaalsmuts
de Pfaffenhütchen n; Spin-
delbaum m
da benved
sv benved
es bonetero m
it evonimo m
Euonymus L.

3555
SPINE; THORN
fr épine f
ne doorn
de Dorn n
da vedtorn; torn
sv tagg; torn
es espina f
it spina f

SPINY, 2927

3556
SPINY SOWTHISTLE
fr laiteron m âpre
ne ruwe melkdistel

de dornige Gänsedistel *f*
da ru svinemælk
sv svintistel
es lechuguilla *f*
it cicerbita *f*; grespino *m*
   Sonchus asper L. Hill.

3557
SPIRAL NEMATODE
(PEA)
fr Rotylenchus *m*
ne vroege vergeling
de Spiralälchen *n*
da spiralnematod
sv spiralnematod
it nematode *m* a spirale
   Rotylenchus robustus

3558
SPIT (SOIL)
fr coup *m* de bêche
ne steek
de Spatenstich *m*
da spadestik *n*
sv spadtag, *n*; spadstick *n*
es palada *f*
it palata *f*

SPLIT, 701

3559
SPLIT PEA
fr pois *m* cassé
ne spliterwt
de geschälte Erbse *f*
da flækært
sv spritärt
es guisante *m* partido
it pisello *m* spaccato

3560
SPLIT UP, TO
fr production *f* des caïeux
ne verklistering; vermoering
de 'Verklistern' *n*
da deler sig for meget
sv lökglomerat
es producción *f* de bolitas
it disfarsi in bulbicini

3561
SPORE
fr spore *f*
ne spore

de Spore *f*
da spore
sv spor
es espora *f*
it spora *f*

3562
SPOROPHYTE; CRYPTO-
GAM
fr cryptogame *f*
ne sporeplant
de Sporenpflanze *f*
da sporeplante
sv kryptogam; sporväxt
es criptøgama *f*
it crittogamo *m*
   Cryptogamae

3563
SPORT; MUTATION
fr 'sport' *m*; mutation *f*
   somatique
ne sport
de Sport *m*
da mutation
sv sport
es mutación *f*
it mutazione *f* spontanea

SPOT, bacterial, 224
—, black, 352, 353
—, Botrytis, 411
—, brown, 450
—, cane, 550
—, chocolate, 673
—, coral, 835, 836
—, Jonathan, 1980
—, leaf and pod, 2070
—, lenticel, 2134
—, ring, 3139, 3140, 3141,
3142
—, septoria, 3348, 3349
—, tar, 3796
see also: leaf spot

3564
SPOTTED; SPECKLED
fr ponctué; moucheté
ne gespikkeld
de getüpfelt
da prikket
sv prickig
es salpicado
it punteggiato

3565
SPOTTED MILLEPEDE
fr blanicule *f* mouchetée
ne miljoenpoot
de Tüpfeltausendfuss *m*
da myriopoda; Blaniulus
   guttulatus
sv fläckig tusenfoting
es ciempiés *m*
it millepiedi *m*; blaniuli *m*
   Blaniulus guttulatus
   Bosc.

3566
SPRAY, TO
fr pulvériser; traiter;
   seringuer
ne bespuiten
de spritzen; bespritzen
da sprøjte
sv bespruta
es pulverizar
it spruzzare

3567
SPRAY BOOM
fr rampe *f*
ne spuitboom
de Spritzgestänge *n*
da sprøjtestang; sprøjtebom
sv sprutramp
es rampa *f* de pulverización
it braccio *m* irratore

3568
SPRAY DAMAGE
fr dégât *m* de pulvérisation
ne spuitbeschadiging
de Spritzschaden *m*
da sprøjteskade
sv sprutskada
es deterioro *m* ocasionado
   par plaguicidas
it ustione *m* de trattamenti
   antiparassitari; danno *m*
   dalla spruzzatura

3569
SPRAYER
fr pulvérisateur *m*
ne spuitwerktuig *n*; sproei-
   apparaat *n*
de Spritzgerät *n*
da sprøjte

sv spruta
es pulverizador *m*
it irroratrice *f*; atomizzatore *m*

SPRAYER,   3780
—, crop,   905
—, motor,   2399
—, propane,   2954
—, p.t.o. driven,   2970
—, speed,   3545
—, tractor mounted,   3900
—, trailed,   3903

3570
SPRAY GUN
fr lance *f*
ne spuitstok
de Spritzlanze *f*
da sprøjtegriffel
sv spridarrör *n*; sprutpistol
es lanza *f* de pulverización
it lancia *f* irroratrice

3571
SPRAYING
fr pulvérisation *f*; seringa-
  ge *m*
ne bespuiting
de Spritzen *n*
da sprøjtning
sv besprutning
es pulverización *f*
it spruzzatura *f*

3572
SPRAY LIQUID
fr solution *f* à pulvériser
ne sproeivloeistof
de Spritzflüssigkeit *f*
da sprøjtevædske
sv besprutningsvätska
es líquido *m*
it miscela *f* per irrorazione

SPREADER,   4132

SPRING,   4124

3573
SPRING CRIMP
NEMATODE (STRAW-
BERRY)
fr anguillule *f* du fraisier
ne aardbeibladaaltje *n*

de Erdbeerblattälchen *n*
da jordbærbladnematod
sv begoniaål
es nematodo *m* de las fresas
it nematode *m* della fragola
  Aphelenchoides fragariae
  (Ritz.-Bos)

3574
SPRING FOLIAGE
fr fronde *m*
ne voorjaarsgroen *n*
de Frühlingsgrün *n*
da forårsgrønt *n*
sv vårgrönska
es follaje *m* de primavera
it fogliame *m* primaverile

3575
SPRING ONION
fr oignon *m* à repiquer
ne pootui
de Setzzwiebel *f*
da planteløg *n*
sv sättlök
es planta *f* de cebolla
it cipolla *f* di trapianto

3576
SPRINGTAIL
fr collembole *m*
ne springstaart
de Springschwanz *m*
da springhale
sv hoppstjärt
es colembolo *m*
it collembolo *m*
  Collembola spp.

3577
SPRING-TINE CULTIVA-
TOR
fr cultivateur *m* à dents
  flexibles; cultivateur *m*
  'canadien'
ne veertand cultivator
de Federzahngrubber *m*
da fjedertands kultivator
sv fjäderpinnkultivator;
  kultivator med fjädrande
  pinnar
es cultivador *m* de dientes
  flexibles
it coltivatore *m* a denti
  plessibili

3578
SPRING WATER
fr eau *f* de source
ne welwater *n*
de Quellwasser *n*
da vældvand *n*; kildevand *n*;
  trykvand *n*
sv källvatten *n*
es agua *f* manantial
it acqua *f* sorgiva

3579
SPRINKLE, TO
fr arroser (par aspersion)
ne besproeien
de besprengen
da oversprøjte
sv strila; spruta
es rociar; regar
it annaffiare; innaffiare

SPRINKLE-INSTALLA-
TION, automatic,   196

3580
SPRINKLER; IRRIGA-
TION EQUIPMENT
fr tuyau *m* d'arrosage
ne regenleiding
de Regenleitung *f*
da dyssestreng; vandspre-
  deledning
sv dysledning; sprinklerslang
es tubería *f* de riego por
  aspersión
it tubatura *f* di pioggia

SPRINKLER, rotary,   3201

3581
SPRINKLER CIRCUIT
fr conduite *f* d'arrosage
ne regenleiding
de Regnerleitung *f*
da vandingsledning
sv bevattningsledning
es cañería *f* de riego
it condotta *f* di pioggia

3582
SPRINKLER SYSTEM
fr système *m* d'arrosage par
  aspersion
ne regeninstallatie

de Beregnungsanlage *f*
da regnvandingsanlæg *n*
sv bevattningsanläggning
es sistema *f* de riego por
   aspersión
it sistema *f* d'aspersione

3583
SPRINKLING
fr arrosage *m* par aspersion
ne beregening; besproeiing
de Beregnung *f*
da vanding; overbrusning
sv bevattning (medelst
   spridare); besprutning
es riego *m* por aspersión
it aspersione *f* da pioggia
   artificiale

SPROUT, 3379

3584
SPROUT, TO
fr émettre des pousses;
   émettre des bourgeons
ne spruiten
de durchtreiben
da skyde
sv spira; gro
es retoñar; brotar
it germogliare; pullulare

3585
SPRUCE
fr épicéa *m*; sapinette *f*
ne spar
de Fichte *f*; Rottanne *f*
da gran
sv gran
es abeto *m*
it abete *m*
   Picea Dietr.

SPRUCE, blue, 386
—, common, 776
—, hemlock, 1776
—, Koster's, 386
—, Norway, 776
—, Sitka, 3426

3586
SPRUCE APHID
fr Elatobium *m* abietinum
ne groene sparreluis

de Fichtenröhrenlaus *f*;
   Sitkafichtenlaus *f*
da Elatobium abietinum
sv Elatobium abietinum
es pulgón *m* verde del abeto
it Elatobium *m* abietinum
   Elatobium abietinum
   Wlk.

3587
SPRUCE 'PINEAPPLE'
GALL ADELGES;
EASTERN SPRUCE GALL
APHID (US)
fr puceron *m* à galle conique de l'épinette;
   chermès *m* du sapin
ne sparappalgalluis
de gelbe Fichtengallenlaus *f*
da Adelges (grangallelus)
sv större granbarrlus
es pulgón *m* de agalla del
   abeto
it cherme *m* dell'abete rosso
   Sacchiphantes abietis L.
   Adelges a. (UK);
   Chermes a. (US)

SPRUCE SPIDER MITE,
807

SPUD, 4112

3588
SPUR
fr éperon *m*
ne spoor
de Sporn *m*
da spore
sv sporre
es espolón *m*
it sperone *m*

3589
SPUR BLIGHT
(RASPBERRY)
fr taches *f pl* noires
ne twijgsterfte
de Himbeersterben *n*
da stængelsyge
sv hallonskottsjuka
es mal *m* de los ramos
it marciume *m* del fusto
   Didymella applanata
   Sacc.

3590
SPURGE
fr euphorbe *f*
ne wolfsmelk
de Wolfsmilch *f*
da vortemælk
sv törel; euforbia
es euforbia *f*
it caracia *f*
   Euphorbia L.

SPURGE, petty, 2695

3591
SQUILL
fr scille *f*
ne Scilla
de Scilla *f*
da skilla
sv blåstjärna; Scilla
es escila *f*
it Scilla *f*
   Scilla L.

3592
STABLE HUMUS
fr humus *m* durable
ne stabiele humus
de Dauerhumus *m*
da stabil humus
sv svårlösliga humusämnen
   npl
es humus *m* estable
it umus *m* stabile

3593
STACK, TO
fr empiler
ne stapelen
de stapeln
da stable
sv stapla
es apilar
it ammassare

3594
STACK, TO; TO BUILD
(MANURE PILE)
fr mettre en tas; faire l'abat-
   tage
ne opzetten
de aufsetzen; aufstocken
da opsætte; sætte op
sv sätta upp

es montar la pila
it fare la stiva

3595
STACKER
fr gerbeur m
ne stapelaar
de Hubstapler m
da stabler
sv stapeltruck
es agavilladora f
it carrello m a piattaforma
sollevabile

3596
STACKING
fr gerbage m
ne stapeling
de Stapelung f
da stabling
sv stapling
es apilado n; amontonado m
it accatastamento m

STACKING, open, 2532

3597
STAGE
fr phase m; stade m
ne stadium n
de Stadium n
da stadium; fase
sv stadium
es estadio m; fase f
it stadio m; periodo m

3598
STAGHORN FERN
fr Platycerium m
ne hertshoornvaren
de Geweihfarn m
da hjortetakbregne
sv älghornsbräken; hjort-
hornsbräken
es platicerio m; cuerno m de
alce
it felce m corno cervino
Platycerium Desv.

3599
STAGING; BENCH
fr tablette f
ne tablet
de Tablette f; Tischbeet n

da bord n
sv bord n
es mesa f; mesita f; tabla f
it tavoletta f

3600
STAG'S HORN SUMAC
fr sumac m de Virginie
ne fluweelboom
de Essigbaum m
da hjortetaktræ n
sv rönnsumak
es zumaque m
it Rhus f typhina
Rhus typhina L.

3601
STAKE; STICK
fr tuteur m
ne bloemstok; steunpaal
de Pflanzenstab m; Baum-
pfahl m
da stok; blomsterpind;
støttepæl
sv blomstöd n; käpp;
stödpåle
es rodrigón m; tutor m
it tutore m da fiore; palo m
di sostegno

STALK,   3634
—, fruit,   1483
—, leaf,   2109
—, seed,   3328

STALKED,   2692

3602
STAMEN
fr étamine f
ne meeldraad
de Staubgefäss n; Staubblatt
n
da støvdrager; støvblad n;
støvbærer
sv ståndasse
es estambre m
it stame m

3603
STAMINATE (bot.);
MALE
fr mâle
ne mannelijk

de männlich
da manlig
sv manlig; maskulin
es masculino
it mascolino

3604
STANDARD (bot.);
VEXILLUM
fr étendard m
ne vlag
de Fahne f
da fane
sv segel
es bandera f
it vessillo m

3605
STANDARD COSTS
fr coûts m pl standard
ne standaardkosten
de Standardkosten f pl
da standardomkostninger
sv normalkostnad
es costes m pl standard
it costi m pl standard

3606
STANDARD ERROR
(OF THE MEAN)
fr erreur f moyenne; écart
(m) moyen
ne middelbare fout; stan-
daardafwijking
de mittlerer Fehler m
da middelfejl; standard-
afvigelse
sv medelfel; standardavvikelse
es error m medio; desviación
f media
it errore m medio; differenza f media

3607
STANDARD FIGURES
fr valeurs f pl normales
ne normaalcijfers n pl
de Normalziffern f pl
da normaltal n pl
sv normaltal
es cifras f pl normales
it valori m pl normali

3608
STANDARDIZATION
fr normalisation *f*; standardisation *f*
ne normalisatie; standaardisatie
de Standardisierung *f*; Normung *f*
da standardisering
sv standardisering; normalisering
es normalización *f*; standardización *f*
it normalizzazione *f*; standardizzazione *f*

3609
STANDARDIZE, TO
fr normaliser
ne normaliseren
de normalisieren
da standardisere
sv standardisera
es normalizar
it normalizzare

3610
STANDARD TREE
fr haute-tige *m*
ne hoogstam
de Hochstamm *m*
da højstammet
sv högstam
es de tronco *m* alto
it albero *m* ad alto fusto

STAND OVER, TO,   3523

3611
STAPLE
fr agrafe *f*
ne niet; kram; haak
de Heftklammer *f*
da hefteklamme
sv häftklammer
es grapa *f*
it graffa *f*

3612
STAPLE, TO
fr agrafer
ne nieten, krammen
de heften
da hefte

sv häfta; nita
es remachar; grapar
it aggraffare

3613
STAPLING MACHINE
fr agrafeuse *f*
ne nietmachine; hechtmachine
de Heftmaschine *f*
da heftemaskine
sv häftmaskin
es grapadora *f*
it graffatrice *f*

3614
STARCH
fr fécule *f*; amidon *m*
ne zetmeel *n*
de Stärke *f*
da stivelse
sv stärkelse
es fécula *f*
it fecola *f*

3615
STARLING
fr sansonnet *m*
ne spreeuw
de Star *m*
da stær *mf*
sv stare
es estornino *m*
it storno *m*
    Sturnus vulgaris L.

3616
STAR OF BETHLEHEM
fr dame-d'onze-heures *f*
ne vogelmelk
de Vogelmilch *f*
da kost-fuglemælk
sv stjärnlök
es leche *f* de gallina
it latte *m* di gallina
    Ornithogalum umbellatum L.

3617
STARTING MATERIAL
fr matériel *m* de base
ne uitgangsmateriaal *n*
de Ausgangsmaterial *n*
da udgangsmateriale *n*; modermateriale *n*

sv utgångsmaterial *n*; modermateriale *n*
es material *m* inicial
it materia *f* d'inizio (d'origine)

3618
STATE OF AGGREGATION
fr état *m* d'aggrégation
ne aggregatietoestand
de Aggregatzustand *m*
da aggregationstilstand
sv aggregationstillstånd
es estado *m* de agregación
it stato *m* d'aggregazione

3619
STATIC LOAD
fr charge *f* statique
ne statische lading
de statische Ladung *f*
da statisk ladning
sv statisk laddning
es carga *f* estática
it carico *m* statico

3620
STATISTICAL ANALYSIS
fr méthodes *f pl* statistiques
ne wiskundige verwerking
de mathematische Bearbeitung *f*
da statistisk bearbejdning
sv statistisk bearbetning
es método *m* estadístico; cálculo *m* estadístico
it analisi *f* matematica

3621
STAY TUBE
fr tube *f* de consolidation
ne steunpijp
de Stutzrohr *n*
da støtterør *n*
sv stödrör *n*
es tubo *m* de apoye
it tubo *m* d'ancoraggio

3622
STEAM, TO (SOIL)
fr stériliser à la vapeur
ne stomen
de dampfen

da dampe
sv ånga
es esterilizar al vapor
it sterilizzare da vapore

3623
STEAM BOILER
fr chaudière *f* à vapeur
ne stoomketel
de Dampfkessel *m*
da dampkedel
sv ångpanna
es caldera *f* de vapor
it caldaia *f* a vapore

3624
STEAM DOME
fr dôme *m* de vapeur
ne stoomdom
de Dampfdom *m*
da damphat
sv ångdome
es vapor *m* 'dome'
it cupola *f* a vapore

3625
STEAMED BONE FLOUR
fr poudre *m* d'os dégélatini-
   sé
ne ontlijmd beendermeel *n*
de entleimtes Knochenmehl *n*
da dampet benmel *n*
sv ångpreparerat benmjöl *n*
es polvo *m* de hueso desgela-
   tinado
it ossa *f* polverizzata dege-
   latinata

3626
STEAMING PLOUGH
fr sous-soleuse *f* pour
   désinfection
ne stoomploeg
de Dämpfpflug *m*
da dampplov
sv ångplog; ångslede
es subroladora *f* para desinfección con vapor
it attrezzo *m* per desinfe-
   zione a vapore

3627
STEAM-PIPE
fr conduite *f* de vapeur
ne stoomleiding

de Dampfleitung *f*
da dampledning
sv ångledning
es tubo *m* para vapor
it condotta *f* da vapore

STEAM PIPE, main,   2248

3628
STEAM PRESSURE
fr pression *f* de vapeur
ne stoomdruk
de Dampfdruck *m*
da damptryk *n*
sv ångtryck
es presión *f* de vapor
it pressione *f* di vapore

3629
STEAM-SHEET
fr bâche *f* à la stérilization
   du sol
ne stoomzeil *n*
de Dampfplane *f*
da dampningsklæde *n*
sv ångningsduk
es cubierta *f* para vapor
it tela *f* da vapore

3630
STEAM SPACE
fr espace *f* de vapeur
ne stoomruimte
de Dampfraum *m*
da damprum *n*
sv ångkammare
es espacio *m* de vapor
it spazio *m* di vapore

3631
STEAM STERILIZATION
fr stérilisation *f* du sol à la
   vapeur
ne grondstomen
de Bodendämpfung *f*
da dampdesinfektion of jord
sv ångning; jorddesinfek-
   tion; jordsterilisering;
es esterilización *f* del suelo
   al vapor
it sterilizzazione *f* del suolo
   da vapore

3632
STEAM TRAP
fr séparteur *m* d'eau
ne condenspot
de Kondenstopf *m*
da kondensvandbeholder
sv kondenssamlare
es separador *m* de agua
it scaricatore *m* di condensa

3633
STEAM-TUBE
fr tuyau *m* à vapeur
ne stoomslang
de Dampfschlauch *m*
da dampslange
sv ångslang
es tubo *m* de goma; tubo *m*
   para vapor
it tubo *m* da vapore

3634
STEM; STALK
fr pédoncule *m*; pétiole *m*;
   tige *f*
ne steel; stengel
de Stiel *m*; Stengel *m*
da stilk; stængel
sv stjälk; stängel
es tallo *m*
it peduncolo *m*

STEM, main,   2249

3635
STEM AND BULB
NEMATODE
fr anguillule *f* des tiges et
   des bulbes
ne hyacintestengelaaltje *n*
de Hyazinthenstengelälchen
da stængelnematod
   (i hyacint)
sv hyacintkål
es anguilula *f* del tallo
it nematode *m* delle stelo
   Ditylenchus dipsaci
   (Kühn)

3636
STEMBASE
fr cavité *f* pédonculaire
ne steelholte
de Stielhöhle *f*

da stilkgrube
sv skafthåla
es cavidad f del pedúnculo
it fossa f picciolare; cavità f peduncolare

3637
STEM CANKER (ROSA)
fr chancre m de la tige
ne stamkanker
de Rindenfleckenkrankheit f
da stængelsvampe
sv stjälkröta
es cáncer m del tallo
it cancro m del fusto
Leptosphaeria coniothyrium (Fuck) Sacc.

STEM CLASPING, 3640

3638
STEM CUTTING
fr bouture f
ne scheutstek n
de Triebsteckling m
da skudstikling
sv skottstickling
es esqueje m de tallo
it talea f erbacea

3639
STEM EELWORM
fr anguillule f de la tige
ne stengelaaltje n
de Stengelälchen n
da stængelål
sv stängelnematod
es anguilula f del tallo
it nematode m delle foglie
Ditylenchus dipsaci (Kühm) Filip.

3640
STEM EMBRACING;
STEM CLASPING
fr embrassant; amplexicaule
ne stengelomvattend
de stengelumfassend
da stængelomfattende
sv stjälkomfattande
es amplexicaule; abrazadoro
it abbracciante

3641
STEM FRUIT
fr fruit m principal
ne stamvrucht
de Stammfrucht f
da stammefrugt
sv stamfrukt
es fruto m tronquero
it frutto m del tronco

STEMLESS, 10

3642
'STEM PARALYSIS'
(GRAPE CULTIVATION)
fr paralysie f des tiges
ne lamsteligheid
de Lahmstieligkeit f
da stilkindtørring
sv stjälklamhet
it paralisi f del peduncolo

3643
STEM ROT
(DIANTHUS CAR.)
fr fusarioses f pl parenchymatiques
ne scheutrot n
de Stengelfäule f
da nellike-fusariose
sv stjälkbrand
es fusariosis f
it fusariosi f del garofano
Fusarium spp.

STEM ROT, didymella, 1054

3644
STEP LADDER
fr échelle f
ne plukladder
de Treppenleiter f
da plukkestige
sv plockningsstege
es escalera f
it scala f a gradini

3645
STERILE
fr stérile
ne onvruchtbaar
de unfruchtbar
da ufrugtbar; steril

sv ofruktbar
es estéril
it sterile

3646
STERILITY
fr stérilité f
ne onvruchtbaarheid
de Unfruchtbarkeit f
da ufrugtbarhed; sterilitet
sv ofruktbarhet
es esterilidad f
it sterilità f

3647
STERILIZE TO;
TO DISINFECT
fr désinfecter; stériliser
ne ontsmetten; steriliseren
de entseuchen; desinfizieren
da desinficere; sterilisere
sv desinficera; sterilisera
es desinfectar; esterilizar
it disinfettare; sterilizzare

3648
STEWED FRUIT
fr compote f
ne compote
de Kompott n
da kompot
sv kompott
es compota f
it composta f

STICK, 3601

3649
STICK SULPHUR
fr soufre m en bâtons
ne pijpzwavel
de Stangenschwefel m
da stangsvovl n
sv stångsvavel n
es azufre m en barras
it solfo m in bastocini

3650
STICKY CLAY SOIL
fr argile f alcaline
ne knikklei; knipklei
de Knick m; dichter Alluvialton m
da knikklæg

sv fast lerskikt
es arcilla *f* alcalina
it argilla *f* alcalina

3651
STIGMA (bot.)
fr stigmate *m*
ne stempel
de Narbe *f*; Stempel *m*
da ar *n*; støvfang *n*
sv märke
es estigma *m*
it stimma *m*; pistillo *m*

3652
STILLAGE
fr chariot *n* susbaissé
ne laadbord *n*
de Rollplattform *f*
da platform
sv lastbord *n*
es caballete *m*
it piattaforma *f*

3653
STINGING HAIR
fr stimule *m*; poil *m* urticant
ne brandhaar *n*
de Brennhaar *n*
da brændhår *n*
sv brännhår *n*
es pelo *m* urticante
it pelo *m* ortico

3654
STINKWEED; FIELD
PENNYCRESS
fr tabouret *m* des champs
ne witte krodde
de Ackerpfennigkraut *n*;
   Hellerkraut *n*
da pengeurt
sv penningört
es hierba *f* del ochavo
it erba *f* storna
   Thlaspi arvense L.

3655
STIPULE
fr stipule *f*
ne steunblad *n*
de Nebenblatt *n*
da akselflig; akselblad

sv stipel
es estipula *f*
it stipola *f*

3656
ST. JOHN'S WORT
fr millepertuis *m*
ne hertshooi
de Johanniskraut *n*; Hartheu *n*
da perikon
sv hyperikum; johannesört
es hipericón *m*; hierba *f* de
   San Juan
it Hypericum *m*
   Hypericum L.

3657
ST. LUCIE CHERRY
fr mahaleb *m*; bois *m* de
   Sainte-Lucie
ne Weichselkers
de Mahalebkirsche *f*;
   Weichselkirsche *f*
da weichel
sv vejksel
es cerezo *m* de Santa Lucia
it ciliegio *m* Santa Lucia;
   Prunus *f* mahaleb
   Prunus mahaleb L.

3658
STOCK
fr giroflée *f*
ne violier
de Levkoje *f*
da levkøj
sv lövkoja
es alhelí *m*
it violacciocca *f*
   Matthiola R. Br.

STOCK PLANT, 2395

3659
STOCK TAKING
fr inventaire *m*; inventarisa-
   tion *f*
ne inventarisatie
de Inventur *f*
da inventaroptælling; status-
   opgørelse
sv inventering
es inventarisación *f*
it far l'inventario

STOLON, 3217

3660
STOMA
fr stomate *m*
ne huidmondje *n*
de Spaltöffnung *f*
da spalteåbning
sv klyvöppning
es estoma *f*
it stoma *m*

3661
STOMACH POISON
fr poison *m* par ingestion
ne maaggift *n*
de Magengift *n*
da mavegift
sv maggift *n* ;
es veneno *m* estomacal
it veleno *m* per ingestione

STONE, 3150

3662
STONECROP
fr orpin *m* brûlant
ne muurpeper
de Mauerpfeffer *m*
da bidende stenurt
sv gul fetknopp
es pan *m* de cuco; siempreviva *f* menor
it erba *f* pignola
   Sedum acre L.

STONE FRUIT, 1146

STOOL, TO, 3853

3663
STORAGE (BULB GROW-
ING)
fr conservation *f* en séchoir
ne schuurbewaring
de Scheunenlagerung *f*
da behandling i opbevarings-
   rum *n*
sv lagerförvaring
es conservación *f* en almacén
it deposito *m*

3664
STORAGE; WARE-
HOUSING
fr entreposage *m*; stockage
   m
ne opslag; bewaring
de Lagerung*f*; Einlagerung*f*
da lager *n*
sv lagring
es almacenamiento *m*
it immagazzinamento *m*

STORAGE, controlled at-
mosphere, 823
—, duration of, 1170
—, gas, 823

3665
STORAGE COSTS
fr frais *m pl* de stockage
ne bewaarkosten
de Lagerungskosten *pl*
da lagringsomkostninger
sv lagringskostnader
es costes *m pl* de almacenamiento
it spese *f pl* d'immagazzinamento

3666
STORAGE DISEASE
fr maladie *f* d'entreposage
ne bewaarziekte
de Lagerschaden *m*; Lager-
   krankheit *f*
da lagersygdom
sv lagringssjukdom
es enfermedad *f* de almacenamiento
it malattia *f* da frigorifero

3667
STORAGE ROT
(GLADIOLUS)
fr moisissure *f*
ne bewaarrot *n*
de Pencillium-Lagerfäule *f*
da gladiolus-penselskimmel
sv grönmögel
es podrido *m* de almacén
it marciume *m* da Penicillium
   Penicillium gladioli
   Mc Cull. et Thom.

3668
STORAGE TEMPERA-
TURE
fr température *f* de conservation
ne bewaartemperatuur
de Lagertemperatur *f*
da opbevaringstemperatur
sv förvaringstemperatur
es temperatura *f* de conservación
it temperatura *f* di deposito

3669
STORE, TO
fr emmagasiner; stocker;
   entreposer
ne opslaan
de lagern; aufspeichern
da lagre; opmagasinere
sv magasinera; lagra
es almacenar
it immagazzinare

3670
STOREHOUSE; STORE
ROOM
fr magasin *m*; depôt *m*
ne bewaarplaats
de Aufbewahrungsort *m*;
   Lager *n*
da opbevaringssted *n*
sv förrådshus *n*; lagerplats
es almacén *m*
it magazzino *m*

3671
STORM DAMAGE
fr dégâts *m pl* causés par la
   tempête
ne stormschade
de Sturmschaden *m*
da stormskade
sv stormskada
es daños *m pl* causados por
   la tempestad
it danni *m pl* della tempesta

3672
STRAIGHT
(SEED GROWING)
fr d'authenticité variétal
ne zaadecht; zaadvast
de samenecht

da frø-ægte
sv fröäkta
es semilla *f* pura; semilla *f*
   genuina; semilla *f* auten-
   tica; semilla *f* original
it puro da seme

3673
STRAIN (BREEDING)
fr souche *f*
ne lijn
de Stamm *m*; Rasse *f*
da linie
sv linje
es linaje *m*; línea *f* genealógica
it linea *f*

STRAINER, 3399

STRANGLING DISEASE
OF CONIFERS, 1058

3674
STRAP SHAPED
fr ligulé
ne lintvormig
de bandförmig
da tungeformet
sv tungformig
es ligulado
it ligulato

3675
STRATIFIED
fr stratifié
ne gelaagd
de geschichtet
da lagdelt
sv varvig; skiktad; varvad
es estratificado
it stratificato

STRATIFORM, 2060

3676
STRATIFY, TO
fr stratifier
ne stratificeren
de stratifizieren
da stratificere
sv stratifiera
es estratificar
it stratificare

3677
STRAW
fr paille *f*
ne stro *n*
de Stroh *n*
da halm
sv strå; halm
es paja *f*
it paglia *f*

3678
STRAW-BALE
fr botte *f* de paille
ne strobaal
de Strohballen *m*
da halmballe
sv halmbal
es fardo *m* da paja
it balla *f* di paglia

3679
STRAWBERRY
fr fraise *f*
ne aardbei
de Erdbeere *f*
da jordbær *n*
sv jordgubbe
es fresa *f*
it fragola *f*

STRAWBERRY, perpetual
fruiting, 2687

3680
STRAWBERRY APHID
fr puceron *m* du frasier
ne aardbeiknotshaarluis
de Erdbeerblattlaus
da jordbærbladlus
sv jordgubbsbladlus
es pulgón *m* del fresal
it afide *m* della fragola
Pentatrichopus fragaefolii
Cock; *Chaetosiphon* (UK)

3681
STRAWBERRY BLOSSOM
WEEVIL
fr anthonome *m* du fraisier
ne aardbeibloesemkever
de Erdbeerblütenstecher *m*
da hindbærsnudebille
sv hallonblomvivel; jord-
gubbsvivel

es antónomo *m* de la fresa
y del frambuesco
it antonomo *m* delle fragole
e dei lamponi
Anthonomus rubi Herbst.

3682
STRAWBERRY MITE;
CYCLAMEN MITE (US)
fr tarsonème *m* du fraisier
ne aardbeimijt; cyclamen-
mijt
de Erdbeermilbe *f*; Cycla-
menmilbe *f*
da jordmærmide; cyclamen
mide
sv jordgubbskvalster *n*;
cyklamenkvalster *n*
es ácaro *m* del fresal; ácaro
m del ciclamen
it acaro *m* delle fragola;
tarsonemide *m* della fragola
Tarsonemus fragariae
Zimm.; *Tars. pallidus*
Banks.; *Steneotarsone-
mus pallidus*

3683
STRAWBERRY PLANT
fr fraisier *m*
ne aardbeiplant
de Erdbeerpflanze *f*
da jordbærplante
sv jordgubbsplanta
es planta *f* de fresa
it fragola *f*
Fragaria spp.

3684
STRAWBERRY
RHYNCHITES
fr charançon *m*
ne aardbeistengelsteker
de Erdbeerstengelstecher *m*
da jordbærsnudebille
sv jordgubbsvivel
es curculiónido *m* del fresal
it punteruolo *m*
Rhynchites germanicus
Herbst.; *Caenorhinus g.*
(UK)

3685
STRAWBERRY SEED
BEETLE
fr anthonome *m* du fraisier
et du framboisier
ne aardbeiloopkever
de behaarter Samenlauf-
käfer *m*
da løbebille
sv hårig jordlöpare
es cárabo *m* del fresal
it antonomo *m* del lampone
e della fragola
Ophonus pubescens Müll.

3686
STRAWBERRY YELLOW
EDGE
fr maladie *f* du bord jaune
du fraisier
ne geelrand (aardbei)
de Blattrandvergilbung *f* der
Erdbeere
da jordbær-gulrandsyge
sv gulkantsjuka (jordgubbar)
es enfermedad *f* del borde
amarillo de la fresera
it giallume *m* marginale
della fragola

3687
STRAWBOARD
fr panneau *m* de fibre
ne vezelplaat
de Faserplatte *f*
da fiberplade
sv fiberplatta
es panel *m* de fibra
it cartone *m* di pasta di
paglia

3688
STRAW LITTER; STRAW
FOR BEDDING
fr litière *f*
ne strooisel *n*
de Streu *f*
da strøelse
sv strö *n*
es paja *f* de cama
it letto *m* di paglia

3689
STREET MARKET
fr négoce *m* de quatre
   saisons
ne straathandel
de Strassenhandel *m*
da gadehandel
sv gatuhandel
es comercio *m* callejero
it bancarella *f*

STRESS, 597

3690
STRETCH, TO
fr étirer
ne rekken
de strecken
da række; strække sig
sv töja (töja ut)
es estirar; prolongar; ahilar
it distendere

3691
STRETCH (WIRE), TO
fr tendre
ne spannen
de spannen
da spænde; trække
sv spänna
es tender
it tendere

3692
STRIATED; STRIPED
fr strié
ne gestreept
de gestreift
da stribet
sv strimmig
es rayado
it striato

3693
STRING; TWINE
fr ficelle *f*
ne bindtouw *n*
de Bindfaden *m*
da sejlgarn *n*; snor
sv bindgarn *n*; snöre
es cuerda *f*; bramante *m*
it spago *m*

STRINGLY MYCELIUM,
2441

3694
STRINGY
fr filandreux; fibreux
ne draderig
de faserig
da trådet; træet
sv trådig
es fibroso; filamentoso
it filamentoso

3695
STRIP, TO (HYACINTH)
fr égrapper
ne ritsen (hyacint)
de ritzen; ausschneiden
da fjerne hyacintblomster
sv repa
es desflorar
it sfilare

STRIPED, 3692

3696
STRUCTURAL PROFILE
fr profil *m* structural
ne structuur profiel *n*
de Strukturprofil *n*
da strukturprofil
sv strukturprofil
es perfil *m* estructural
it profilo *m* della struttura

3697
STRUCTURE
fr structure *f*
ne structuur
de Struktur *f*
da struktur
sv struktur
es estructura *f*
it struttura *f*

3698
STRUT
fr contre-fiche *f*; entretoise *f*
ne schoor
de Stütze *f*
da stiver; støtte
sv sträva
es puntal *m*
it puntone *m*

3699
STUMP (MUSHROOM
GROWING)
fr bout *m* de queue
ne stomp; voet
de Fussende *n*; Stielende *n*
da stokende
sv fotände
es punta *f*
it torso *m*

3700
STUMP
fr souche *f*; chicot *m*
ne stronk
de Strunk *m*
da stub
sv stubbe
es tronco *m*; troncho *m*
it ceppo *m*

STUNTING, 1183

3701
STYLE (bot.)
fr style *m*
ne stijl
de Griffel *m*
da griffel
sv stift
es estilo *m*
it stilo *m*

3702
STYLET
fr stylet
ne mondstekel
de Mundstachel *m*
da brod
sv muntagg
es estilete *m*
it stiletto *m*

3703
SUBLIMATE
fr sublimé *m*
ne sublimaat *n*
de Sublimat *n*
da sublimat
sv sublimat
es sublimado *m*
it sublimato *m*

3704
SUBSIDIARY EFFECT
fr effet *m* secondaire
ne nevenwerking
de Nebenwirkung *f*
da tågevirkning
sv biverkning; sidoverkning
es efecto *f* secundario
it effetto *m* sussidiario

3705
SUBSOIL
fr sous-sol *m*
ne ondergrond
de Untergrund *m*; Unter-
boden *m*
da undergrund
sv alv
es subsuelo *m*
it sottosuolo *m*

3706
SUBSOILER
fr sous-souleuse *f*
ne woeler
de Untergrundlockerer *m*
da jordperforator
sv alvluckrare
es subsoladora *f*
it attrezzo *m* per ripuntatu-
ra; aratro *m* sotterraneo

3707
SUBSOILING WITH SOIL
MIXING
fr fouiller
ne mengwoelen
de mengwühlen
da blandegrubbe
sv blandkultivera; bland-
grubba
es subsolar; mezclar
it grufolare *f* da mescolanza

SUBSOIL PLOUGHING,
1009

3708
SUBTERRANEAN
fr souterrain
ne onderaards
de unterirdisch
da underjordisk
sv underjordisk

es subterráneo
it sotterraneo

3709
SUBTROPICAL FRUIT
fr fruits *m pl* tropicaux;
fruits *m pl* exotiques;
agrumes *m pl*
ne zuidvruchten
de Südfrüchte *f pl*
da sydfrugter
sv sydfrukt
es frutos *m pl* meridionales
it frutte *f pl* del mezzo
giorno

3710
SUBULATE; AWL-
SHAPED
fr subulé; pointu
ne priemvormig
de pfriemförmig
da sylformet
sv sylformig
es acuminado
it lesiniforme

3711
SUCCULENT PLANT
fr plante *f* grasse
ne vetplant; succulent
de Sukkulente *f*; Fettpflanze
f
da sukkulent
sv suckulent; fetbladsväxt
es planta *f* crasa; succulenta
f
it pianta *f* grassa

3712
SUCKER; PSYLLID (US)
fr psylle *m*
ne bladvlo
de Blattfloh *m*
da bladloppe
sv bladloppa
es psila *f*; mieleta *f*
it psilla *f*
Psylla spp.

SUCKER, apple, 135
—, box, 417
—, pear, 2648
—, root, 3183

3713
SUCKERS
fr drageons *mpl*
ne wortelopslag
de Stockausschlag *m*
da rodskud *n*
sv rotskott *n*
es retoños *m pl*
it pollone *m* radicato

3714
SUCKING ROOT
fr racine *f* absorbante
ne zuigwortel
de Saugwurzel *f*
da sugerod
sv sugrot
es raíz *f* chupona
it radice *f* assorbente

3715
SUCK TO THE SURFACE,
TO (BULB GROWING)
fr renouvellement *m* de la
terre par aspersion
ne omspuiten
de tief umspülen
da flytte egnet sandlag op
til overfladen
sv jordvändning med sand-
sugare
es renovación *f* de la tierra
por aspersión
it schizzare in alto (terra)

3716
SUCTION; SOIL-
MOISTURE TENSION
fr tension *f*; suction *f* d'eau
ne zuigspanning
de Saugspannung *f*
da sugekraft
sv sugspänning; sugkraft
es tensión *f* de succión
it tensione *f* d'assorbimento

3717
SUCTION DRAIN
fr petit drain *m*
ne zuigdrain
de Saugdrän *m*; Sauger *m*
da sugedræn
sv (dränerings) sugrör *n*

es drenaje *m* secundario;
  affluente *m*
it drenaggio *m* d'assorbimento

3718
SUCTION TUBE
fr pipe *f* d'admission
ne zuigbuis
de Saugrohr *n*
da sugerør *n*
sv sugrör *n*
es tubo *m* de admisión
it tubo *m* d'assorbimento

3719
SUGAR
fr sucre *m*
ne suiker
de Zucker *m*
da sukker *n*
sv socker *n*
es azúcar *m*
it zucchero *m*

3720
SUGAR PEA
fr pois *m* mangetout
ne peul (groente)
de Zuckererbse *f*
da sukkerært *m*
sv sockerärt
es guisante *m* azucarado;
  mollar *m*; tirabeque *m*
it pisello *m* mangiatutto
  Pisum sativum L.
  cv. Saccharatum (Ser.)
  Alef.

3721
SULPHATE
fr sulfate *m*
ne sulfaat *n*
de Sulfat *n*
da sulfat *n*
sv sulfat *n*
es sulfato *m*
it solfato *m*

3722
SULPHATE OF
AMMONIA
fr sulfate *m* d'ammoniaque
ne zwavelzure ammoniak

de schwefelsaures Ammo-
  niak *n*
da svovlsur ammoniak
sv svavelsyrad ammoniak
es sulfato amónico *m*
it ammoniaca *f* solforica

3723
SULPHATE OF COPPER
fr sulfate *m* de cuivre
ne kopersulfaat *n*
de Kupfersulfat *n*
da kobbersulfat *n*
sv kopparsulfat *n*
es sulfato *m* de cobre
it solfato *m* di rame

3724
SULPHATE OF POTASH
fr sulfate *m* de potasse
ne zwavelzure kali
de schwefelsaures Kali *n*
da svovlsur kali *n*
sv svavelsyrad kali *n*
es sulfato *m* de potasa
it solfato *m* di potassa

3725
SULPHATE OF POTASH-
MAGNESIA
fr sulfate *m* double de potasse et de magnésie
ne patentkali
de Patentkali *n*
da patentkali *n*
sv kalimagnesia
es sulfato *m* doble de potasio y magnesio
it solfato *m* raddoppiate
  di potassio

3726
SULPHITE, TO
fr soufrer
ne sulfiteren
de schwefeln
da svovle
sv svavla
es sulfitar
it solfare; solforare

3727
SULPHUR
fr soufre *m*
ne zwavel

de Schwefel *m*
da svovl *n*
sv svavel *n*
es azufre *m*
it zolfo *m*; solfo *m*

SULPHURIZE, TO, 1496

3728
SUMMER CHAFER
fr hanneton *m* de la St.-Jean
ne junikever
de Junikäfer *m*; Brachkäfer *m*
da St. Hans-oldenborre;
  brandenborre
sv pingborre
es escarabajo *m* san-juanero
it amfimallo *m*
  Amphimallon solstitialis L.

3729
SUMMER FRUIT
TORTRIX MOTH
fr tordeuse *f* verte
ne vruchtbladroller
de Fruchtschalenwickler *m*
da knopvikler
sv knoppvecklare
es tortrix *m* de los frutales
it tortrice *f* delle gemme
  Adoxophyes reticulana
  Hb.; *A. orana* (UK)

3730
SUMMER HOUSE
fr tonnelle *f*
ne tuinhuisje *n*
de Gartenhäuschen *n*;
  Laube *f*
da lysthus *n*; havehus *n*
sv lusthus *n*
es caseta *f* de jardín
it pergolato *m*

3731
SUMMER HYACINTH
fr jacinthe *f* du Cap
ne Kaapse hyacint
de Sommerhyazinthe *f*
da sommerhyacinth
sv kaphyacint
es jacinto *m* del Cabo
it giacinto *m* del Capo
  Galtonia candicans Dcs.

3732
SUMMER SAVORY
fr sarriette *f* commune
ne bonenkruid *n*
de Bohnenkraut *n*
da sår
sv kyndel
es ajedrea *f* de jardín
it santoreggia *f*
    Satureia hortensis L.

3733
SUNFLOWER
fr soleil *m*; tournesol *m*
ne zonnebloem
de Sonnenblume *f*
da solsikke
sv solros
es girasol *m*
it girasole *m*
    Helianthus L.

3734
SUNKEN
fr creux
ne ingezonken
de eingesunken
da indsunket
sv insjunken
es hundido
it immerso; affondato

3735
SUNK INVESTMENT
fr investissement *m* de pro-
    ductivité
ne diepte-investering
de Rationalisierungs-
    investitionen *f pl*
da (basis)investering
sv grundinvestering;
    botteninvestering
it investimenti *m pl* pro-
    duttivi (passati)

3736
SUN ROSE
fr hélianthème *m*
ne zonneroosje *n*
de Sonnenröschen *n*
da soløje *n*
sv solvända
es rosa *f* de sol

it Helianthemum *m*
    Helianthemum Mill.

3737
SUNSCORCH
fr coup *m* de soleil
ne zonnebrand
de Sonnenbrand *m*
da brændt af solen
sv solbränd
es quemadura *f* del sol
it colpo *m* di sole

3738
SUPERFICIAL SCALD
fr échaudure *f*
ne scald *n*
de Schalenbräune *f*
da skold
sv skalbränna
es escaldadura *f*
it riscaldo *m*; 'scald' *m*

3739
SUPERIOR (bot.)
fr supère
ne bovenstandig
de obenständig
da oversædig
sv översittande
es superior
it superiore

3740
SUPERMARKET
fr supermarché *m*
ne supermarkt
de Supermarkt *m*
da supermarked
sv supermarket; stormark-
    nad
es supermercado *m*
it supermercato *m*

3741
SUPERHOSPHATE
fr superphosphate *m*
ne superfosfaat *n*
de Superphosphat *n*
da superfosfat *n*
sv superfosfat *n*
es superfosfato *m*
it superfosfato *m*

3742
SUPPLEMENTARY
DRAINAGE
fr drainage *m* souterrain
    supplémentaire
ne onderbemaling
de Teilentwässerung *f*
    (mit Schöpfwerk)
da underafvanding
sv kompletterande avlopp;
    underavflytning
es drenaje *m* supletorio
it drenaggio *m* supplemen-
    tario

SUPPLEMENTARY
LIGHTING, 26

3743
SUPPLY
fr arrivage *m*; offre *f*
ne aanvoer; aanbod *n*
de Zufuhr *f*; Anfuhr *f*; An-
    lieferung*f*; Angebot *n*
da tilførsel; udbud
sv tillförsel; utbud; tillgång;
    offert
es arribo *m*; oferta *f*
it fornitura *f*; approvvigio-
    namento *m*; offerta *f*

3744
SUPPLY LINE
fr ligne *f* d'arrivée; conduite
    f d'admission
ne aanvoerleiding
de Zuleitung *f*; Zuleitungs-
    rohr *n*
da tilførselsledning
sv tilloppsledning
es caño *m* de traída; tubería
    f de abastecimiento
it tubo *m* conduttore *f*

3745
SURFACE
fr surface *f*
ne oppervlakte
de Oberfläche *f*
da overflade
sv yta
es superficie *f*
it superficie *f*

SURFACE DRAIN, 1090

3746
SURFACE DRAINING;
DITCHING
fr creuser des fossés
   (d'assainissement)
ne greppelen
de Wasserfurchen ziehen;
   entwässern
da grøftegravning
sv dika; dikesgrävning
es abrir zanjas
it drenaggio m superficiale

3747
SURFACE PANNING
(SOIL); COPPING (SOIL)
fr battre; plomber
ne dichtslaan
de verdichten
da klappe
sv klappa
es apelmazarse
it sbattere

SURVEY, TO, 2270

3748
SURVEYING (LAND)
fr arpentage m
ne landmeting
de Feldmessung f
da landmåling
sv lantmäteri
es agrimensura f
it agrimensura f

3749
SURVEYING STAFF
fr jalon m
ne jalon
de Jalon m; Absteckpfahl m
da mærkepæl; højdemarke-
   ring
sv båk; märkpåle
es jalón m
it biffa f

3750
SUSCEPTIBILITY
fr susceptibilité f
ne vatbaarheid
de Anfälligkeit f; Empfäng-
   lichkeit f
da modtagelighed

sv mottaglighet
es susceptibilidad f
it suscettibilità f; recettivi-
   tà f

3751
SUSCEPTIBLE
fr sensible; prédisposé
ne vatbaar
de empfänglich
da modtagelig
sv mottaglig
es sensible; susceptible;
   predispuesto
it suscettibile

3752
SUSCEPTIBLE TO FROST
fr sensible à la gelée
ne vorstgevoelig
de frostempfindlich
da frostfølsom
sv frostöm
es sensible al frío
it sensibile al gelo

3753
SUSPENSION HEIGHT
fr hauteur f de suspension
ne ophanghoogte
de Aufhängehöhe f
da ophængshøjde
sv upphängningshöjd
es altura f de suspensión
it altezza f d'attacco

3754
S-VIRUS DISEASE
(POTATO)
fr maladie f à virus S de la
   pomme de terre
ne S-virusziekte (aardappel)
de S-Viruskrankheit f der
   Kartoffel
da S-virose (kartoffel)
sv S-virus sjuka (potatis)
es virus m S de la patata
it virus m S

3755
SWAMP; BOG; MARSH
fr marais m; marécage m
ne moeras n
de Sumpf m; Morast m

da morads m; sump
sv moras n; kärr n
es pantano m
it palude f

3756
SWAMP BLUEBERRY
fr myrtille f cultivée
ne bosbes (blauwe bes)
de (Kultur) Heidelbeere f
da storfrugtet blåbær n
sv amerikanskt blåbär n
es arándano m americano
it mirtillo m; bacca f silvestre
   Vaccinium corymbosum L.

3757
SWAMP CYPRESS
fr cyprès m chauve; cyprès
   m de Louisiane
ne moerascypres
de Sumpfzypresse f
da sumpcypres
sv sumpcypress
es taxodio m
it cipresso m palustre
   Taxodium Rich.

3758
SWARD
fr pelouse f
ne grasmat
de Grasnarbe f
da grønsvær
sv grästorv; gräsvall
es césped m; peluse f
it prato m

3759
SWARM OF BEES
fr essaim m; jetée f
ne bijenzwerm
de Bienenschwarm m
da bisværm
sv bisvärm
es enjambre m
it sciame m

SWEAT, TO, 2632

3760
SWEDE; RUTABAGA
fr chou-navet m
ne koolraap
de Kohlrübe f; Steckrübe f

da kålrabi; kålroe
sv kålrot
es colinabo *m*
it cavolo *m* rapa
   Brassica napus L. cv.
   Napobrassica L. Rehb.

3761
SWEDE MIDGE
fr cécidomyie *f* du chou-
   fleur
ne koolgalmug
de Drehherzmücke *f*;
   Kräuselgallmücke *f*
da krusesygegalmyg
sv kålgallmygga
es cecidomia *f* de la col;
   mosquito *m* de la col;
   mosca *f* de la coliflor
it cecidomia *f* del cavolo;
   cecidomia *f* delle crucifere
   Contarinia nasturtii
   Kieffer

3762
SWEETBRIAR
fr églantier *m* odorant
ne eglantier
de schottische Zaunrose *f*;
   Weinrose *f*
da æblerose
sv äppelros
es escaramujo *m*
it rosaio *m* selvatico;
   canino *m*
   Rosa rubiginosa L.

3763
SWEET CORN; MAIZE
fr maïs *m* sucré
ne suikermaïs
de Zuckermais *m*
da sukkermajs
sv sockermajs
es maíz *m* dulce
it granturco *m* sucroso
   Zea mays L. cv. Saccha-
   rata (Koern.) Asch. et
   Graebn.

3764
SWEET GALE
fr galé *m*
ne gagel
de Gagel *m*

da pors
sv pors
es mirto *m* de Brabante
it mirica *f*
   Myrica gale L.

3765
SWEET GUM
fr copal *m* d'Amérique
ne amberboom
de Amberbaum *m*
da ambratræ *n*
sv ambraträd *n*
es liquidámbar *m*
it liquidambra *m*
   Liquidambar styraciflua L.

3766
SWEET MUST
fr moût *m*
ne zoete most
de Süssmost *m*
da sødmost
sv sötmust
es mosto *m* dulce
it mosto *m* dolce

3767
SWEET PEA
fr pois *m* de senteur
ne pronkerwt; lathyrus
de Edelwicke *f*
da ærteblomst
sv luktärt
es guisante *m* de olor
it pisello *m* odoroso
   Lathyrus odoratus L.

3768
SWEET ROCKET
fr julienne *f* des jardins
ne damastbloem
de Nachtviole *f*
da aftenstjerne
sv trädgårdsnattviol
es juliana *f* matronal
it violacciocca *f*
   Hesperis matronatis L.

3769
SWEET SCENTED VER-
NAL GRASS
fr flouve *f* odorante
ne reukgras *n*

de Ruchgras *n*
da gulaks *n*
sv vårbrodd
es grama *f* de olor
it Anthoxanthum *m*
   Anthoxanthum odoratum
   L.

3770
SWEET SEDGE
fr acore *m* odorant
ne kalmoes
de Kalmus *m*
da almindelig kalmus
sv kalmus
es cálamo *m* aromático
it calamo *m* aromatico
   Acorus calamus L.

SWEET SULTAN, 238

3771
SWEET WILLIAM
fr oeillet *m* de poète
ne duizendschoon
de Bartnelke *f*
da studenternellike
sv borstnejlika
es minutisa *f*
it garofano *m* barbuto
   Dianthus barbatus L.

3772
SWEET WOODRUFF
fr aspérule *f* odorante
ne lieve-vrouwe-bedstro *n*
de Waldmeister *m*
da skovmærke *n*
sv mysk; myskmadra
es asperilla *f* olorosa
it stellina *f* odorosa
   Asperula odorata L.

3773
SWELL, TO (BUDS);
TO GROW
fr gonfler
ne zwellen
de schwellen
da svulme
sv svälla
es hincharse; crecer
it gonfiare

3774
SWINE-CRESS; WART
CRESS
fr corne-de-cerf *f*; pied-de-
corneille *m*
ne varkenskers
de niederliegender Krähen-
fuss *m*
da ravnefod
sv kråkkrasse
es cerezo *m* de cerdo
it coclearia *f*
Coronopus squamatus
(Forsk.) Aschrs.

SWING-BACK ARM,
4135

SWISS CHARD, 3552

3775
SWITCH
fr interrupteur *m*
ne schakelaar
de Schalter *m*
da omskifter
sv strömställare
es interruptor *m*
it interrutore *m*

3776
SWORD SHAPED
fr ensiforme
ne zwaardvormig
de schwertförmig
da sværdformet
sv svärdlik
es ensiforme
it ensiforme

3777
SYMPTOM
fr symptôme *m*
ne symptoom *n*
de Symptom *m*; Krank-
heitserscheinung *f*
da symptom *n*
sv symptom *n*
es síntoma *m*
it sintomo *m*

SYMPTOM, disease, 1084

3778
SYNDROME
fr syndrome *m*
ne ziektebeeld *n*; syndroom
de Syndrom *n*; Krankheits-
bild *n*
da syndrom
sv sjukdomsbild; syndrom
es sindrome *m*
it sindrome *f*

3779
SYNTHETICS (MUSH-
ROOM GROWING); SYN-
THETIC COMPOST;
ARTIFICIAL MANURE
fr compost *m* synthétique
ne kunstmatige (syntheti-
sche) compost
de synthetischer Nährboden
da syntetisk kompost
sv syntetisk kompost
es estiércol *m* artificial;
compost *m*
it letame *m* artificale

3780
SYRINGE; SPRAYER
fr seringue *f*
ne tuinspuit
de Gartenspritze *f*
da havesprøjte; sprøjte
sv (trädgårds) spruta
es manga *f* de riego; man-
guera *f*
it siringa *f*

3781
SYRUP
fr sirop *m*
ne siroop
de eingedickter Fruchtsaft *m*;
Sirup *m*
da indkogt frugsaft; frugt-
sirup
sv saft
es jarabe *m*
it sciroppo *m*

T

3782
TABLE OF MIXTURE;
TABLE OF COMPATIBI-
LITIES
fr tableau *m* de mélange
ne mengtabel
de Mischungstafel *f*
da blandingstabel
sv blandningsschema *n*
es tabla *f* de mezclas
it tabella *f* di mescolanza

TAKE A SAMPLE, TO,
3241

3783
TAKE CUTTINGS, TO
fr bouturer
ne stekken
de durch Stecklinge ver-
     mehren; stecken
da stikke; stiklingsformering
sv sticklingsföröka; sticka
es multiplicar por esquejar;
     desquejar
it moltiplicare da talea

3784
TAKE ROOT, TO
fr s'enraciner
ne bewortelen
de bewurzeln
da rode; slå rod
sv rota sig
es barbar
it radicarsi

3785
TAKING ROOT
fr enracinement *m*
ne beworteling
de Bewurzelung *f*
da roddannelse
sv rotbildning
es enraizamiento *m*
it radicazione *f*

3786
TAKING UP;
ASSIMILATION
fr assimilation *f*
ne opneming
de Aufnahme *f*
da optagelse
sv upptagning
es asimilación *f*
it assimilazione *f*

3787
TALCUM POWDER
fr poudre *f* de talc
ne talkpoeder *n*
de Talkpulver *n*; Talkpuder
     m
da talkumpudder *n*
sv talkpuder *n*
es talco *m*; polvo *m* de talco
it polvere *f* di talco

3788
TAMARISK
fr tamaris *m*
ne tamarisk
de Tamariske *f*
da tamarisk
sv tamarisk
es tamarisco *m*; taray *m*
it Tamarix *m*
     Tamarix L.

3789
TAMPING TOOL
fr batte *f*
ne zodenklopper
de Grassodenstampfer *m*
da stamper
sv klappbräda
es pisón *m* para céspedes
it pestone *m*

3790
TANGERINE
fr mandarine *f*
ne mandarijn
de Mandarine *f*
da mandarin

sv mandarin
es mandarina *f*
it mandarino *m*

TAPE MEASURE, 2317

3791
TAP ROOT
fr pivot *m*; racine v pivo-
     tante
ne penwortel
de Pfahlwurzel *f*
da pælerod
sv pålrot
es raíz *f* penetrante; raíz *f*
     eje
it fittone *m*

3792
TAP WATER
fr eau *f* de conduite
ne leidingwater *n*
de Leitungswasser *n*
da ledningsvand *n*
sv vattenledningsvatten *n*
es agua *f* corriente
it acqua *f* potabile

3793
TARGET PRICE
fr prix *m* d'objectif
ne richtprijs
de Richtpreis *m*
da vejledende pris
sv riktpris *n*
es precio *m* indicativo
it prezzo *m* indicativo

3794
TAR OIL SPRAY; TAR
WASH
fr huile *f* d'anthracène
ne vruchtboomcarbolineum
de Obstbaumkarbolineum *n*
da frugttrækarbolineum
sv fruktträdskarbolineum
es carbolíneo *m* para fruta-
     les
it olio *m* d'antracene

3795
TARRAGON
fr estragon *m*
ne dragon
de Estragon *m*
da esdragon
sv dragon
es estragón *m*
it targone *m*
Artemisia dracunculus L.

3796
TAR SPOT (ACER)
fr croûtes *f pl* noires de
l'érable
ne inktvlekkenziekte
de Teerfleckigkeit *f*;
Schwarzfleckigkeit *f*;
Ahornrunzelschorf *m*
da ahorn-rynkeplet
sv tärfläcksjuka
es tinta *f*
it croste *f pl* nere delle
foglie
Rhytisma acerinum
(Pers.) Fr.

TAR WASH, 3794

3797
TASK SETTING
fr répartition *f* des tâches
ne taakstelling
de Aufgabenstellung *f*
da procesopstilling
sv ackordssättning; arbets-
instruktion
es reparto *m* de las tareas
de los trabajos
it distribuzione *f* del lavoro

3798
TEAM WORK
fr travail *m* d'équipe
ne groepswerk *n*
de Gruppenarbeit *f*
da gruppearbejde *n*
sv lagarbete *n*
es trabajo *m* de equipo
it lavoro *m* di gruppo

3799
TEAR STRENGTH
fr résistance *f* à la déchirure
ne scheursterkte

de Zerreissfestigkeit *f*
da revnefasthed
sv rivhållfasthet
es resistencia *f* al desgarro
it resistenza *f* alla rottura

TECHNIQUE OF CULTI-
VATION, 1684

3800
TECHNOLOGICAL
fr technologique
ne technologisch
de technologisch
da teknologisk
sv teknologisk
es tecnológico
it tecnologico

3801
TEGUMENT; SEED COAT
fr tégument *m*
ne zaadhuid
de Samenhaut *f*
da frøhinde
sv fröhinna
es tegumento *m*
it tegumento *m*

3802
TEMPERATE GLASS-
HOUSE; COOL GLASS-
HOUSE
fr serre *f* tempérée
ne gematigde kas
de temperiertes Gewächs-
haus *n*
da tempereret hus *n*
sv tempererat växthus *n*
es invernadero *m* templado
it serra *f* temperata

3803
TEMPERATURE RULE
fr régime *m* des températures
ne temperatuurregiem *n*
de Temperaturregel *f*
da temperaturforhold *n*
sv temperaturregim;
temperaturbehärskning
es régimen *m* de temperatura
it governo *m* di temperatura

3804
TENANT
fr fermier *m*
ne pachter
de Pächter *m*
da forpagter
sv arrendator
es arrendatario *m*
it affittuario *m*

3805
TENDERNESS
fr tendreté *f*
ne malsheid
de Zartheit *f*
da blødhed
sv sprödhet
es ternura *f*
it tenerezza *f*

3806
TENDRIL
fr vrille *f*
ne rank
de Ranke *f*
da slyngtråd; ranke
sv ranka; klänge
es sarmiento *m*; pámpano *m*
it tralcio *m*

3807
TENSILE STRENGTH
fr effort *m* de traction
ne treksterkte
de Zugkraft *f*
da trækstyrke
sv draghållfasthet
es esfuerzo *m* de tracción
it sforzo *m* di trazione

3808
TENSIOMETER
fr tensiomètre *m*
ne tensiometer
de Tensiometer *n*;
Dampfspannungsmeter *n*
da tensiometer
sv tensiometer
es tensiometro *m*
it tensiometro *m*

3809
TENSION; VOLTAGE
fr tension *f*; voltage *m*
ne spanning

de Spannung *f*
da spænding
sv spänning
es tensión *f*; voltage *m*
it voltaggio *m*

3810
TENSION ROD
fr aiguille *f* pendante
ne ophangstaaf
de Tragstab *m*
da trækstang
sv dragstag
es varilla *f* de tensión
it tirante *m*

3811
TERMINAL
fr terminal
ne eindstandig
de endständig
da endestillet
sv toppställd
es terminal
it terminale

3812
TERMINAL BUD
fr bourgeon *m* terminal
ne eindknop
de Endknospe *f*
da endeknop
sv terminalknopp; ändknopp
es yema *f* terminal
it gemma *f* terminale

3813
TERMINAL FLOWER
fr fleur *f* terminale
ne topbloem
de Endblüte *f*
da endeblomst
sv toppblomma
es flor *f* terminal
it fiore *m* terminale

3814
TERMINAL GROWTH
fr point *m* de croissance
ne groeitop
de Wachstumsspitze *f*;
   Triebspitze *f*
da vækstpunkt
sv skott-spets

es ápice *m* vegetativo
it apice *m* vegetativo

3815
TERNATE
fr terné
ne drietallig
de dreizählig
da trekoblet
sv trefingrad
es terno
it ternario

3816
TERRACE
fr terrasse *f*
ne terras *n*
de Terrasse *f*
da terrasse
sv terrass
es terraza *f*
it terrazza *f*

TEST, 1259

3817
TEST CROP; EXPERI-
MENTAL CROP
fr culture *f* d'essais
ne proefgewas *n*
de Versuchsgewächs *n*
da forsøgsafgrøde
sv försöksdrift
es plantas *fpl* experimentales
it piantata *f* sperimentale

3818
TEST PLANT;
DIFFERENTIAL
fr plante *f* témoin; plante-
   test *f*
ne toetsplant
de Indikatorpflanze *f*;
   Nachweispflanze *f*;
   Testpflanze
da prøveplante; testplante
sv indikatorplanta; test-
   planta
es planta *f* testigo
it pianta *f* provatore

3819
TEXTURE
fr texture *f*
ne textuur

de Textur *f*
da tekstur
sv textur
es textura *f*
it granulazione *f* costitutiva

3820
THALE CRESS
fr arabidopsis *m*
ne zandraket
de Schmalwand *f*
da gåsemad
sv backtrav
es jaramago *m* de las
   arenas
it pelosella *f*
   Arabidopsis thaliana L.
   Heynh.

3821
THAW, TO
fr dégeler
ne ontdooien
de auftauen
da tø op
sv tina (upp)
es deshelar
it disgelare

3822
THERMOCOUPLE
fr couple *m* thermo-électri-
   que
ne thermokoppel
de Thermokuppel *f*
da termoelement *n*
sv termoelement *n*
es par *m* termo eléctrico
it coppia *f* termo-elettrico

3823
THERMOGRAPH
fr thermographe *m*
ne thermograaf
de Thermograph *n*;
   Wärmeschreiber *m*
da termograf
sv termograf
es termografo *m*
it termografo *m*

3824
THICKENER;
THICKENING AGENT
(PROCESSING)
fr épaissant *m*
ne verdikkingsmiddel *n*
de Verdickungsmittel *n*
da jævningsmiddel *n*
sv förtjockningsmedel *n*
es agente *m* de condensación
it modo *m* per condensamento

3825
THIN, TO
fr éclaircir
ne uitdunnen
de auslichten; vereinzeln
da udtynde
sv gallra
es aclarar; ralear
it diradare

3826
THIN GRAPES, TO
fr éclaircir
ne krenten
de ausdünnen; ausbrechen
da klaseudtynding
sv druvgallring
es aclarar; ralear bayas
it diradamento *m*

3827
THIN WALLED
fr de l'épicarpe mince
ne dunwandig
de dünnwändig
da tyndvægget
sv tunnväggig
es de pericarpo delgada
it con parete sottile

3828
THIRD COUNTRIES
fr pays *mpl* tiers
ne derde landen
de Drittländer *pl*
da tredielande *n pl*
sv tredje länder *n pl*
es países *m pl* terceros
it pæsi *m pl* terzi

3829
THISTLE
fr chardon *m*
ne distel
de Distel *f*
da tidsel
sv tistel
es cardo *m*
it cardo *m*
   Carduus L.

THISTLE, Carline,   587
—, creeping,   886
—, globe,   1599

THORN,   3555

3830
THORNED
fr épineux
ne gedoornd
de gedornt; dornig
da tornet
sv taggig; tornig
es espinoso
it spinato

THORNY,   2927

3831
THREE-JOINT TRUSS
fr ferme *f* à trois charnières
ne driescharnierspant
de Dreigelenkbinder *m*
da tre-charnier spær
sv 3-ledsbåge
es cercha *f* de tres charnelas
it capriata *f* a tre snodi

THREE-LEAVES (IRIS),
365

3832
THREE POINT LINKAGE
fr attelage *m* 'trois points'
ne driepuntsbevestiging
de Dreipunktanschluss *m*
da trepunkttilslut *n*
sv trepunktsupphänging;
   trepunktsanfästning
es enganche *m* triple
it attacco *m* a tre punti

3833
THREE SPAN
fr serre *f* à trois chapelles
ne driekapper
de dreischiffiges Gewächshaus *n*
da treblok
sv 3-hus block
es invernadero *m* de tres unidades
it serra *f* a tre campate

3834
THREE WAY VALVE
fr vanne *f* à trois voies
ne driewegafsluiter
de Wechselventil *n*
da tre-vejs ventil
sv trevägsventil
es válvula *f* de tres vías
it valvola *f* a tre vie

3835
THRIFT
fr gazon *m* d'Espagne
ne Engels gras
de Grasnelke *f*
da fåreleger; Engelsk græs *n*
sv trift
es clavel *m* silvestre
it Armeria *f*
   Armeria Willd.

3836
THRIPS
fr thrips *m*
ne trips
de Blasenfuss *m*; Thrips *m*
da thrips
sv trips
es thrips *m*
it tisanotteri *f*; tripide *m*
   Thysanoptera

THRIPS, cabbage,   519
—, glasshouse,   1588
—, pea,   2655

3837
THROAT (bot.)
fr gorge *f*
ne keel
de Schlund *m*
da svælg *n*

sv svalg
es garganta *f*
it gola *f*

3838
THROTTLE VALVE
fr étrangleur *m*
ne smoorklep
de Drosselklappe *f*
da drosselventil
sv spjäll *n*; bladventil
es válvula *f* de estrangulamiento
it valvola *f* a farfalla

3839
THUJOPSIS
fr hiba *m*
ne Thujopsis
de Hibalebensbaum *m*
da hønsebonstræ *n*
sv hiba
es hiba *f*
it hiba *f*
   Thujopsis Sieb. et Zucc.

3840
THYME
fr thym *m*
ne tijm
de Thymian *m*
da timian
sv timjan
es tomillo *m*
it timo *m*
   Thymus vulgaris L.

THYME, common, 778
—, grey, 1664

TICK, 2356

3841
TIE
fr tirant *m*
ne trekstang
de Zugband *m*
da trækstang
sv dragband *n*
es tirante *m*
it tirante *m*

3842
TIE, TO
fr tuteurer; palisser
ne aanbinden

de anbinden
da binde op
sv binda upp
es atar
it attaccare

3843
TIE MATERIAL
fr matériel *m* pour lier
ne bindmateriaal *n*
de Bindematerial *n*
da bindemateriale *n*
sv bindmaterial *n*
es material *m* a sujetar
it materiale *m* per legature

3844
TIER; RACK (MUSH-
ROOM GROWING)
fr étagère *f*
ne stellage; stelling
de Stellage *f*; Gestell *n*
da hyldestativ
sv 'stellage'; ställning
es estante *m*
it palco *m*

3845
TIE UP, TO
fr attacher
ne opbinden
de anbinden; aufbinden
da binde op
sv uppbinda
es atar
it legare; attaccare

3846
TIGER LILY
fr lis *m* tigré
ne tijgerlelie
de Tigerlilie *f*
da tigerlilje
sv tigerlilja
es azucena *f* tigrina
it giglio *m* della Cina
   Lilium tigrinum Grol.

3847
TIGHTEN, TO
fr étancher; étanchéiser
ne afdichten
de abdichten
da tætte

sv täta; dikta
es tapar
it tappare

3848
TILE DRAIN
fr fil *m* de drain
ne drainreeks
de Dränstrang *m*
da drænsystem *n*
sv dränsystem *n*; rörströrsträngar
es drenaje *m* en serie
it serie *f* della fognatura

3849
TILE DRAINAGE INTO
PITS
fr drainage *m* à circuit fermé
ne gesloten drainage
de Systemdränung *f*
da rørdræn *n*
sv slutet dräneringssystem;
   oavhängig dränering
es drenaje *m* en circuito
   cerrado
it drenaggio *m* chiuso

3850
TILE DRAINING
fr drainage *m* par tuyaux
ne buisdrainage
de Röhrendränierung *f*
da rørdræning
sv täckdikning
es drenaje *m* de tubos
it drenaggio *m* da tuba

3851
TILL, TO (SOIL);
TO CULTIVATE
fr labourer
ne bewerken
de bearbeiten
da bearbejde
sv bearbeta
es labrar; cultivar
it coltivare

3852
TILLAGE
fr labourage *m*
ne grondbewerking
de Bearbeitung *f*
da bearbejdning

sv bearbetning
es labranza f; laborero m
it coltivazione f

3853
TILLER, TO; TO STOOL
fr taller; drageonner
ne uitstoelen
de sich bestocken
da skyde rodskud
sv skjuta rotskott
es ahijar; macollar
it tallire

3854
TILTED CAP (MUSH-
ROOM)
fr chapeau m incliné
ne scheve hoed
de schiefer Hut m
da skaev hat
sv skev hatt; sned hatt
es cabeza f inclinada
it capello m storto (inchi-
nato)

3855
TILTH
fr couche f arable
ne bouwvoor
de Pflugfurche f
da muldlag n; madjord
sv matjord
es surco m
it solco m

TIME OF PLANTING,
2773

TIME OF SOWING,   3522

3856
TIME STUDY
fr étude f de temps
ne tijdstudie
de Zeitstudie f
da tidsstudie
sv tidsstudie
es estudio m de tiempos
it etudio m di tempo

3857
TIME SWITCH
fr contacteur m horaire
ne tijdschakelaar
de Zeitschalter m

da tidsur n
sv kopplingsur n
es contactor m horario
it contatore m orario

3858
TIME WAGE
fr payement m au temps;
salaire m fixe
ne tijdloon n
de Zeitlohn m
da tidsløn
sv tidlön
es salario m por periodo
it paga f per tempo

3859
TIMOTHY
fr fléole f des prés
ne timotheegras n
de Timotheegras n
da timoté; eng. rottehale
sv timotej
es fleo m
it Phleum m
    Phleum pratense L.

TIN,   544

3860
TIN, TO; TO CAN
fr emboîter
ne inblikken
de eindosen
da sætte på dåser
sv konservera
es poner en latas
it mettere in scatola

TINNED FRUIT,   558

TINNED VEGETABLES,
559

3861
TIPBURN (LETTUCE)
fr jaunissement m des bords:
nécrose f des bords des
feuilles
ne rand
de Blattrandkrankheit f
da bladrandsyge ('tipburn')
sv vissna bladkanter

es podredrumbre m margi-
nal
it margine m disseccato

3862
TIPPING POT TRAY
ne kantelpotblad n
de Kipptopf m
da vende-potteblade
sv multipot
es estante m revolvible para
macetas
it vaso m roversciabile

3863
TIP UP CART; TIP UP
TRUCK
fr tombereau m
ne kipkar
de Kippkarren m
da tipvogn
sv tippvagn
es volquete m
it carriuola f a bilico

3864
TISSUE (bot.)
fr tissu m
ne weefsel n
de Gewebe n
da væv n
sv vävnad
es tejido m
it tessuto m

TISSUE, leaf,   2112

3865
TISSUE PAPER
fr papier m de soie
ne zijdepapier n
de Seidenpapier n
da silkepapir n
sv silkespapper n
es papel m de seda
it carta f velina

3866
TOAD
fr crapaud m
ne pad (dier)
de Kröte f
da tudse
sv padda

es sapo *m*
it rospo *m*
  Bufo spp.

3867
TOADSTOOL
fr champignon *m* à chapeau
ne paddestoel
de Pilz *m*
da paddehat
sv svamp
es hongo *m*
it fungo *n*

TOBACCO, 2472

3868
TOBACCO MOSAIC
VIRUS
fr virus *m* de la mosaïque
  du tabac
ne tabaksmozaïekvirus *n*
de Tabakmosaikvirus *n*
da tobak-mosaik-virus
sv tobaksmosaikvirus
es virus *m* del mosaico del
  tabaco
it virus *m* del mosaico del
  tabacco

3869
TOBACCO NECROSIS
(TULIP)
fr nécrose *f* de la tulipe
ne augustaziekte
de Augustakrankheit *f*
da augustasyge
sv augustasjuka
es necrosis *f* del tabaco
it necrosi *f* del tulipano
  Nicotiana-virus II

3870
TOBACCO NECROSIS
VIRUS
fr virus *m* de la nécrose du
  tabac
ne tabaksnecrosevirus *n*
de Tabaknekrosevirus *n*
da tobak-nekrose-virus
sv tobaksnekrosvirus
es virus *m* de la necrosis del
  tabaco

it virus *m* della necrosi del
  tabacco

3871
TOLERANCE
fr tolérance *f*
ne tolerantie
de Toleranz *f*
da tolerance
sv tolerans
es tolerancia *f*
it tolleranza *f*

3872
TOMATO
fr tomate *f*
ne tomaat
de Tomate *f*
da tomat
sv tomat
es tomate *m*
it pomodoro *m*

3873
TOMATO DOUBLE
STREAK
fr maladie *f* complexe des
  stries nécrotiques (tomate)
ne complexe strepenziekte
  (tomaat)
de schwere Strichelkrank-
  heit *f* (Tomate)
da tomat-stribesyge
sv komplex strimsjuka
  (tomat)
es estriado *m* necrótico
  (tomate)
it striatura *f* necrotica
  complessa (pomodoro)

3874
TOMATO MOTH
fr noctuelle *f* potagère
ne groente-uil
de Gemüsseeule *f*
da haveugle
sv grönsaksfly
es nóctua *f* de las legum-
  bres
it nottua *f* degli orti;
  erbaggivora *f*
  Diataraxia oleracea L.
  Mamestra ol.

3875
TOMATO NARROW
LEAF; FERN LEAF
fr maladie *f* filiforme
  (tomate)
ne naaldblad *n* (tomaat)
de Fadenblättrigkeit *f*;
  Farnblättrigkeit *f*
  (Tomate)
da bregneblad *n*
sv ormbunksblad *n*; tråd-
  bladighet
es hoja *f* filiforme (tomate)
it laciniatura *f* delle foglie;
  nematofillia *f* (pomodoro)

3876
TOMATO PASTE
fr purée *f* de tomates
ne tomatenpuree
de Tomatenpuree
da tomatpuré
sv tomatpuré
es puré *m* de tomates
it puré *m* di pomodoro;
  passato *m* di pomodoro

3877
TOMATO SPOTTED WILT
fr maladie *f* bronzée
  (tomate)
ne bronsvlekkenziekte
de Bronzefleckenkrankheit *f*
  (Tomate)
da bronzetop
sv ringfläcksjuka; brons-
  fläcksjuka
es bronceado *m* (tomate)
it bronzatura *f* (pomodoro)

3878
TOMATO STREAK
fr maladie *f* des stries nécrotiques de la tomate
ne strepenziekte (tomaat)
de Strichelkrankheit *f*
da tomat-stribesyge
sv strimsjuka (tomat)
es rayado *m* necrótico del
  tomate
it striatura *f* necrotica del
  pomodoro

3879
TOMENTOSE
fr tomenteux
ne viltig
de filzig
da filtet
sv filtad
es afieltrado
it feltrino

3880
TOOL BAR
fr barre *m* porte-outils
ne werktuigraam *n*
de Vielfachgerät *n*
da redskabsophaeng
(traktor)
sv verktygshållare
es porta-útiles *f*
it barra *f* porta-attrezzi

TOOTHED,   1036

3881
TOP; RIDGE
fr cime *f*; sommet *m*
ne kruin
de Wipfel *m*
da trætop
sv topp; spets; trädtopp
es cima *f*
it cima *f*

3882
TOP, TO (ONIONS)
fr défeuiller
ne afstaarten
de entlauben
da afløve; afblade
sv beskära; toppa
es 'derribar' cebollas
it sfogliare

3883
TOP, TO; TO HEAD
DOWN
fr pincer
ne toppen
de entspitzen; pinzieren
da toppe; knibe
sv toppa
es despuntar; descabezar
it spuntare

3884
TOP CUTTING
fr bouture *f* de tête
ne kopstek
de Kopfsteckling *m*
da topstikling
sv toppstickling
es esqueje *m* terminal
it talea *f* erbacea

3885
TOP DRESSING
fr engrais *m* en couverture
ne overbemesting
de Kopfdüngung *f*
da top dressing
sv övergödsling
es abono *m* de cobertura
it concimazione *f* di copertura

3886
TOP GRAFT, TO;
TO TOP WORK
fr surgreffer
ne omenten
de umpfropfen
da ompodning
sv omympa
es reinjertar
it reinnestare

3887
TOP NECROSIS
fr nécrose *f* apicale
ne topsterfte; topnecrose
de Spitzennekrose *f*
da top-nekrose
sv toppnekros
es necrosis *m* apical
it necrosi *f* apicale;
acronecrosi *f*

3888
TOPPED CARROT
fr carottes *f pl* en vrac
ne breekpeen
de Brechmöhren *f pl*;
geköpfte Herbstmöhren
f pl
da aftoppede karotte;
brækgulerod
sv toppad morötter
es zanahoria *f* quebradiza
it carota *f* a svettare

3889
TOPPLE, TO (BULB
GROWING)
fr fléchissement *m*
ne kiepen
de Kippen *n*; Umfallen *n*
da faldesyge
sv fallsjuka
es derribarse
it rovesciarsi

3890
TOP SOIL
fr couche *f* arable
ne teellaag
de Ackerkrume *f*
da madjordslag *n*; muldlag *n*
sv matjordslager *n*
es capa *f* arable
it strato *m* arabile

TOP WORK, TO,   3886

3891
TORTRIX MOTH;
LEAF ROLLER MOTH
(US) (larvae = leaf rollers)
fr tordeuse *f*
ne bladroller
de Blattwickler *m*
da vikler
sv (blad) vecklare
es tortrix *m*
it tortrice *f*
Tortricidae

TORTRIX MOTH,
cherry-bark,   650
—, green oak,   1654
—, rose,   3191
—, summer fruit,   3729

TORUS,   3051

TOTAL SOLUBLE SAND,
1160

3892
TOWER GLASSHOUSE
fr serre *f* tour
ne torenkas
de Turmgewächshaus *n*
da tårnvæksthus *n*

sv tornväxthus *n*
es invernadero *m* torre
it serra *f* torre

3893
**TOWN REFUSE COM-
POST**
fr compost *m* d'ordures
ne stadsvuilcompost
de Stadtabfallkompost *m*
da dagrenovation; byaffald
n; skrald
sv avfallskompost
es compost *m* de basuras
públicas
it concime *m* di cascame
comunale

3894
**TOWN WASTE PEAT
COMPOST**
fr détritus *m*; compost *m* de
détritus
ne stadsvuil-veencompost
de Kompost *m* aus Stadt-
mull und Torf
da tørveblandet dagrenovation
sv 'Dano-kompost'
es mezcla *f* de basuras con
turba detritus
it cascame *m* della città

3895
TRACE ELEMENT
fr élément *m* mineur
ne sporeëlement *n*
de Spurenelement *n*
da sporelement *n*;
mikronæringsstof *n*
sv mikroelement *n*;
mikronäringsämne *n*
es elemento *m* menor;
microelemento *m*
it microelemento *m*

3896
TRACER
fr élément *m* marque
ne tracer; merkatoom *n*
de Indikator *m*; Tracer *m*
da indikator; tracer
sv indikator; tracer
es marcador *m*; indicador *m*
it elemento *m* indicatrice

3897
**TRACK-LAYING TRAC-
TOR; TRACK TRACTOR**
fr chenillard *m*
ne rupstrekker
de Raupenschlepper *m*;
Kettenschlepper *m*
da larvefods traktor
sv bandtraktor
es tractor *m*
it trattrice *f* a congoli

3898
TRACK WIDTH
fr voie *f*
ne spoorbreedte
de Spurweite *f*
da sporvidde
sv spårvidd
es vía *f*
it carreggiata *f*

3899
**TRACTOR MOUNTED
FERTILIZER BROAD-
CASTER**
fr épandeur *m* à la volée
porté
ne aanbouwcentrifugaal-
strooier
de Schlepper-Anbauschleu-
derstreuer *m*
da ophængt centrifugal
kunstgødningsspreder
sv traktormonterad centrifugal gödselspridare
es repartidor *m* de abonos a
voles traccionado
it spandiconcime *m* a
spaglio montatato

3900
**TRACTOR MOUNTED
SPRAYER**
fr pulvérisateur *m* porté
ne aanbouwspuit
de Anbauspritze *f*
da ophængt sprøjte
sv traktormonterad spruta
es pulverizador *m* para
acoplar al tractor
it polverizzatore *m* portato

3901
TRADE VALUE
fr valeur *f* marchande
ne handelswaarde
de Handelswert *m*
da handelsværdi
sv handelsvärde
es valor *m* comercial
it valore *m* commerciale

3902
TRAILED PLOUGH
fr charrue *f* simple trainée
ne rondgaande ploeg
de Beetpflug *m*
da bugseret plov
sv bogserad plog för
tegplöjning
es arado *m* simple traccionado
it aratro *m* trainato

3903
TRAILED SPRAYER
fr pulvérisateur *m* trainé
ne aanhangspuit
de Anhängespritze *f*
da bugseret sprøjte
sv bogserad spruta
es pulverizador *m* de arrastrio
it polverizzatore *m* trainato

3904
TRAILER DRAWBAR
fr attelage *m* de remorque
ne trekhaak
de Anhängerkupplung *f*
da tilslutningskobling
sv dragkrok
es enganche *m* de remolque
it barra *f* di traino del
rimorchio

3905
TRAINING STAKE
fr tuteur *m*; piquet *m*
ne stok voor vormbomen
de Stütze *f* für Formobst
da stok
sv stöv
es rodrigón *m*; tutor *m*
it tutore *m*

3906
TRANSFORMATION
fr transformation f; conversion f
ne omzetting
de Umsetzung f
da omdannelse; omsætning
sv omvandling; omsättning
es transformación f
it trasformazione f

3907
TRANSFORMER
fr transformateur m
ne transformator
de Transformator m
da transformator
sv transformator
es transformador m
it trasformatore m

3908
TRANSITION PEAT
fr tourbière f intermédiaire
ne moerasveen n
de Übergangsmoor n; Sumpfmoor n
da sumpmose
sv kärrtorv
es turba f de transición
it torba f transitora

3909
TRANSLUCENT
fr translucide
ne lichtdoorlatend
de lichtdurchlässig
da gennemsigtig
sv genomskinlig
es transluciente; transparente
it traslucido

3910
TRANSPLANT, TO
fr transplanter
ne verplanten; verpoten
de verpflanzen
da omplante
sv flytta; omplantera
es trasplantar
it trapiantare

3911
TRANSPLANTER
fr planteuse f; repiqueuse f
ne plantmachine
de Pflanzmaschine f
da plantemaskine
sv planteringsmaskin
es sembradora f; repicadora f
it trapiantatrice f

3912
TRANSPORT ROAD;
TRANSPORT STRIP
fr voie f de pulvérisation
ne rijbaan
de Reitbreite f; Reitbahn
da kørebane
sv körbana
es calle f
it striscia f stradale

3913
TRASHING; ROOTING
(MUSHROOM GROWING); CORING;
CHOGGING
fr dessoucher; nettoyer
ne schoonmaken van het bed; stompen uithalen
de Pilzstümpfe m pl entfernen
da oprense
sv putsa (bäddarna)
es limpiar; sacar los muertos; sacar raices
it pulire

3914
TRAY; BOX
fr caisse f
ne kist
de Anbaukiste f
da kasse
sv låda
es caja f
it cassa f

3915
TREE
fr arbre m
ne boom
de Baum m
da træ n
sv träd n

es árbol m
it albero m

3916
TREE FORM
fr forme f de l'arbre
ne boomvorm
de Baumform f
da træform
sv trädform
es forma f del árbol
it forma f d'alberi

3917
TREE LOPPER
fr fourche f d'élagage
ne snoeihoutschuif
de Schnittholzschieber m
da grenknuser; grenmulder
sv grensamlare
es horquilla f de limpieza
it potatore m

3918
TREE OF HEAVEN
fr vernis m du Japon; ailanthe m
ne hemelboom
de Götterbaum m
da skyrækker
sv gudaträd n
es ailanto m
it Ailanthus m
Ailanthus altissima (Mill) Swingle

3919
TREE PEONY
fr pivoine f en arbre
ne boompioen
de Baumpäonie f
da træpaeon
sv (vanlig) buskpion
es peonía f arbórea
it peonia f arborea
Paeonia suffruticosa Andr.

3920
TREE PLANTER
fr plantoir m
ne plantboor
de Pflanzbohrer m
da plantebor n

sv planteringsborr
es plantador *m* a barena
it trivella *f* da piantare

TREES AND SHRUBS,
4201

3921
TREE SCRAPER
fr grattoir-émoussoir *m* à
   écorce
ne boomkrabber
de Baumkratzer *m*; Baum-
   scharre *f*
da skrabejern *n*
sv barkskrapa
es rascador *m* de cortezas
it raschietto *m* (d'albero)

3922
TREE STUMP
fr souche *f*
ne boomstronk
de Baumstumpf *m*
da træstub
sv trädstubbe
es tocón *m*
it ceppo *m*

3923
TREE SUPPORT; STAKE;
POST; POLE
fr tuteur *m*
ne boompaal
de Baumpfahl *m*
da støttepæle
sv trädstör
es tutor *m*
it tutore *m*

3924
TREE TIE
fr lien *m* pour arbres
ne boomband *n*
de Baumband *n*
da bånd *n*
sv trädbindmaterial *n*
es bandaje *m* para árboles
it legaccio *m*

TREE-TIE, plastic, 2791

3925
TRELLIS
fr lattis *m*; treillis *m*
ne latwerk

de Lattenwerk *n*
da tremmeværk *n*
sv spaljé; läktverk *n*
es vallado *f* de listones
it graticolato *m*

3926
TRENCH
fr tranchée *f* de drainage
ne drainsleuf
de Drängraben *m*
da drænfure; drænrende
sv dränränna
es zanja *m* de drenage
it scannellatura *f* della
   fognatura

TRENCH, forcing,   1430

3927
TRENCH, TO;
TO DOUBLE DIG
fr défoncer; bêcher en profondeur
ne diepspitten
de rigolen
da kulegrave; reolgrave
sv djupgräva
es desfondar
it vangare a fondo

TRENCHING PLOUGH,
1091

TRIAL, 1259

TRIAL GARDEN, 1262

TRIAL PLOT, 1261

TRIAL RESULTS, 3114

3928
TRIANGULAR
fr triangulaire
ne driehoekig
de dreieckig
da trekantet
es triangulär
sv triangular
it triangolare

3929
TRIANGULAR PLANT-
ING
fr plantation *f* en quinconce
ne driehoeksverband *n*

de Dreiecksverband *m*
da forbundtplanting
sv plantering i förband;
   triangel plantering
es plantación *f* al tresbolillo
it piantatura *f* in triangolo

3930
TRIBASIC
fr tribasique
ne driebasisch
de dreibasisch
da trebasisk
sv trebasisk
es tribásico
it tribasico

3931
TRICKLE IRRIGATION
fr irrigation *f* par aspersion
ne druppelbevloeiing
de Tropfenbewässerung *f*
da drypvanding
sv droppbevattning
es riego *m* de gotas
it irrigazione *f* da goccia

TRIM, TO, 2963

3932
TROLLEY
fr diable *m* à caisses
ne steekwagen
de Kistenkarre *f*
da kassevogn
sv gaffeltruck
es carrito *m* con un gancho
   especial
it carello *m* sollevatore

TRUE BUG, 472

3933
TRUE-TO-TYPE
fr de race pure
ne raszuiver
de sortenrein
da sortsægte
sv rasren; sortäkta
es de pure raza; genuino
it di razza pura

3934
TRUMPET CREEPER
fr bignone *f*
ne trompetbloem

de Jasmintrompete *f*;
   Klettertrompete *f*
da trompetblomst
sv trumpetranka
es flor *f* de trompeta;
   bignonia *f*
it Campsis *m*
   Campsis Lour.

3935
TRUMPET FLOWER
fr Incarvillea *f*
ne tuingloxinia
de Staudengloxinie *f*
da havegloxinia
sv Incarvillea
es Incarvillea *f*
it Incarvillea *f*
   Incarvillea Juss.

3936
TRUMPET NARCISSUS
fr narcisse *m* à trompette
ne trompetnarcis
de Trompetennarzisse *f*
da trompetnarcis
sv påsklilja
es narciso *m* de trompeth
it trombone *m*

3937
TRUNK
fr tronc
ne stam
de Stamm *m*
da stamme
sv stam
es tronco *m*
it tronco *m*

3938
TRUSS
fr ferme *f*
ne spant
de Binder *m*
da spærfag *n*
sv takstol
es cabrio *m*; armazón *f*
it capriata *f*; trave *f*; travatura *f*

3939
TRUSS SIZE
fr grandeur *f* de la grappe
ne trosgrootte

de Blütenstandgrösse *f*
da klasestørrelse
sv klasstorlek
es volumen *m* del racimo
it grandezza *f* del grappolo

3940
TRUSS VIBRATOR
fr vibrateur *m* pour tomates
ne tomatentriller
de Tomatenrüttelgerät *n*
da tomat-vibrator
sv tomatskakare
es vibrador *m* para tomates
it vibratore *m* per fecondazione

3941
TUB
fr bac *m*
ne kuip
de Kübel *m*
da balje
sv balja
es cuba *f*
it secchio *m*

3942
TUBE
fr tube *m*
ne buis
de Röhre *f*
da rør *n*
sv rör *n*
es tubo *m*
it tubo *m*

TUBE HEATING, 2748

3943
TUBER
fr tubercule *m*
ne wortelknol
de Wurzelknolle *f*
da rodknold
sv rotknöl
es tubérculo *m*
it tubero *m*

3944
TUBERCLE
fr nodosité *f* de la racine
ne wortelknolletje *n*
de Wurzelknöllchen *n*

da rodgalle
sv bakterieknöll
es nudosidad *f*
it tuberino *m* radicale

3945
TUBEROSE
fr tubéreuse *f*
ne tuberoos
de Tuberose *f*
da tuberose
sv tuberos
es nardo *m*
it tuberosa *f*
   Polianthes tuberosa L.

3946
TUBEROUS BEGONIA
fr bégonia *m* tubéreux
ne knolbegonia
de Knollenbegonie *f*
da knoldbegonie
sv knölbegonia
es Begonia *f* tuberosa
it Begonia *f* tuberosa
   Begonia tuberhybrida
   Voss.

3947
TUBEROUS PLANT
fr plante *f* tubéreuse;
   plante tuberculifère
ne knolgewas *n*
de Knollengewächs *n*
da knoldvækst
sv knölväxt
es planta *f* tuberosa
it pianta *f* tuberosa

3948
TUBER ROT (DAHLIA)
fr pourriture *f* brune du
   Dahlia
ne bruinrot *n*
de Trockenfäule *f* der Knol-
   len
da brunråd
sv vissnesjuka
es podredumbre *f* bruna
it putrefazione *f* bruna

3949
TUB PLANT
fr plante *f* en bac
ne kuipplant

de Kübelpflanze *f*
da baljeplante
sv balja
es planta *f* de cuba
it tino *m*; tinozza *f*

3950
TUBULAR
fr tubuleux; tubulaire
ne buisvormig
de röhrig; röhrenförmig
da rørformet
sv rörformig
es tubiforme
it tubolare

3951
TUBULAR FLORET
fr fleuron *m*
ne buisbloem
de Röhrenblüte *f*
da skiveblomst (rørformet)
sv rörformig blomma;
   diskblomma
es tubuliflora *f*; flor *f* tubiforme
it fiore *m* tubolare

3952
TUFTED VETCH
fr vesceron *m*; jarosse *f*
ne vogelwikke
de Vogelwicke *f*
da muse-vikke
sv kråkvicker
es veza *f* cracca
it veccia *f* montanina:
   cracca *f*
   Vicia cracca L.

3953
TULIP
fr tulipe *f*
ne tulp
de Tulpe *f*
da tulipan
sv tulpan
es tulipán *m*
it tulipano *m*
   Tulipa L.

3954
TULIP BREAKING
fr panachure *f* de la fleur
ne breken

de Buntstreifigkeit *f*;
   'brechen' von Tulpen
da 'bræking'
sv 'breaking'
es variegación *f* infecciosa
   del tulipán
it rottura *f* del colore;
   variegatura *f* infecciosa
   del tulipano
   *Marmor tulipae (Tulipa
   virus I)*

3955
TULIP MOSAIC
fr panachure *f* de la tulipe
ne tulpemozaïek *n*
de Buntstreifigkeit *f* der
   Tulpe
da tulipan-mosaik
sv tulpanmosaik
es mosaico *m* del tulipán
it mosaico *m* del tulipano;
   variegatura *f* del tulipano

3956
TUMO(U)R
fr tumeur *f*
ne gezwel *n*
de Tumor *m*
da svulst; tumor
sv tumör; utväxt
es tumor *m*
it tumore *m*

3957
TUNIC (BULB)
fr pelure *f*
ne bolhuid
de Zwiebelhaut *f*
da løgskal
sv lökskal
es túnica *f*
it tunica *f* del bulbo

TURF,   3460

TURF, TO,   875

3958
TURGIDITY
fr turgescence *f*
ne turgor
de Turgor *m*
da saftspænding; turgortryk *n*

sv turgor; saftspänning
es turgescencia *f*
it turgescenza *f*

3959
TURNIP
fr navet *m* potager
ne raap; knol
de Weissrübe *f*; Wasser-
   rübe *f*
da majroe; hvideroe
sv rova; knöl
es nabo
it cavolo *m* rapa
   Brassica campestris L.
   cv. Rapa

3960
TURNIP GALL WEEVIL
fr charançon *m* gallicole du
   chou
ne galboorsnuitkever
de Kohlgallenrüssler *m*
da kålgallesnudebille
sv kålvivel
es agalla *f* de la col
it punteruolo *m* galligeno
   del cavolo
   *Ceuthorrhynchus pleuro-
   stigma* Marsh.

3961
TURNIP SAWFLY
fr tenthrède *m* de la rave
ne knollebladwesp
de Kohlrübenblattwespe *f*
da kålhveps
sv kålbladsstekel
es falsa oruga *f* de los nabos
   y coles
it tendredine *f* delle rape
   e dei navoni; tentredine *f*
   delle crucifere
   Athalia rosae L.

3962
TURNOVER TAX
fr taxe *f* sur le chiffre
   d'affaires
ne omzetbelasting
de Umsatzsteuer *f*
da omsætningsafgift
sv omsättningsskatt

es impuesto *m* sobre las
ventas
it tassa *f* sul volume d'affari

3963
TWIG
fr rameau *m*
ne twijg
de Zweig *m*
da kvist
sv skott *n*; gren; kvist
es rama *f*
it ramincello *m*

3964
TWIG SCAB
fr gale *f* des rameaux
ne takschurft
de Zweigschorf *m*
da grenskurv
sv grenskorv
es sarna *f* de las ramas
it rogna *f* dei rami

TWINE, 3693

3965
TWINING (bot.)
fr volubile
ne windend
de windend
da snoende
sv slingrande
es voluble
it volubile

TWITCH GRASS, 867

3966
TWO-JOINT TRUSS
fr ferme *f* à deux charnières
ne tweescharnierspant
de Zweigelenkbinder *m*
da to-charnier spær
sv 2-ledsbåge
es cercha *f* de dos charnelas
it capriata *f* a due snodi

TWO-LIPPED, 2007

3967
TWO PASS BOILER
fr chaudière *f* à double
circuit
ne tweetreksketel
de Zweizugkessel *m*
da kedel med to røgslag
sv tvåstegspanna
es caldera *f* de doble circuito
it caldaia *f* a doppio circuito

TWO-SPOTTED SPIDER
MITE, 3071

3968
TYPE OF SOIL
fr type *m* de sol
ne grondsoort
de Bodenart *f*
da jordtype
sv jordart
es tipo *m* de suelo
it tipo *m* di suolo

U

3969
UMBEL
fr ombelle *f*
ne scherm *n*
de Dolde *f*
da skærm
sv (blom-)flock
es umbela *f*
it umbella *f*

3970
UMBRELLA PLANT
fr Sciadopitys *m*
ne Japanse parasolden
de Schirmtanne *f*
da parasolgran
sv solfjädertall
es pino *m* quitasol;
piñonero *m*
it pina *f* umbella
Sciadopitys Sieb. et Zucc.

3971
UNDAMAGED
fr entier et sain
ne gaaf
de gesund; unbeschädigt
da hel; ubeskadiget
sv hel; oskadad
es entero; sano
it intero

3972
UNDERCROPPING;
INTERCROPPING
fr culture *f* intercalaire
ne onderbeplanting
de Unterpflanzung *f*
da underplantning
sv underkultur
es cultivo *m* intercalado;
cultivo *m* bajo
it coltura *f* intercalante

3973
UNDERGROUND
fr souterrain
ne ondergronds
de unterirdisch

da underjordisk
sv underjordisk
es subterráneo
it sotterraneo

UNDERGROUND
WATER, level of, 2145

UNEVEN COLOURED,
378

3974
UNHEATED
fr non chauffé
ne onverwarmd
de ungeheizt
da uopvarmet
sv oeldad; ouppvärmd
es sin calefacción
it non scaldato

3975
UNICELLULAR
fr unicellulaire
ne eencellig
de einzellig
da encellet
sv encellig
es unicelular
it unicellulare

3976
UNILOCULAR;
ONE-CELLED
fr uniloculaire
ne eenhokkig
de einfächerig
da enrummet
sv enrummig
es unilocular
it uniloculare

3977
UNISEXUAL
fr unisexué
ne eenslachtig
de eingeschlechtig
da enkønnet; særkønnet
sv enkönad

es unisexual
it unisessuale

3978
UNIT COSTS
fr prix *m* de revient;
coûts *m pl* unitaires
ne kostprijs
de Gestehungskosten *f pl*;
Stückkosten *f pl*
da fremstillingsomkostning
sv självkostnad
es costes *m pl* de producción
it costi unitari *m pl* di produzione

3979
UNIT COSTS CALCULA-
TION
fr calcul *f* de prix de revient
ne kostprijsberekening
de Gestehungskostenkalku-
lation *f*; Selbstkostenkal-
kulation *f*
da kalkulation af frem-
stillingsomkostningerne
sv självkostnadskalkyl
es cálculo *m* de costes de
producción
it calcolo *m* dei costi unitari

3980
UNIT COSTS REDUC-
TION
fr diminution *f* du prix de
revient
ne kostprijsverlaging
de Gestehungskostensen-
kung *f*; Selbstkostensen-
kung *f*
da reduktion af fremstillings
omkostningerne
sv sänkning av självkost-
nader
es disminución *f* del coste de
producción
it riduzione *f* dei costi unitari

3981
UNRIPE; IMMATURE
fr vert; non mûr
ne onrijp
de unreif
da umoden
sv omogen
es verde; inmaturo
it immaturo

3982
UNSALABLE; DUMPED;
UNSOLD
fr invendu
ne doorgedraaid
de unverkauft
da usolgt
sv osäljbar
es invendible; no vendido
it invendibile

3983
UNSTRATIFIED
fr typhonien; non stratifié
ne ongelaagd
de ungeschichtet
da ikke lagdelt

sv oskiktad; olagrad
es no estratificado
it non stratificato

3984
UNTREATED
fr non traité
ne onbehandeld
de unbehandelt
da ubehandlet
sv obehandlad
es sin tratar; testigo m
it non trattato

3985
UREA
fr urée f
ne ureum n
de Harnstoff m
da urinstof n
sv urinämne n
es urea f
it urea f

3986
UREA FORMALDEHYDE
fr urée-formol f

ne ureum-formaldehyde
de Harnstoff-formal m
da urea formaldehyd
sv urinformaldehyd
es urea-formol m
it formaldeide f ureica

3987
URINE
fr urine f
ne urine
de Harn m
da urin
sv urin
es orina f
it urina f

3988
URN-SHAPED
fr urcéolé
ne kroesvormig
de becherförmig
da bægerformet
sv bågformig
es vasiforme
it ciotoliforme

V

3989
VACUUM COOLING
fr réfrigération *f* sous vide
ne vacuümkoeling
de Vakuumkühlung *f*
da vacumkøling
sv vakumkylning
es refrigeración *f* al vacío
it raffreddamento *m* sotto
   vuoto

3990
VALERIAN
fr valériane *f*
ne valeriaan
de Baldrian *m*
da baldrian
sv vänderot
es valeriana *f*
it valeriana *f*
   Valeriana L.

3991
VALLEY GUTTER
fr gouttière *f* en *V*
ne zakgoot
de Sackrinne *f*
da skotrende
sv ränndal
es canal *m*
it grondaia *f* posta a valle

3992
VALVE
fr soupape *f*
ne afsluiter
de Ventil *n*
da ventil
sv ventil; avstängningsven-
   til
es válvula *f*
it valvola *f*

VALVE, ball, 236
—, globe, 1600

3993
VAPORIZATION
fr vaporisation *f*
ne vergassing

de Vergasung *f*; Begasung *f*
da forgasse
sv förgasning
es vaporación *f*
it vaporizzazione

3994
VAPORIZING BURNER
fr brûleur *m* à vaporisation
ne bakvergassingsbrander
de Verdampfungsbrenner *m*
da fordampningsfyr *n*
sv förgasningsbrännare
es quemador *m* de vaporización
it bruciatore *m* vaporizzatore

3995
VAPOROUS
fr gazeux
ne gasvormig
de gasförmig
da gasformig
sv gasformig
es gaseiforme; gaseoso
it gassoso

3996
VAPOUR
fr vapeur *f* d'eau
ne waterdamp
de Wasserdampf *m*
da vanddamp
sv vattenånga
es vapor *m* de agua
it vapore *m* acqueo

3997
VAPOUR BARRIER
(MUSHROOM GROWING)
fr couche *f* étanche à la vapeur; écran *m* de vapeur
ne dampdichte laag
de Dampfsperre *f*
da dampspærre
sv ångspärr

es estanco al vapor; pantalla
   f contra el vapor
it barriera *f* di vapore

3998
VARIABILITY
fr variabilité *f*
ne variabiliteit
de Variabilität *f*; Veränder-
   lichkeit *f*
da variabilitet
sv variabilitet; föränderlig-
   het
es variabilidad *f*
it variabilità *f*

3999
VARIABLE COSTS
fr charges *m* variables
ne variabele kosten
de variabele Kosten *f pl*
da variable omkostninger
sv rörliga (variabla) kostnader
es costes *m pl* variables
it costi *m pl* variabili

4000
VARIETAL PURITY;
TRUENESS-TO-TYPE
fr pureté *f* variétale
ne soortechtheid
de Sortenechtheit *f*
da sortsægthed
sv sortäkthet
es pureza *f* de la variedad
   (especie)
it purezza *f* varietale

4001
VARIETY
fr variété *f*
ne variëteit; ras *n*
de Varietät *f*; Abart *f*;
   Spielart *f*
da varietet; afart
sv varietet
es variedad *f*
es varietà *f*

4002
VARIETY CHOICE
fr choix *m* des variétés
ne rassenkeuze
de Sortenwahl *f*
da sortsvalg *n*
sv sortval *n*
es elección *f* de variedades
it scelta *f* della varietà

4003
VARIETY-TRIAL
fr test *m* variétal; épreuve *f* variétale
ne rassenproef
de Sortenversuch *m*
da prøvedyrkning; sortsforsøg *n*
sv sortförsök *n*
es ensayo *m* de variedades
it esperimento *m* di razze

4004
VASCULAR BUNDLE
fr faisceau *m* vasculaire
ne vaatbundel
de Gefässbündel *n*
da karstreng
sv kärlknippe
es haz *f* vascular
it nucleo *m* vascolare

4005
VASCULAR DISEASE
fr maladie *f* vasculaire
ne vaatziekte
de Gefässkrankheit *f*
da karsygdom
sv vissnesjuka; kärlmykos
es enfermedad *f* vascular
it malattia *f* vascolare

4006
VECTOR
fr vecteur *m*
ne vector
de Vektor *m*
da vektor; overfører
sv vektor
es vector *m*
it vettore *m*

4007
VEGETABLE (adj.)
fr végétal
ne plantaardig

de pflanzlich
da vegetabilsk
sv vegetabilisk
es vegetal
it vegetale

4008
VEGETABLE GARDEN; KITCHEN GARDEN
fr jardin *m* potager
ne groentetuin
de Gemüsegarten *m*
da køkkenhave
sv köksträdgård
es huerta *f*
it orto *m*

4009
VEGETABLE GROWING
fr culture *f* maraîchère; maraîchage *m*
ne groenteteelt
de Gemüsebau *m*; Gemüsekultur *f*
da grønsagskultur; køkkenurtedyrkning
sv grönsaksodling; köksväxtodling
es cultivo *m* de hortalizas; horticultura *f*
it coltura *f* di verdura

VEGETABLE KINGDOM, 2775

VEGETABLE MILK, 2046

4010
VEGETABLES
fr légume *m*
ne groente
de Gemüse *n*
da grønsager; køkkenurter
sv grönsaker; koksväxter
es hortaliza *f*; verdura *f*
it verdura *f*; ortaggio *m*

VEGETABLES, deep frozen, 1008
—, dehydrated, 1029
—, fresh, 1458
—, leaf, 2113

4011
VEGETABLE SEEDS
fr semences *f pl* potagères
ne groentezaden
de Gemüsesamen *m pl*
da grønsagsfrø *n pl*; køkkenurtfrø *n pl*
sv köksväxtfrö *n pl*; grönsaksfrö *n pl*
es semillas *f pl* hortalizadas
it semi *m pl* di verdura

4012
VEGETABLES IN BRINE
fr légumes *m pl* conservés au sel; légumes *m pl* en saumure
ne gezouten groenten
de eingesalzene Gemüse *n pl*
da saltede grønsager
sv insaltade grönsaker
es hortalizas *f pl* en salmuera
it ortaggi *m pl* salati

4013
VEGETATION
fr végétation *f*
ne plantengroei; vegetatie
de Pflanzenwuchs *m*; Vegetation *f*
da vegetation; plantevækst
sv vegetation; växtlighet
es vegetación *f*
it vegetazione *f*

4014
VEGETATIVE
fr végétatif
ne vegetatief
de vegetativ
da vegetativ
sv vegetativ
es vegetativo
it vegetativo

4015
VEIL (MUSHROOM GROWING)
fr voile *f*
ne (champignon) vlies *n*
de Velum *n*; Schleier *m*
da slør *n*
sv slöja; segel *n*
es velo *m*
it velo *m*

VEIN, 2465

4016
VEIN MOSAIC
fr mosaïque f des nervures
ne nerfmozaïek n
de Adernmosaik n
da nerve-mosaik
sv nervmosaik
es mosaico m de la nerva-
dura
it mosaico m nervale

VENETIAN SUMACH,
4137

VENATION, 2464

4017
VENLO HOUSE
fr serre f Venlo métallique
ne Venlokas
de Venlogewächshaus n
da Venlo-væksthus n
sv Venlo-växthus n
es invernadero m tipo Venlo
it serra f di Venlo; Venlo-
block m

4018
VENTILATE, TO
fr ventiler; aérer
ne luchten; ventileren
de lüften; ventilieren
da luften
sv ventilera; lufta
es ventilar
it ventilare

4019
VENTILATION
fr ventilation f
ne luchting
de Lüftung f
da ventilation
sv ventilation; luftning
es ventilación f
it ventilazione f

4020
VENTILATOR
fr ventilateur m
ne ventilator; luchtraam n
de Ventilator m; Lüftung f

da ventilator; luftvindue
sv ventilator; luftfönster
es ventilador m
it ventilatore m; aeratore m

4021
VENT PLUG
fr robinet m d'échappement
d'air
ne ontluchtingskraan
de Entlüftungshahn m
da udluftningsventil
sv luftningsventil; luftbok-
rur
es grifo m de escape de aire
it apertura f di sfoge

4022
VENTRAL SUTURE (bot.)
fr suture f ventrale
ne buiknaad
de Bauchnaht f
da bugsøm m
sv buksöm
es sutura f ventral
it sutura f ventrale

4023
VERBENA
fr verveine f
ne Verbena
de Eisenkraut n
da jærnurt
sv Verbena
es Verbena f
it Verbena f
Verbena L.

VERDIGRIS, 4226

4024
VERNALISATION;
YAROVISATION
fr jarovisation f; vernalisation f
ne jarowisatie; vernalizatie
de Blühbeeinflussung f nach
Jarow; Vernalisation f
da jarowisation; vernalise-
ring
sv jarovisation; vernalise-
ring

es yarovización f; vernali-
zación f
it raffreddamento m di semi
germinanti; iarovisazione
f; vernalizzazione f

VERONICA, 3546

4025
VERTICILLIUM WILT;
WILT DISEASE
fr flétrissement m; verticilliose f; trachéomycose f
ne verwelkingsziekte
de Welkekrankheit f
da kransskimmel; visnesyge
sv kransmögel; vissnesjuka
es marchitado m; enferme-
dad f de marchitamiento
it verticilliosi f; malattia f
d'appassimento
Verticillium; Fusarium

4026
VERTICILLIUM WILT
(CUCUMBER)
fr verticilliose f
ne slaapziekte
de Welkekrankheit f
da kransskimmel
sv Verticillium
es marchitez m
it verticilliosi f
Verticillium

4027
VERTICILLIUM WILT
(POTATO)
fr verticilliose f; trachéomycose f
ne ringvuur n
de Pilzringfäule f
da kransskimmel
sv vissnesjuka
es traqueo-verticilosis f
it tracheo-verticilliosi f
Verticillium albo-atrum
Reinke et Berth.

4028
VETCH
fr vesce f
ne wikke
de Wicke f
da vikke

sv vicker
es algarroba f; Vicia f
it veccia f
   Vicia L.

VETCH, common, 779
—, hairy, 1709
—, tufted, 3952

VEXILLUM, 3604

VIABILITY, 1565

4029
VIABLE
fr germinatif
ne kiemkrachtig
de keimfähig
da spiredygtig; spirekraftig
sv groningsduglig
es viable
it germinativo

4030
VIBRATE, TO
fr secouer
ne trillen
de trillen; vibrieren
da kunstigt bestøve
sv vibrera
es vibrar
it tremare

4031
VIBRATING SCREEN;
VIBRATING RIDDLE
fr trieur m à vibrations
ne trilzeef
de Rüttelsieb n
da rystesold n
sv skaksåll n
es clasificador m de vibra-
   ción
it vaglio m vibratore; vibro-
   vaglio m

4032
VIBRATING TINE
CULTIVATOR
fr vibroculteur m
ne triltand cultivator
de Feingrubber m
da finkultivator

sv fjäderpinnkultivator med
   smala fjädrande
es vibrocultivador m
it coltivatore m a denti
   vibranti

4033
VIGOROUS ROOTSTOCK
fr porte-greffe m fort
ne sterke onderstam
de starke Unterlage f
da kraftig grundstamme
sv starkt växande underlag n
es patrón m de vigor
it soggetto m vigoroso

4034
VIGOUR
fr vigueur f
ne groeikracht
de Wuchskraft f
da vækstkraft
sv växtkraft
es poder m vegetativo;
   vigor m vegetativo
it vigoria f; forza f vegetativa

4035
VINE
fr vigne f
ne wijnstok
de Rebe f; Weinstock m
da vinstok
sv vinstock
es vid f
it vite f
   Vitis vinifera L.

4036
VINE CULTURE
fr viticulture f
ne wijnbouw
de Weinbau m
da vindyrkning
sv vinodling
es viticultura f
it viticultura f

4037
VINE WEEVIL
fr otiorrhynque m de la
   vigne
ne lapsnuittor
de Lappenrüssler m

da øresnudebille
sv öronvivel
es curculiónido m de la vid
it gorgoglione m
   Otiorrhynchus spp.

4038
VIOLET; VIOLA
fr violette f
ne viooltje n
de Veilchen n
da viol
sv viol
es violeta f
it violetta f
   Viola L.

VIOLET, South African,
3518

4039
VIOLET ROOT ROT
(WITLOOF CHICORY)
fr rhizoctone m violet;
   maladie f de la mort du
   safron
ne violet wortelrot n
de violetter Wurzeltöter m
da violet rodfiltsvamp
sv rotfiltsjuka
es rhizoctonia f violacea;
   mal m vinoso
it mal m vinato
   Helicobasidium purpureum
   Pat.

4040
VIRGINIAN CREEPER
fr vigne f vierge
ne wilde wingerd
de wilder Wein m
da klatrevildvin
sv vildvin
es viña f virgen
it vite f vergine
   Parthenocissus
   quinquefolia Planch.

4041
VIROLOGY
fr virologie f
ne virologie
de Virologie f; Viruskunde f
da virologi

sv virologi
es virología *f*
it virologia *f*

4042
VIRUS-ATTACK
fr atteinte *f* virologique;
    attaque *f* virologique
ne virusaantasting
de Virusberührung *f*
da virusangreb *n*
sv virusangrepp *n*
es ataque *m* de virus
it infezione *f* da virosi

4043
VIRUS DISEASE
fr maladie *f* à virus; virose *f*
ne virusziekte
de Viruskrankheit *f*; Virose *f*
da virussygdomme
sv virussjukdom; viros
es enfermedad *f* virósica
it malattia *f* da virus

4044
VISIBLE RADIATION
fr radiation *f* visible
ne zichtbare straling
de sichtbare Strahlung *f*
da synlig stråling
sv synlig strålning
es radiación *f* visible
it radiazione *m* visible

VITALITY, 1682

4045
VITAMINS
fr vitamines *f pl*
ne vitaminen
de Vitaminen *f pl*
da vitaminer *n pl*
sv vitaminer *n pl*
es vitaminas *f pl*
it vitamine *f pl*

4046
VOLATILE
fr volatile
ne vluchtig
de flüchtig
da flygtig
sv flyktig; eterisk
es volátil
it volatile

4047
VOLATILIZE, TO
fr se volatiliser
ne vervluchtigen
de verflüchtigen
da forflygtige
sv förflyktiga
es volatilizarse
it volatilizzarsi

4048
VOLE
fr campagnol *m*
ne woelmuis
de Wühlmaus
da markmus; stadsmus
sv sork
es ratón *m* de campo
it topo *m* campagnolo

VOLTAGE, 3809

4049
VOLUME
fr volume
ne volume *n*
de Volumen *n*
da volumen *n*
sv volym
es volumen *m*
it volume *m*

4050
VOLUNTEER BULBS;
GROUND KEEPERS
fr bulbes *m pl* oubliés
ne opslag
de 'Aufschlag'; beim Roden
    im Vorjahr übersehene
    Zwiebeln
da at fjerne glemte løg
sv kvarglömda lökar
es retoños *m pl*
it emergenza *f* falsa

W

4051
WADERS (BOOTS)
fr bottes *f pl* d'égoutier
ne baggerlaarzen
de Baggerstiefel *m pl*
da fiskestøvler; vadestøvler
sv mudderstövlar; vadar-
    stövlar
es botas *f pl* de pocero
it stivali *m pl*

4052
WAGE IN KIND
fr rémunération *f* en natu-
    re
ne loon *n* in natura
de Naturallohn *m*
da naturallønning
sv naturalön *n*; lön i natura
es remuneración *f* en especie
it salario *m* in natura

4053
WAGES
fr salaire *m*
ne arbeidsloon *n*
de Arbeitslohn *m*
da arbejdsløn
sv arbetslön
es salario *m*
it salario *m*

4054
WALKING TRACTOR
fr motoculteur *m*
ne tweewielige trekker
de Einachsschlepper *m*
da tohjulet traktor
sv tvåhjulig traktor
es tractor *m*
it trattrice *f* a congoli

4055
WALL
fr mur *m*
ne muur
de Mauer *f*
da mur
sv mur

es muro *m*
it muro *m*

4056
WALL BRACING
fr renforcement *m* d'une
    façade
ne gevelverstijving
de Giebelversteifung *f*
da gavlafstivning
sv gavelförstyvning
es refuerzo *m* de una facha-
    da
it rinforzo *m* di parete

4057
WALLFLOWER
fr giroflée *f*
ne muurbloem
de Goldlack *m*
da gyldenlak
sv gyllenlack
es alhelí *m* amarillo
it viola *f* gialla
    Cheiranthus cheiri L.

4058
WALL PLATE
fr sablière *f*
ne muurplaat
de Schwellholz *n*
da vandplanke
sv syll
es placa *f* de pared
it piano *m* chi posa

4059
WALL SPEEDWELL
fr véronique *f* des champs
ne veldereprijs
de Feld-Ehrenpreis *m*
da mark-ærenpris
sv fältveronika
es verónica *f*
it ederella *f*
    Veronica arvensis L.

4060
WALNUT
fr noyer *m*
ne walnoot

de Walnuss *f*
da valnød
sv valnöt
es nogal *m*
it noce *m*
    Juglans L.

WALNUT, common,   780

WAREHOUSING,   3664

4061
WARE POTATO
fr pomme *f* de terre de con-
    sommation
ne eetaardappel
de Speisekartoffel *f*
da spisekartoffel
sv matpotatis
es papa *f* comestible
it patata *f* mangabile
    Solanum tuberosum L.

4062
WARM-WATER TREAT-
MENT
fr traitement à l'eau chaude
ne warmwaterbehandeling
de Warmwasser-behandlung,
da varmtvandsbehandling
sv varmvattenbehandling
es tratamiento *m* con agua
    templada
it trattamento *m* con acqua
    calda

WART CRESS,   3774

4063
WART DISEASE
(POTATO)
fr maladie *f* verruqueuse;
    gale *f* noire
ne wratziekte
de Kartoffelkrebs *m*
da kartoffelbrok
sv potatiskräfta
es sarna *f* verrugosa;
    sarna *f* negra

it rogna *f* nera
    Synchtrium endobioticum
    (Schilb.) Perc.

4064
WASH, TO; TO RINSE
fr laver; rincer
ne wassen; spoelen
de waschen; spülen
da vaske; spule
sv tvätta; spola
es lavar; limpiar
it lavare; risciacquare

4065
WASH IN, TO
fr jeter; pousser
ne inspoelen
de einspülen
da tilvande
sv tillvattna
es lavar; enaguazar
it iniettare

WASP, common, 781
—, German, 1563
see also; gall wasp

4066
WASTE
fr déchet *m*
ne afval *n*; uitval *n*
de Abfall *m*
da affald *n*
sv avfall *n*
es desperdicio *m*
it quasto *m* (scarto)

4067
WASTE LAND
fr terre *f* inculte
ne woeste grond
de Ödland *n*
da øde land *n*
sv vildmark
es tierra *f* incultivable
it terreno *m* deserto

4068
WASTE LIME
fr chaux *f* de déchet
ne afvalkalk
de Abfallkalk *m*
da affaldskalk

sv avfallkalk
es cal *f* de desperdicio
it calce *f* rimasuglia

4069
WATER, TO
fr arroser
ne begieten
de begiessen
da vande
sv vattna
es regar
it annaffiare

4070
WATER BALANCE
fr balance *f* hydrique
ne waterbalans
de Wasserbilanz *f*
da vandbalance
sv vattenbalans
es equilibro *m* hidrométrico
it equilibro *m* d'acqua

4071
WATER COLUMN
fr colonne *f* d'eau
ne waterkolom
de Wassersäule *f*
da vandsøjle
sv vattenpelare
es columna *f* de agua
it colonna *f* d'acqua

4072
WATER CONSERVATION
fr bilan *m* hydrique
ne waterhuishouding
de Wasserhaushalt *m*
da vandhusholdning
sv vattenhushållning
es conservación *f* del agua
it equilibro *m* d'acqua;
    conservazione *f* d'acqua

4073
WATER CONSUMPTION
fr consommation *f* d'eau
ne waterverbruik
de Wasserverbrauch *m*
da vandforbrug *n*
sv vattenförbrukning
es consumo *m* del agua
it consumo *m* d'acqua

4074
WATER CONTROL
fr économie *f* des eaux
ne waterbeheersing
de Wasserwirtschaft *f*
da vandkontrol
sv reglering av vattentillför-
    sel
es control *m* de aguas
it economia *f* d'acqua

4075
WATER CORE
fr maladie *f* vitreuse; vitres-
    cence *f*
ne glazigheid
de Glasigkeit *f*
da glasagtighed; vanddruk-
    kenhed
sv glasighet
es vidriosidad *f*; brillo *m*
    vidrioso
it vetrosità *f*

4076
WATER CRESS
fr cresson *m* de fontaine
ne waterkers
de Brunnenkresse *f*
da brøndkarse
sv källkrasse
es berro *m*
it nasturzo *m*
    Nasturtium officinale R.
    Br.

4077
WATER CULTURE
fr hydroculture *f*
ne watercultuur
de Wasserkultur *f*; Hydro-
    kultur *f*
da vandkultur
sv vattenkultur
es cultivo *m* hidropónico
it idrocoltura *f*

4078
WATER DISCHARGE
fr écoulement *m* d'eau
ne waterafvoer
de Entwässerung *f*; Wasser-
    abführung *f*
da vandafledning

sv vattenavlopp *n*
es desagüe *m*
it scarico *m* dell'acqua

4079
**WATER DISTRIBUTING
SYSTEM**
fr conduite *f* d'eau
ne waterleiding
de Wasserleitung *f*
da vandledning
sv vattenledning
es conducto *m* de agua;
cañería *f* de agua
it conduttura *f* d'acqua

4080
WATERDRAINAGE
fr évacuation *f* d'eau
ne waterafvoer
de Wasserabfluss *m*
da vandafledning
sv (vatten-)avlopp *n*
es desagüe *m*
it scolo *m*; scarico *m* dell'-
acqua

4081
WATER HAMMER
fr coup *m* de bélier
ne waterslag
de Wasserschlag *m*
da vandslag *n*
sv vattenstöl; ventilslag *n*
es avenida *f* de agua
it colpo *m* d'arieta

WATERING, 1959

4082
WATERING CAN
fr arrosoir *m*
ne gieter
de Giesskanne *f*
da vandkande
sv vattenkanna
es regadera *f*; rociadera *f*;
aspersor *m*
it annaffiatoio *m*

4083
WATER LEVEL
fr plan *m* d'eau
ne waterstand; waterpeil *n*
de Wasserstand *m*

da vandstand
sv vattenstånd *n*
es plano *m* de agua
it livello *m* d'acqua

4084
WATERLEVEL GAUGE
fr tube *f* d'indicateur de
niveau
ne peilglas *n*
de Wasserstandsrohr *n*
da vandstandsglas *n*
sv vattenståndsrör *n*
es tubo *m* indicador de nivel
it indicatore *m* di levello

4085
WATER-LILY
fr nénuphar *m*
ne waterlelie
de Seerose *f*
da nøkkerose
sv näckros
es ninfea *f*
it ninfea *f*
Nymphaea L.

4086
WATERLOGGING
fr sursaturation *f* en eau
ne wateroverlast
de Ubersättigung *f* mit
Wasser; Vernässung *f*
da overmætning med vand
sv vattenöverskott *n*
es saturación *f* en agua
it saturazione *f* d'acqua

4087
WATER MELON
fr melon *m* d'eau;
pastèque *f*
ne watermeloen
de Wassermelone *f*
da vandmelon
sv vattenmelon
es sandía *f*
it cocomero *m*
Citrullus vulgaris L.

4088
WATER MOVEMENT
fr mouvement *m* de l'eau
ne waterbeweging
de Wasserbewegung *m*

da vandbevægelse
sv vattenrörelse
es movimiento *m* del agua
it movimento *m* dell'acqua

4089
**WATERNECK
(BULB DISEASE)**
fr tulipe *f* infléchie
ne kieper
de Kipper *m*
da faldesyg tulipan
sv fallsjuk tulpan
it pianta *f* bulbosa a peduncolo acquoso

4090
WATER PRESSURE
fr tension *f* d'eau
ne waterspanning
de Wasserdruck *m*
da vandstryk *n*
sv vattentryck *n*
es presión *f* de agua
ti pressione *f* d'acqua

4091
WATERPROOF
fr impermeable; étanche à
l'eau
ne waterdicht
de wasserdicht
da vandtæt
sv vattentät
es impermeable
it a prova d'acqua

4092
WATERPROOF FITTING
fr appareil *m* d'éclairage
étanché
ne waterdichte armatuur
de wasserdichte Leuchte *f*
da vandtæt armatur
sv vattentät armatur
es armadura *f* a prueba de
agua
it apparecchio *m* d'illuminazione impermeabile

4093
WATER PURIFICATION
fr filtrage *m* d'eau
ne waterzuivering
de Wasserreinigung *f*

da vandrensning
sv vattenrening
es purificación f de aguas
it chiarificazione f dell'acqua

4094
WATER SEAL
fr sceau m d'eau
ne waterzegel n
de Wassersiegel n
da vandsegl
sv vattensigill n
es cerrado m hidráulico
it sigillo m d'acqua

WATER SHOOT, 4096

4095
WATER SHORTAGE
fr carence f d'eau
ne watertekort n
de Wassermangel m
da vandmangel
sv vattenbrist
es carencia f del agua
it deficienza f d'acqua

4096
WATER SPROUT;
WATER SHOOT
fr gourmand m
ne waterloot
de Geiltrieb m; Wasser-
  schoss m
da vanskud n
sv vattenskott
es mamón m; chupón m
it succhione m

4097
WATER STORAGE
CAPACITY
fr capacité f de stockage
  d'eau; capacité f d'emmagasiner l'eau
ne waterbergend vermogen n
de Wasseraufnahmevermö-
  gen n
da vandholdende evne
sv vattenhållande förmåga;
  vattenkapacitet
es capacidad f de retención
it capacità f di ricuperazio-
  ne d'acqua

4098
WATER SUPPLY
fr approvisionnement m en
  eau
ne watervoorziening
de Wasserversorgung f
da vandforsyning
sv vattenförsörjning
es aprovisionamiento m en
  agua
it approvvigionamento m
  d'acqua

4099
WATER TABLE;
GROUNDWATER LEVEL
fr plan m d'eau(souterraine)
  nappe f phréatique
ne grondwaterspiegel;
  freatisch vlak
de Grundwasserspiegel m
da grundvandsspejl n
sv grundvattenyta; grund-
  vattenspegel
es capa f freática; plano m
  del agua subterranea
it livello m della falda freatica; livello m dell'acqua
  sotterrana

4100
WATERTANK
fr réservoir m d'eau
ne waterbak
de Wasserbehälter m
da vandbeholder
sv reservoar; vattenbehålla-
  re
es depósito m de agua
it serbatoio m d'acqua

4101
WATER-VOLE
fr rat m d'eau
ne waterrat; woelrat
de Schermaus f; grosse
  Wühlmaus f
da vandrotte
sv vattensork
es rata f de agua
it topo m d'acqua
  Arvicoli sp.

4102
WAVELENGTH
fr longueur f d'onde
ne golflengte
de Wellenlänge f
da bølgelængde
sv våglängd
es longitud f de onda
it lunghezza f d'onda

WAVY, 3425

4103
WAX PLANT
fr Hoya m
ne wasbloem
de Wachsblume f
da voksblomst
sv porslinsblomma
es flor f de la cera
it Hoya f carnosa
  Hoya carnosa R. Br.

WAY OF GROWING,
1679

4104
WEAK ROOTSTOCK
fr porte-greffe m faible
ne zwakke onderstam
de schwache Unterlage f
da svag grundstamme
sv svagt växande underlag n
es patrón m débil
it soggetto m debole

4105
WEAR AND TEAR
fr usure f technique
ne technische slijtage
de technische Entwertung f;
  Verschleiss m; Abnut-
  zung f
da teknisk slitage
sv tekniskt slitage n
es envejecimiento m técnico
it logorio m; consumo m
  tecnico

4106
WEASEL'S SNOUT
fr muflier m tête de mort;
  gueule-de-loup f
ne akkerleeuwebek

de Ackerlöwenmaul *m*
da ager-løvemund
sv kalvnos
es hierba *f* becerra
it gallinella *f*; bocca *f* di leone selvatica
   Antirrhinum orontium L.

4107
WEATHER, TO
fr se désagréger; décomposer
ne verweren
de verwittern
da forvitre
sv förvittra
es descomponerse
it decomporsi

WEBWORM, juniper, 1987
—, parsnip, 2613

4108
WEDGE
fr coin *m*
ne wig; keil
de Keil *m*
da kile
sv kil
es rincón *m*
it cuneo *m*

WEDGE GRAFTING, 702

WEDGE SHAPED, 934

4109
WEED
fr plante *f* adventice; mauvaise herbe *f*
ne onkruid *n*
de Unkraut *n*
da ukrudt *n*
sv ogräs *n*
es mala hierba *f*
it erba *f* infestante; malerba *f*

WEED, annual, 112
—, bishop's, 1668
—, butterfly, 506
—, Kew, 1995

—, pineapple, 3048
—, perennial, 2675
—, willow, 3070

4110
WEED, TO
fr désherber; sarcler
ne wieden
de jäten
da luge
sv rensa (bort)
es escardar; desmalezar
it diserbare; sarchiare; scerbare

4111
WEED CONTROL
fr lutte *f* contre les mauvaises herbes
ne onkruidbestrijding
de Unkrautbekämpfung *f*
da ukrudtsbekæmpelse
sv ogräsbekämpning
es lucha *f* contra las malas hierbas; escarda *f*
it lotta *f* contro le erbe infestanti

4112
WEEDER; SPUD
fr pique-ivraie *m*
ne onkruidsteker
de Unkrautstecher *m*
da tidseljern *n*
sv ogrässtickare; tisteljarn *n*
es arrancador *m* de malas hierbas
it sarchio *m*; zappetto *m*

4113
WEEDING FORK
fr petite serfouette *f*
ne wiedvorkje *n*
de Jätekralle *f*
da lugeklo
sv rensklo
es escardadera *f*
it forchettina *f* a sarchiare

4114
WEEDING-HOOK
fr sarcloir
ne schrepel
de Jäthäckchen *n*

da skrabejern *n*
sv skrapa; handhacka
es escardillo *m*; garabato *m*
it raschiatoio *m*

4115
WEEDING MACHINE
fr sarcleuse *f*; désherbeuse *f*
ne wiedmachine
de Jätmaschine *f*
da lugemaskin
sv rensmaskin
es escardadera *f*
it macchina *f* di sarchiatura

4116
WEED KILLER; HERBICIDE
fr herbicide *f*
ne onkruidbestrijdings-middel *n*
de Unkrautbekämpfungs-mittel *n*
da ukrudtsmiddel *n*
sv ogräsmedel *n*
es herbicida *m*
it erbicido *m*

4117
WEEPING DISEASE (MUSHROOM GROWING)
fr maladie *f* de la goutte
ne traanziekte
de Tränenkrankheit *f*; Tropfenkrankheit *f*
da grædesygdom
sv gråtsjuka
es enfermedad *f* de la gota
it infezione *f* gocciolosa

4118
WEEPING TREE
fr arbre *m* pleureur
ne treurboom
de Trauerbaum *m*
da hængetræ *n*
sv hängträd *n*
es arbol *m* llorón
it albero *m* piangente

4119
WEEVIL (ONION); CEUTORRYNCHUS SUTURALIS
fr Ceutorrhynchus *m* suturalis

ne uieboorsnuitkever
de Zwiebelrüssler *m*
da løgsnudebille
sv lökvivel
es hernia *f* de la cebolla
it Ceutorrynchus *m* suturalis
Ceuthorrhynchus suturalis
F.

WEEVIL, apple, 138
—, apple blossom, 128
—, apple bud, 129
—, cabbage seed, 516
—, cabbage seedpot, 516
—, cabbage stem, 517
—, clay-coloured, 695
—, clover seed, 718
—, curculionid, 938
—, hazel leaf roller, 1743
—, leaf eating, 447
—, pea and bean, 2621
—, pine, 2737
—, sand, 3243
—, strawberry blossom,
3681
—, turnip gall, 3960
—, vine, 4037
see also: bean weevil;
leaf weevil

4120
WEIGHT
fr poids *m*
ne gewicht *n*
de Gewicht *n*
da vægt
sv vikt
es peso *m*
it peso *m*

4121
WEIGHT LOSS
fr perte *f* de poids
ne gewichtsverlies *n*
de Gewichtsverlust *m*
da vægttab *n*
sv viktförlust
es pérdida *f* de peso
it calo *m*; perdita *f* di peso

4122
WELD, TO
fr souder
ne lassen
de schweissen

da svejse
sv svetsa
es soldar
it saldare

4123
WELL
fr puits *m*
ne put
de Brunnen *m*
da brønd
sv brunn
es pozo *m*
it pozzo *m*

4124
WELL; SPRING
fr source *f*
ne wel
de Brunnen *m*
da kilde
sv brunn
es manantial *m*
it sorgente *f*; fonte *f*

WELL, negative, 2462

4125
WELL DECAYED DUNG;
WELL ROTTED MANU-
RE
fr fumier *m* décomposé
ne verrotte mest
de verrotteter Dünger *m*
da velomsat gødning
sv förmultnad gödsel
es estiércol *m* decompuesto
it concime *m* putrefatto

4126
WELL-DRAIN
fr épuisement *m* en puits
ne bronbemaling
de Brunnendrän *m*
es bomba *f* achicadora
it prosciugamento *m* da
sorgente

4127
WELLINGTONIA;
MAMMOTH TREE
fr Sequoia *m*
ne Mammoetboom
de Mammutbaum *m*
da kalifornisk kæmpefyr
sv mammutträd *n*

es Sequoia *m* gigante
it Sequoia *f* gigantea
Sequoia gigantea DC

4128
WELL WATER
fr eau *f* de suintement;
eau *f* de pression
ne drangwater
de Druckwasser *n*
es agua *m* viva
it acqua *f* di pressa

4129
WESTLAND BLOCK
fr Westland-warenhuis
ne Westland warenhuis *n*
de Westland-Block *m*
da Westland- block
sv växthus *n* Westland typ
es warenhuis Westland
it Westland-block

4130
WET BUBBLE
(MUSHROOM GROWING)
fr môle *f*; Mycogone *f*
ne natte mol
de nasse Molle *f*; Weichfäule *f*
da Mycogone
sv deformeringsröte;
Mycogone
es mole *m*; micogone *m*
it Mycogone *m*
Mycogone perniciosa
Magn.

4131
WET SOIL
fr sol *m* humide
ne natte grond
de nasser Boden *m*
da våd jord
sv fuktig jord; vattensjuk
jord
es suelo *m* húmedo
it suolo *m* umido

4132
WETTER; SPREADER
fr (agent) mouillant *m*
ne uitvloeier
de Netzmittel *n*; Anfeuchter
m

da spredemiddel *n*
sv spridningsmedel *n*
es mojante *f*
it bagnante *m*

4133
WHEAT CURL MITE (US)
fr phytopte *m* de la tulipe
ne tulpegalmijt
de Tulpengallmilbe *f*
da tulipangalmide
sv tulpangallkvalster
es ácaro *m* de agalla del
  tulipán
it eriofide *m* del tulipano
  Eriophyes tulipae Keifer;
  Aceria t. (UK)

WHEEL, TO, 254

4134
WHEELBARROW
fr brouette *f*
ne kruiwagen
de Schubkarren *m*
da trillebør; hjulbør
sv skottkärra
es carretilla *f*
it carriuola *f*

4135
WHEELED FEELER
ARM; SWING-BACK
ARM
fr disque *m* décavaillonneur
ne zwenkende schijf
de Schwingscheibe *f*
da drejeskive
sv kännararm med hjul
es disco *m* giratorio
it forcella *f* oscillante

4136
WHEEL HAND HOE
fr binette *f* à bras
ne handschoffelmachine
de Radhacke *f*
da håndradrenser
sv handhacka
es binadora *f* de mano
it sarchiatrice *f* a mano

4137
WHIG TREE;
VENETIAN SUMACH
fr arbre *m* à la perruque
ne pruikeboom
de Perückenbaum *m*
da paryktræ *n*
sv perukbuske
es arból *m* de las pelucas
it sommacco *m*
  Cotinus coggygria Mill.

4138
WHIP AND TONGUE
GRAFTING
fr greffer à l'anglaise
ne copuleren
de kopulieren
da kopulere
sv kopulera
es injertar
it innestare a doppio

4139
WHIPTAIL
fr coeur *m* grippé
ne klemhart *n*
de Klemmherz *n*
da hjerteløshed
sv hjärtlöshet
es cogollo *m* estrangulado
it cuore *m* serrato

4140
WHITEBEAM
fr alisier *m*
ne meelbes
de Mehlbeere *f*
da akselrøn
sv vitoxel
es serbal *m*
it Sorbus *m* aria
  Sorbus aria L.

4141
WHITE CABBAGE
fr chou *m* blanc
ne witte kool
de Weisskraut *n*
da hvidkål
sv vitkål
es col *f* blanca; repollo *m*
jt cavolo *m* bianco
  Brassica oleracea L.
  cv. Alba DC.

4142
WHITE CLOVER
fr trèfle *m* blanc
ne witte klaver
de Weissklee *m*
da hvidkløver
sv vitklöver
es trébol *m* blanco
it trifoglio *m* bianco
  Trifolium repens L.

WHITEFLY, glasshouse,
1590

WHITE GRUB, 640, 731

4143
WHITE MOULD
(LATHYRUS)
fr ramulariose *f*
ne bladvlekkenziekte
de Weissfleckenkrankheit *f*
da bladskimmel
sv bladmögel
es viruelas *f*
it ramulariosi
  Ramularia deusta (Fuck)
  Back.

4144
WHITE PLASTER
MOULD
fr plâtre *m* blanc; plâtre *m*
  farineux
ne witte kalkschimmel
de weisser Gips *m*; Gipskrankheit *f*
da hvid gipssvamp
sv vit gipssvamp
es yeso *m* blanco
it muffa *f* bianca
  Scopulariopsis fimicola
  Arn. et Barth.

4145
WHITE POPLAR
fr peuplier *m* blanc
ne witte abeel
de Silberpappel *f*
da sølvpoppel
sv silverpoppel
es chopo *m* blanco; álamo
  m blanco
it pioppo *m* bianco
  Populus alba L.

WHITE ROT, 3438

4146
WHITE ROT (ONION)
fr pourriture *f* blanche de
l'oignon
ne witrot *n*
de Mehlkrankheit *f*
da hvidråd
sv vitröta
es podredumbre *f* blanca
it marciume *m* bianco
Sclerotium cepivorum
Bork.

4147
WHITE RUST (ARABIS)
fr rouille *f* blanche
ne witte roest
de weisser Rost *m*
da korsblomsternes hvidrust
sv vitröst
es roya *f* blanca
it ruggine *f* bianca
Albugo candida (Gray)
Kze.

4148
WHITE TIP (LEEK)
fr mildiou *m*
ne papiervlekkenziekte
de Papierfleckenkrankheit *f*
da porreskimmel
sv bladmögel
es mildio *m*
it Phytophthora *f*
Phytophthora porri
Foister

4149
WHITEWASH
fr lait *m* de chaux; blanc *m*
d'Espagne
ne krijtwit *n*
de Kreideweiss *n*
da skyggevædske
sv kalk för skuggning; krita
es blanco *m* de España
it creta *f*

4150
WHITEWASH, TO
fr badigeonner à la chaux
ne krijten

de kalken
da skygge; kridte
sv skugga med krita
es blanquear
it inbiancare

4151
WHITLOW GRASS
fr drave *f* printanière
ne vroegeling
de Frühlingshunger-
blümchen *n*
da vår-gæslingeblomst
sv nagelört
es erofila *f* temprana
it stellaria *f*
Erophila verna L.
Chevalier

4152
WHOLESALE MARKET
fr marché *m* de gros
ne groothandelsmarkt
de Grossmarkt *m*
da engrosmarked *n*
sv partihandel
es mercado *m* al por mayor
it mercato *m* all'ingrosso

4153
WHOLESALER
fr grossiste *m*
ne groothandelaar; grossier
de Grosshändler *m*
da grossist
sv grossist
es mayorista *m*
it grossista *m*

4154
WHOLESALE TRADE
fr commerce *m* de gros
ne groothandel
de Grosshandel *m pl*
da engros-handel
sv grosshandel; partihandel
es comercio *m* al por mayor
it commercio *m* all'ingrosso

4155
WHORLED
fr verticillé
ne kransstandig
de quirlständig
da kransstillede

sv kransställd
es verticilado
it verticillato

WHORTLEBERRY, 311

4156
WIDEN, TO; TO BELL
fr élargir
ne optrompen (pijp)
de aufdornen; aufweiten
da udvide
sv vidga
es ensanchar
it ampliare

4157
WILD CARROT
fr carotte *f* sauvage
ne wilde peen
de wilde gelbe Rübe *f*
da vild gulerod
sv vildmorot
es zanahoria *f* silvestre
it carota *f*
Daucus carota L.

4158
WILD CHERRY TREE;
GEAN
fr merisier *m*
ne zoete kers; kriek
de Vogelkirsche *f*; Süss-
kirsche *f*
da kirsebær *n*
sv sötkörsbär *n*; fågelbar *n*
es cerezo *m* dulce
it ciliegio *m* dolce
Prunus avium L.

4159
WILD RADISH; RUNCH
fr ravenelle *f*; radis *m* sauvage; raveluche *f*
ne knopherik
de Hederich *f*; wilder Rettich
m
da kiddike
sv åkerrättika
es rabaniza *f*; rabanillo *m*;
labrestos *m*
it rafanistro *m*
Raphanus raphanistrum L.

4160
WILD SHOOTS
fr rejeton *m* radical; pousse
    f radicale; rejet *m* de ra-
    cine; drageon *m*
ne wortelopslag *n*
de Wurzelaustrieb *f*
da rodskud *n*
sv rotskott *n*
es brote *m*; yema *f*; cierzas
    fpl
it pollone *m*

4161
WILLOW
fr saule *m*
ne wilg
de Weide *f*
da pil
sv vide
es sauce *m*
it salcio *m*
    Salix L.

WILLOW, basket, 263
—, creeping, 887
—, golden, 1608

4162
WILLOW BEETLE;
POPLAR AND WILLOW
BORER (US)
fr charançon *m*; cryptorhy-
    ne *f* de l'aulne
ne elzesnuitkever
de Erlenwürger *m*
da ellesnudebille
sv alvivel
es gorgojo *m* del álamo y
    del sauce
it punteruolo *m* del salice e
    del pioppo; crittorrinco *m*
    Cryptor(r)ynchus lapathi
    L.; *Sternochetus l.* (US)

WILLOW BEETLE, blue,
781
—, brassy, 425

4163
WILLOW SAWFLY
fr mouche *f* à scie; tenthrède
    f
ne bladwesp
de Blattwespe *f*

da bladhveps
sv bladstekel
es avispa *f*
it pontania *f*
    Pontania spp.

WILLOW WEED, 3070

4164
WILT (DIANTHUS CAR.)
fr Phialophora *f*; maladie *f*
    de flétrissement
ne vaatziekte
de Welkekrankheit *f*
da nellike vifteskimmel
sv vissnesjuka
es grasa *f*
it tracheo-fialoforosi *f*
    i.a. Phialophora

WILT, blossom, 374
—, fusarium, 1514, 1515,
1516
—, shoot, 3382
—, tomato spotted, 3877
—, verticillium, 4025, 4026,
4027

WILT, TO, 1278

WILT DISEASE, 4025

4165
WILTING
fr flétrissement *m*
ne verwelking
de Verwelken *n*; Welken *n*;
    Welke *f*
da visnen
sv vissnande; slokning
es marchitarse
it avvizzimento *m*

4166
WILTING POINT
fr point *m* de flétrissement
ne verwelkingspunt
de permanenter Welke-
    punkt *m*
da visnegrænse
sv vissnepunkt
es punto *m* de marchita-
    miento
it punto *m* dell'appassimen-
    to

4167
WINCH
fr treuil *m*
ne lier
de Seilwinde *f*
da spil
sv vinsch; lintrumma
es torno *m*
it verricello *m*

4168
WIND BRACING
fr contreventement *m*
ne windverband *n*
de Windverband *m*
da vindafstivning
sv vindförband *n*
es paravientos *m pl*
it controventatura *f*

WIND BREAK, 3375

4169
WIND-EROSIVE SOIL
fr terre *f* susceptible à
    s'envoler
ne stuivende grond
de verwehender Boden *m*
da vindømfindtlig jord
sv vindömtålig jord
es suelo *m* erosivo
it suolo *m* erosivo

4170
WIND PRESSURE
fr pression *f* du vent
ne winddruk
de Winddruck *m*
da vindtryk *n*
sv vindtryck *n*
es presión *f* del vento
it pressione *f* del vento

WIND SCREEN, 3375

4171
WING
fr aile *f*
ne zwaard *n*
de Flügel *m*
da vinge
sv vinge
es ala *f*
it ala *f*

WING, lace, 2020

4172
WINGED
fr ailé
ne gevleugeld
de geflügelt
da vinget
sv vingad
es alado
it alato

4173
WINGED FRUIT
fr fruit *m* ailé; samare *f*
ne gevleugelde vrucht
de Flügelfrucht *f*
da vingefrugt
sv vingfrukt
es samara *f*; fruto *m* alado
it frutto *m* alato

4174
WING NUT
fr pterocaryer *m*
ne vleugelnoot
de Flügelnuss *f*
da vingelvalnød
sv vingnöt
es nuez *f* alada
it Pterocarya *f*
   Pterocarya Kunth.

WINTER, TO,   1793

4175
WINTERACONITE
fr Eranthis *f*
ne winteraconiet
de Winterling *m*
da eranthis
sv vintergäck
es eranta *f*
it aconito *m* d'inverno
   Eranthis Salisb.

4176
WINTER CAULIFLOWER;
BROCCOLI
fr chou-brocoli *m*
ne winterbloemkool
de Winterblumenkohl
da vinterblomkål
sv vinterblomkål
es brécoles *m pl*; coliflor *f*
   de invierno
it cavolfiore *m* invernale

WINTER DORMANCY,
4180

WINTER GRAIN MITE,
3066

4177
WINTER JASMINE
fr jasmin *m*
ne winterjasmijn
de Winterjasmin *m*
da vinterjasmin
sv vinterjasmin
es jazmín *m* nodiflora
it gelsomino *m* d'inverno;
   gelsomino *m* fiorente
   nudamente
   Jasminum nudiflorum Ldl.

4178
WINTER MOTH
fr phalène *f* hiémale;
   cheimatobie *f*
ne kleine wintervlinder
de kleiner Frostspanner *m*
da lille frostmåler
sv frostfjäril
es mariposa *f* nocturna de
   invierno; falena *f* invernal de los frutales
it falena *f* invernale; falena
   f degli alberi de frutto
   Operophtera brumata L.

4179
WINTER PURSLANE
fr pourpier *m* d'hiver;
   claytone *f* de Cuba
ne winterpostelein
de Winterportulak *m*
da vinterportulak
sv västindisk spenat
es verdolaga *f* de invierno
it portulaca *f* d'inverno
   Claytonia perfoliata
   Donn.

4180
WINTER REST PERIOD;
WINTE DORMANCY
fr repos *m* hivernal
ne winterrust
de Winterruhe *f*
da vinterhvile; dvale

sv vintervila
es reposo *m* invernal
it riposo *m* invernale

4181
WINTER SAVORY
fr sarriette *f* vivace
ne winterbonenkruid *n*
de Winterbohnenkraut *n*
da vintersar
sv vinterkyndel
es ajedrea *f*
it santoreggia *f* invernale
   Satureia montana L.

4182
WIRE BASKET
fr pot-treillis *m*
ne mandpot
de Gittertopf *m*
da netpotte
sv nätkruka; nätkorg
es bote *m* de malla; tiesto *m*
   de malla; cesto *m* de malla
it castino *m* in filo metallico

4183
WIRE CUTTERS
fr cisailles *f pl* à fil de fer
ne draadschaar
de Drahtschere *f*
da trådsaks
sv trådsax
es cizallas *f pl*
it cesoie *f pl*

4184
WIRE NETTING
fr treillis *m* de poulailler
ne kippengaas *n*
de Maschendraht *m*
da trådnet *n*
sv nät *n* (av metalltråd);
   hönsnät
es malla *f* de gallinero
it rete *f* metallica

4185
WIREWORM
fr larve *f* du taupin; larve *f*
   fil de fer
ne ritnaald; koperworm
de Drahtwurm *m*

da smelderlarve
sv (sädes-) knäpparlarv
es doradilla *f*; gusano *m*
   de alambre
it ferretto *m*; bissola *f*;
   Agriotes spp. *(larvae)*

4186
WISTARIA
fr glycine *f*
ne blauwe regen
de Glyzine *f*
da blåregn
sv blåregn *n*
es glicina *f*
it glicine *m*
   Wisteria Nutt.

4187
WITCHES' BROOMS
(BETULA)
fr balai *m* de sorcière
ne heksenbezem
de Hexenbesen *m*
da heksekost
sv häxkvast
es escoba *f* de bruja
it scopazzi *f*
   Taphrina betulina Rostr.

4188
WITCHES' BROOM AND
LEAF CURL (CHERRY)
fr balai *m* de sorcière; enroulement *m*; cloque *f* du
   pêcher et de l'amandier
ne heksenbezem; krulziekte
de Hexenbesen *m*; Kräuselkrankheit *f*
da heksekost
sv häxkvast; krussjuka
es escoba *f* de bruja; lepra *f*
   de las hojas
it bolla *f* delle foglie;
   accartocciamento *m* delle
   foglie; scopazzi *f*
   Taphrina cerasi (Fuck)
   Sad.

4189
WITCHES' BROOM
DISEASE
fr balai *m* de sorcière
ne heksenbezemziekte
   (appel, framboos)

de Hexenbesenkrankheit *f*
da heksekost
sv häxkvast
es escoba *f* de bruja
it scopazzi; cladomania *f*

4190
WITCH HAZEL
fr hamamélis *m*
ne toverhazelaar
de Zaubernuss *f*
da troldnød
sv trollhassel
es Hamamelis
it Hamamelis
   Hamamelis L.

WITHER, TO, 1158

4191
WITHERED;
SHRIVELLED
fr défraîchi
ne verlept
de welk
da rynket
sv vissen
es ajado; marchito
it appassito; avvizzito

4192
WITHERING
fr dessèchement *m*
ne verdorring
de Verdorrung *f*
da nedvisning; udtørring
sv nervissning; uttorkning
es marchitez *f*
it avvizzimento *m*

4193
WITLOOF CHICORY
fr chicorée-witloof *m*;
   chicorée *f* de Bruxelles
ne witlof
de Zichoriensalat *m*
da julesalat
sv cikorisallad
es achicoria *f* de barba
   gruesa; achicoria *f* de
   Bruselas
it cicoria *f*
   Cichorium intybus L.

4194
WOOD; FOREST
fr bois *m*; forêt *f*
ne bos *n*
de Wald *m*
da skov
sv skog
es bosque *m*
it foresta *f*; selva *f*

4195
WOODEN STRIP
LATTICE
fr grillage *m* en lattis
ne lattenrooster
de Lattengitter *n*
da tremmerist
sv trägaller *n*
es rejilla *f* de madera
it graticolato *m*

4196
WOODLOUSE;
SOWBUG (US)
fr cloporte *m*
ne pissebed
de Assel *f*
da bænkebider
sv gråsugga
es cochinilla *f*
it isopodo *m*
   Oniscus spp.

4197
WOOD MEADOW
GRASS
fr paturin *m* des bois
ne bosbeemdgras *n*
de Hainrispengras *n*
da lund-rapgræs *n*
sv lundgröe
es Poa *f* espiguilla
it Poa *f*
   Poa nemoralis L.

4198
WOOD PIGEON
fr pigeon *m* ramier; palom-
   be *f*
ne houtduif
de Ringeltaube *f*
da skovdue; ringdue
sv ringduva
es paloma *f* torcaz

it colombo *m* selvatico
Columba palumbus L.

4199
WOOD PRESERVATION
fr conservation *f* du bois
ne houtimpregnering
de Holzkonservierung *f*
da træimpregnering
sv träimpegnering
es conservación *f* de madera
it protezione *f* per il legno

4200
WOODY
fr ligneux
ne houtachtig
de holzig
da træagtig
sv träaktig
es leñoso
it legnoso

4201
WOODY PLANTS;
TREES AND SHRUBS
fr plantes *f pl* ligneuses
ne houtige gewassen *n pl*
de Gehölze *n pl*; holzartige
Gewächse *n pl*
da træagtige planter
sv vedartade växter
es plantas *f pl* leñosas
it piante *f pl* legnose

4202
WOOLLY APHID;
WOOLLY APPLE APHID
(US)
fr puceron *m* lanigère
ne appelbloedluis; bloedluis
de Blutlaus *f*
da æbleblodlus; blodlus
sv blodlus
es pulgón *m* lanigero del
manzano
it afide *m* lanigero del melo;
afide *m* sanguigno del
melo
Eriosoma lanigerum
Hausm.

4203
WOOLLY LARCH APHID;
LARCH ADELGES
fr puceron *m* lanigère du
mélèze
ne larikswolluis
de rote Fichtengallenlaus *f*
da Adelgea (aldlus)
sv lärkträdsbarrlus
es pulgón *m* lanigero del
alerce
it cherme *f* del larice
Adelges laricis Vallot

4204
WORK CONTENTS
fr quantité *f* de travail
ne werkhoeveelheid
de Arbeitsanfall *m*; Arbeitsmenge *f*; Arbeitsumfang
m
da arbejdsmængde; arbejds-
omfang *n*
sv arbetsmängd; arbets-
volym
es contenido *m* de trabajo;
tipo *m* de trabajo
it fabbisogno *m* di lavoro

4205
WORKER BEE
fr (abeille) ouvrière *f*
ne werkbij
de Arbeitsbiene *f*
da arbejder (bi)
sv arbetsbi
es abeja *f* abrera
it ape *f* operaia
Apis mellifera L.

4206
WORKING-CAPITAL
fr actifs *m pl* circulants
ne vlottende produktie-
middelen *n pl*
de umlaufende Produktionsmittel *n pl*; Umlaufkapi-
tal *n*; Verbrauchsver-
mögen *n*
da cirkulerende produk-
tionsmiddel *n pl*
sv rörliga produktions-
medel *n pl*

es medios *m pl* de produc-
cíon circulantes
it mezzi *m pl* produttivi
circolanti

4207
WORKING PRESSURE
fr pression *f* de service
ne werkdruk
de Arbeitsdruck *m*
da arbejdstryk *n*
sv arbetstryck *n*
es présion *f* de trabajo
it pressione *f* di lavoro

4208
WORK PLAN
fr plan *m* de travail
ne werkplan *n*
de Arbeitsplan *m*
da arbejdsplan
sv arbetsplan; arbetsssche-
ma *n*
es plan *m* de trabajo
it piano *m* di lavoro

WORK SEQUENCE, 2543

4209
WORK SHED
fr hangar-atelier *m*; salle *f*
de travail
ne bedrijfsschuur
de Wirtschaftsgebäude *n*
da arbejdsskur *n*
sv arbetsskjul *n*
es pabellón *m* de trabajo
it granaio *m* aziendale

4210
WORK SIMPLIFICATION
fr simplification *f* du travail
ne werkvereenvoudiging
de Arbeitsvereinfachung *f*
da arbejdsforenkling
sv arbetsförenkling
es simplificación *f* del trabajo
it semplificazione *f* del lavoro

4211
WORK SPACE; WORK
PLACE
fr poste *m* de travail
ne werkplek

de Arbeitsplatz *m*
da arbejdsplads
sv arbetsplats
es lugar *m* de trabajo
it posto *m* di lavoro

4212
WORM
fr ver *m*
ne worm
de Wurm *m*
da orm
sv mask
es gusano *m*; lombriz *f*
it verme *m*

WORM, earth, 1194
—, eel, 1200
—, imported cabbage, 3445
—, imported currant, 770

4213
WORM CHANNEL
fr galerie *f* de lombric
ne wormgang
de Wurmgang *m*
da ormegang; ormehul
sv maskgång; maskväg
es galería *f* de lombriz
it galleria *f* di verme

4214
WORM EATEN;
WORMY;
MAGGOTTY
fr véreux
ne wormstekig
de wurmstichig; madig
da ormstukken; ormædt
sv maskstungen
es carcomido
it vermicato

4215
WORMWOOD
fr absinthe *f*
ne alsem
de Wermut *m*
da malurt
sv malört
es ajenjo *m*
it assenzio *m*
Artemisia absinthium L.

WORMY, 4214

4216
WRAP IN GAUZE, TO
fr emballer en toile
ne ingazen
de einballieren
da klumpindbinding
sv inhölja (in gas)
es enfundar en gasa
it velare

4217
WRAP UP, TO; TO PACK
fr empaqueter; emballer
ne inpakken
de einpacken
da pakke ind
sv packa in
es empaquetar; embalar
it imballare

4218
WREATH
fr couronne *f* de fleurs
ne bloemkrans
de Blumenkranz *m*
da blomsterkrans
sv blomkrans
es corona *f* de flores
it corona *f*

4219
WRINKLE, TO
fr rider
ne rimpelen
de runzeln
da rynke
sv rynka
es arrugarse
it incresparsi

WRINKLED, 3216

4220
WRINKLED SEEDED
PEA
fr petit pois *m* ridé
ne kreukzadige erwt
de Markerbse *f*
da markært
sv märgärta
es guisante *m* verde arrugado
it pisello *m* a grano rugoso

4221
WRITE OFF, TO
fr amortir
ne afschrijven
de abschreiben
da afskrive
sv avskriva
es amortizar
it amortizzare

4222
WYCH ELM; SCOTCH
ELM
fr orme *m* de montagne
ne bergiep
de Bergulme *f*; Bergrüster *f*
da bjergelm
sv skogsalm
es sandalo *m* falso
it Ulmus *m* glabra
Ulmus glabra Huds.

X

4223
XYLEM
fr vaisseau *m* ligneux

ne houtvat *n*
de Holzgefäss *n*
da vedkar *n*

sv (ved-)kärl
es vaso *m* leñoso
it vaso *m* legnoso

Y

YAROVISATION, 4024

YARROW, 2342

4224
YELLOWING
fr jaunissement *m*
ne vergeling
de Vergilbung *f*
da gulning
sv kloros
es amarilleo *m*
it ingiallimento *m*

4225
YELLOW LEAF BLISTER (POPULUS)
fr cloque *f* du peuplier
ne bobbelziekte
de Pappelblasenkrankheit *f*; Taphrina *f*
es lepra *f* del chopo
it bolla *f* del pioppo
  Taphrina populina Grev. ex. Fr.

4226
YELLOW MOULD; VERDIGRIS
fr moisissure *f* jaune; vert *m* de gris
ne gele schimmel
de gelber Schimmel *m*; Grünspan *m*
da vert de gris
sv vert de gris
es cardenillo *m*; vert *m* de gris; moho *m* amarillo
it muffa *f* gialla
  Myceliophthora lutea Cost.

4227
YELLOW SAND
fr sable *m* jaune
ne geelzand *n*
de Gelbsand *m*
da bakkesand *n*; gultsand *n*
sv gul sand *n*
es arena *f* amarilla
it sabbia *f* gialla

4228
YELLOWS OF HYACINTH
fr maladie *f* bactérienne de la jacinthe
ne geelziek van Hyacinthus
de Hyacinthus Gelbfäule *f*
da hyacinth-gulbakteriose
sv gulbakterie
es amarillez *m* bacteriana del jacinto
it marciume *m* gallo dei giacinti
  Xanthomonas hyacinthi (Wakk.) Dows.

4229
YELLOW STRIPS (NARCISSUS)
fr mosaique *f*
ne grijs *n*
de Mosaikkrankheit *f*
da mosaiksyge
sv mosaik
es mosaico *m*
it malattia *f* mosaica

4230
YELLOW-TAILED MOTH; GOLD-TAILED MOTH
fr bombyx *m* cul doré
ne donsvlinder
de heller Goldafter *m*
da guldhale
sv guldsvansspinnare
es lipáride *m*
it portesia *f* simile; bombice *m* dal ventro dorato
  Euproctis similis Fuess.

4231
YEW
fr if *m*
ne Taxus
de Eibe *f*
da taks
sv idegran
es tejo *m*
it tasso *m*
  Taxus L.

4232
YIELD; RETURN
fr rendement
ne opbrengst
de Ertrag *m*
da udbytte *n*
sv produktion; avkastning
es rendimiento *m*
it rendita *f*

4233
YORKSHIRE FOG
fr houlque *f* laineuse
ne witbol
de Honiggras *n*
da fløjelsgræs *n*
sv luddtåtel
es holco *m* lanoso
it Holcus *m* lanatus
  Holcus lanatus L.

YOUNG LETTUCE, 1190

4234
YOUNG PLANTS
fr jeunes plantes *f pl*; plants *m pl*
ne jonge planten; plantgoed *n*
de Jungpflanzen *f pl*; Pflanzgut *n*
da ungplanter; plantemateriale
sv småplantor; plantmaterial
es plantas *f pl* jóvenes
it piante *f pl* giovani

4235
Y-VIRUS DISEASE (POTATO)
fr maladie *f* à virus Y (pomme de terre)
ne Y-virusziekte (aardappel)
de Y-Viruskrankheit *f* (Kartoffel)
da Y-virose (kartoffel)
sv Y-virus sjuka (potatis)
es virus *m* Y (patata)
it malattia *f* da virus Y (patata)

Z

4236
ZIG-ZAG HARROW
fr herse *f* zig-zag framée
ne zigzag-eg
de Zickzackegge *f*
da zig-zag-harve
sv lättharv; zig-zag harv
es grada *f* zig-zag de arras-
   tie
it erpice *m* a zig-zag

4237
ZINC
fr zinc *m*
ne zink *n*
de Zink *n*
da zink
sv zink

es zinc *m*
it zinco *n*

4238
ZINC-DAMAGE
fr dégat(s) *m* de zinc
ne zinkschade
de Zinkschaden *m*
da zinkskade
sv zinkskada
es damnificación *f* por zinc
it danno *m* causato dallo
   zinco

4239
ZINNIA
fr Zinnia *m*
ne Zinnia

de Zinnie *f*
da Zinnia; frøkenhat
sv Zinnia
es zinnia *f*
it Zinnia *f*
   Zinnia L.

4240
ZONAL PELARGONIUM
fr géranium *m* des horticulteurs
ne geranium
de Zonalpelargonie *f*; Geranie *f*
da pelargonie
sv pelargon
es geranio *m*
it geranio *m*
   Pelargonium zonale Ait.

INDEXES

FRANÇAIS

charrue à bêches, 3529
— balance, 233
— monosoc, 3420
— polysoc, 2423
— portée de tour, 2407
— reversible demi-tour, 1711
— rigoleuse, 1091
— simple trainée, 3902
— taupe, 2374
châssis, 1449, 3249
— double, 1100, 3533
— hollandais, 1179
châtaignier, 655
chaton, 608
chaudière, 391
— à double circuit, 3967
— à haute pression, 1796
— à vapeur, 3623
— d'eau chaude, 1836
— marine, 2278
— sectionnée, 3305
chauffage du sol, 3484
— électrique du sol, 1207
— par radiateurs, 2748
— par tuyaux, 2748
chauffer, 1754
— à sec, 1762
chauler, 2171
chaume, 925
chaux, 2170
— calcinée, 3007
— carbonatée agricole, 1671
— de déchet, 4068
— éteinte, 3435
— magnésienne carbonatée, 578
— magnésienne en poudre, 1672
— sili-potassique, 531
— vive, 3007
chef de multiplication, 1583
cheimatobie, 4178
cheminée, 661
chêne, 2505
— d'Amérique, 3067
— pédonculé, 772
— rouvre, 1172
chenillard, 3897
chenille, 607
— arpenteuse, 2212
— mineuse du cytise, 2019
chénopode, 1299
chercheur scientifique, 3106
chermès des conifères, 2733
— du pin Weymouth, 2733
— du sapin, 3587

chevelu, 1705
cheveux de Vénus, 2239
chèvrefeuille, 1813
chevron, 3017
chicorée, 658
— de Bruxelles, 4193
— frisée, 1224
chicorée-witloof, 4193
chicot, 3700
chiendent ordinaire, 867
chimère, 1622
Chionodoxa, 1601
chlorate de soude, 3462
chlore, 669
—, sans, 1451
chlorophylle, 670
chloropicrine, 671
chlorose, 672
— cicatrique, 1624
chlorure, 668
— de polyvinyle, 2839
— de potassium, 2429
choix des variétés, 4002
chou, 509
— blanc, 4141
chou-borgne, 364
chou-brocoli, 439, 4176
chou cabus, 1746
— cabus de printemps, 2577
— cabus frisé, 3255
choucroute, 3254
chou de Bruxelles, 455
— de Chine, 663
— de Milan, 3255
— feuillé, 402
chou-fleur, 611
chou frisé, 401
— marin, 3297
— monté, 402
chou-navet, 3760
chou-rave, 2005
chou-rouge, 3058
chou vert, 1991
chrysalide, 2982
chrysalider, se, 2983
chrysanthème, 680
— frutescent, 2276
Chrysanthemum maximum, 3367
chrysomèle de l'osier, 425, 781
— du peuplier, 3069
chrysorrhée, 451
chute, 1280
— des aiguilles, 2459

— des boutons, 1145
— des feuilles, 1019
— des fruits, 1985
ciboulette, 667
cicadelle, 682
— du rosier, 3188
cicatrice, 3267
— foliaire, 2094
cidre, 683
— doux, 131
cigarier, 1743
cilié, 684
cime, 3881
cimicifuge, 473
cinéraire hybride, 685
circuit de distribution, 1089
circulaire, 688
circulation, 689
— de la sève, 1407
cisailles à fil de fer, 4183
— à haies, 1772
ciselage, 1484
citron, 2128
citronnier, 2129
citrouille, 1612
cladonie crénelée, 691
cladosporiose, 1698, 2089, 2107
claie, 1852
— à ombrer, 3359
clair, 432
clapet de retenue, 2489
clause de sauvegarde, 1240
claviforme, 720
clayette, 1367
claytone de Cuba, 4179
clé de robinet, 732
clematite, 703
climat, 705
climatisation, 52, 706
cloche, 713
cloison cellulaire, 629
clone, 715
cloporte, 4196
cloque de l'azalée, 209
— du pêcher et de l'amandier, 4188
— du peuplier, 4225
clôture, 1223, 1306
clôturer, 1307
cneorrhinus plagiatus, 3243
coalescence, 722
cobalt, 727
coccinelle, 2073
cochenille, 3266

— des grains, 2738
— d'humus, 1851
forme de bouton, en, 466
— d'écusson, en, 3377
— de l'arbre, 3916
— de plume, en, 2745
— de rostre, en, 270
— du limbe, 2097
formé juvénile, 1989
formulation, 1441
Forsythia, 1442
fossé, 1090
fosse, 2751
— d'aisance, 639
fossé d'écoulement, 605
— de drainage, 1120
fosse de profil, 2946
fougère, 1312
fouiller, 1694, 3707
fourche à fumier, 1436, 2263
— d'élagage, 3917
— élévatrice, 1438
fourmi, 114
fournisseur de graines, 3314
fourrière, 1748
foyer, 1505
fragon, 503
frais, 862
— de production, 861
— de stockage, 3665
fraise, 3679
fraiser, 3202
fraisier, 3683
— des quatre saisons, 2687
framboise, 3033
framboisier, 3036
frangé, 1461
Freesia, 1452
freinage de la croissance, 1688
freinant, 648
frelon, 1817
frêne, 166
friable, 1460
friche, 1283
—, en, 1282
frisé, 893, 942
friser, 939
frisolée, 892
fritillaire, 1462
fronde, 3574
fructification, 1475, 2738
fruit, 1469, 1470
— ailé, 4173
— à pépins, 2840

— atrophié, 646
— cotelé, 3205
— déhiscent, 1027
fruiterie, 1482
fruit fascié, 3205
fruitier, 1482
fruit principal, 3641
fruits à pépins, 1723
— confits, 546
— exotiques, 3709
— frais, 1457
— tropicaux, 3709
frutescent, 3394
fumagine, 341, 3511
fumer, 2262
fumeterre officinal, 1497
fumier, 2261, 2544
— de cheval, 1821
— décomposé, 4125
— de ferme, 1297
— de mouton, 3369
— de porc, 2725
— de poule, 656
— de réchaud, 1759
— de vache, 877
fumigant, 1492
fumigation, faire une, 1493
fumure, 2269
— carbonique, 738
— de fond, 258
— d'investissement, 257
— en ligne, 1323
— en sillon, 1323
funkia, 1503, 2756
fusain, 3554
fusariose, 1153, 1510, 1511,
    1512, 1513, 1515, 3180, 3515
— du narcisse, 256
Fusarioses parenchymatiques,
    3643
fuseau, 3553
fût, 602

gadoues, 3355
gage, 2792
gaillarde, 1517
gaillet, 700
gaine, 3368
gainier, 1982
galé, 3764
gale, 775
— argentée, 3413
— de la pomme de terre, 2873
— des rameaux, 3964

galène, 3412
gale noire, 4063
— poudreuse, 2887
galerie de lombric, 4213
galéruque de la boule de neige,
    879
galéruque de l'aulne, 66
galet, 2662
galinsoga, 1995
galle commune, 3259
galles foliaires, 1518, 2116
— foliaires de l'azalée, 209
galline, 2883
galvanisé à chaud, 1834
galvaniser, 1522
gamète, 1523
gamopétale, 1524, 2381
gamosépale, 1525
garantie, 2792, 3306
— d'un crédit, 883
garnitures de chaudière, 392
gaz carbonique, 580
gazeux, 3995
gaz fumigène, 1409
— naturel, 2451
gazon, 2051
— d'Espagne, 3835
gazonnant, 638
gazonner, 875
geai, 1975
gelée, 1976
— nocturne, 2474
geler, 1453
gène, 1551
générateur d'air chaud, 1832
génération, 1552
genêt, 1556
genévrier, 1986
génotype, 1557
genre, 1559
gentiane, 1558
géoligique, 1560
geranium, 881, 1561
— à feuilles de lierre, 1966
— des horticulteurs, 4240
géranium-lierre, 1966
gerbage, 3596
gerbeur, 2587, 3595
germe, 1562
germer, 1564
germes de malt, 2255
germinatif, 4029
germination, 1566
— de pollen, 2823

luxuriant, 8
luzerne, 2224
— lupaline, 340
lyophilisation, 1454

mâche, 851
machine à effeuiller, 1018
— à faire des mottes, 3475
— à retourner, 2267
— à sérancer, 1704
— draineuse, 1124
— frigorifique, 3086
magasin, 3670
magnésie, 2230
magnésium, 2231
magnolier, 2235
mahaleb, 3657
Mahonia, 2238
main d'oeuvre, 2260
— disponible, 2011
— familiale, 1290
— nombreuse, demandant une, 2012
— salariée, 2561
maïs sucré, 3763
malade, 1083
— de l'eau, 379
maladie, 1081
— à virus, 2440, 4043
— à virus S, 3754
— à virus Y, 2440, 4235
— bactérienne, 215, 216, 217, 222, 223
— bactérienne de la jacinthe, 4228
— bactérienne des taches, 213, 214
— bronzée, 3877
— causée par le Botrytis, 409
— complexe des stries nécrotiques, 3873
— criblée du cerisier, 3387
— cryptogamique, 1499, 1502, 2403
— de carence, 1012
— de flétrissement, 1516, 4164
— de la goutte, 4117
— de la momification, 2428
— de la mort du safran, 4039
— de la pelure, 3433
— de l'orme, 1177
— d'entreposage, 3666
— de Pfeffingen, 1195
— des collerettes, 350

— des grosses nervures, 2140
— des plantes, 2768
— des stries nécrotiques de la tomate, 3878
— des taches, 410
— des taches amères, 323
— des taches noires, 352
— des taches pourpres, 2096
— des taches rouges, 3063
— de taches en anneau, 3142
— digitoire, 719
— du bord jaune du fraisier, 3686
— du collet des plantes d'épicéa et de sapin, 1058
— du corail, 836
— du flétrissement, 1514
— du rattle, 3043
— du rouge du pin, 2458
— du sol, 3479
— filiforme, 3875
— vasculaire, 4005
— verruqueuse, 4063
— vitreuse, 4075
mâle, 3603
malformation, 2252
maltose, 2254
manche à eau, 1830
mandarine, 3790
manganèse, 2256
mange-tout, 3452
manomètre, 2920
manque d'azot, 2483
— d'eau, 2369
— de lumière, 2152
— de pluie, 1014
manteau-de Notre-Dame, 2022
maraîchage, 4009
maraîcher, 1677
marais, 3755
marbrure, 2401
marché, 2282
— de gros, 4152
marcottage aérien, 57, 3138
marcotte, 2056
marcotter, 2058
marécage, 3755
marjolaine, 2280
marmelade, 1473
marne, 2291
— calcaire, 2174
marronnier d'Inde, 1819
masque à gaz, 1545
massette, 3077

mastic, 2995
— à greffer, 1626
mastiquer, 630
matériau isolant, 1930
matériel de base, 3617
— de couches, 1429
— de couverture, 871
— pour lier, 3843
matière active, 22
— colorante, 756
— inerte, 1897
— inorganique, 1918
— négligeable, 1897
— organique, 2545
— plastique, 2782
— sèche, 1152
matières premières, 3045
matricaire, 3048
maturation, 3147
— complémentaire, 45
maturité, 3146
mauvaise herbe, 2675, 4109
— à graines, 112
mauvaise récolte, 899
mauve, 2253
— en arbre, 3395
— frisée, 941
mazout, 2516
mécanisation, 2318
meiose, 3075
mélange, 3155
mélanger, 2357
mélange terreux, 3490
mêler, 2357
mélèze, 2030
méligèthe du colza, 373
mélisse citronnelle, 237
— officinale, 237
melon, 2325
— d'eau, 4087
membrane cellulaire, 629
membraneux, 2326
menthe crépue, 3541
— des champs, 1331
— poivrée, 2667
mercure, 2328
merisier, 4158
méristème, 2330
merle, 329
mésocarpe, 1474
mésophylle, 2112
mesure de lutte, 824
mesures de protection, 2958
métamorphose, 2333

Physostegia, 2508
Phytomyza ilicis, 1805
phytopte, 2077, 2804
— de la tulipe, 4133
— de l'azalée, 208
— du noisetier, 2499
phytotron, 1686, 2714
Picea pungens 'Glauca', 386
pie, 2236
pied d'alouette, 2039
pied-de-coq, 253
pied-de-corneille, 3774
pied d'oiseau, 3351
pied-mère, 2395
pied noir, 337, 338
piéride de la rave, 3445
— du chou, 520, 2037
— du navet, 1655
pierre à chaux, 2176
— calcaire, 2176
piétin, 3179
pieu, 2819
pigamon, 2311
pigeon ramier, 4198
pileux, 1706
piment, 3068
pimprenelle, 499
pin, 3283
— cembro, 151
pincer, 2730, 3883
pin de l'Himalaja, 307
— des montagnes, 2406
— maritime, 2279
— noir, 195
Pinus peuce, 2228
pipe d'admission, 3718
piqué, 1903
pique-ivraie, 4112
piquet, 2819, 3905
pissenlit, 982
pistil, 2749
pivoine, 2666
— en arbre, 3919
pivot, 3791
placer, 1948
plaie de taille, 2964
planche, 279, 389
— de semis, 3312
plancher de séchage, 1151
plan d'eau, 4083, 4099
— de commercialisation, 2286
— de culture, 902, 2772
— de travail, 4208
planificateur, 2709

planning, 2758
plantain, 2761
plantation, 1531, 2763
— en blocs, 367
— en quinconce, 3929
— en rangées, 3211
plant de pomme de terre, 3324
plante, 2759
— à balcon, 234
— adventice, 4109
— à feuillage ornemental, 1415
— alpine, 73
— aquatique, 141
— bulbeuse, 487
— cultivée, 928
— d'appartement, 1842
plante de jour court, 3383
— de jour long, 2207
— de marécage, 2298
— de rocaille, 73
— de semis, 3320
— de serre, 1585
— d'ornement, 1005, 2555
— en bac, 3949
— en motte, 2878
— en pot, 2880
— grasse, 3711
— grimpante, 709
— héméropériodique, 2207
plante-hôte, 1831
pante indicatrice, 1892
— nyctipériodique, 3383
— parasite, 2605
— portant des chatons, 81
planter, 2760
— en demeure, 2452
— pour le forçage, 1368
plantes condimentaires, 924
— ligneuses, 4201
— médicinales, 2319
— montant à graine, 393
— supérieures, 1794
— vivaces, 1784, 2673
plante témoin, 3818
plante-test, 3818
plante tuberculifère, 3947
— tubéreuse, 3947
planteuse, 3911
— de pommes de terre, 2872
plantoir, 3920
plants, 2774, 2776, 4234
— de choux hivernés, 2571
— forestiers et ornementaux, 1434
plantule, 3320

plaque ondulée, 855
plastilien, 2791
plastique armé, 3092
platane, 2757
plateau à pots, 2882
— du bulbe, 255
plate-bande, 279, 400, 1388
plate-forme, 2718, 2721
plâtre, 1702
— blanc, 4144
— brun, 448
— farineux, 4144
— rouge, 448
Platycerium, 3598
plein de sève, 1984
plier, 3889
plomber, 3298, 3747
pluie, 3018
plumule, 2807
pluriloculaire, 2424
plus d'un an, de 2384
pluviomètre, 3019
poêle à l'huile, 2518
poids, 4120
— atomique, 183
— brut, 1666
— net, 2468
— propre, 994
poils absorbants, 3173
poil sécréteur, 1577
poils radiculaires, 3173
poilu, 1706
poil urticant, 3653
poinsettie, 2811
point de croissance, 3814
— de flétrissement, 4166
— de satiété, 3253
— de saturation, 3253
— de végétation, 1681
pointeau de fermeture, 2461
pointe au travail, 2634
pointillé, 1106
pointu, 3710
poire, 2638
poireau, 2124
poirée, 3552
poire pierreuse, 2647
poirier, 2649
pois, 2620
— à cosses violettes, 2294
— à écosser, 1536
— cassé, 3559
— de senteur, 3767
— gris, 1333

rainure, 1665
raisin, 1635
rame, 276
rameau, 3963
— à bois, 2211
rames à pois, 2650
ramification, 3026
ramifié, 422, 3027
ramifier, se, 3028
rampant, 3029
rampe, 3567
ramulariose, 4143
rangée, 3210
râpe, 1645
raphia, 3016
rasette, 3431
— latérale, 100
rat, 3039
ratatiner, 3392
rat d'eau, 4101
râteau, 3022
— à herbe, 2054
râteler, 3025
ratisser, 3023
ratissoire, 1178
rat musqué, 2433
raveluche, 4159
ravenelle, 645, 4159
ray-grass, 3226
— anglais, 2672
— d'Italie, 1963
rayon médullaire, 2754
rayonnement, 3012
— d'origine thermique, 330
— ionisant, 1950
rayonner sur, 1955
rayonneur, 1136, 2281
recéper, 954
réceptacle, 3051
récessif, 3052
recherche, 3104
— des bulbes malades, 1926
récolte, 895, 1733
— déficitaire, 899
récolter, 1734
récolteuse de bulbes, 484
— de choux de Bruxelles, 456
— de haricots verts, 274
recouvrement, 2029
recouvrir, 869
recrû, 3099
rectinervé, 2603
recuire, 828
réducteur, 1910

réduction, 3074
— de croissance, 1689
— du jour, 3385
réduire, 3072
réflecteur, 3083
réflexion, 3082
réfrigération, 831, 1143
— sous vide, 3989
refroidir, 829
refroidissement, 1143
— par eau glacée, 1859
refroidisseur, 830
refroidisseur à air, 53
régime, 1472
— des températures, 3803
région bulbicole, 478
— horticole, 1824
registre de cheminée, 978
règle graduée, 2316
règles de normalisation, 1616
règne végétal, 2775
régulation des prix, 2922
reine, 3004
reine-marguerite, 662
rejet, 3217, 3379
— de racine, 4160
rejeton, 3379
— radical, 4160
relais, 3095
relevage hydraulique, 1857
rembourrage, 2184
remède, 3096
remembrement, 2027
remontant, 2686
remorque, 1485
rempoter, 2879
rémunération en nature, 4052
rendement, 4232
— brut, 1667
— moyen, 202
renforcement d'une façade, 4056
réniforme, 3100
renoncule, 504, 3032
renouée, 2003
— amphibie, 90
— des oiseaux, 2001
— persicaire, 3070
renouvellement de la terre par
  aspersion, 3715
répartition de la lumière, 2154
— des tâches, 3797
repiquer, 2777, 2928
repiqueuse, 3911
repos hivernal, 4180

réproduction sexuée, par, 1553
réséda, 2339
réservoir d'eau, 4100
— de détente, 1258
résidu, 3108
résine époxyde, 1235
résistance, 3110
— à la déchirure, 3799
— au champ, 1334
résistant, 3111
— au choc, 1874
— au froid, 1728
respiration, 427, 3112
restaurer, 3113
résultats d'analyse, 93, 3472
— expérimentaux, 3114
retard, 3116
retardant, 648, 1691
retardation, de récolte, 901
retarder, 3117
retenant l'humidité, 2370
retombant, 1141, 2665
retombée radioactive, 3014
retourner, 2800
rétrécir, se, 3390
rétrécissement du limbe, 2090
revêtement de planches, 2729
— de plastique, 2783
rhizoctone noir, 350
— violet, 4039
rhizoglyphe commun, 485
rhizome, 3122
rhomboïdal, 3124
rhubarbe, 3125
— de Tartarie, 664
Ribes alpinum, 2405
ridé, 3216
rider, 4219
rièble, 700
rigoler, 1507
rincer, 698, 1378, 3143, 4064
robinet, 730
— d'échappement d'air, 4021
— d'évacuation, 383
— de vidange, 1123
— purgeur, 360
robinier, 1284
rocaille, 3151
roches, 3150
roche sédimentaire, 3309
ronce du Japon, 1974
roncer, 304
ronger le mésophylle, 3430
rongeur, 3154

NEDERLANDS

NEDERLANDS

bonk, 3205
bonte bessevlinder, 2237
boom, 3915
boomband 2791, 3924
boomgaard, 2541
boomgaardzwam, 1419
boomkrabber, 3921
boomkweker, 2496
boomkwekerij, 2494
boomkwekerijprodukten, 2497
boompaal, 3923
boompioen, 3919
boomstronk, 3922
boomteelt, 144
boomvorm, 3916
boon, 271
—, bruine, 1996
—, witte 1996
boordbed, 400
boorder, 402
boorsnuitkever, 516
borax, 398
bordeauxse pap, 399
border, 400
boren, 841
borgstelling, 883
borium, 403
boriumgebrek, 404
borstelen, 454
borstelpoetsmachine, 1148
borstverstuiver, 3198
bos, 493, 4194
bosbeemdgras, 4197
bosbes, 3756
—, blauwe, 311
—, gewone, 311
bos- en haagplantsoen, 1434
bosgrond, 1432
bospeen, 495
bossen, 494
bosstrooisel, 1431
bostel, 430
botanisch, 407
boterbloem, 504
boterbloemluis, 1586
Botrytisstip, 411
Botrytisziekte, 409
bourgondische pap, 497
bouwland, 143
bouwvoor, 2797, 3855
bovengronds, 36
bovenstandig, 3739
braak, 1282
braakland, 1283

braak liggen, 2149
braam, 326
braamstruik, 327
brand, 117, 3451
brander, 498
brandhaar, 3653
brandnetel, 2467
brandnetelbladziekte, 335
brandplek, 3279
brandplekken, 1356
brandstof, 1489
brandvlekkenziekte, 118
brandziekte, 117
breedbladige weegbree, 774
breedstraler, 1062
breedwerpig, 435
breekpeen, 3888
breken, 3954
brem, 442
brijnkraan, 360
broccoli, 439
broeibak, 1367, 1833
broeicompost, 1427
broeien, 1425
broeikist, 1367
broeimateriaal, 1429
broeimest, 1759
broeiveur, 2531
broeivoor, 1430
broes, 3186
bronbemaling, 4126
bronsvlekkenziekte, 3877
bronzen wilgehaantje, 425
bruidsbloem, 1047
bruidsboeket, 431
bruin, inwendig, 842
bruine boon, 1996
— kalkschimmel, 448
bruine-stengelvlekkenziekte,
  2988
bruine vlekken, 443
bruingeel, 446
bruinrood, 445
bruinrot, 3948
brutogewicht, 1666
bruto-opbrengst, 1667
buffer, 471
buigzaam, 1377
buiknaad, 4022
buikziek, 2568
buis, 3942
buisbloem, 3951
buisdrainage, 3850
buisfolie, 2061

buisverwarming, 2748
buisvormig, 3950
buitentarief, 1272
bundelen, 496

cactusaaltje, 521
calcium, 528
Californische pap, 2177
callus, 535
Camassia, 540
cambium, 541
camelia, 542
cantharel, 644
capillair, 565
capillaire opstijging, 569
— stijghoogte, 567
— werking, 566
— zone, 568
capillariteit, 564
capucijners, 1333
carbonaat, 577
ceder, 616
cel, 621
celdeling, 623
celkern, 2493
cellulose, 627
cellulose-acetaat, 628
celstof, 627
celvocht, 625
celvorming, 624
celwand, 629
celweefsel, 626
centrifugaalpomp, 635
centrifugaalstrooier, 1321
centrifugaalventilator, 634
Chamaecyparis, 1285
champignon, gekweekte, 927
—, gesloten, 507
—, gevliesde, 936
—, halfopen, 936
—, open, 1369
champignonbroed, 2432
champignonmest, 2430
champignonvlieg, 2701
champignonvlies, 4015
chemische verbinding, 649
chilisalpeter, 659
Chinees klokje, 1442
Chinese aster, 662
— kool, 663
— rabarber, 664
— roos, 3189
chloor, 669
chloorarm, 2221

dopluis, 3266
doppen, 3373
dopvrucht, 16
dor, 148
dosering, 1105, 3040
dotterbloem, 2297
douane-unie, 952
Douglasspar, 1111
dovenetel, 993
draadschaar, 4183
draadvormig, 1343
draagstof, 590
draaiende sproeier, 3201
draaihartigheid, 364
draderig, 3694
dragon, 3795
drainafstand, 3525
drainage, 1118
—, gesloten, 3849
—, open, 2530
drainagediepte, 1041
draineerbuis, 1121, 1126
draineermachine, 1124, 1127
draineren, 1117
drainreeks, 3848
drainsleuf, 3926
drangwater, 2669, 4128
dravik, zachte, 440
driebasisch, 3930
driehoekig, 3928
driehoeksverband, 3929
driekapper, 3833
driekleurig viooltje, 2597
drielobbige amandel, 1181, 1398
driepuntsbevestiging, 3832
driescharnierspant, 3831
drietallig, 3815
driewegafsluiter, 3834
droge mol, 1149
drogen, 1028, 1147
droge stof, 1152
dromedarisluis, 2038
droogbloeier, 2312
droogleggen, 1117, 1125, 3056
droogmakerij, 2817
droogontsmetting, 1150
droogrand, 1157
droogrot, 1154, 1155
droogstoken, 1762
droogverkoop, 1156
droogvloer, 1151
druif, 1635
druifhyacint, 1638
druifluis, 1640

druipwaterdichte armatuur,
  1138
druivenkas, 1637
druivesap, 1639
drukverstuivingsbrander, 2919
druppelbevloeiing, 3931
druppelvorming, 1142
dubbele bak, 1109, 3533
dubbel gevind, 316
dubbelkalkfosfaat, 1053
dubbelneus, 1108
dubbelsuperfosfaat, 1110
duindoorn, 3293
duinzand, 1163
duivekervel, 1497
duizendblad, 2342
duizendknoop, 2003
duizendpoot, 631
duizendschoon, 3771
dunwandig, 3827
dunsel, 1190
duurzaamheid, 1166
duurzame produktiemiddelen,
  1167
dwaling, 3155
dwerggroei, 1183
dwergziekte, 2626

echte Christusdoorn, 914
Eckelraderziekte, 1195
economische slijtage, 2512
edelweis, 1196
EEG, 1199
eekhoorntjesbrood, 636
eenbasisch, 2378
eenbladig, 2381
eencellig, 3975
eendekroos, 1161
eenfase, 3422
eenhokkig, 3976
eenhuizig, 2380
eenjarig, 106
eenjarige hardbloem, 108
eenmalige verpakking, 2488
eenmansbedrijf, 2522
eenruiter, 1179
eenscharige ploeg, 3420
eenslachtig, 3977
eenwielige trekker, 3424
één zetten, op, 3419
eetaardappel, 4061
eetbaar, 1198
eg, 1730
egel, 1769

eggen, 1731
eglantier, 3762
egtand, 1732
ei, 1203
eiafzetting, 1204
eicel, 2529
eidodend, 2572
eierplant, 1205
eigen arbeidskrachten, 1290
— gewicht, 994
— vermogen, 2575
eik, 2505
eikebladpatroon, 2507
eikebladroller, 1654
eikehakhout, 2506
eikemeeldauw, 2886
eikvaren, 2834
eindbuis, 1227
eindknop, 3812
eindstandig, 3811
eirond, 2564
eitje, 2574
eiwit, 63
ekster, 2236
elektrische grondverwarming,
  1207
elektrisch vermogen, 2890
elektronenmicroscopie, 1209
elektronisch blad, 1208
element, 1210
elliptisch, 1212
els, 64
elzebastaardrups, 67
elzehaan, 66
elzesnuitkever, 4162
emelt, 880
emulgeerbaar, 1219
emulsie, 1221
emulsiebrander, 1220
enatiemozaïek van erwt, 2630
energie-golflengte-kromme, 3543
Engels gras, 3835
— raaigras, 2672
engerling, 640
enkelvoudig, 3418
ent, 1620, 3274
entbastaard, 1622
entchlorose, 1624
enten, 1621, 3539
— ter zijde, 3397
enting, 1623
entkever, 138
entmes, 1625
entomoloog, 1229

juffertje in 't groen, 2220
junikever, 3728
junival, 1985
jute, 1988

kaal, 1573
Kaapse hyacint, 3731
Kaaps viooltje, 3518
kaasjeskruid, 2253
kafje, 1603
kainiet, 1990
kalf, 2185
kali, 2856
—, zwavelzure, 3724
kalialuin, 2857
kalibreren, 532
kalifixatie, 2863
kaligebrek, 2862
kaligehalte, 2861
kalikiezelkalk, 531
kaliloog, 613
kalimeststof, 2858
kalisalpeter, 2479
kalium, 2860
kaliumcarbonaat, 2856
kalizout, 2859
kalk, 2170
kalkammonsalpeter, 529
kalkarm, 1016
kalkbodem, 524
kalkfactor, 2173
kalkgehalte, 2172
kalkhoudend, 523
kalkmergel, 2174
kalkmeststof, 2179
kalkmijdend, 527
kalkminnend, 526
kalksalpeter, 2478
kalkschimmel, bruine, 448
—, witte, 4144
kalkschuwend, 527
kalksteen, 2176
kalkstikstof, 965
kalktoestand, 2175
kalmia, 533
kalmoes, 3770
kamerlinde, 3535
kamerplant, 1842
kamgras, 890
kamille, 2307
kamperfoelie, 1813
kanaal, 545
kanker, 551, 1054
—, zwarte, 331

kantelpotblad, 3862
kantig, 103
kap, 269, 3531
kapitaalgoederen, 570
kapmes, 676
kardinaalsmuts, 3554
kardoen, 586
karteren, 2270
karton, 584
kartonnen doos, 585
karwij, 575
kas, 1579
—, tweezijdige, 3532
—, warme, 1755
kasbaas, 1583
kasbedding, 1580
kascultuur, 1581 *
kasgeld, 3049
kasgrond, 1587
kasplant, 1585
kastanje, tamme, 655
kastrips, 1588
kastype, 1589
katje, 608
katjesdrager, 81
katteklei, 606
kattekruid, 609
kattestaart, 9, 610, 2216
kavel, 2218
keel, 3837
keerklep, 2489
kegelvormig, 804
keil, 4108
keileem, 413
keizerskroon, 1462
kelk, 538
kelkblad, 3346
kentia, 1993
kernhout, 1753
kernrot, 469
Kerria, 1994
kers, zoete, 4158
—, zure, 3517
kersevlieg, 651
kerspruim, 653
kerstcactus, 677
kerstroos, 678
kerstster, 2811
kervel, 654
ketel, 391
ketelappendages, 392
kettingzaag, 641
keukenkruiden, 924
keukenzoutgehalte, 1998

keuring, 1924
keuringsdienst, 1927
keurmeester, 1928
kever, 294
kiel, 1992
kiem, 1562
kiemblad 866, 3319
kieming, 1566
kiemkracht, 1565
kiemkrachtig, 4029
kiemplant, 3320
kiemschimmel, 979, 1658
kiemsnelheid, 3041
kiemtoestel, 1567
kiemvermogen, 1568
kiemwit, 1226
kiepen, 3889
kieper, 4089
kieseriet, 1997
kiezel, 3401
kiezelhoudend, 3403
kiezelsteen, 2662
kiezelzuur, 3402
kipkar, 3863
kippemest, 656
kippengaas, 4184
kipploeg, 233
kist, 3914
kistenlediger, 1162
kisten op stellages, 2532
kit, 2995
kitten, 630
klaproos, 850
klaver, 717
—, rode, 3059
—, witte, 4142
klaverzuring, 2576
kleefkruid, 700
klei, 694
kleiachtig, 696
kleigrond, 697
kleine frambozeluis, 3034
— iepespintkever, 3441
— narcisvlieg, 3444
— veldkers, 1707
— wintervlinder, 4178
klein fruit, 3443
kleinhandelaar, 3115
klein hoefblad, 757
— koolwitje, 3445
kleinkronige narcis, 3440
kleinste rozebladwesp, 2093
kleinverpakking, 812
kleinvruchtigheid, 646

lichtdoorlatend, 3909
lichtdoorlatendheid, 2162
lichten, 3020, 3021
lichtgebrek, 2152
lichtintensiteit, 2158
lichtinval, 2155
lichtsterkte, 2158
lichtsterktekromme, 2159
lichtverdeling, 2154
lichtverlies, 2217
lichtvoorziening, 2161
lid, 1944
lidcactus, 677
lidrus, 2295
lier, 4167
lieveheersbeestje, 2073
lieve-vrouwe-bedstro, 3772
liggend, 2939
liggende vetmuur, 2643
liguster, 2934
ligusterpijlstaart, 2935
lijn, 3673
lijnvormig, 2181
lijsterbes, 2404
lijsterbesmot, 130
linde, 2178
lintbloem, 2163
lintvormig, 3674
lip, 2186
lipbloem, 2008
lippenstiftschimmel, 2187
liquide middelen, 2189
liquiditeit, 2192
lisdodde, 3077
litteken, 3267
Lobelia, 2204
lokmiddel, 231
longkruid, 2225
loodarsenaat, 2065
loodglans, 3412
loodkruid, 2067
loodzand, 359
loof, 1414
loofboom, 437
loofklapper, 1739
looftrekker, 1738
loofverbruining, 74
loonbedrijf, 820
loon in natura, 4052
loonwerk, 821
loot, 3379
lopende band, 1225
lork, 2030
los, 2213

losmaken, 2215
löss, 2206
lucerne, 2224
luchtafvoer, 54
luchtafvoerkanaal, 55
luchtdicht, 61
luchtdroog, 56
luchtdruk, 182
luchten, 50, 4018
luchting, 4019
luchtkartering, 39
luchtkoeler, 53
luchtraam, 4020
luchttoevoer, 60
luchtverontreiniging, 59
luchtverstuivingsbrander, 51
luchtvochtigheid, 181
luchtvochtigheidsmeter, 1862
luchtwasser, 62
luchtwortel, 37
luis, 125
lupine, 2226
luxmeter, 2227

maagdepalm, 2682
maaggift, 3661
maaibalk, 958
maaien, 2411
maailader, 959
maaimachine, 2412
maaiveld, 3498
maandbloeier, 2687
maat, 3427
maatsortering, 3429
made, 2229
madeliefje, 975
magnesia, 2230
magnesiapoederkalk, 1672
magnesium, 2231
magnesiumgebrek, 2232
magnesiumkalk, koolzure, 578
Magnolia, 2235
mahonia, 2238
majoraan, 2280
malsheid, 3805
malve, 2253
Mammoetboom, 4127
mand, 262
mandarijn, 3790
mandpot, 4182
mangaan, 2256
mangaangebrek, 2257
mangaanovermaat, 1254
mangat, 2258

mannelijk 3603
manometer, 2920
mantelkoeling, 1968
manuur, 2259
marcotteren, 57, 3138
maretak, 2354
margriet, 2276
—, grootbloemige, 3367
markeur, 2281
marktbericht, 2288
marktordening, 2286
marktprijs, 2287
marktstabilisatie, 2289
marktstructuur, 2290
marktverstoring, 2283
marmelade, 2293
maskerbloem, 2376
massaselectie, 2302
matig zwakke onderstam, 2322
maximumprijs, 2305
maximum-temperatuur, 2306
mechanisatie, 2318
meelbes, 4140
meeldauw, 2340, 2885
—, valse, 1115
meeldraad, 3602
meerhokkig, 2424
meermalig fust, 3119
meermolm, 2413
meeropbrengst, 1890
meerscharige ploeg, 2423
meetlat, 2316
meetlint, 2317
meidoorn, 1741
meikever, 731
meiziekte, 1662
melde, 2537
—, uitstaande 773
melig, 2313
melige koolluis, 510
melkdistel, 111
—, ruwe, 3556
melkdistelluis, groene, 948
melksap, 2046
meloen, 2325
membraanpomp, 1051
mengen, 2357
mengklep, 2360
mengmeststof, 792
mengploegen, 2796
mengtabel, 3782
mengventilator, 58
mengwoelen, 2359, 3707
menie, 3065

merel, 329
merg, 2753
mergel, 2291
mergelgrond, 2292
mergstraal, 2754
meristeem, 2330
merkatoom, 3896
meskouter, 2000
mest, 2261
—, afgedragen, 3547
—, organische, 2544
—, verrotte, 4125
mestbehoefte, 1325
mesthoop, 1164, 2264
mestmachine, 2265, 2267
meststof, 1320
—, volledige, 785
mestvaalt, 1164
mestverspreider, 2266
mestvork, 2263
metselspecie, 2386
microben, 2335
microklimaat, 2336
micropyle, 1798
microscopisch 2337
middagbloem, 2331
middel, 3096
middelbare fout, 3606
middellandsezeevlieg, 2320
middennerf, 2338
middenprijs, 201, 2315
mier, 114
mierikswortel, 1822
mijt, 2356
milde humus, 2341
milieu, 1231
miljoenpoot, 3565
milligram-equivalent, 2343
mimosa, 2344
minderwaardig produkt, 1903
mineervlieg, 2087
mineraal, 2347
mineralen, 2349
mineralisatie, 2348
mineren, 2346
minimumprijs, 2350
minimum-temperatuur, 2351
mirt, 2444
miskleurig, 229
misoogst, 899
mispel, 2323, 2324
misvormig, 1020
misvorming, 2252
modder, 2414

modificatie, 2364
moederbol, 2392
moedercel, 2393
moederplant, 2395
moederplantje, 2394
moeras, 3755
moerasandoorn, 2299
moerascypres, 3757
moeraskers, 2300
moerasplant, 2298
moerassige grond, 2301
moerasveen, 3908
moerbei, 2418
moerbeiboom 2419
moerplant, 2395
mol, 2371
—, natte, 4130
moldrainage, 2373
molm, 1634
molploeg, 2374
molybdeen, 2375
mondstekel, 3702
monilia-rot, 449
monnikskap, 21, 2377
monocultuur, 2379
monster, 3240
monsters nemen, 3241
morel, 2383
moscovische mat, 267
most, 2435
mosveen, 2391
mot, 716
motorhak, 2397
motormaaier, 2398
motorrugnevelspuit, 2889
motorrugverstuiver, 2888
motorspuit, 2399
mousseren, 1202
moutkiemen, 2255
moutsuiker, 2254
mozaïek, 2388
—, tussennervig, 1946
mozaïekziekte, 2390
muis, 2408
muizetarwe, 3042
mul, 1460
multipot, 2426
mummieziekte, 2428
mus, 3536
mutageen, 2436
mutant, 2437
mutatie, 2438
mutatieveredeling, 2439
muur, 657, 4055

muurbloem, 4057
muurkas, 2118
muurpeper, 3662
muurplaat, 4058
M-virusziekte, 2440
mycelium, 2442
myceliumgroei, 2443

naakte slak, 1335
naald, 2457
naaldafsluiter, 2461
naaldblad, 2447, 3875
naaldboom, 805
naaldengrond, 806, 2734
naaldvormig, 2460
na-calculatie, 1269
nachtonderbreking, 2475
nachtschade, zwarte, 342
nachtvorst, 2474
NaCl-gehalte, 1998
nagel, 300, 693
nagelkruid, 199
nakomelingschap, 2948
na-opkomst-toepassing, 2852
napje, 935
narcis, 973
—, kleinkronige, 3440
narcismijt, 489
narcisvlieg, grote, 2445
—, kleine, 3444
narijping, 45
nateelt, 3303
natontsmetter, 2190
natrium, 3461
natriumchloraat, 3462
natronloog, 614
natronsalpeter, 3463
natrot, 3468, 3469
natte grond, 4131
natte mol, 4130
natuurlijke vijand, 2450
natuurlijk fosfaat, 2703
navel, 1798
na-zaaien-toepassing, 2853
necrose, 2455
neerslag, 2895, 3018
neerslagtekort, 1014
negatieve bron, 2462
nektarklier, 2456
nematicide, 2463
nerf, 2465
nerfmozaïek, 4016
nervatuur, 2464
net, 2466

schieters, 393
schijfcactus, 2536
schijfkamille, 3048
schijfkouter, 1078
schijnkrans, 1287
schijnvrucht, 2969
schijveneg, 1079
schil, 3432
schildje, 3289
schildluis, 3266
schildvormig, 3377
schimmel, gele, 4226
—, grauwe, 1657, 1658, 1659
—, olijfgroene, 2521
schimmelziekte, 1499, 2403
schoffel, 1178
schoffelen, 3287
schoffelmachine, 3288
schokker, 2294
schoonmaken van het bed, 3913
schoor, 3698
schoorsteen, 661
schoorsteenschuif, 978
schop, 3388
schors, 250
schorsboorder, 650
schorsbrand, 555
schorseneer, 3282
schorskever, 251
schrepel, 4114
schroef, 1774
schub, 3264
schuifafsluiter, 3437
schuimbeestje, 769
schuimplastic, 2784
schuimziekte, 3277
schuine palmet, 2509
schurft, 3260, 3261
schutblad, 418
schutting, 1306
schuur, 252
schuurbewaring, 3663
Scilla, 3591
sclerotiën, 3275
sclerotiënrot, 3276
scrubber, 3285
sedimentaire afzetting, 3308
sedimentgesteente, 3309
seizoenarbeid, 3300
selderij, 619
selderijvlieg, 620
selecteren, 3333
selectie, 3334
selectief, 3335

sering, 2164
seringemot, 2165
serologie, 3350
serradella, 3351
sierappel, 1399
siergewas, 2555
sierheester, 2556
sierkers, Japanse, 1972
siertabak, 2472
sierteeltgewas, 1005
siertuin, 2554
sigarenmaker, 1743
Silene, 603
siliconhars, 3405
sinaasappel, 2538
sinaasappelboom, 2540
siroop, 3781
Sitkaspar, 3426
sjalot, 3361
sjalotteluis, 3362
skeletteermot, 132
skeletteren, 3430
sla, 2139
slaapmutsje, 534
slaapziekte, 4026
slaboon, 1455, 1996, 3452
slagvast, 1874
slak, 3439
—, naakte, 1335
slakvormige bastaardrups, 2646
slamozaïek, 2141
slang, 1830
slapend oog, 1103
slapers, 1104
sleedoorn, 354
slempgevoelig, 3344
slempig, 2416
sleutelbloem, 2931
slib, 3407
slijk, 2414
slijkgat, 2417
slingerdunner, 3054
slip, 2203
slof, 666
sloot, 1090
slootwaterpeil, 1092
sluipwesp, 1869
sluitkool, 1746
smaakafwijking, 2513
smakelijkheid, 2583
smalbladig, 2448
smalbladigheid, 2090
smeul, 979, 1658, 3450
smoorklep, 3838

snavelvormig, 270
snede, 953
sneeuwbal, 3453
sneeuwbalhaan, 879
sneeuwbes, 3454
sneeuwklokje, 3455
sneeuwroem, 1601
snelkoppelbuis, 3005
snelspuit, 3545
snelvriezen, 3006
snelweger, 3544
snijbiet, 3552
snijbloem, 955
snijboon, 3436
snijden, 908
snijgroen, 956
snijkant, 962
snij-orchidee, 2598
snijselderij, 619
snijsla, 963
snoeien, 2963
snoeihout, 2966
snoeihoutschuif, 3917
snoeihoutversnipperaar, 3286
snoeimes, 2965
snoeischaar, 2968, 3302
snoeiwond, 2964
snoeizaag, 2967
snoer, 839
snotkoker, 3157
snuitkever, 718, 938
—, behaarde, 447
sociale lasten, 3459
solvabiliteit, 3507
soort, 3542
soortechtheid, 4000
sorteermachine, 1614, 3514
sorteerruimte, 1617
sorteren, 3428, 3513
sortering, 1615
sorteringsvoorschrift, 1616
sortiment, 178
spaan, 665, 2034
Spaanse peper, 3068
spade, 3528
spannen, 3691
spanning, 3809
spanrups, 2212
spant, 3017, 3938
spantafstand, 3526
spar, 3585
sparappelgalluis, 3587
sparreluis, groene, 3586
sparrespintmijt, 807

vergeet-mij-niet, 1435
vergeling, 4224
—, vroege, 3557
vergelingsziekte, 297
vergif, 2813
vergiftig, 2816
vergiftigen, 2814
vergiftiging, 2815
vergroeidbladig, 1524, 1525
vergroeien, 1693
vergroeiing, 722
vergrootglas, 2234
verjongen, 3113
verkalking, 642
verkavelen, 2607
verkleuring, 643, 1080
verklistering, 3560
verkoopprijs, 3341
verkoopwaarde, 3233
verkruimelen, 916
verkurken, 1439
verlaten, 3117
verlept, 4191
vermeerderen, 2949
vermenigvuldigen, 2949
vermenigvuldiging, 2425
vermoeidheid, 1300
vermoering, 3560
vermogensbehoefte, 571
vermolmd, 1000
vernalisatie, 4024
vernevelen, 184
verontreiniging, 2828
veroudering, 46, 2512
verpakking, 2579
—, eenmalige, 2488
verplanten, 3910
verpoppen, zich, 2983
verpoten, 3910
verpotten, 2879
verrotte mest, 4125
verrotten, 3195
verschrompelen, 3392
verse groente, 1458
verseluchtkap, 1456
vers fruit, 1457
verslijmen, 1569
verspeenhout, 2925
verspenen, 2928
verspreid, 76
verstek, 2514
verstuiven, 2976
verstuiverbeker, 185
verstuivingsbrander, roterende,

3196
vertakken, zich, 3028
vertakt, 422, 3027
verteren, 2993
vervangingswaarde, 3101
vervluchtigen, 4047
vervroegen, 1425
vervuiling, 2828, 3486
verwarmen, 1754
verwarmingskabel, 1758
verwelken, 1278
verwelking, 4165
verwelkingspunt, 4166
verwelkingsziekte, 4025
verweren, 4107
verwerking, 2937
verwerkingsindustrie, 2938
verwilderen, 2452
verzadigen, 3251
verzadiging, 3252
verzadigingsgraad, 1026
verzadigingspunt, 3253
verzamelrooier, 2869
verzilten, 2569
verzinken, 1522
verzuren, 19
vetkruid, 3310
vetmuur, liggende, 2643
vetplant, 3711
vetvlekkenziekte, 1712
veur, 1506
vezelig, 1327
vezelplaat, 3687
vierwielige trekker, 1443
vijg, 1340
vijgeboom, 1341
vijver, 2842
viltig, 3879
vingerhoedskruid, 1444
violet wortelrot, 4039
violier, 3658
viooltje, 4038
—, driekleurig, 2597
—, Kaaps, 3518
Virginiaanse vogelkers, 675
virologie, 4041
virusaantasting, 4042
virusvlekkenziekte, 3126
virusziekte, 4043
vitaminen, 4045
vlaamse gaai, 1975
vlag, 3604
vlakwortelend, 3365
vlambloem, 2700

vlamkast, 761
vlammenwerper, 1366
vlampijp, 1359
vlas, 1373
vlekkenziekte, 119, 2070
vlekkerigheid, 2401
vleugelnoot, 4174
vlezig, 1376
vlieg, 1412
—, Japanse, 3123
—, witte, 1590
vlier, 1206
vlies, open, 1724
vliezig, 2326
vlijtig liesje, 238
vlinder, 505
vlinderbloemigen, 2600
vlindervormig, 2599
vloeibare brandstof, 2191
vloeistof, 2188
vlottende produktiemiddelen,
4206
vlucht, 1411
vluchtig, 4046
vochtgehalte, 2368
vochthoudend, 2370
vochtig, 2365
vochtigheid, absolute, 3
—, relatieve, 3094
vochtigheidsgraad, 1025
vochtspanning, 3491
vochttekort, 2369
voederwikke, 779
voeding, 2502
voedingsbodem, 2321, 2500
voedingsoplossing, 2501
voedingspijp, 1301
voedingspomp, 1302
voedingswaarde, 2504
voedingsziekte, 1012
voedsel, 1420
voedselopneming, 7
voedzaam, 2503
voet, 3699
voetrot, 1423, 1512
voetziekte, 3180
vogelkers, 1243
—, Virginiaanse, 675
vogellijm, 2354
vogelmelk, 3616
vogelmest, 2883
vogelschade, 976
vogelwikke, 3952
volautomatisch, 1491

welriekend, 1446
welwater, 3578
wendakker, 1748
wentelploeg, 3121
werkbij, 4205
werkdruk, 4207
werkgever, 1217
werkhoeveelheid, 4204
werkhouding, 2854
werknemer, 1216
werkplan, 4208
werkplek, 4211
werktuigraam 3880
werkvereenvoudiging, 4210
werkvolgorde, 2543
wesp, 1563
Westlandwarenhuis, 4129
wieden, 4110
wiedmachine, 4115
wiedvorkje, 4113
wier, 3301
wig, 4108
wigvormig, 934
wijker, 1344
wijnbes, Japanse, 1974
wijnbouw, 4036
wijnstok, 4035
wijzer, 2812
wijzerplaat, 1049
wikke, 4028
wildafweermiddelen, 122
wilde peen, 4157
— wingerd, 4040
wildschade, 977
wilg, 4161
wilgehaan, 781
wilgehaantje, bronzen, 425
wilgehoutrups, 1604
winddruk, 4170
windend, 3965
windscherm, 3375
windverband, 4168
wingerd, wilde, 4040
winteraconiet, 4175
winterbloemkool, 4176
winterbonenkruid, 4181
wintereik, 1172
winterhard, 1728
winterjasmijn, 4177
winterpostelein, 4179
winterrust, 4180
wintervlinder, grote, 2400
— kleine, 4178
wiskundige verwerking, 3620

witbol, 4233
witlof, 4193
witrot, 4146
witsnot, 3438
witte abeel, 4145
— boon, 1996
— kalkschimmel, 4144
— klaver, 4142
— kool, 4141
— krodde, 3654
— roest, 4147
witte-vlekkenziekte, 2108
witte vlieg, 1590
woekerplant, 2605
woelen, 1694
woeler, 1695, 3706
woelmuis, 4048
woelrat, 4101
woeste grond, 4067
wol, 2567
woldopluis, 865
wolf, 1116
wolfsmelk, 3590
wolluis, 2314, 2733
wolverlei, 150
worm, 4212
wormgang, 4213
wormstekig, 4214
wortel, 3165
wortelaaltje, 2310
—, vrijlevend, 2138
wortelactiviteit, 3167
wortelbeeld, 3177, 3184
wortelboorder, 1538
wortelbrand, 337, 3179
wortelduizendpoot, 3290
wortelgestel, 3184
wortelhals, 3170
wortelharen, 3173
wortelkluit, 3168
wortelknobbel, 912
wortelknobbelaaltje, 3175
wortelknol, 3943
wortelknolletje, 3944
wortelknop, 3172
wortelluis, 2142
wortelmutsje, 3169
wortelonkruid, 2675
wortelopslag, 3713, 4160
wortelrooier, 596
wortelrot, 345, 2713, 3179
—, violet, 4039
wortelscheut, 3183
wortelschieten, 3166

wortelstek, 3171
wortelstok, 3122
wortelvlieg, 595
woudboom, 1433
wratten, 2117
wratziekte, 4063

Y-virusziekte, 4235

zaad, 3311
—, origineel, 2552
zaadbehandeling, 3332
zaadcontrole, 3329
zaaddoes, 3323
zaaddrager, 3322
zaadecht, 3672
zaadhandel, 3330
zaadhuid, 3801
zaadkever, 2975
zaadleverancier, 3314
zaadmonster, 3326
zaadonkruid, 112
zaadontsmetting, 3316
zaadontsmettingsmiddel, 3315
zaadplant, 3322
zaadpluis, 2601
zaadreiniger, 3313
zaadstengel, 3328
zaad schieten, in, 3219
zaadteelt, 3318
zaadvast, 3672
zaadwet, 3325
zaadwinkel, 1529
zaagwesp, 3256
zaaibakje, 3331
zaaibed, 3312
zaaidatum, 3521
zaaien, 3519
zaaiing, 3320
zaaimachine, 3317
zaaipan, 3321
zaaisel, 3520
zaaitijd, 3522
zachte dravik, 440
zacht fruit, 3467
zak, 230
zakbuis, 1112
zakgoot, 3991
zand, 3242
zandgrond, 3246
zandig, 3244
zandraket, 3820
zandwikke, 1709
zavel, 3245

DEUTSCH

DEUTSCH

Amortisation, 89
Ampfer, stumpfblättriger, 436
Ampferblattlaus, 1095
Ampferblattwespe, 1096
Amsel, 329
anaerob, 91
Analyse, 92
Analysenergebnis, 93
Ananas, 2732
Anbau, 929
Anbaudrehpflug, 1711
anbauen, 926
Anbaufläche, 147
Anbauhygiene, 931
Anbaukiste, 3914
Anbauplan, 902
Anbaurecht, 1678
Anbauspritze, 3900
Anbautechnik, 1684
Anbauwinkelpflug, 2407
Andienungspflicht, 795
Anemone, 97
Anerkennungsbedingungen,
	3103
Anerkennungsdienst, 1927
Anerkennungsstelle, 1927
Anfälligkeit, 3750
anfeuchten, 2366
Anfeuchter, 4132
Anfuhr, 3743
Angebot, 3743
Angelika, 98
Angriff, 2755
Anhängerkupplung, 3904
Anhängespritze, 3903
anhäufeln, 1193
Anheftung, 186
Anlage, 1531
Anlagekapital, 1167
Anlagekosten, 864, 2551
Anlieferung, 3743
Annuität, 113
anorganisch, 1917
anorganischer Stoff, 1918
anpflanzen, 2760
Anpflanzung, 2763
Anschaffungswert, 2553
anschlämmen, 3408
Ansteckbarkeit, 815
Ansteckung, 1900
Antagonismus, 115
antasten, 188
Antherenbrand, 117
Anthracnose, 2082, 2095

Anwachsraum, 3540
Anzuchtbeet, 2495
Anzuchtshaus, 2951
Apfel, 126
Apfelbaum, 137
Apfelbeere, 674
Apfelblattlaus, grüne, 1650
—, rosige, 3193
Apfelblattmotte, 132
Apfelblattsauger, gemeiner, 135
Apfelblütenstecher, 128
Apfelmus, 133
Apfelsaft, 131
Apfelsägewespe, 134
Apfelsine, 2538
Apfelsinenbaum, 2540
Apfelsirup, 136
Apfelwein, 136
Apfelwickler, 736
Aprikose, 139
Aprikosenbaum, 140
Araukarie, 660
Arbeiter in Dauerstellung,
	3090
Arbeitgeber, 1217
Arbeitsablauf, 2543
Arbeitsanfall, 4204
Arbeitsaufzeichnungen, 2014
Arbeitsbedarf, 2016
Arbeitsbiene, 4205
Arbeitsbuchführung, 2014, 2015
Arbeitsdruck, 4207
Arbeitserledigungskosten, 2533
Arbeitsersparnis, 2017
Arbeitsfähigkeit, 2854
Arbeitsfolge, 2543
arbeitsintensiv, 2012
Arbeitskosten, 2010
Arbeitskräftebesatz, 2011
Arbeitslohn, 4053
Arbeitsmenge, 4204
Arbeitsnehmer, 1216
Arbeitsordnung, 2543
Arbeitsplan, 4208
Arbeitsplatz, 4211
Arbeitsproduktivität, 2013
Arbeitsspitze, 2634
Arbeitsstunde, 2259
Arbeitsteilung, 1094
Arbeitsumfang, 4204
Arbeitsvereinfachung, 4210
Arbeitsvoranschlag, 2009
Areal, 147
Arom 1372

Aroma, 1372
aromatisch, 152
Arsenverbindung, 155
Art, 3542
Artischocke, 1597
Arve, 151
Ascochyta-Fleckenkrankheit,
	375
Ascochyta-Krankheit, 3046
Aspergillus, 174
Assel, 4196
Assimilate, 177
assimilieren, 176
Ast, 421
Aststerben, 996
Atemwurzel, 428
ätherische Oele, 1242
Atmosphäre, 180
Atmung, 427, 3112
Atomgewicht, 183
Aucuba-Mosaik, 2865
Aufastungsschere, 2968
Aufbereitung und Verpackung,
	1717
Aufbewahrungsort, 3670
aufbinden, 3845
aufbrechen, 500
aufdornen, 4156
aufforsten, 43
Aufgabenstellung, 3797
Aufguss, 870
Aufhängehöhe, 3753
Aufkippvorrichtung, 1162
auflagern, 416
auflösen, 1086
Aufnahme, 3786
aufnehmbar, 198
aufpflanzen, 1368
aufpumpen, 2981
aufrechtstehend, 1238
aufreissen, 2800
Aufschlag, 4050
aufschwemmen, 3408
aufsetzen, 3594
aufspeichern, 3669
aufsteigend, 163
aufstocken, 3594
auftauen, 3821
Aufwand, 862
aufweiten, 4156
Auge, 1276
äugeln, 459
Augensteckling, 1277
Augustakrankheit, 3869

Beisskohl, 3552
Bekämpfung, 822
Bekämpfungsmassnahme, 824
Bekämpfungsmittel, 2690, 2778
bekleiden, 2727
Belastungsfähigkeit, 597
beleuchten, 1870
Beleuchtung, 160, 2156
Beleuchtungsmesser, 2227
Beleuchtungsstärke, 1934
Belichtung, 160
Belichtungszeit, 1168, 1871
Bepflanzungsdichte, 2767
Bepflanzungsplan, 2772
Berater, 34
Beratung, 33
Beratungsdienst, 35
Berberitze, 249
Beregnung, 3583
Beregnungsanlage, 3582
Beregnungsautomat, 196, 1960
bereift, 2962
Bergföhre, 2406
Bergkiefer, 2406
Berglorbeer, 533
Bergrüster, 4222
Bergulme, 4222
Bergwohlverleih, 150
Berieselung, 1959
Berufskraut, 1374
berühren, 188
Besandung, 873
beschatten, 3358
Beschichtung, 726
beschneiden, 2963
Beschreikraut, 1374
Besenginster, 1556
Besenheide, 2183
Besichtiger, 1928
besprengen, 3579
bespritzen, 3566
bestäuben, 1174, 2825
Bestäuber, 2827
Bestäubung, 2826
bestellte Fläche, 475
bestocken, sich, 3853
bestrahlen, 1955
Bestrahlung, 1956
Bestrahlungsstärke, 1935
Betonbalken, 799
Betrieb, 1801
Betriebsabrechnung, 2534
Betriebsaufwand, 2533
Betriebsertrag, 1667

Betriebsorganisation, 2547
Betriebsstruktur, 1295
Betriebstyp, 1296
Betriebsvergleich, 783
Betriebswirtschaft, 1294
betriebswirtschaftliche Buchführung, 502
bewässern, 1958
Bewässerung, 1959
Bewegungsstudie, 2396
bewurzeln, 3784
Bewurzelung, 3174, 3184, 3785
Bibernelle, 499
Bickbeere, 311
biegsam, 1377
Biene, 283
Bienenkönigin, 3004
Bienenkorb, 289
Bienenschwarm, 3759
Bienenvolk, 288
Bienenzucht, 291
Bienenzüchter, 290
Bildungsabweichung, 2252
Bilsenkraut, 1780
Bindematerial, 3843
Binder, 3938
Binderabstand, 3526
Bindesalat, 857
Bindfaden, 3693
Binnenseepolder, 2817
Binse, 3220
Biologie, 315
biologische Bekämpfung, 314
Birke, 317
Birkenblattwespe, breitfüssige, 67
Birnbaum, 2649
Birne, 2638
Birnenblattgallmücke, 2642
Birnenblattlaus, braune, 2640
—, mehlige, 2639
Birnenblattsäuger, gelber, 2648
Birnengallmücke, 2644
Birnengittersort, 3223
Birnenknospenstecher, 129
Birnensägewespe, 2645
Birnenwurzellaus, 3272
Bisamratte, 2433
bitter, 321
bitterfrei, 322
Bittersalz, 2233
Blankoversuch, 2511
Blasebalg, 302
Blasenfuss, 3836

Blasenstrauch, 355
Blatt, 2068
Blattacksel, 2071
Blattälchen, 460
Blattanalyse, 2069
Blattbegonie, 299
Blattbeschädigung, 2081
Blattbrand, 376
Blattdorn, 2098
Blätter, 2023
Blätterkohl, 401
Blattfall, 1019
Blattfallkrankheit, 2078, 2106
Blattfleckenfäule, bakterielle, 222
Blattfleckenkrankheit, 120, 410, 450, 2074, 2075, 2082, 2095, 2099, 2100, 2101, 2102, 2103, 2107, 3139, 3349
—, bakterielle, 223
—, eckige, 104
— und Fäulnis, 3141
Blattflecken- und Stengelfäule, bakterielle, 216
Blattfloh, 3712
Blattform, 2097
Blattgallen, 2083
Blattgemüse 2113
Blattgewebe, 2112
Blattgrün, 670
Blattgrund, 2072
Blattkaktus, 1234
Blattknospe, 2079
Blattlaus, 125
—, schwarze, 325
blattlos, 2084
Blattminierfliege, 1805
Blattnarbe, 2094
Blattoberfläche, 2110
Blattpflanze, 1415
Blattrand, 2086
Blattrandkäfer, gestreifter, 2621
Blattrandkrankheit, 3861
Blattrandvergilbung der Erdbeere, 3686
blättrige, Galle, 2116, 2117
Blattrollen, 2092
Blattrollkrankheit, 2658, 2870
Blattscheide, 3368
Blattspitze, 2111
Blattspreite, 356
Blattsteckling, 2080
Blattstellung, 154
Blattstiel, 2109

Blattwespe, 3256, 4163
Blattwickler, 2091, 2850, 3891
Blaubeere, 311
blauer Erlenblattkäfer, 66
— Weidenblattkäfer, 781
blaufärben, 384
Blaufichte, 386
'blaugewachsene' Tulpen-
zwiebeln, 385
blauhülsige Auskernerbse, 2294
Blausäuregas, 966
Blausieb, 2135
Blechbüchse, 544
Bleiarsenat, 2065
bleibend, 2688
Bleiber, 2683
bleichen, 357
Bleichsellerie, 358
Bleiglanz, 3412
Bleisand, 359
Bleiwurz, 2067
Blindwuchspflanze, 2004
Blockpflanzung, 367
Blühbeeinflussung, 1908
— nach Jarow, 4024
blühen, 370, 1384
Blume, 1383
Blumenanlage, 1401
Blumenbeet, 1388
Blumen binden, 1385
Blumenbinder, 1381
Blumenbinderei, 1382
Blumenduft, 1445
Blumengarten, 1394
Blumengeschäft, 1403
Blumenlagerhaus, 490
Blumenkohl, 611
Blumenkohlkrankheit, 2116
Blumenkohlmosaik, 612
Blumenkorb, 1386
Blumenkorso, 1392
Blumenkranz, 4218
Blumenkrone, 852
Blumennessel, 746
Blumenteppich, 1390
Blumentopf, 1402
Blumenvase, 1406
Blumenzwiebel, 476
Blumenzwiebelanbaugebiet, 478
Blumenzwiebelerntemaschine,
484
Blumenzwiebelfeld, 479
Blumenzwiebelkultur, 482
Blumenzwiebelscheune, 490

Blumenzwiebelsortiermaschine,
480
Blumenzwiebelzüchter, 481
Blüte, 371, 372, 1383, 1397
Blütenbildung, 1393
Blütenboden, 3051
Blütenendfäule, 380
Blütenhülle, 2679
Blütenknospe, 1389
Blütenstand, 1907
Blütenstandgrösse, 3939
Blütenstaub, 2822
Blütenstengel, 2664
Blütenstiel, 1405
Blütentaubheit, 365
blütentragend, 1387
Blütentraube, 1391
Blütezeit, 1400
Blutlaus, 4202
Blutmehl, 368
Bockkäfer, 2209
Boden, 1192, 2025
Bodenanalyse, 3471
Bodenart, 3502, 3963
Bodenbedeckung, 2420, 3478
Bodenbefund, 3472
Bodenbehandlung, 3501
Bodenbelüftung, 3470
Bodenbeschaffenheit, 2453,
3477
Bodenbildung, 2663
Bodendämpfung, 3496, 3631
Bodenentseuchung, 3480
Bodenfruchtbarkeit, 3482
Bodenheizung, 3484
Bodenkarte, 3489
Bodenkartierung, 3499
Bodenkrankheit, 3479
Bodenkunde, 3495
Bodenmüdigkeit, 3481
Bodenoberfläche, 3498
Bodenprobe, 3494
Bodenprofil, 3493
Bodenquerschnitt, 3493
Bodenschicht, 3487
bodenübertragbares Virus, 3476
Bodenuntersuchung, 3500
Bodenverarmung, 1877
Bodenverbesserung, 3485
Bodenverhagerung, 1877
Bodenwärme, 3483
Bodenwasserspannung, 3491
Bohne, 271
Bohnengelbmosaik, 278

Bohnenkraut, 373
Bohnenmosaik, gewöhnliches,
766
Bohnenpflückmaschine, 274
Bohnenspinnmilbe, 3071
Bohnenstange, 27
Bor, 403
Borax, 398
Bordeauxbrühe, 399
Boretsch, 397
Borke, 250, 264
Borkenkäfer, 251
Bormangel, 404
Böschungsmäher, 246
Botanik, 480
botanisch, 407
Botrytis-Fäule, 1662
Botrytisflecke, 411
Botrytiskrankheit, 409, 410
'Bovist, 1149
brach, 1282
Brachkäfer, 3728
Brachland, 1283
brach liegen, 2149
Brandfleckenkrankheit, 118
Brandkrankheit, 3451
Brandmal, 3279
Brandstelle, 3279
braune Achateule, 102
— Birnenblattlaus, 2640
— Bohne, 1996
brauner Gips, 448
braune Spinnmilbe, 127
Braunfleckigkeit, 443, 1698,
2089
braungelb, 446
braunrot, 445
Brause, 3186
Brautstrauss, 431
Brechbohne, 1455, 3452
brechen von Tulpen, 3954
Brechmöhren, 3888
Brei, 2985
Breitadrigkeit, 2140
breitfüssige Birkenblattwespe,
67
Breitlauch, 2124
Breitmilbe, 438
Breitstrahler, 1062
breitwürfig, 435
Brenner, 498
Brennessel, 2467
Brennfleckenkrankheit, 119, 550,
2070

Dränierung, 1118
Dränmaschine 1124, 1127
Dränrohr, 1121
Dränröhre, 1126
Dränstrang, 3848
Dräntiefe, 1041
Drehbecher, 185
Drehbecherbrenner, 3196
Drehherz, 364
Drehherzmücke, 3761
Drehstrahlregner, 3201
dreibasisch, 3930
dreieckig, 3928
Dreiecksverband, 3929
Dreigelenkbinder, 3831
dreilappige Mandel, 1398
Dreipunktanschluss, 3832
dreischiffiges Gewächshaus,
3833
dreizählig, 3815
Drillmaschine, 3317
Drillschar, 1135
Drittländer, 3828
Drohne, 1140
Drosselklappe, 3838
Druckluftzerstäuberbrenner, 51
Druckölbrenner, 2919
Druckstelle, 452
Druckwasser, 2669, 4128
Druckzug, 1426
Drüse, 1575
Drüsenhaar, 1577
drüsig, 1576
Dünensand, 1163
Dung, 2261
düngen, 2262
Dünger, 2261
—, organischer, 2544
Düngeraufbereitungsmaschine,
2265
Düngerbedürfnis, 1325
Düngerhaufen, 2264
Düngerwender, 2267
Dunggabel, 1436
Dunghaufen, 1164
Düngung, 2269
Düngungsempfehlung, 1324
Düngungsversuch, 2268
dünnwändig, 3827
Durchgangsventil, 1600
durchgefrorener Schwarztorf,
1466
durchlässig, 2685
Durchlässigkeit, 2684

durchschiessen, 3219
Durchschnittsertrag, 202
durchsickern, 2668
durchspülen, 1378, 3143
durchtreiben, 3584
durchwachsen, 2676
durchwässern, 1378
Durchwuchs, 419
Durchwurzelung, 3178
dürr, 148
Dürrfleckenkrankheit, 75
Düse, 2492
Duwock, 2295

Eberesche, 2404
Ebereschenmotte, 130
Eberwurz, 587
ebnen, 2144
echte Motte, 716
eckig, 103
eckige Blattfleckenkrankheit,
104
Edeldistel, 3296
Edelkastanie, 655
Edelpelargonie, 3088
Edelpilz, 927
Edelreis, 1620, 3274
Edelweiss, 1196
Edelwicke, 3767
Efeu, 1964
efeublättriger Ehrenpreis, 1967
Efeupelargonie, 1966
Egge, 1730
eggen, 1731
Eggenzinken, 1732
Ehrenpreis, 3546
—, efeublättriger, 1967
Ei, 1203
Eiabsatz, 1204
Eibe, 4231
Eibisch, 2296, 3189
—, syrischer, 3395
Eiche, 2505
Eichen, 2574
Eichenblattmuster, 2507
Eichenmehltau, 2886
Eichenschälwald, 2506
Eichenwickler, grüner, 1654
Eierfrucht, 1205
Eierschwamm, 644
eiförmig, 2564
Eigengewicht, 994
Eigenkapital, 2575
Einachsschlepper, 4054

einballieren, 4216
einbasisch, 2378
einblättrig, 2381
eindeichen, 1068, 1218
eindosen, 3860
einfach, 3418
einfächerig, 3976
Einfassung, 400
einfriedigen, 1307
Einfriedigung, 1223
eingedickter Fruchtsaft, 3781
eingesalzene Gemüse, 4012
eingeschlechtig, 3977
eingesunken, 3734
eingewöhnen, 11
eingiessen, 2884
einhäusig, 2380
einheimisch, 1893
einjährig, 106
einjähriger Knäuel, 108
einjähriges Rispengras, 109
einkochen, 2911
Einlagepapier, 872
Einlagerung, 3664
Einlegegurke, 3133
einmachen, 2911
Einmachessig, 2723
einmieten, 2991
einpacken, 4217
Einphasen (Strom), 3422
einpoldern, 1218
Einradschlepper, 3424
einrecken, 3024
Einscharpflug, 3420
Einschlag, 2062, 2809
einschlagen, 2992
Einschlagplatz, 2062
Einschnitt, 953, 1887
einschnüren, 312
Einschnürungskrankheit, 1058
einschrumpfen, 3389
einsenken, 2991
einspritzen, 1911
einspülen, 4065
einstutzen, 954
eintopfen, 2855
eintrocknen, 1044
einwurzeln, 3166
Einzeldränung, 2530
Einzelhändler, 3115
Einzelkorndrillgerät, 3327
Einzelkornstruktur, 3421
Einzelkosten, 1073
einzellig, 3975

Frostringe, 1465
Frostschaden, 1463
Frostspanner, grosser, 2400
—, kleiner, 4178
Frucht, 1470
Fruchtansatz, 1481
fruchtbar, 1315
Fruchtbarkeit, 1316
Fruchtbarkeitsverlauf, 1317
Fruchtbildung, 1475
Fruchtblatt, 589
Fruchtfleisch, 1474
Fruchtfleischbräune, 2223
Fruchtknäuel, 1472
Fruchtknoten 2563
Fruchtpulpe, 1480
Fruchtsaft, 1478
Fruchtschalenwickler, 3729
Fruchtstiel, 1483
Fruchtwechsel, 904
Fruchtzweig, 1471
Frühbeet, 1833
Frühblüher, 402
Frühe, 1187
frühe Bräune, 1188
Frühkartoffeln, 1191
Frühlingsgrün, 3574
Frühlingshungerblümchen, 4151
Frühtreiberei, 1189
Fuchsschwanz, 610
Füller, 1344
Fundament, 1422
Funkenfanger, 3534
Funkie, 1503, 2756
Furche, 1506
Fusariumfäule, 1153, 1511
Fusarium-Krankheit, 1510
Fusariumwelke 1514, 1516
Fusarium-Zwiebeltrockenfäule, 1513
Fussende, 3699
Fusskrankheit, 1423, 1512, 3180
Futterwicke, 779

Gabelhubwagen, 2587
Gabelstapler, 1438, 1718
Gagel, 3764
Gallapfel, 1519
Galle, blättrige, 2116, 2117
Gallmilbe, 2077, 2804
Gallmücke, 334, 1520
Gallwespe, 1521
Gamete, 1523
Gammaeule, 3417

Gänsedistel, dornige, 3556
Gänsedistellaus, grünliche, 948
Gänsefingerkraut, 3416
Gänsefuss, 2, 1299
ganzrandig, 1228
garantierter Mindestpreis, 1697
gären, 1309
Gärkompost, 1427
Gärmaterial, 1429
Garten, 1527
Gartenabfälle, 1537
Gartenbau, 1827
Gartenbauberater, 1823
Gartenbaubetrieb, 2284
Gartenbaufachschule, 3273
Gartenbaugebiet, 1824
Gartenbaukundiger, 1829
Gartenbauschule, 1826
Gartenbau unter Glas, 1828
Gartenbauzentrum, 633
Gartenerde, 1535
Gartenhäuschen, 3730
Gartenhippe, 2965
Gartenkresse 889
Gartenlaubkäfer, 1530
Gartenmargerite, 3367
Gartensäge, 2967
Gartenschere, 2968, 3302
Gartenschlauch, 1533
Gartenspritze, 3780
Gartentorf, 1468
Gartenwolfsmilch, 2695
Gärtner, 1532
Gärtnerei, 2284, 2494
gärtnern, 1528
Gärung, 1310
Gärungsvorgang, 1311
Gasbrenner, 1541
Gasentladungslampe, 1543
gasförmig, 3995
gasförmiger Brennstoff, 1544
Gaslagerraum, 1547
Gaslagerung, 823
Gasmaske, 1545
Gasschäden, 1542
Gasvergiftung, 1546
Gattung, 1559
Gattungsname, 1555
Gauklerblume 2376
gebogenes Spezialmesser, 950
gebrannter Kalk, 3007
Gebrauchsanweisung, 1075
Gebrauchswert, 3050
gedornt, 3830

Geestboden, 1550
Gefässbündel, 4004
Gefässkrankheit, 4005
gefiedert, 2741
gefleckter Kohltriebrüssler, 517
Geflügelmist, 2883
geflügelt, 4172
gefranst, 1461
gefrieren, 1453
Gefriertrocknung, 1454
Gefriertruhe, 1806
gefüllt, 1107, 1490
gefurcht, 1508
Gegendumgestaltung, 3089
Gegendverbesserung, 3089
gegenständig, 2535
Gegenwartswert, 2907
gegliedert, 157
Gehalt, 817
Gehölze, 4201
Gehölzzucht, 144
geil, 8
Geiltrieb, 4096
Geissfuss, 1668
Geissklee, 442
gekerbt, 888
geköpfte Herbstmöhren, 3888
gekühlte Schauvitrine, 3084
gelbe Fichtengallenlaus, 3587
— Pflaumensägewespe, 2805
gelber Birnenblattsäuger, 2648
— Schimmel, 4226
gelbe Stachelbeerblattwespe, 770
Gelbfleckigkeit, 2389
Gelbsand, 4227
Gelbstreifigkeit der Zwiebel, 2528
Gelee, 1976
geleimter Holzbinder, 2024
Gelenkblume, 2508
Gelenkwellenschutz, 2892
gemeine Akazie, 1284
— Austernschildlaus, 2578
— Buschhornblattwespe, 2735
— Kiefer, 3283
— Melde, 773
gemeiner Apfelblattsauger, 135
— Erdrauch, 1497
— Kommaschildlaus, 2434
— Maikäfer, 731
— Ohrwurm, 768
gemeine Rosengallwespe, 282
gemeines Greiskraut, 1673

gemeine Spinnmilbe, 3071
gemeines Rispengras, 3208
gemeine Wespe, 1563
Gemeinkosten, 1895
— der Kuppelproduktion, 1979
Gemeinschaftsgefrieranlage, 2205
Gemeinschaftswerbung, 750
gemischte Knospe, 2358
Gemswurz, 2136
Gemüse 4010
Gemüseampfer, 1787
Gemüsebau, 4009
Gemüseeule, 3874
Gemüsegarten, 4008
Gemüsekultur, 4009
Gemüsesamen, 4011
Gemüsetrockenanstalt, 1031
Gemüsewanze, 1246
Gemüsewurzelfliege, 275
Gen, 1551
Generation, 1552
generativ, 1553
generative Vermehrung, 3357
Genotypus, 1557
geologisch, 1560
Geranie, 4240
gerippt, 3127
Geruch, 3446
geruchlos, 3271
gesagt, 3352
geschäftsanlagen, 1236
geschälte Erbse, 3559
geschichtet, 3675
Geschiebelehm, 413
geschlechtliche Vermehrung, 1554
geschlossener Champignon, 507
Geschmack, 2583
Geschmacksfehler, 2513
gespalten, 701
Gespinstmotte, 3442
Gestehungskosten, 3978
Gestehungskostenkalkulation, 3979
Gestehungskostensenkung, 3980
Gesteine, 3150
Gestell, 3844
gestielt, 2692
gestreift, 3692
gestreifter Blattrandkäfer, 2621
gesund, 1749, 3971
geteilt, 2614
getüpfelt, 3564

Gewächs, 2759
Gewächshaus, 1579
Gewächshausblock, 2427
Gewächshausboden, 1587
Gewächshauskultur, 1581
Gewächshauspflanze, 1585
Gewächshaustyp, 1589
Gewächshausvormann, 1583
Gewächskontrolle, 897
Gewährleistung, 3306
Gewebe, 2332, 3864
Geweihfarn, 3598
Gewicht, 4120
Gewichtssortierung, 1619
Gewichtsverlust, 4121
gewimpert, 684
gewinnen, 3056
gewöhnliches Bohnenmosaik, 766
— Erbsenmosaik, 2636
gewölbt, 825
Gewürzkraut, 153
gezähnt, 1036
Gibberellin, 1571
Giebelversteifung, 4056
Giersch, 1668
Giesskanne, 4082
Giesswasser, 1961
Gift, 2813
giftig, 2816
Ginkgobaum 2240
Ginster, 442
Gips, 1702
—, brauner, 448
Gipskrankheit, 4144
Gitterrad, 522
Gittertopf, 4182
Gladiole, 1574
Glasbreite, 2595
Glasfaser, 1578
Glasigkeit, 4075
Glasscheibe, 2594
Glasverschmutzung, 1591
Glaswolle, 1593
glatt, 3448
Glattwalze, 1370
Gläubiger, 2130
Glied, 1944
Gliederfüssler, 156
Gliederkessel, 3305
Gliedertiere, 156
Glockenblume, 301
glockenförmig, 303
Glockenheide, 1756

glockig, 303
Gloeosporium-Fäule, 324
Gloxinie, 1602
Glühlampe, 1884
Glührest, 1160
Glührückstand, 3109
Glühverlust, 1763
Glyzine, 4186
Goldafter, 451
—, heller, 4230
Goldbandlilie, 1606
Goldglöckchen, 1442
Gold-Johannisbeere, 1605
Goldlack, 4057
Goldmohn, 534
Goldregen, 2018
Goldregenminiermotte, 2019
Goldrute, 1607
Götterbaum 3918
Graben, 1090
graben, 1063
Grabenpflug, 1091
Grabenräumer, 1091
Grabenwasserstand, 1092
Grabkranz, 1498
Grad, 1023
Granatapfelbaum, 2841
Granulat, 1632
granulieren, 1633
Gras, 1641
Grashalm, 1642
Grasharke, 2054
Grasnarbe, 3758
Grasnelke, 3835
Grassodenstampfer, 3789
Grasstreifen, 1643
graue Erbsen, 1333
grauer Kugelrüssler, 3243
Graufäule, 1661
Graupappel, 1663
Grauschimmel, 1657, 1658, 1660
Grauschimmelfäule, 2454
Grauschimmelkrankheit, 409, 1352, 1353, 2168
Greiskraut, gemeines, 1673
Grenzbetrieb, 2274
Grenzkosten, 2272, 2273
Grenznutzen, 2275
Griffel, 3701
Grindel, 2794
Grindkraut, 3263
grob, 724
grobkörnig, 725
grossblättrig, 2033

grosse Brombeerblattlaus, 3215
— Narzissenfliege, 2445
Grössenmass, 3427
Grössensortierung, 3429
grosser Frostspanner, 2400
— Kohlweissling, 2037
— Obstbaumsplintkäfer, 2032
— Pappelbock, 2035
— schwarzer Rüsselkäfer, 2737
— Ulmensplintkäfer, 1214
— Wegerich, 774
— Weidenrindenlaus, 2038
— Wühlmaus, 4101
Grosshandel, 4154
Grosshändler, 4153
Grosskiste, 491, 2586
Grossmarkt, 4152
Grossraumladen, 1037
Grubber, 1695, 2359
— mit starren Zinken, 3136
Grube, 2751
Grundanalyse, 3471
Grundbedeckung, 3478
Grundbezitzabgaben, 2026
Grunddüngung, 258
grundieren, 2930
Grundstoffe, 3045
Gründung, 1422
Gründüngung, 1653
Gründüngungspflanze, 1652
Grundwasser, 1674
Grundwasserspiegel, 4099
Grundwasserstand, 2145
grüne Apfelblattlaus, 1650
— Erbsenblattlaus, 2622
— Pfirsichlaus, 2625
grüner Eichenwickler, 1654
Grünkohl, 1991, 3255
Grünkragen, 1651
grünliche Gänsedistellaus, 948
Grünrüssler, 2114
Grünspan, 4226
Gruppenarbeit, 3798
Guano, 1696
Gummiholzkrankheit, 3214
Gundelrebe, 1669
Gundermann, 1669
Günsel, 474
Gurke, 920
Gurkenkrätze, 1698
Gurkenmosaik, 921
Gurkennekrose, 922
Gurkentreibfurche, 1430
Gurkenvirus, 923

Gürtelschorf, 3259
Güteklasse, 1613
Gütesortierung, 1618

Haarkrone, 2601
Haarmücke, 308
Haarwurzel, 1705
Hacke, 1800
hacken, 3287
Hackmaschine, 3288
Hackmesser, 676
Haferwurz, 3236
Haftsumme, 3306
Haftwurzel, 38
Hahn, 730
Hahnenbalken, 748
Hahnenfuss, 504
Hahnenkamm, 733
Hahnschlüssel, 732
Hainbuche, 1815
Hainrispengras, 4197
halbautomatisch, 3342
halboffener Champignon, 936
Halbstamm, 1710
Hallimasch, 149
Halm, 925
Halsfäule, 1659
haltbar, 2042
Haltbarkeit, 1165
Handarbeit, 2260
Handdrillmaschine, 1719
Handelsdünger, 161
Handelspolitik, 764
Handelswert, 3901
handförmig, 1064
— gelappt, 2590
— gespalten, 2589
— geteilt, 2591
Handhubwagen, 2587
handnervig, 2588
Hanf, 1777
Harke, 3022
harken, 3023
Harn, 3987
Harnstoff, 3985
Harnstoff-formal, 3986
Härte, 1725
harter Schwingel, 1722
Hartfäule, 1726
Hartheu, 3656
hartlamellig, 1724
Hartriegel, 1098
hartschalig, 1720
Hase, 1729

Haselnuss, 1744
Haselnussknospengallmilbe, 2499
Haselnusstrauch, 1742
Haspel, 3080
Haube, 713
Haubendämpfung, 3372
Häufelhacke, 1178
häufeln, 1193
Häufler, 3135
Hauptabsperrventil, 2250
Hauptast, 2242, 2245
Hauptdampfrohr, 2248
Hauptrippe, 2338
Hauptrohr, 2246
Hauptstengel, 2249
Hauptstrang, 2244
Haupttrieb, 2066
Hauptwurzel, 2247
Hausabfallkompost, 1100
Haushaltgefriermöbel, 1806
Hauskehricht, 1840
Hauslauch, 1841
Hausmüll, 1840
Haussperling, 1843
Haustreiberei, 1896
Hauswurz, 1841
häutig, 2326
Hechelmaschine, 1704
Hecke, 1767
Heckenkirsche, 1813
Heckenschere, 1772
Heckensystem, 1771
Heckenwickler, 3191
Heckstapler, 1437, 1438
Hederich, 4159
heften, 3612
Heftklammer, 3611
Heftmaschine, 3613
Heideerde, 1757
Heidelbeere, 311, 3756
Heilkräuter, 2319
heissbehandelt, 1765
heizen, 1754
Heizinstallation, 1760
Heizkabel, 1758
Heizkörper, 3013
Heizrohr, 1359
Heizungstechniker, 1761
Heliotrop, 1775
hell, 432
heller Goldafter, 4230
Hellerkraut, 3654
Hemlocktanne, 1776

Kiesel, 3401
kieselhaltig, 3403
Kieselsäure, 3402
Kieselstein, 2662
Kieserit, 1997
Kieskultur, 1648
Kippen, 3889
Kipper, 4089
Kippflug, 233
Kippkarren, 3863
Kipptopf, 3862
Kirschblattwespe, schwarze,
    2646
Kirschfruchtfliege, 651
Kirschlorbeer, 652
Kirschpflaume, 653
Kistenkarre, 3932
Kitt, 2995
kitten, 630
Kittfalz, 1592
Klarglas, 1713
Klärschlamm, 3355, 3356
Klatschmolm, 850
Klebemittel, 27
klebendes Labkraut, 700
Klebkraut, 700
Klee, 717
Kleie, 420
kleinblütiges Franzosenkraut,
    1995
kleine Himbeerblattlaus, 3034
kleiner Apfelglasflügler, 1173
— Frostspanner, 4178
— Holzbohrer, 79
— Sauerampfer, 3371
— Ulmensplintkäfer, 3441
Kleinfrüchtigkeit des Apfels, 646
Kleingarten, 69
kleinste Rosenblattrollwespe,
    2093
Kleinverpackung, 812
Kleinzikade, 682
Klemmherz, 4139
Klemmkarre, 692
kletternd, 710
Kletterpflanze, 709
Klettertrompete, 3934
Klima, 705
Klimabeherrschung, 706
Klimagewächshaus, 2714
Klimakammer, 1686
Klimatisierung, 52
Klon, 715
Klumpen, 721

Klumpenblätterkrankheit, 209,
    1518
Knallapparat, 318
Knäuel, einjähriger, 108
Knaulgras, 734
Knick, 3650
Knickschicht, 782
Knoblauch, 1540
Knochenmehl, 394
Knöllchenbakterien, 2487
Knollenbegonie, 3946
Knollengewächs, 3947
Knollensellerie, 618
Knopfkraut, 1995
Knospe, 458
Knospenbildung, 464
Knospenfall, 1145
Knospenschuppe, 470
Knospensterben, 219
Knospenwickler, roter, 467
knospig, 466
Knoten, 1978
Knotenblech, 1699
Knöterich, 2003
knotig, 2002
Kobalt, 727
kochen, 827
Kohl, 509
Kohlblattlaus, mehlige, 510
Kohlenhydrat, 576
Kohlensäure, 583
Kohlensäureassimilation, 582
Kohlensäuregas, 580
Kohlensäuregasdüngung, 581
Kohlensäuregasverabreichung,
    581
kohlensaurer Kalk, 530
Kohlenstoff, 579
Kohleule, 514
Kohlfliege, 515
Kohlgallenrüssler, 511, 516,
    3960
Kohlgallmücke, 424
Kohl-Gänsedistel, 111
Kohlhernie, 719
Kohlrabi, 2005
Kohlrübe 3760
Kohlrübenblasenfuss, 519
Kohlrübenblattwespe, 3961
Kohlschabe, 1050
Kohlschotenmücke, 424
Kohlschotenrüssler, 516
Kohlschwärze, 75, 986
Kohlstrunk, 518

Kohltriebrüssler, gefleckter, 517
Kohlwanze, 1246
Kohlweissling, 520, 3445
—, grosser, 2037
Kohlzystenälchen, 512
Kokardenblume, 1517
Kokon, 735
Kolben, 3530
Kolbenpumpe, 2750
kolloidal, 752
Kolloide, 751
Kommaschildlaus, gemeiner,
    2434
Kommissionär, 765
Kompost, 786
—, abgetragener, 3547
— aus Stadtmull und Torf, 3894
Komposthaufen, 788
kompostieren, 787
Kompostierungsbetrieb, 789
Kompostierungsplatz, 790
Kompott, 3648
Kompressor, 794
Kondensator, 802
kondensieren, 801
Kondenstopf, 3632
Kondenswasser, 800
Konifere, 805
Königsfarn, 3213
Königskerze, 2422
Königslilie, 3087
Konnektiv, 808
Konserven, 2914
Konservenindustrie, 2915
Konservierung, 1479, 2908
Konservierungsmittel, 2910
Konsistenz, 809
Konsument, 810
Kontaktgif, 814
Kontaktwirkung, 813
Kontokorrentkredit, 31
Kontrollbesichtigung, 1924
kontrollierte Erhitzung, 2632
Konzentrat, 797
Konzentration, 798
Konzentrationsmesser, 1322
Konzentrationsmessgerät, 1070
Kopf, 1745
Kopfbildung, 1750
Kopfbruststück, 637
Köpfchen, 572
Kopfdüngung, 3885
köpfen, 1747
Kopfformung, 1750

Saatkiste, 3331
Saatreiniger, 3313
Saatschale, 3321
Saatzeit, 3522
Sack, 230
Säckelblume, 615
Sackkarre, 3227
Sackrinne, 3991
Säedatum, 3521
säen, 3519
Saft, 1983
saftig, 1984
Saftstrom, 1407
Sägewespe, 3256
Saisonarbeit, 3300
Salat, 2139
Salatbohne, 1455
Salaterdlaus, 2142
Salatmosaik, 2141
Salatrübe, 292
Salatwurzellaus, 2142
Salbei, 3230
Saldo, 232
Salmiakgeist, 85
Salpeter, 3235
Salz, 3237
Salzgehalt, 3238
Salzsäure, 1858
Salzverhältnis, 3234
Same, 3311
Samenbau, 3318
Samenbeizung, 3316
samenecht, 3672
Samenhandel, 3330
Samengeschäft, 1529
Samenhaut, 3801
Samenkäfer, 2975
Samenkapsel, 3323
Samenlaufkäfer, behaarter, 3685
Samenlieferant, 3314
Samenmuster, 3326
Samenpflanze, 3322
Samenprobe, 3326
Samenprüfung, 3329
Samenstengel, 3328
Samenträger, 3322
Samenunkraut, 112
Sämereienhandel, 3330
Sämling, 3320
Sammelroder, 2869
Samtblume, 44
Samtfleckenkrankheit, 2089
Sand, 3242
— abgraben, 1066

Sandboden, 3246
Sanddorn, 3293
Sandgraurüssler, 3243
sandig, 3244
sandiger Tonboden, 3245
Sandwicke, 1709
Saprophyten, 3247
sättigen, 3251
Sättigung, 3252
Sättigungsgrad, 1026
Sättigungspunkt, 3253
Saubohne, 434
sauer, 17
Sauer, 3515
Sauerampfer, 3512
Sauerkirsche, 2383, 3517
Sauerklee, 2576
Sauerkraut, 3254
Saugdrän, 3717
Sauger, 3717
Saugrohr, 3718
Saugspannung, 3716
Saugwurzel, 3714
Säule, 759
Säuregrad, 20
saurer Humus, 18
saure Tulpen, 3515
schachbrettartiges Stapeln, 647
Schadbild, 2619
Schaden durch Übermass, 2698
— durch Vögel, 976
schädlich, 2490
schädliche Verunreinigung, 1912
Schädlinge, 2491
Schafgarbe, 2342
Schafmist, 3369
Schafschwingel, 3370
Schale, 1844, 3432
Schalenbräune, 3738
Schalerbse, 1536
Schalotte, 3361
Schalottenlaus, 3362
Schalter, 3775
scharfes Adernmosaik der Erbse,
    2630
scharlachrot, 3268
Schattendecke, 3359
Schaufel, 1178, 3388
Schaumbeton, 1413
Schaumkraut, behaartes, 1707
Schaumzikade, 769
Schaumzirpe, 769
Scheibenegge, 1079
Scheibensech, 1078

Scheide, 3537
scheierlos, 1724
Scheinfrucht, 2969
Scheinquirl, 1287
Scheinzypresse, 1285
Schermaus, 4101
Scheune, 252
Scheunenlagerung, 3663
Schicht, 2057
schieben, 254
Schieber, 3437
schiefer Hut, 3854
Schiesser, 393
Schiff, 3531
Schildchen, 3289
schildförmig, 3377
Schildlaus, 3266
—, austernförmige, 2578
Schilf, 3076
Schilfmatte, 3078
Schilftorf, 3079
Schimmel, gelber, 4226
—, olivgrüner, 2521
Schimmelpilz, 1499
Schingpflanze, 709
Schirmtanne, 3970
schlafendes Auge, 1103
Schlafmützchen, 534
schlagfest, 1874
Schlamm, 1130, 3407
schlammig, 2416
Schlammloch, 2417
Schlammventil, 383
Schlauch, 1830
Schlauchfolie, 2061
Schlehe, 354
Schleier, 4015
Schleierkraut, 1701
Schleifenblume, 548
Schlepper, 1443
Schlepper-Anbauschleuder-
    streuer, 3899
Schleuderdüngerstreuer, 1321
Schlick, 1284
Schliessfrucht, 16
Schlingpflanze, 709
Schlund, 3837
schlüpfen, 1737
Schlupfwespe, 1869
Schlüsselblume, 2931
Schmalbauch, 447
schmalblättrig, 2448
Schmalblättrigkeit, 2090
Schmalwand, 3820

verwesen, 1003, 3195
Verwesung, 426
verwildern, 2452
verwittern, 4107
verzehren, 2993
verzinken, 1522
verzweigen, sich, 3028
verzweigt, 422, 3027
Verzwergung, 1183
vibrieren, 4030
vielblättrig, 2833, 2836
Vielfachgerät, 3880
violetter Wurzeltöter, 4039
Virologie, 4041
Virose, 4043
viröser Atavimus, 335
Virusberührung, 4042
Viruskrankheit, 4043
Viruskunde, 4041
Vitaminen, 4045
Vogelkirsche, 4158
Vogelknöterich, 2001
Vogelmiere, 657
Vogelmilch, 3616
Vogelwicke, 3952
vollautomatisch, 1491
Volldünger, 785
Vollkosten, 1932
Volumen, 4049
Voranschlag, 465
Vor-Auflaufbehandlung, 2853, 2898
Vorfluter, 605
Vorfrucht, 2894
Vorgewende, 1748
Vorkeim, 2960
vorkeimen, 2899
Vorkühlung, 2896
Vorkultur, 1360, 2443
Vorratsdüngung, 257
Vorratsroder, 2868
Vor-Saatbehandlung, 2917
Vorschäler, 3431
Vorschätzung, 465
Vorschrift, 1074
Vorverpackung, 2903
Vorwärmer, 2901
vorweichen, 2916

Wacholder, 1986
Wachsblume, 4103
wachsen, 1675
Wachsschicht, 369
Wachstum, 1685

Wachstumsfaktor, 1687
Wachstumsforderung, 1692
Wachstumshemmung, 1688, 1689
Wachstumsspitze, 3814
Wachstumsstockung, 1945
Wachstumszeit, 1680
Wald, 4194
Waldbaum, 1433
Waldboden, 1432
Waldmeister, 3772
Waldrebe, 703
Waldsträucher und Heckengehölze, 1434
Waldstreu, 1431, 2734
Wall, 244
Walnuss, 780, 4060
Walnussblattlaus, 1245
Walnussbrand, 217
Walze, 3159
walzen, 3158
wandernde Wurzelnematode, 2138
Wanze, 472
Wanzenblume, 843
Warmbeet, 1833
Wärmeaustauscher, 536
Wärmebehandlung, 1766
Wärmeschreiber, 3823
Wärmestrahlung, 330
Wärmetherapie, 1766
Warmhaus, 1755, 1835
Warmluftofen, 1832
Warmwasserbehälter zum Kochen, 1837
Warmwasserbehandlung, 1838, 4062
Warmwasserkessel, 1836
waschen, 4064
Wasserabfluss, 4080
Wasserabführung, 4078
Wasserauffanggraben, 605
Wasseraufnahmevermögen, 4097
Wasserbehälter, 4100
Wasserbewegung, 4088
Wasserbilanz, 4070
Wasserdampf, 3996
wasserdicht, 4091
wasserdichte Leuchte, 4092
Wasserdost, 1778
Wasserdruck, 4090
Wasserentziehung, 1030
Wasserentzug, 1030

Wasserflecken, 377
Wasserfurchen ziehen, 3746
Wassergehalt, 2368
Wasserhaushalt, 4072
Wasserkapazität, 1330
Wasserknöterich, 90
wasserkrank, 379
Wasserkultur, 4077
Wasserleitung, 4079
Wasserlinse, 1161
Wassermangel, 4095
Wassermelone, 4087
Wasserpflanze, 141
Wasserreinigung, 4093
Wasserrübe, 3959
Wassersäule, 4071
Wasserschlag, 4081
Wasserschoss, 4096
Wassersiegel, 4094
Wasserstand, 4083
Wasserstandsrohr, 4084
Wasserstoff, 1860
Wasser- und Bodenverband, 1119
Wasserverband, 1119
Wasserverbrauch, 4073
Wasserversorgung, 4098
Wasserwirtschaft, 4074
wechselständig, 76
Wechselventil, 3834
Wegerich, 2761
—, grosser, 774
Wegranke, 1770
weiblich, 1305
weich, 3464
weiche Trespe, 440
Weichfäule, 4130
Weichobst, 3467
Weichselkirsche, 3657
Weide, 4161
Weidelgras, 3226
Weidenblattkäfer, 425
—, blauer, 781
Weidenbohrer, 1604
Weidenkorb, 1715
Weidenrindenlaus, grosse, 2038
Weidenspinner, 3250
Weiderich, 2216
Weinachtskaktus, 677
Weihnachtsstern, 2811
Weinbau, 4036
Weinbeere, japanische, 1974
Weinrose, 3762
Weinstock, 4035

DANSK

DANSK

abe, 3205
abeblomst, 2376
abetræ, 660
abrikos, 139
abrikostræ, 140
absolut fuktighed, 3
absorbere, 4
absorbtion, 5
absorbtionsevne, 6
acklimatisere, 11
actinomyceter, 1356
additiver, 1421
Adelgea, 4203
Adelges (grangallelus), 3587
adgangsbetingelser, 3103
adsorbere, 28
adsorbtion, 29
adsorbtionsevne, 30
adventivknop, 32
afart, 4001
afbanke, 3265
afblade, 1017, 3882
afbladningsmaskine, 1018
afblomstre, 1278, 2123
afdrevet kompost, 3547
afdrivningsrum, 903
afdække, 869
affald, 1903, 4066
affaldskalk, 4068
affarve, 1002
affarvning, 1080
afgrave, 1065
— sand, 1066
afhugge, 957
afhænde, 1721
afkalke, 998
afklipning, 2966
afkom, 2948
afkøling, 1143
aflade, 1371
aflang, 2510
aflægge, 2058
aflægger, 2056
aflægning, 57, 3138
afløb, 1119
afløbsrende, 1120
afløbsrør, 2560
afløve, 3882
afløver, 1739

afmodne, 3145
afsalte, 1042
afskalle, 3373
afskare, 957
afskrive, 4221
afskrivning, 1039
afskærme, 3284
afsmag, 2513
afsondring, 3304
afstamning, 1043
afsuge, 1882
afsvampningsmiddel, 3315
afsætningsfonds, 2285
afsætningskanal, 1089
afsætte, 1038
aftagelig, 3097
aftapningshane, 1123
aftenstjerne, 3768
aftoppede karotte, 3888
aftopper, 1739
aftrækskanal, 55
afvande, 1117
afvanding, 2977
afvigende, 1061
afviger, 3093
afvikling, 89
agatugle, brun, 102
ager, 1328
agerbrug, 142
agerfure, 2797
ager galtetand, 1339
agerjord, 143, 1535
ager-løvemund, 4106
ager-mynte, 1331
ager-padderokke, 771
ager sennep, 645
agersnegl, 1335
ager-snerle, 1329
ager-stedmoderblomst, 1332
ager-svinemælk, 2674
ager-tidsel, 886
aggregat, 47
aggregationstilstand, 3618
agrar, 49
agurk, 920, 3133
agurk-mosaik, 921
agurkmosaikvirus, 923
agurk-nekrose, 922
ahorn, 2271

ahorn-rynkeplet, 3796
ajle, 2193
ajlebeholder, 639
akeleje, 758
akkordløn, 2724
aks, 3549
akse, 206
aksel, 203
akselblad, 3655
akselflig, 3655
akselknop, 205
akselrøn, 4140
akselstillet, 204
aktive bestanddele, 22
aktiver, 175
al, 1954
aldring, 46
alge, 3301
alkalisk, 68
allétræ, 200
alm 2342
— eg, 772
almindelig, brandbæger, 1673
— dværgløvefod, 2611
— firling, 2643
— hanekro, 1779
— kalmus, 3770
— oldenborre, 731
— rajgræs, 2672
— rapgræs, 3208
— svinemælk, 111
alm-løvefod, 2022
aloe, 72
alpeplante, 73
alperibs, 2405
alpeviol, 967
altanplante, 234
alun, 77
amarant, 610
ambratræ, 3765
Amerikansk rødeg, 3067
— tranebær, 83
ammoniak, 85
ammoniakalun, 87
ammoniak-base, 105
ammonium, 86
ammonsalpeter, 88
amortisation, 89
anaerob, 91

cellevæv, 626
cellulose, 627
cellulose-acetat, 628
cembrafyr, 151
centerråd, 842
centnergræskar, 2980
centrifugalpumpe, 635
centrifugalventilator, 634
Chaetomium, 2521
champignon, 927
champignonflue, 2701
champignongødning, 2430
Chermes nordmannianae, 3411
chilesalpeter, 659
chrysanthemumflue, 681
chrysanthemum-sortråd, 3046
cider, 683
cikade, 682
cikorie, 658
cineraria, 685
cinnobersvamp, 835
cirkelrund, 688
cirkulation, 689
cirkulerende produktions-
   middel, 4206
citron, 2128
citronmelisse, 237
citronsyre, 690
citrontræ, 2129
Clematis, 703
CO₂-gødning, 723
conserveringsmiddel, 2913
cordon, 839
Corynebacterium, 2083
CO₂-tilförsel, 723, 738
cyclamen, 967
cyclamenmide, 3682
cyclisk belysning, 968
cyste, 971
cystenematod, 972

Dactylium, 729
daddel, 988
daddelpalme, 989
dagforkortning, 3385
dagforlaengelse, 2131
daglilie, 991
daglængde, 2132
dagrenovation, 1100, 3893
dagslys, 990
dagsværdi, 2907
dahlie, 974
dam, 2842
dampdesinfektion af jord, 3631

dampe, 3622
dampet benmel, 3625
dampharve, 2362
damphat, 3624
dampkedel, 3623
dampledning, 3627
dampningsklæde, 3629
dampning under plastsejl, 3372
dampplov, 3626
damprist, 1799
damprum, 3630
dampsejl, 2789
dampslange, 3633
dampspærre, 2367, 3997
damptryk, 3628
d-b-syge, 2431
debitor, 406
degeneration, 1022
degenerationssyge, 1057
degenerere, 1021
dehydrering, 1030
dekomponere, 2993
dele, 1093
deler sig for meget, 3560
del i udbytte, 395
delt, 701, 2614
deltaformet, 1033
demineralisation, 1034
demonstrationsgartneri, 1263
denitrificere, 1035
desinfektion, 1085
desinficere, 3647
destillat, 1087
detailhandler, 3115
Deutzia, 1047
die-back, 1057
differensomkostninger, 2272
differentiering, 1060
diffus-stråler, 1062
dild, 1069
direkte omkostninger, 1073
dissociationsgrad, 1024
dobbelkalkfosfat, 1053
dobbelsuperfosfat, 1110
dobbelt, 1490
— bænk, 1109
— bænkevindue, 3533
dobbeltfinnet, 316
dobbeltnæse, 1108
dolomit, 1099
dominans, 1101
dosering, 1105, 3040
douglasgran, 1111
drejeblomst, 2508

drejeskive, 4135
driftsherregevinst, 1230
driftsorganisation, 2547
driftsregnskab, 2534
driftssammenligning, 783
driftsstruktur, 1295
driftstype, 1296
driftsøkonomie, 1294
driftsøkonomisk regnskabsføring, 502
drivbuske, 3396
drivbænk, 1833
drivbænke, 1449
drive, 1425
drivhus, 1428, 1579
drivhuskultur, 1581, 1582
drivhusplante, 1585
drivhus til vindyrkning, 1637
drivkasse, 1367
drivløg, 3232
drivmateriale, 1429
drone, 1140
drosselventil, 3838
drue, 1635
druehyacinth, 1638
druesaft, 1639
drypvanding, 3931
dræn, 1118
dræne, 1117
drænemaskine, 1124
drænfure, 3926
drængrøft, 605
dræningsmaskine, 1127
drænrende, 3926
drænrør, 1121, 1126
drænrørafstand, 3525
drænrørsdybde, 1041
drænsystem, 3848
dråbedannelse, 1142
duft, 3446
dugget, 2962
dunet, 1114
dunhammer, 3077
dvale, 4180
dværgcypres, 1285
dværgvækst, 1183
dybfrosne grønsager, 1008
dybfryse, 1007
dybpløje, 1009
dybrodende, 1010
dyndet, 2416
dyrke, 926
— på rygge, at, 3134
dyrker, 429, 1677, 2765

SVENSKA

SVENSKA

lövfall, 1019
lövjord, 2088
lövkoja, 3658
lövräfsa, 2054
lövskogsnunna, 1572
lövträd, 437
lövvivel, 2114
—, avlång, 447

magasinera, 3669
maggift, 3661
magnesia, 2230
magnesium, 2231
magnesiumbrist, 2232
magnesiumsulfat, 2233
Magnolia, 2235
Mahonia, 2238
majskolvsavfall, 848
makedonisk tall, 2228
mal, 716
maltgroddar, 2255
maltsocker, 2254
Malva, 2253
malört, 4215
mammutträd, 4127
mandarin, 3790
mandelträd, 71
mangan, 2256
manganbrist, 2257
manganöverskott, 1254
mangold, 3552
manlig, 3603
manlucka, 2258
manometer, 2920
mantelkylning, 1968
manuellt arbete, 2260
marginalföretag, 2274
marginalkostnad, 2273
marginalnytta, 2275
marginellt självkostnadspris,
    2272
marinpanna, 2278
mark, 2025
markbehandling, 3501
markbeskaffenhet, 2453
markkartering, 3499
marklära, 3495
marknad, 2282
marknadspris, 2287
marknadsrapport, 2288
marknadsreglering, 2286
marknadsstabilisering, 2289
marknadsstruktur, 2290
marknadesstörning, 2283

markprofil, 3493
marktäckning, 2420
markvård, 3501
markyta, 3498
markör, 1136, 2281
marmelad, 2293
martorn, 3296
mask, 2229, 4212
maskgång, 4213
maskin för plantering och ut-
    läggning av folie, 2421
— för utläggning av folie, 2787
maskinstation, 820
maskros, 982
maskstungen, 4214
maskulin, 3603
maskväg, 4213
massreklam, 750
massurval, 2302
matarpump, 1302
matarrör, 1301
matjord, 1535, 3855
matjordslager, 3890
matpotatis, 4061
matpumpa, 1612
mattram, 2997
maximipris, 2305
maximitemperatur, 2306
me, 2343
med blad, 1416, 2115
medel, 3096
medelavkastning, 202
medelfel, 3606
Medelhavsfruktfluga, 2320
medelintäkt, 202
medelpris, 201, 2315
medicinalväxter, 1788, 2319
medicinalväxtodling, 1785
meja, 2411
mejram, 2280
mekanisering, 2318
mellanförädling, 1939
mellankultur, 1937
melon, 2325
membranpump, 1051
meristem, 2330
merutbyte, 1890
mesembrianthemum, 2331
metamorfos, 2333
micropyle, 1798
middagsblomma, 2331
mikrober, 2335
mikroelement, 3895
mikroklimat, 2336

mikronäringsämne, 3895
mikroskopisk, 2337
mild humus, 2341
miljö, 1231
milliekvivalent, 2343
mindre knoppvecklare, 467
— krysantemumfluga, 681
— narcissfluga, 3444
minera, 2346
mineraler, 2349
mineralisering, 2348
mineralisk, 2347
minerarfluga, 2087
miniaturväxthus, 713
minimipris, 2350
minimitemperatur, 2351
minska, 3072
minutförpackning, 812
mispel, 2323
mispelträd, 2324
missbildning, 2252
missformad, 1020
missfärgad, 229, 378
missväxt, 899
mistel, 2354
mittaxel, 632
mittnerv, 2338
mittpris, 2315
mittstamm, 632
mjuk, 3464
mjukplister, 1781
mjöldagg, 2340, 2885, 2886
mjölig, 2313
mjölksaft, 2046
mjölktistel, 111
mjölktistelbladlus, 948
modercell, 2393
moderlök, 2392
modermaterial, 3617
moderplanta, 2395
modifikation, 2364
mogen, 3144
mogna, 2304
mognad, 3146
mogning, 3147
mogningsrum, 3148
molybden, 2375
monokultur, 2379
moras, 3755
morell, 2383
morot, 593
morotcystnematod, 594
morotsfluga, 595
morotsupptagare, 596

pumpstation, 2979
punkterad, 1106
puppa, 2982
puré, 2985
purjo, 2124
purjolök, 2124
purpurröd, 3061
putsa, 3913
putsningsmaskin, 1148
p.v.c., 2839
påhängväxelplog, 1711, 2407
påle, 2819
pålrot, 3791
påsklilja, 3936
pärlhyacint, 1638
pärllök, 3414
päron, 2638
päronbladgallmygga, 2642
päronbladloppa, 2648
päronbladlus, 2639, 2640
päronblomvivel, 129
pärongallmygga, 2644
päronrost, 3223
päronstekel, 2645
päronträd, 2649

rabarber, 3125
rabarbermosaik, 3126
rabatt, 400, 1388
rad, 3210
radavstånd, 3212, 3527
radgödsling, 1323
radiator, 3013
radioaktivt nedfall, 3014
radplantering, 3211
radsådd, 1137
radsåningsmaskin, 3317
(raffia-) bast, 3016
rajgräs, 2672, 3226
—, engelskt, 2672
—, italienskt, 1963
rand, 400
randig ärtvivel, 2621
ranka, 423, 3806
ranunkel, 3032
rappa av vatten, 382
rapsbagge, 373
rapsfjäril, 1655
rapssugare, 1246
rapsvivel, blygrå, 516
rasp, 1645
raspblad, 1195
rasren, 3933
rattle, 3043

rattlevirus, 3044
realisationsvärde, 600
recessiv, 3052
redovisning, 12
reducera, 3072
reduceringsventil, 3073
reduktion, 3074
reduktionsdelning, 3075
referenspris, 3081
reflektor, 3083
reflexion, 3082
regional förbättringskampanj,
    3089
registrering, 2014
reglering av vattentillförsel,
    4074
regnbrist, 1014
regnmätare, 3019
relativ fuktighet, 3094
relä, 3095
remonterande, 2686
renhet, 2986
renkultur, 2984
rensa, 698, 4110
— lök, 699
— upp, 1131
rensklo, 4113
renslucka, 2417
rensmaskin, 4115
rensningsmaskin, 3313
rep, 2180
repa, 3695
Reseda, 2339
reservoar, 4100
resistens, 3110
resistent, 3111
respiration, 427, 3112
rest, 3108
restvärde, 3107
resultatjämförelse, 783
retardera, 3117
retarderade lökar, 3118
returemballage, 3119
returrör, 1112
reversion, 335
rhododendron-nätstinkfly, 3123
ribba, 2047
riddarsporre, 2039
rikblommig, 1450
rikblomstrande, 1450
riktpris, 3793
ringblomma, 2277
—, afrikansk, 563
ringduva, 4198

ringfläcksjuka, 3142, 3877
ringspinnare, 2021
ris, 2966
risknippa, 1279
rista för befruktning, 3360
rivhållfasthet, 3799
robinia, vanlig, 1284
rodnad, 388
romersk sallad, 857
ros, 3185
rosbuske, 3187
rosenbladrullstekel, liten, 2093
rosenbladstekel, vitgördlad, 243
rosenböna, 3218
rosenkrage, 2997
rosenkvitten, liten, 1973
rosenmandel, 1181, 1398
rosenskottstekel, 146
rosenstav, 2147
rosenstrit, 3188
rosettbildning, 3192
rosling, 96
rosmosaik, 3190
rost, 1646, 3222
rostig, 3225
rostringar, 3137
roststav, 1644
rot, 3165
rotaktivitet, 3167
rota sig, 3166, 3784
rotationsbrännare, 3196
rotbildning, 3785
rotbrand, 337, 345, 979
roterande bevattnare, 3201
rotfiltsjuka, 980, 4039
rotgallnematod, 3175
rothals, 3170
rothalsröta, 1423, 3468
rothugga, 3021
rothår, 3173
rotklump, 3168
rotknopp, 3172
rotknöl, 3943
rotkräfta, 912
rotlus, 2142
rotmönster, 3177
rotmössa, 3169
rotning, 3174
rotorklippare, 3199
rotröta, 2713, 3179
ıotselleri, 618
rotskott, 3183, 3713, 4160
rotstickling, 3171
rotstock, 3122

socker, 3719
sockermajs, 3763
sockerärt, 3720
solbränd, 3737
solfjäderformig, 1365
solfjädertall, 3970
solros, 3733
solvens, 3507
solvända, 3736
sork, 4048
sort, 3542
sortera, 3513
sortering, 1615
— efter kvalitet, 1618
— efter storlek, 3429
— efter vikt, 1619
sorteringsband, 1925
sorteringsmaskin, 1614, 3514
— för blomsterlökar, 480
sorteringsregler, 1616
sorteringsrum, 1617
sorteringssåll, 3129
sortförsök, 4003
sortiment, 178
sortlista, 2194
sorttorv, 336
sortval, 4002
sortäkta, 3933
sortäkthet, 4000
sot, 117, 3451, 3508
sotdagg, 341, 3511
sotfläck, 3510
sotlucka, 3509
sovande knölar, 1104
— öga, 1103
spade, 3528
spadmaskin, 3529
spadstick, 3558
spadtag, 3558
spaljé, 3925
spaljéträd, 1241
spansk peppar, 3068
sparris, 168
sparrisbagg, 169
sparrisböna, 1455
sparrisfluga, 170
sparrisjärn, 171
sparriskniv, 171
sparrissängplov, 172
Sparrmannia, 3535
sparv, 3536
spatelformig, 3538
spenat, 3551
spets, 3881

spetskål, 2577
spetsvivel, 718
spindelnät, 728
spindelpyramid, 3553
spindelväv, 728
spindelvävssjuka, 729
spinnmal, 3442
spira, 1564, 3584
spiralnematod, 3557
spiskummin, 933
spjutformig, 1736
spjäla, 2047
spjäll, 978, 3838
splintborre, 251
splintved, 3248
spola, 4064
— ur, 2063
spor, 3561
sporre, 3588
sport, 3563
sporväxt, 3562
spottning, 2214
spottstrit, vanlig, 769
spricka, 539, 878
sprida, 3270
spridare, 2492
spridarrör, 3570
spridningsmedel, 2905, 4132
spridningsvinkel, 101
sprinklerslang, 3580
sprita, 3373
spritärt, 1536, 3559
spruta, 3569, 3579
sprutmunstycke, 2492
sprutpistol, 3570
sprutramp, 3567
sprutskada, 3568
sprödhet, 3805
spröjs, 247, 1596
spånkorg, 666
spårvidd, 3898
späcka, 3539
spänna, 3691
spänning, 3809
spännvidd, 2565
stadium, 3597
stadiumundersökning, 1942
stallgödsel, 1297
stam, 1710, 3937
stambasröta, 749
stamdike, 2244
stamfläcksjuka, 2988
stamfrukt, 3641
stamknopp, 2807

stamknöl, 846
stamledning, 2246
stam- och bladbakterios, 220,
221
stamurval, 3423
standardavvikelse, 3606
standardisera, 3609
standardisering, 3608
stapellåda, 491, 2586
stapeltruck, 3595
stapla, 3593
stapling, 3596
stare, 3615
starkt humifierad vitmosstorv,
1466
— växande underlag, 4033
starrtorv, 3307
statisk laddning, 3619
statistisk bearbetning, 3620
stellage, 3844
stenfrukt, 1146
stengelbont, 2875
stenmögel, 3180
stenparti, 3151
stenpartiväxt, 73
stensjuka, 2647
stensöta, 2834
sterilisation, 743
sterilisera, 3647
sticka, 3783
stickande, 2927
stickling, 960
— med klack, 964
sticklingsbord, 961
sticklingsföröka, 3783
sticklingsförökning i vatten-
dimma, 2355
sticklök, 2526
stickmyrten, 503
stickprov, 3030, 3240
stift, 3701
stigande grundvatten i en polder,
2669
stinkfly, 472
stipel, 3655
stippelstreep, 277
stjälk, 3634
stjälkbakterios, 338, 339
stjälkbrand, 3643
stjälklamhet, 3642
stjälklos, 10
stjälkomfattande, 3640
stjälkröta, 3637
stjärnbuske, 1047

ESPAÑOL

ESPAÑOL

— de gas carbónico, 582
asimilar, 176
aspereza de la piel, 3207
— de los frutos, 3221
aspergilo, 174
asperifolia, 173
asperilla olorosa, 3772
áspero, 3204
aspersor, 4082
— rotativo, 3201
aster, 2334
atacar, 188
ataque, 187, 1904
— de virus, 4042
atar, 496, 1362, 3842, 3845
atmósfera, 180
atomizador, 2353
— de espalda, 1999, 2889
atomizar, 184
aulaga, 1509
autofecundación, 3338
autofértil, 3337
automático, 1491
automatización, 197
autoservicio, 3339
autosteril, 3340
avellana, 1744
avellano, 1742
avenida de agua, 4081
avispa, 1563, 4163
— de agallas, 1521
— del abeto, 2735
— de las hojas del fresal, 243
— del manzano, 134
axila, 203
— de la hoja, 2071
axilar, 204
azada, 1800, 3528
azalea, 207
— del Japón, 1971
azascón, 3065
azúcar, 3719
azucena, 2166
— aúrea, 1606
— real, 3087
— tigrina, 3846
azufrar, 1496
azufre, 3727
— en barras, 3649
— en flor, 1404
azular, 384
azularse, 385
azul celeste, 3434
azulejo, 849

babosa, 1335
babosilla del peral, 2646
bacifero, 211
baciforme, 212
bacteria, 228
bacterias aerobias, 41
— nodulares, 2487
bactericido, 225
bacteriófago, 227
bacteriología, 226
bacteriosis de las lilas, 215
bagazo, 430
baldío, 1282
balsamina nicaragua, 238
ballico, 3226
— italiano, 1963
— perenne, 2672
bambú, 239
banco de crédito agrícola, 1292, 1293
banda, 1225
— de pasto, 1643
— de plástico, 2790
bandaje para árboles, 3924
bandera, 3604
barbar, 3784
barbecho, 1282, 1283
barra, 247
— de grilla, 1644
— guadanadora, 958
— para poner vidrios, 1596
barrenillo, 79, 251
— del olmo, 1214
— grande del manzano, 2032
— pequeño del olmo, 3441
barreno de la raíz, de la peonía, 1538
— de los retoños del meloco-tonero, 2628
— de los retoños del rosal, 146
— de suelo, 3473
barrera antivapor, 2367
barro, 2200, 3407
báscula, 3544
base, 2321
— del bulbo, 255
— del limbo, 2072
básico, 68
basquet, 666
bastardear, 1021
bastardo, 265
basura casera, 1100
basuras de ciudad, 1840
batalla de flores, 1392

baya, 306
— de turbera, 1244
bayas, 305, 3443
bebida refrescante, 3465
Begonia, 298
— de hojas ornamentales, 299
— rex, 299
— tuberosa, 3946
beleño negro, 1780
belladona, 992
beneficio del empresario, 1230
berengena, 1205
berro, 4076
— falso, 2300
berza, 509
betún de injertar, 1626
bibio de las huertas, 308
bienal, 309
bienes de capital, 570
— líquidos, 2189
bignonia, 3934
bilabial, 2007
binadora, 1113, 1800
— de acción selectiva, 3288
— de mano, 4136
— de mano rodado, 3162
binar, 2215
biología, 315
bipinado, 316
bisexual, 320
bistorta, 2003
bisuperfosfato, 1110
— cálcico, 1053
bitter pit, 323
blanco de España, 4149
blando, 3464
blanquear, 357, 4150
blenocampa chiquita, 2093
boca de dragón, 123
bodega, 2345
— refrigerada, 3085
bohordo, 3328
boj, 415
bola de nieve, 3453
boletín de análisis, 93, 3472
bolitas de nieve, 3454
bolsa de pastor, 3376
bomba achicadora, 4126
— centrífuga, 635
— de alimentación, 1302
— de membrana, 1051
— de pistón, 2750
— de pozo, 2977
— de sumersión, 2808

campanula, 301, 543
campaña, 713
— agrícola, 1735
campinera de multiplicación, 2950
campo, 1328
— de bulbos, 479
— experimental, 1261
canal, 545, 1700, 3991
— de aireación, 55
— de distribución, 1089
— madre, 605
canasta, 666
canastillo, 2764
cáncer, 551, 554, 555
— bacteriano, 220, 221
— del tallo, 3637
— negro, 331
cancro, 552
cantera, 2345
caña, 925
— común, 1570
— de bambú, 240
— de las Indias, 557
cáñamo, 1777
cañería de agua, 4079
— de riego, 3581
cañizo, 1852
caño de traída, 3744
cañuela de ovejas, 3370
— durilla, 1722
capa, 872, 2057
— arable, 3890
capacidad de campo, 1330
— de retención, 4097
capa de arcilla alcalina, 782
— de hierro, 1954
— de humus, 1847
— de limo, 2201
— de tierra, 2059, 3487
— de tierra arable, 2797
— freática, 4099
— perturbante, 1938
caparrosa, 834
capas, en, 2060
capilar, 565
capilaridad, 564
capilla (cuerpo de invernadero), 3531
capital ajeno, 405
— propio, 2575
capítulo, 572
capricornio, 2209
cápsula, 574

— de la semilla, 3323
capuchina, 2449
capullo, 735
cárabo del fresal, 3685
caracol, 3439
características del suelo, 3477
Caragana, 2660
carbolínco para frutales, 3794
carbón, 117, 3451
carbonato, 577
— cálcico, 530, 1671
— cálcico-magnésico, 578
carbono, 579
carburador de petroles, 2693
carcomido, 1000, 4214
cardenillo, 4226
cardo, 586, 1599, 3829
— corredor, 3296
— cundidor, 886
— plateado, 587
carencia de boro, 404
— de humedad, 2369
— del agua, 4095
— de potasa, 2862
— en cobre, 833
— en fósforo, 1013
— en hierro, 1953
— en magnesio, 2232
— en manganeso, 2257
careta, 1545
carga, 597
cargadora de patatas, 2869
carga estática, 3619
cargas sociales, 3459
caries del tronco, 1419
cariofilada, 199
cariopside, 598
carnet de gestión, 502
carnoso, 1376
carpe, 1815
carpelo, 589
carraspique, 548
carretilla, 3227, 4134
carrito con un gancho especial, 3932
carro elevador, 1438
cartela, 1699
carter de protección, 1447
cartografía aérea, 39
— del suelo, 3499
cartografiar, 2270
cartón, 584
— embreado, 3164
— ondulado, 854

cartucho fumigante, 1494
casa contratante, 820
casaruelos, 1673
cascajo, 3401
cascajoso, 3403
cáscara, 1844, 3432
— dura, de, 1720
cáscaras, 1853
caseta de jardín, 3730
castaño, 655
— de Indias, 1819
castras plateadas, 3413
categoría, 1613
cavadora-mezcladora, 2359
cavar, 1063, 1694, 3021
cavidad del pedúnculo, 3636
ceanoto, 615
cebo, 231
cebolla, 2523
cebolleta, 667
cebollino, 3414
cebollita para plantar, 2526
cecidomia de la col, 3761
— de las peras, 2644
cedro, 616
C.E.E., 1199
cefalotorax, 637
célula, 621
— huevo, 2529
— madre, 2393
celulosa, 627
cenagoso, 2416
cenizas, 1160, 3109
central contable, 13
centro hortícola, 633
cep, 636
cepillar, 454
cepellón champa, 3168
cepillo mecánico, 1148
cerastino, 2409
cerca, 1223
cercar, 1307
— con diques, 1068
cercospora, 2101
cercha de dos charnelas, 3966
— de malla, 2049
— de tres charnelas, 3831
cereza ácida, 3517
cerezo de adorno japonés, 1972
— de cerdo, 3774
— de Santa Lucia, 3657
— de Virginia, 675
— dulce, 4158
cerrado hidráulico, 4094

cuba, 3941
cubierta, 1885
— de arena, 873
— de invernadero, 269
— para esterilización del suelo
con vapor, 2789
— para vapor, 3629
cubierto de helada blanca, 2962
cubo para plantas, 2781
cubrir, 599, 869
— con césped, 875
— de barro, 3409
— de injertar, 462, 1625
cuchilla del arado, 2000
— para ahuecar jacintos, 950
— para cortar espárragos, 171
cuello de la raíz, 3170
— verde, 1651
cuenta de la explotación, 2534
cuerda, 2180, 3693
cuerno de alce, 3598
cuerpo sólido, 3503
cueva, 2345
culantrillo de poza, 2239
cultivador, 429, 930, 1677, 2765
— comercial, 762
— de bulbos, 481
— de dientes flexibles, 3577
— de dientes rígidos, 3136
— rotativo, 3197
cultivar, 926, 1676, 3851
cultivo, 929
— bajo, 3972
— consecutivo, 3303
— de abono en verde, 1652
— de bulbos, 482
— de hortalizas, 4009
— de invernadero, 1581
— de plantas, 2766
— de plantas medicinales, 1785
— de semillas, 3318
— en camas, 280
— en hileras, 281
— en invernadero, 1582
— en lomos, 3134
— esquilmante, 2566
— hidropónico, 4077
— intercalado, 1937, 3972
— precedente, 2894
— puro, 2984
cultivos al aire libre, 2559
cultivo sin suelo, 1648, 3488
— sobre grava, 1648
cuneiforme, 934

cuneta, 1090, 1120
cúpula, 935
cupuliforme, 937
curculiónido, 138, 938
— de la endivia, 718
— de las orquídeas, 24
— de la vid, 695, 4037
— del fresal, 3684
curva de distribución de intensidad de una fuente luminosa,
2159
curvinervado, 951
chaetomio, 2521
chalote, 3361
Chamecyparis, 1285
champiñon abierto, 1369
— a punto de abrirse, 936
— cerrado, 507
— de Paris, 927
champiñones abiertos, 1724
chamuscado, 2101
— de las raíces, 337
chancro, 551, 1054
chaparral, 2506
chimenea, 661
chinche, 472
— de la col, 1246
— del bujo, 3123
chirivita, 975
chopo blanco, 4145
choucroute, 3254
chucrut, 3254
chupón, 2045, 4096

dactilio, 729
dactilo aglomerado, 734
dafne, 983
dalia, 974
damasquina, 44
damnificacíon por zine, 4238
daño causado por animales de
caza, 977
— causado por pájaros, 976
— de gas, 1542
— de heladas, 1463
— de insectos masticadores,
1921
daños causados por la tempestad, 3671
dardo, 3386
dar sombra, 3358
dátil, 988
datos de cosecha, 898
decoloración, 1080

decolorar, 1002
decusado, 1006
defensa biológica, 314
déficit, 1011
defoliación, 1019
— del abeto Douglas, 2459
deforme, 1020
degeneración, 1022
degenerar, 1021
degenerarse, 2452
dejar de florecer, 2123
deltoideo, 1033
dentado, 888, 1036
dentelaria, 2067
deportado, 2514
depositarse, 1038, 3354
depósito de agua, 4100
— sedimentario, 3308
depreciación, 1039
depresión del terreno, 1040
derecho de cultivo, 1678
— de exportación, 1267
— privado, de, 2933
— público, de, 2972
'derribar' cebollas, 3882
derribarse, 3889
desagüe, 4078, 4080
desalar, 1042
desarenar, 1066
desarollo del micelio, 2443
— longitudinal, 2210
desbotonar, 1077
descabezar, 1747, 3883
descalcificar, 998
descargador, 1162
descendencia, 1043, 2948
descogollado, 767
descolorido, 229, 378
descomponer, 3195
descomponerse, 1003, 2993, 4107
descomposición, 426, 1004
— interna, 1941
— interna de las frutas por
effecto del frío, 2223
descompuesto, 2994
desecación, 1045
desecar, 1044, 1117, 1125, 1762
desenrolladora situadora de
película plástica, 2421
desenrollador de película, 2787
desenterrar, 3021
desflorar, 3695
desfondar, 1065, 3927
desgranar, 3373

encalar, 2171
encargado de invernadero, 1583
encendido, 410, 836, 1352, 1353
 1698, 2168, 3141
encina, 2505
encorgerse, 3392
encrespar, 939
endogamía, 1883
endospermo, 1226
endrino, 354
endurecer, 1721
endurecerse, 3354
enebro, 1986
eneldo, 1069
enemigo natural, 2450
enfermedad, 1081
— bacteriana, 217, 352
— carencial, 1012
— criptogámica, 1499, 1502
— de almacenamiento, 3666
— de la gota, 4117
— de la madera gomosa, 3214
— de la momificación, 2428
— de la piel dermatosis, 3433
— de las plantas, 2768
— de las ramas, 556
— del borde amarillo de la
 fresera, 3686
— del mosaico, 2390
— del suelo, 3479
— de manchas de la grasa, 1712
— de marchitamiento, 4025
— holandesa del olmo, 1177
— por anguilulas, 1201
— por hongo, 2403
— vascular, 4005
— virósica, 4043
enfermo, 1083
enfriamiento, 1143
— del agua, 1859
enfriar, 829
enfundar en gasa, 4216
enganche de remolque, 3904
— triple, 3832
enjambre, 3759
enjuagar, 1378, 3143
enlucido, 726
enmienda calcárea, 2179
énotera, 1250
enraizamiento, 3785
— profundo, de, 1010
enredadera, 1329
— de campanilla, 2385
enriquecimiento con ácido

carbónico, 581
— en carbónico, 738
enrollado de la hoja del guisante,
 2658
enrollamiento de hojas, 2092
— de las hojas, 2870
enroya, 2101
ensanchar, 4156
ensayo, 1259
— de abonos, 2268
— de uniformidad, 2511
— de variedades, 4003
— práctico, 2893
ensiforme, 3776
ensuciado del cristal, 1591
ensuciamiento, 2828, 3486
entarimado, 3163
entero, 1228, 3971
enterrar, 2991, 2992
— con arado, 2795
entomólogo, 1229
entrada del aire, 60
entrega, 1032
— obligatoria, 795
entrenudo, 1944
entresacar, 2928
envejecimiento, 46
— económico, 2512
— técnico, 4105
envenenamiento, 2815
envenenar, 2814
envoltura, 1949
epidemiología, 1232
epidermis, 1233
epifilo, 1234
época de floración, 1400
— de plantación, 2773
— de recolección, 2719
— de siembra, 3522
equilibro hidrométrico, 4070
equipo, 1236
equiseto, 2295
— menor, 771
eranta, 4175
erecto, 1238
erguido, 1238
érica, 1756
erigerón, 1374
erismo, 1770
erizo, 1769
erodio, 1790
erofila temprana, 4151
erosión, 1239
error medio, 3606

escabiosa, 3262, 3263
escala de Jacob, 1969
escaldadura, 3738
escalera, 3644
— para cosechar frutas, 2721
escalona para plantar, 3363
escama, 3264
— de bulbo, 488
— de la yema, 470
escarabajo, 294, 591, 704
— de la patata, 753
— de la remolacha, 2726
— de San Juan, 731
— gris de jeta redonda, 3243
— saltarín, 286
— san-juanero, 3728
escaramujo, 1097, 3762
escarchada, 2331
escarda, 4111
escardadera, 2, 100, 4113, 4115
— de un arado, 3431
escardar, 3287, 4110
escardillo, 1178, 4114
escarlata, 3268
escarola, 1224
escavar, 1065
escila, 3591
escleranto anual, 108
esclerocio, 1656
esclerocios, 3275
escoba de bruja, 4189
escoba de retamas, 2052
escoba de bruja, 4187, 4188
escogadora, 3514
escolítido, 251
escorias de cobre molidas, 1351
— Thomas, 259
escorpioide, 3281
escorzonera, 3282
escuela de especialistas en
 horticultura, 3273
— de horticultura, 1826
escutelo, 3289
esfagno, 2391, 3548
esfera, 1049
esfinge, 1740
— del aligustre, 2935
esfuerzo de traccíon, 3807
espacio de vapor, 3630
espantalobos, 355
esparcidor de estiércol, 2266
esparcir, 3270
esparmania, 3535
espárrago, 168

laminilla, 2023
lámpara de descarga, 1543
— de incandescencia, 1884
— de vapor de mercurio de alta
  presión, 1797
— fluorescente, 1410
— hortícola, 1825
lana de vidrio, 1593
lanceolado, 1736
— de las hojas, 2090
lancha dragadora, 1133, 2415
languidez, 1057
— de la seta, 2431
lanza de pulverización, 3570
lanza-llamas, 1366
larva, 2040
— del abejorro, 640
larvicido, 2041
lata, 544
latex, 2046
laurel, 268
— cerezo, 652
— real, 652
lavado, 2064
lavador de aire, 62
lavándula, 2050
lavar, 4064, 4065
lebrillo germinador, 3321
lecanino, 3266
leche de gallina, 3616
lecho forestal de coníferas, 2734
lechuga, 2085, 2139
— crespa de hielo, 1866
— de campo, 851
— de cortar, 963
— flamenga, 513
— romana, 857
— tierna, 1190
lechuguilla, 2674, 3556
leguminosas, 2127
lengua de buey, 95
— de vaca, 940
lente de aumento, 2234
lenteja de agua, 1161
lentejilla, 2133
lenticula, 2133
leña proveniente de la poda,
  2966
leñoso, 4200
lepidio, 889
lepra de las hojas, 4188
— del chopo, 4225
leptosferiosis, 553
lesión, 2137

levantar, 3021
ley de arriendos, 2121
leyes protectoras del seleccionador, 3325
liberalización, 2148
libocedro, 1886
libre de cloro, 1451
liebre, 1729
liebrecilla, 849
ligamaza mielada, 1811
ligulado, 3674
ligustro, 2934
lila, 2164
limaco, 1335
limbo, 356
limo, 3407
— de alcantarilla purificado,
  3356
limón, 2128
limonada, 3465
limonero, 2129
limoso, 2416
limpiadora de simientes, 3313
limpiar, 454, 698, 3913, 4064
— bulbos, 699
linaje, 3673
línea, 3210
— genealógica, 3673
lineal, 2181
lino, 1373
liofilización, 1454
lipáride, 4230
liquidámbar, 3765
liquidez, 2192
líquido, 2188, 3572
— complementario, 870
— desinfectante, 2190
lirio, 2166
— de los valles, 2169
— de San Juan, 991
lisimaquia roja, 2216
liso, 3448
lista de precios, 2924
listón, 247, 2047
litera, 2196
lixiviación, 2064
lixiviar, 2198
Lobelia, 2204
lóbulo, 2203, 3319
lodo, 2414
— de ciudades, 3355
loess, 2206
lofiro, 2735
lombarda, 3058

lombriz, 1194, 4212
longitud de onda, 4102
lonja de clasificación, 1617
lote, 2218
lucha, 822
— contra enfermedades de
  plantas, 1082
— contra las malas hierbas, 4111
— contra los insectos, 1920
lugar de clasificación, 1617
— de trabajo, 4211
— de ventas, 2282
Lunaria, 1809
lupino, 2226
lupulina, 340
luxámetro, 2227
luz artificial, 159
— coloreada, 755
— diurna, 990
llantén, 2761
llave de grifo, 732
— falsa, 1303
lleno, 1490
lluvia, 3018
— de oro, 2018

maceta, 1402
— de plástico, 2788
— de turba prensada, 2659
macollar, 3853
machete, 676
madreselva, 1813
maduración, 3147
— artificial, 45
madurar, 2304, 3145
madurez, 3146
maduro, 3144
— prematuro, 2902
magnesia, 2230
magnesio, 2231
— de cal en polvo, 1672
Magnolia, 2235
magulladura, 452
magullar, 453
Mahonia, 2238
maíz dulce, 3763
mala hierba, 112, 2675, 4109
mal blanco, 2886
— de la goma, 3387
— del cuello de las plantas
  forestales, 1058
— del esclerocio, 673, 1658,
  1659, 2454
— de los ramos, 3589

porta-útiles, 3880
posición, 2851
— del cuerpo, 2854
postillo de chimenea, 978
potasa, 2856
potasio, 2860
potencia, 2890
Potentilla, 686
potra de la col, 517
pozo, 4123
— absorbente, 2462
— negro, 604, 639
pre-calentador, 2901
precalentar, 2900
precintar, 3298
precio de mercado, 2287
— de orientación, 2549
— de referencia, 3081
— de venta, 3341
— garantizado, 1697
— indicativo, 3793
— máximo, 2305
— medio, 201, 2315
— mínimo, 2350
precipitación atmosférica, 3018
— radioactiva, 3014
— sucia, 1076
precipitado, 2895
precocidad, 1187
precultivo, 1360
predator, 2897
predispuesto, 3751
preembalaje, 2903
prenda, 2792
prendedor de chispa, 3534
preparación, 1717
— de basuras, 787
preparado, 2904
preparar, 2906
prerefrigeración, 2896
preservando, 2913
presión atmosférica, 182
— de agua, 4090
— del vento, 4170
— de trabajo, 4207
— de vapor, 3628
prestamista, 2130
préstamo, 2202
prestatario, 406
presupuesto de trabajo, 2009
previsiones, 465
primavera, 2931
primordio de flor, 1401
Primula, 2931

proceso de formentación, 1311
producción, 895
— alterna, 310
— bruta, 1667
— de bolitas, 3560
— superior, 1890
producir, 2940
productividad, 2943
— del trabajo, 2013
producto, 2936
— congelado, 1467
— de protección, 2909
— en polvo, 1176
productos accessorios, 2352
— anejos, 2352
— de asimilación, 177
— del vivero, 2497
— perecederos, 2681
profundidad del drenaje, 1041
profundo, 796
progenie, 2948
programa, 2758
— de rutina, 1408
programador de riego, 1960
programa permanente de base,
1408
prolongar, 3690
propano, 2953
prótalo, 2960
protandria, 2956
protección de la toma de fuerza,
2892
proteger, 2957, 3284
protegido de la helada, 2959
protoginia, 2961
protoplasma, 625
proveedor de semillas, 3314
prueba, 1259
pruina, 369
Prunus triloba, 1181, 1398
pseudo verticilo, 1287
psila, 3712
— del boj, 417
— del manzano, 135
— del peral, 2648
pubescente, 1114
publicidad en masa, 750
pudrir, 3195
puente, 2185
— de carga, 2199
puerro, 2124
puesta de huevos, 1204
puerta de la fogaina de la
cámara de combustión, 1355

pulgón, 125
— amarillo del grosellero, 3060
— amarillo del nogal, 1245
— ceroso de la col, 510
— de agalla del abeto, 3587
— de la acedera, 1095
— de la frambuesa, 3034, 3215
— de la lechuga, 948
— de la patata, 2864
— de la raíz de la lechuga, 2142
— de las habas, 325
— de las hojas, 285
— del fresal, 3680
— del sauce, 2038
— lanigero, 947, 2314, 3411
— lanigero de la corteza del pino,
2733
— lanigero del alerce, 4203
— lanigero del manzano, 4202
— lanoso del haya, 287
— negro, 325
— negro del peral, 2640
— rosa del peral, 2639
— rosado del manzano, 3193
— verde del abeto, 3586
— verde de la patata, 1586
— verde de las ramas del manzano, 1650
— verde del guisante, 2622
— verde del melocotonero, 2625
Pulmonaria, 2225
pulpa, 1474, 1480, 2974
— de frutas, 1473
pulverización, 3571
pulverizador, 3569
— acelerado potente, 3545
— de arrastrio, 3903
— de cultivo, 905
— de motor, 2399
— de propano, 2954
— para acoplar al tractor, 3900
— sobre toma de fuerza, 2970
pulverizar, 1174, 2976, 3566
punta, 3699
— de cierre, 2461
— de trabajo, 2634
puntal, 3698
punteado, 214, 216, 1106, 3140
— de las hojas 104, 120, 223,
450, 2074, 2078, 2099, 2102,
2107, 3139, 3349
— gris, 411
— rojo, 2096
punto de estrangulamiento, 412

ITALIANO

alga, 3301
allacciare, 312
allegazione, 1481
allontamento dei bottoni, 1077
alloro, 268
allume, 77
— ammoniacato, 87
— di potassa, 2857
allungamento, 1215
alluvione marittima, 3294
alno, 64
— nero, 65
aloe, 72
alopecuro 2309
Alstroemeria, 2689
altea, 1804, 3395
alternanza di produzione, 310
alternariosi, 986
— delle foglie, 74
alternato, 76
altezza capillare, 567
— d'attacco, 3753
altica, 1375
alveare, 288
alzamento di terra, 244
amaranto, 610
amarasco, 2383
amarilli, 78
amaro, 321
ambiente, 1231
— nutritivo, 2500
Amelanchier, 3456
amelmarsi, 3409
amento, 608
amfimallo, 3728
ammaccare, 453
ammalato, 1083
— dall'acqua, 379
— nel involucro, 3433
ammassare, 3593
ammasso, 721
amministrazione delle acqua, 1119
ammollare, 2916
ammoniaca, 105
— liquida, 85
— solforica, 3722
ammonium, 86
ammortamento, 89
amorino, 2339
amortizzare, 4221
amovibile, 3097
ampliare, 4156
anaerobe, 91

anagallide, 3269
analisi, 92
— del suolo, 3471
— fogliale, 2069
— matematica, 3620
— vegetale, 896
ananasso, 2732
anarsia, 2628
ancorare, 94
ancusa, 95
androgino, 320
Andromeda, 96
anelli da gelo, 1465
Anemone, 97
aneto, 1069
Angelica, 98
angiosperme, 99
angolato, 103
angolo di divergenza, 101
angoloso, 103
anguillula, 1200
anguillulas dei prati, 3176
anile, 1894
animali dannosi, 2491
annaffiare, 3579, 4069
annaffiatoio, 4082
annata agraria, 1735
— contabile, 14
annerimento basale, 273
— delle piante ortensi, 410
annualità, 113
annuo, 106
antagonismo, 115
antera, 116
Anthoxanthum, 3769
anticipare, 1425
anticipazione della raccolta, 900
antigelo, 2959
antirrino, 123
antisettico, 124
antonomo del lampone e della fragola, 3685
— delle fragole e dei lamponi, 3681
— del melo, 128
— del pero, 129
antracnosi, 118, 119, 120, 375, 550, 2070
anturia, 121
aparine, 700
ape, 283
— artificiale, 158
— operaia, 4205
— regina, 3004

aperto, 1814
apertura di sfoge, 4021
apice, 1681
— fogliale, 2111
— vegetativo, 3814
apicoltore, 290
apicoltura, 291
apione dei capolini del trifoglio, 718
apparato d'Orsat, 2557
apparecchio da germinare, 1567
— di diffusione, 1062
— d'illuminazione, 2157
— d'illuminazione a tenuta stagna, 1138
— d'illuminazione impermeabile, 4092
appassire, 1278
appassito, 4191
appezzamento, 2218, 2606
appiattire, 1371
applicazione d'acido carbonico, 581
— in post-emergenza, 2852
— in post-semina, 2853
— in pre-emergenza, 2898
— in pre-semina, 2917
approvvigionamento, 3743
— aereo, 60
— d'acqua, 4098
Aquilegia, 758
arancia, 2538
aranciera, 2539
arancio, 2540
arare, 2793
— a fondo, 1009
— da mescolanza, 2796
aratro a bilanciere, 233
— di fognatura, 1091
— monovomere, 3420
— multivomere, 2423
— portato a 1/2 di giro, 1711
— portato a 1/4 di giro, 2407
— rincalzatora per asperagi, 172
— sotterraneo, 3706
— trainato, 3902
— talpa, 2374
— voltolante 3121
Araucaria, 660
arboricoltura, 144
arbusto, 3393
— da forzatura, 3396
— di ribes, 944

218, 222, 223
— dei rami e frutti degli agrumi, 219
— del fagiolo, 1712
— delle crocifere, 347
Begonia, 298
— rex, 299
— tuberosa, 3946
belfiore, 2277
beni deperibili, 2681
berbero, 249
bernoccoluto, 2973
berrettina radicale, 3169
betulla, 317
bianco, 2432
biancospino, 1741
bianco su grano, 1630
bibio, 308
biennale, 309
bietola capuccio, 3552
biffa, 3749
bifosfato calcario, 1053
bilancia, 3544
bilancio preventivo, 465
billeri, 1707
bilobo, 2007
biologia, 315
bipennato, 316
bisessuale, 320
bismalva, 2296
bissola, 4185
bisuperfosfato, 1110
black root, 273
blaniuli, 3565
blocchiera per terreno, 3475
bocca di leone selvatica, 4106
boccia, 1389
'boccio sputato', 2214
bolla delle foglie, 4188
— del pioppo, 4225
bollettino di mercato, 2288
bombice bianco del salice, 3250
— dal ventro bruno, 451
— dal ventro dorato, 4230
— dispari, 1572
— gallonato, 2021
bombo, 492
bonificare, 1117, 1125, 1218
borace, 398
bordo di fiori, 400
boro, 403
borrana, 397
borse del pastore, 3376
bosco ceduo di quercia, 2506

bostrico, 79
botanica, 408
botanico, 407
Botrytis, 1352, 1353, 2168, 2929
bottega di fiori, 1403
bottoli, 848
bottone principale, 2243
bovina, 877
bozzolo, 735
braccio irratore, 3567
branca centrale, 632
— primaria, 2242
brattea, 418
brentoli, 2183
briobia dell'edera, 1965
— delle piante da frutto, 127
— del ribes, 1610
broccolo, 439
Bromus, 440
bronzatura (pomodoro), 3877
bruciatore, 498
— ad atomizzatore pneumatico, 51
— ad emulsione, 1220
— a olio, 2517
— a gas, 1541
— atomizzatore a pressione, 2919
— atomizzatore rotativo, 3196
— vaporizzatore, 3994
bruciatura, 3279
brucio, 2975
bruco, 607, 736, 2229
— delle fave, 272
bruco peloso degli alberi da frutto, 451
brughiera, 1757
buca da profilo, 2947
bucaneve, 3455
buccia, 3432
buco, 2751
— da piantare, 2770
Buddleia, 463
buina, 877
bulbicine, 847
bulbicino, 2515
bulbiforme, 486
bulbillo, 483
bulbi ritardati, 3118
bulbo, 476
— a doppio rampollo, 1108
— da fiore, 476
— di cipolla, 3363
— laterale, 665, 2034

— madre, 2392
bure, 2794
Buxus, 415

cabina da gas, 1547
cacecia, 3191
cacto articolato, 677
— foglioso, 1234
cainito, 1990
calabrone, 1817
calamo aromatico, 3770
calcareo, 523
calce, 2170
— magnesiaca polverizzata, 1672
Calceolaria, 525
— rimasuglia, 4068
— silicica di potassa, 531
— spenta, 3435
calcestruzzo cellulare, 1413
— precompresso, 2921
calce viva, 3007
calcicole, 526
calcifuggente, 527
calcina conchifera, 3374
calcio, 528
calcolo dei costi, 860
— dei costi unitari, 3979
— ex-post, 1269
caldaia, 391
— ad acqua calda, 1836
— ad alta pressione, 1796
— a doppio circuito, 3967
— a sezione, 3305
— a vapore, 3623
— marina, 2278
calderone, 1837
calensola piccola, 2595
calibrare, 532, 3428
calibratrice, 1614
— in base alla qualità, 1618
— in base al peso, 1619
— per bulbi, 480
calibrazione, 3429
calice, 538
— accessorio, 537
caliciforme, 937
callo, 535
Calluna, 2183
calo, 4121
calore d'irraggiamento, 1764
— terrestre, 3483
calta, 2297
calzinazione, 642

coltello a scavare, 950
— a sparagio, 171
— da innestare, 1625
— da inoculare, 462
— da bordos, 1197
coltivare, 926, 1676, 3851
coltivatore, 429, 930, 1677, 2765
— a denti plessibili, 3577
— a denti rigidi, 3136
— a denti vibranti, 4032
— con utensili rotanti, 3197
— da alberi, 2496
— di bulbi, 481
coltivazione, 929, 3852
— di bulbi, 482
— di sementi, 3318
— in elevati strati di terra, 3134
— in file, 281
— senza terra, 1648, 3488
coltro, a coltello, 2000
— a disco, 1078
coltura, 929
— all'aperto, 2559
— bis, 3303
— da serra, 1581
— d'erbe, 1785
— di verdura, 4009
— in eccesso, 2566
— in serra, 1582
— intercalante, 3972
— interposta, 1937
— precorrente, 1360
— pura, 2984
— su quadri, 280
Colutea, 355
combustibile, 1489
— gassoso, 1544
— liquido, 2191
— solido, 3504
comino, 933
commercio all'ingrosso, 4154
commissionario, 765
compagnia di vita, 313
compartecipazione, 3366
compattezza, 1725
composito, 791
composizione chimica, 649
composizioni chimique di
    mercurio, 2329
composta, 3648
composto arsenicale, 155
comunità Economica Europea,
    1199
concavo, 796

concentrazione, 797, 798
— per sublimazione, 1454
concimaia, 1164
concimare, 787, 2262
— in aggiunta, 1319
concimazione, 2269
— carbonica, 738
— con acido carbonico, 581,
    723
— di copertura, 3885
— di provvisione, 257
— filatamente, 1323
— fondamentale, 258
concime, 786, 1320, 2261
— ausiliario, 1320
— azotato, 2485
— calcareo, 2179
— chimico, 161
— completo, 785
— da fungo, 2430
— della immondezza, 1100
— di cascame comunale, 3893
— di potassa, 2858
— di torba, 2653
— mescolato, 792
— organico, 2544
— putrefatto, 4125
— riscaldante, 1427
concrescimento, 722
condensamento, 793
condensare, 801
condensatore, 802
condire, 3299
condizionamento dell' aria, 52
condizione del suolo, 3477
condizioni atmosferiche, 707
condotta aerea, 54
— da vapore, 3627
— di pioggia, 3581
conduttura, 687
— d'acqua, 4079
confezionamento, 2579
confronto dei risultati aziendali,
    783
congelatore domestico, 1806
congelazione a basse tempera-
    tura, 1007
— rapide, 3006
conico, 804
conifero, 805
coniforme, 804
coniglio selvatico, 3009
connettivo, 808
consegna, 1032

— obbligatoria, 795
conservando l'umidità, 2370
conservare, 2911
conservazione, 2908
— d'acqua, 4072
— frigorifica, 744
— in atmosfera controllata, 823
conserve, 2914
— di frutti, 558, 2912
consistenza, 809
consultore, 34
— orticolo, 1823
consumatore, 810
consumo d'acqua, 4073
— tecnico, 4105
contabilità aziendale, 12, 2534
— di gestione, 502
— fiscale, 1361
contagiare, 1898
contagio, 1900
contagiosità, 815
contatore di concentrazione,
    1322
— orario, 3857
contenante silicio, 3403
contenuto, 817, 818
— d'umidità, 2368
— d'umus, 1850
contratto d'affitto, 2120
controllo ambientale, 706
— dei prezzi, 2922
— delle semenze, 3329
— di qualità, 2998
— qualitativo, 2999
— sulla pianta, 897
— sull'esportazione, 1266
controventatura, 4168
convolvulo nero, 328
coperta del suolo, 3478
copertura, 726, 868
— del suolo, 2420
— di colmo, 3132
— d'umus, 1847
coppia termo-elettrico, 3822
coprino, 1913
coprire, 869
— da vetri, 1594
corda a piantare, 2180
— da piantare, 1534
cordiforme, 838
cordone, 839
coreggiola, 2001
— minore, 2295
coreopsis, 843

gas, 1544
— d'acido carbonico, 580
— da fumo, 1409
gassoso, 3995
gattaia, 609
gazza, 2236
gelare, 1453
gelatina, 1976
gelo notturno, 2474
gelso, 2419
gelsomino, 2363
— d'inverno, 4177
— fiorente nudamente, 4177
gemma, 458, 2079
— ascellare, 205
— avventizia, 32
— promiscua, 2358
— terminale, 3812
gemmazione, 464
gemmiforme, 466
generatore d'aria calda, 1832
generazione, 1552
genere, 1559
geno, 1551
genotipo, 1557
genziana, 1558
geologico, 1560
geometra, 2212
— del ribes, 2237
geranio, 881, 1561, 4240
— ederaceo, 1966
— francese, 3088
germe, 1562
germinare, 1564
germinativo, 4029
germinazione, 1566
— del polline, 2823
germinello, 3320
germogliare, 3584
germogli di malto, 2255
germoglio, 3380
— ad occhio, 1277
— del portinnesto, 3182
— laterale, 2045
— ripiegato, 2004
gesso, 1702
ghiaccio, 1865
ghiaia, 1647
ghiandaia, 1975
giacinto, 1854
— a grappolo, 1638
— del Capo, 3731
giagiolo, 1574
giallo brunetto, 446

giallume, 1510
— della barbabietola, 297
— marginale della fragola, 3686
giardiniere, 1532
giardino, 1527
— di fiori, 1394
— di rocca, 3152
— d'ornamento, 2554
— popolare, 69
giavone, 253
gibberelline, 1571
giglio, 2166
— aureo, 1606
— della Cina, 3846
— diurno, 991
— regale, 3087
ginepro, 1986
ginestra a spina, 1509
— di brughiera, 1556
ginestro, 442
Ginkgo biloba, 2240
gipsofila, 1701
girasole, 3733
giunco, 3220
giusquiamo, 1780
glabro, 1573
glandola, 1575
glandoloso, 1576
Gleditsia, 1812
glicine, 4186
glossinia, 1602
gola, 3837
gomena di riscaldamento, 1758
gonfiare, 3773
gorgoglione, 4037
governo di temperatura, 3803
grado, 1023
— d'acidità, 20
— di dissociazione, 1024
— di saturazione, 1026
— d'umidità, 1025
— qualitativo, 1613
graduamento, 1615, 3002
graffa, 3611
graffatrice, 3613
grafioso dell'olmo, 1177
gramigna, 867
graminaceo, 1641
grana fine, di, 1350
granaio aziendale, 4209
grande magazzino, 1037
grandezza del grappolo, 3939
grandifoglia, 2033
granello, 1628

— di polline, 2824
granturco sucroso, 3763
granulare, 1633
granulato, 1632
granulazione costitutiva, 3819
granuli grossolani, a, 725
granuloso, 1631
grappolo, 3010
— di fiori, 1391
— di frutti, 1472
graticcio, 1852
graticolato, 3925, 4195
grattugia, 1645
grespino, 3556
— dei campi, 2674
grillotalpa, 2372
grinzoso, 3216
grondaia, 1700
— posta a valle, 3991
grossezza del granello, 1629
grossista, 4153
grosso, 724
grotta, 2345
grufolare, 1694
— da mescolanza, 3707
grufolatrice a mescolare, 2359
guaina d'una foglia, 3368
guano, 1696
guarnizioni per caldaia, 392
guscio, 2810
gusto anomalo, 2513

Hamamelis, 4190
Helianthemum, 3736
Hepatica, 1782, 2197
heterosis, 1791
heterozygote, 1792
hiba, 3839
Holcus lanatus, 4233
homozygote, 1808
hormone, 1690
— d'impedimento, 1691, 1910
Hosta, 1503
Hoya carnosa, 4103
Hypericum, 3656

iarovisazione, 4024
Iberis, 548
ibrido, 1855
— d'innesto, 1622
icneumonide, 1869
idrato di carbonio, 576
idrocoltura, 1648, 4077
idrogeno, 1860

— per l'uso, 1075
iuta, 1988

Kalmia, 533
kentia, 1993
Kerria, 1994

labbro, 2186
laciniatura delle foglie, 3875
lacno del salice, 2038
lama, 2
— ad angolo, 100
lamburda, 2211
lamella, 2023
lampada ad incandascenza, 1884
— a scarica nei gas, 1543
— a vapore di mercurio ad alta pressione, 1797
lampada per illuminazione di piante, 1825
lampone 3033, 3036
lana di vetro, 1593
lancetta, 2812
lanciafiamma, 1366
lancia irroratrice, 3570
lanciforme, 1736
lanuginoso, 1706
larghezza della lastra di vetro, 2595
— totale della campata, 2565
larice, 2030
larva, 2040
— melolontoide, 640
larvicido, 2041
lastra, 3163
— di vetro, 2594
latta ondulata, 855
latte di gallina, 3616
lattuga, 2139
— cappuccia, 513
— da taglio, 963, 2085
— primaticcia, 1190
— romana, 857
lauro, 268
lauro ceraso, 652
lavanda, 2050
lavare, 4064
lavatore d'aria, 62
lavorare nel giardino, 1528
lavoratore, 1216
lavorazione, 1717
lavoro di gruppo, 3798
— familiare, 1289

— salariato, 821
— stagionale, 3300
legaccio, 3924
— di plastica, 2790
legare, 496, 3845
legge d'affito, 2121
— sulla commercializzazione di sementi, 3325
legno durno, 1753
— di potatura, 2966
legnoso, 4200
legumi, 2127
leguminose, 2600
lemna, 1161
lente d'ingrandimento, 2234
lenticellula, 2133
lepre, 1729
lesiniforme, 3710
lesione, 452, 2137
— di potatura, 2964
letame artificiale, 3779
— di gallina, 2883
— di pollo, 656
— equino, 1820
— liquido, 2193
— usato, 3547
lettiera, 2196
— di conifere, 806
— forestale, 1431
letto, 1448, 2950
— bilaterale, 3533
— caldo, 1833
— di paglia, 3688
— freddo, 741
— gemello, 1109
levamento dei minerali, 1034
levare il vetro, 3020
levistico, 2219
Liatris, 2147
liberalizzazione, 2148
libocedro, 1886
libro, 264
licenza colturale, 1678
ligulato, 3674
ligustro, 2934
lilla, 2164
limaccia grigia, 1335
limaccioso, 2416
limacina del pero, 2646
limantria, 1572
limbo, 356, 2086
limonate, 3465
limone, 2128, 2129
limonite, 1954

linea, 3673
— bianca, 3415
lineare, 2181
lineolatura, 2182
lino, 1373
liofilizzazione, 1454
liquidambra, 3765
liquidità, 2192
liquido, 2188
— disinfettante, 2190
liquore concentrato, 797
liscio, 3448
lisciva di potassa, 613
lisciviare, 2198
litiasi contagiosa, 2647
livellare, 2144
livello, 2143
— ascencionale, 1773
— d'acqua, 4083
— d'acqua di fossato, 1092
— dell'acqua sotterrana, 2145, 4099
— della falda freatica, 4099
lobato palmatiforme, 2590
— penniforme, 2742
Lobelia, 2204
lobo, 2203
locale refrigerante, 2205
loess, 2206
lofiro, 2735
loglio, 2672, 3226
— italiano, 1963
logorio, 4105
lombrico, 1194
loppa, 1603
lotta, 822
— biologica, 314
— contro gli insetti, 1920
— contro le erbe infestanti, 4111
— contro le malattie, 1082
luce artificiale, 159
— colorata, 755
— diurna, 990
lumaca, 3439
luminoso, 432
Lunaria, 1809
lunghezza d'onda, 4102
luogo a scavare, 2952
— d'assortimento, 1617
— di combustione, 760
— d'innesto, 1627
lupinello, 340
lupino, 2226

luxmetro, 2227

macchia delle lenticellule, 2134
macchie brunastre, 443
— d'acqua, 377
— d'amaro, 323
— di Jonathan, 1980
— fogliari, 2075, 2076, 2096, 2099, 3139
— gialle, 2103
— nere, 3510
— rosse, 3063
macchietta da Botrytis, 411
macchina a grufolare, 1695
— a sfogliare, 1018
— di fognatura, 1124
— di Sarchiatura, 4115
— frigorifera, 3086
— per girare letame, 2267
— per il raccolto per piselli, 2631
— per selezione, 1614
— per selezione di patates, 2874
— per situare e stendere il foglio di plastica, 2421
— prosciugatrice, 2979
maculatura, 2401
— a foglia di quercia, 2507
— amara, 323
— anulare, 3142
— anulare tuberosa, 3137
— batterica, 224
— delle foglie, 2095
— lineare, 2182
madreselva, 1813
magazzino, 3670
— a doppia parete, 1968
maggese, 1283
maggiolino, 731
maggiorana, 2280
magnesia, 2230
magnesio, 2231
Magnolia, 2235
Mahonia, 2238
malacosoma, 2021
malattia, 1081
— botrytis, 409
— crittogamica, 1502
— da frigorifero, 3666
— da muffa, 2403
— d'appassimento, 1514, 4025
— da virus, 4043
— da virus Y, 4235

— delle piante, 2768
— del rattle, 3043
— del suolo, 3479
— di deficienza, 1012
— di macchietta negra, 353
— di muffa, 1499
— di Pfeffing, 1195
— di piombo, 3412
— mosaica, 2389, 4229
— orecchietta, 209
— per carenza, 1012
— petecchiale di batterio, 213
— vascolare, 4005
mal bianco dell'uva spina, 84
— caucciù, 3214
— del colletto delle piantine forestali, 1058
— del cuore, 337
— della gomma dei rameti del pesco, 3387
— della tela, 729
— dell'inchiostro, 1914
— dello sclerozio, 915, 1656
— tartufo, 1286
— del piede, 337
malerba, 4109
— perenne, 112
malformazione, 2252
mallo duro, di, 1720
mal secco, 217, 352
malta, 2386
maltosio, 2254
Malus, 1399
Malva, 2253
— crispa, 941
mal vinato, 4039
mandarino, 3790
mandorla, 70
mandorlo, 71
— triloba, 1398
manganese, 2256
mangiabile, 1198
manica, 1830
manicotto, 3005
mano d'opera, 2260
— d'opera familiare, 1290
— d'opera salariata, 2561
manometro, 2920
mappa del suolo, 3489
marcatore, 2281
marcimento secco, 1153
marcio grigio, 1657
marcitura al piede, 1423
marciume, 1659, 2454, 3141

— amaro, 324
— apicale, 380
— bianco, 4146
— bruno, 449
— da fomes, 1418
— da Penicillium, 3667
— da sclerotinia, 3276, 3277
— del colletto, 749, 1424
— del cuore, 1752
— del fusto, 1054, 1660, 3589
— delle pinatine dei semenzai, 979
— gallo dei giacinti, 4228
— molle, 3468, 3469
— nero, 346, 348, 980
— pedale, 338
— radicale, 149, 345, 2713, 3179
— secco, 1510
— secco dei tuberi, 1155
margherita, 2276
— a fiore grande, 3367
margheritina, 975
margine disseccato, 1157, 3861
— di sicurezza, 3228
margotta, 2056
margottare, 57, 2058, 3138
marmellata, 1970, 2293
marna, 2291
— calcarea, 2174
maschera antigas, 1545
mascolino, 3603
mastice, 2995
materia colorante, 756
— da copertura, 871
materia d'inizio, 3617
materiale a piantare, 2774
— per legature, 3843
— organica, 2545
— riscaldante, 1429
— secca, 1152
— secretoria, 3304
— solida, 3503
materie da ingrasso fosfate, 2704
— prime, 3045
mattone refrattario, 1354
maturare, 2304, 3145
maturazione, 3147
— complementare, 45
maturità, 3146
maturo, 3144
mazza di bambu, 240
mazzo di fiori, 414, 1379
meccanizzazione, 2318
meiose, 3075

LATIN

LA

LATIN

Abies Mill., 3410
Abraxas grossulariata, 2237
Acacia spec., 2344
Acalypha, 9
Acanthoscelides obtectus Say, 1134
Acarina Spp., 2356
Acer, 2271
Aceria gracilis, 3038
— tulipae, 4133
Acer palmatum cv. Atropurpureum, 2989
Acherontia atropos, 997
Achillea millefolium, 2342
Aconitum, 21, 2377
Acorus calamus, 3770
Acrolepia assectella Zell., 2125
Acyrthosiphon pisum Harris, 2622
Acythopeus aterrimus Waterh., 24
Adelges abietis, 3587
— cooleyi, 2733
— laricis Vallot, 4203
— nuesslini, 3411
Adiantum, 2239
Adoxophyes orana, 3729
— reticulana Hb., 3729
Aegeria tipuliformis, 945
Aegopodium podagraria, 1668
Aesculus, 1819
Agaricus bisporus Sing., 927
Agelastica alni, 66
Agriotes spp., 4185
Agrobacterium tumefaciens (Sm. et Town.) Conn., 912
Agropyrum repens Pal., 867
Agrostis stolonifera, 885
Agrotidae, 2486
Agrotis spp., 987
Ailanthus altissima (Mill.) Swingle, 3918
Ajuga, 474
Albugo candida (Gray) Kze., 4147
Alchemilla arvensis, 2611
— vulgaris, 2022
Allantus cinctus, 243
Allium ampeloprasum, 3414

— ascalonicum, 3361
— cepa, 2523
— sativum, 1540
— schoenoprasum, 667
Alnus Mill., 64
Aloë, 72
Alopecurus pratensis, 2309
Alstroemeria, 2689
Alternaria spp., 2099
— brassicae (Beck.) Sacc., 986
— dauci (Kühn) Graves et Skolko, 74
— solani (E. et M.) J. et Gr., 75
Althaea officinalis, 2296
— rosea Cav., 1804
Amaranthus, 610
Amelanchier Med., 3456
Ametastegia glabrata Fall., 1096
Amorpha, 266
Amphimallon solstitialis, 3728
Amphorophora rubi Kltb., 3215
Anagallis arvensis, 3269
Ananas comosus Merr., 2732
Anarsia lineatella Zoll., 2628
Anchusa, 95
Andromeda, 96
Anemone, 97
Anethum graveolens, 1069
Angelica archangelica, 98
Anthonomus cinctus Koll., 129
— pomorum, 128
— pyri, 129
— rubi Herbst, 3681
Anthoxanthum odoratum, 3769
Anthriscus cerefolium Hoffm., 654
Anthurium scherzerianum Schott., 121
Antirrhinum, 123
— orontium, 4106
Antophila pariana, 132
Aphelenchoides fragariae (Ritz.-Bos), 3573
— spp., 460
Aphididae, 125
Aphis fabae Scop., 325
— idaei v.d. Goot, 3034
— pomi de G., 1650

Aphis rumicis, 1095
Apion assimile Kirby, 718
Apis mellifera, 283, 4205
— mellifica, 283
Apium graveolens, 358, 618, 619
Aquilegia, 758
Arabidopsis thaliana Heynh., 3820
Araucaria araucana Koch, 660
Archips rosana, 3191
Ardis brunniventris Htg., 146
Arge ochropus Gm., 2036
— rosae, 2036
Argyresthia conjugella Zell., 130
Arion rufus, 1335
Aristolochia, 319
Armeria Willd., 3835
Armillaria mellea (Vahl) Quél., 149
Armoracia rusticana G. M. Sch., 1822
Arnica montana, 150
Aronia Pers., 674
Artemisia absinthium, 4215
— dracunculus, 3795
Arthropoda, 156
Arunde donax, 1570
Arvicoli sp., 4101
Asclepias, 506
Ascochyta phaseolorum Sacc., 375
— pinodella L. K. Jones, 2070
— pisi Lib., 2070
Asparagus officinalis, 168
Asperula odorata, 3772
Aster, 179, 2671
— spec., 2334
Athalia rosae, 3961
Atomaria linearis Steph., 2726
Atriplex, 2537
— patula, 773
Atropa belladonna, 992
Aulacorthum solani Kltb., 1586
Autographa gamma, 3417

Bambusa Schreb., 239
Barathra brassicae, 514
Begonia, 298
— rex hybriden, 299

Colletotrichum fuscum Laub., 120
— lindemuthianum (Sacc. et Magn.) Bri. et Cav., 119
Columbia palumbus, 4198
Colutea, 355
Coniothyrium hellibori Cke. et Massee, 2104
Contarinia nasturtii Kieffer, 364, 3761
— pisi Winn., 2635
— pyrivora Riley, 2644
Convallaria majalis, 2169
Convolvulus arvensis, 1329
Coprinus spp., 1913
Coreopsis, 843
Cornus, 1098
Coronopus squamatus (Forsk.) Aschrs., 3774
Cortaderia selloana A. et G., 2593
Corticium solani, 350
Corvus, 910
Corylus avellana, 1742
Corynebacterium fascians (Tilford) Dowson, 2083, 2116, 2117
Corynespora melonis (Cke) Lindau, 376
Cosmos Cav., 858
Cossus cossus, 1604
Cotinus coggygria Mill., 4137
Crambe maritima, 3297
Crataegus, 1741
Crioceris asparagi, 169
— lilii Scop., 2167
— spp., 169
Crocus, 894
Croesus septentrionalis, 67
Cryptococcus fagi Brsp., 287
Cryptodiaporthe populea (Sacc.) Butris, 555
Cryptogamae, 918, 3562
Cryptomyzus ribis, 3060
Cryptor(r)ynchus lapathi, 4162
Cucumus Sativus, 920, 3009, 3133
Cucurbita maxima Des., 2980
— pepo, 1612
Cuminum cyminum, 933
Curculionidae, 938
Cyclamen, 967
Cydonia Mill., 3008
Cylindrocarpon radicicola Wr.,

3179
Cynara cardunculus, 586
— scolymus Pers., 1597
Cynipidae, 1521
Cynipoidea, 1521
Cynosurus cristatus, 890
Cypripedium, 970
Cytisus, 442

Dactylis glomerata, 734
Dactylium dendroides Bull., 729
Dahlia Cav., 974
Daphne mezereum, 983
Dasyneura brassicae Winn., 424
— pyri Bouché, 2642
— spp., 1520
— tetensi Ruebs, 334
Daucus carota, 593, 4157
Delia antiqua, 2524
— platura, 275
Delphinium, 2039
Depressaria heracliana, 2613
Deroceras reticulatum Müll., 1335
Deutzia Thunbg., 1047
Dianthus barbatus, 3771
— caryophyllus, 588
— plumarius, 2740
Diataraxia oleracea, 3874
Dicentra Bernnk., 361
Dichomeris marginellus F., 1987
Didymascella thyjina (Durand) Maire, 2076
Didymella applanata Sacc., 3589
— ligulicola, 3046
— lycopersici Kleb., 1054, 1423
— pinodes, 2070
Didymellina dianthi C.C. Burt., 3140
Diehlioyces microspora (Diehl et Lambert) Gilkey, 1286
Digitalis, 1444
Dimorphoteca aurantiaca DC., 563
Diplocarpon earlianum (Ell. et Everh.) Wolf, 2096
— maculatum, 2074
— rosae Wolf, 353
Diplolepus rosae, 282
Diprion pini, 2735
Ditylenchus destructor Thorne, 2877
— dipsaci (Kühn), 3635
— dipsaci (Kühn) Filip, 3639

Doralis fabae, 325
— rumicis Leach, 1095
Doronicum, 2136
Deprano peziza ribis, 2106
Dreyfusia nordmannianae Ratzb., 3411
Dulce, 358
Dysaphis plantaginea, 3193
— pyri, 2639

Echinochloa crus-galli (L. P.B.), 253
Echinops, 1599
Elateridae, 704
Elatobium abietinum Wlk., 3586
Elsinoë veneta Jenk., 550
Emphytus cinctus, 243
Enarmonia formosana Scop., 650
— funebrana Tr., 2803
— nigricana F., 2637
— pomonella, 736
Entyloma dahliae Syd., 2102
Epiphyllum Haw., 1234
Equisetum arvense, 771
— palustre, 2295
Erannis defoliaria Clerck, 2400
Eranthis Salisb., 4175
Erica, 1756
Erigeron, 1374
Erinaceus europaeus, 1769
Erioischia brassicae, 515
Eriophyes, 2499
— tulipae Keifer, 4133
Eriosoma lanigerum Hausm., 4202
— lanuginosum Htg., 3272
— ulmi, 947
Erodium cicutarium l'Hérit. ex. Hait., 777
— l'Hérit., 1790
Erophila verna Chevalier, 4151
Erwinia atroseptica (van Hall) Jenn., 338
— carotovora (Jones) Holland, 3469
— carotovora f.sp. parthenii Starr., 3468
Eryngium, 3296
Erythrina, 837
Eschscholtzia Cham., 534
Eumerus strigatus Fall., 3444
Eunonymus, 3554
Eupatorium, 1778